THE FUNGI

THE FUNGI

THIRD EDITION

Sarah C. Watkinson
University of Oxford, Oxford, UK

Lynne Boddy
Cardiff University, Cardiff, UK

Nicholas P. Money
Miami University, Oxford, OH, USA

ELSEVIER

AMSTERDAM • BOSTON • HEIDELBERG • LONDON
NEW YORK • OXFORD • PARIS • SAN DIEGO
SAN FRANCISCO • SINGAPORE • SYDNEY • TOKYO
Academic Press is an imprint of Elsevier

Academic Press is an imprint of Elsevier
225 Wyman Street, Waltham, MA 02451, USA
525 B Street, Suite 1800, San Diego, CA 92101–4495, USA
The Boulevard, Langford Lane, Kidlington, Oxford OX5 1GB, UK
125 London Wall, London, EC2Y 5AS, UK

Notices

Knowledge and best practice in this field are constantly changing. As new research and experience broaden our understanding, changes in research methods, professional practices, or medical treatment may become necessary.

Practitioners and researchers must always rely on their own experience and knowledge in evaluating and using any information, methods, compounds, or experiments described herein. In using such information or methods they should be mindful of their own safety and the safety of others, including parties for whom they have a professional responsibility.

To the fullest extent of the law, neither the Publisher nor the authors, contributors, or editors, assume any liability for any injury and/or damage to persons or property as a matter of products liability, negligence or otherwise, or from any use or operation of any methods, products, instructions, or deas contained in the material herein.

Library of Congress Cataloging-in-Publication Data
A catalog record for this book is available from the Library of Congress

British Library Cataloguing in Publication Data
A catalogue record for this book is available from the British Library

For information on all Academic Press publications
visit our website at http://store.elsevier.com/

Publisher: Sara Tenney
Acquisition Editor: Linda Versteeg-Buschmann
Editorial Project Manager: Mary Preap
Production Project Manager: Chris Wortley
Designer: Alan Studholme

Printed in the United States of America

ISBN: 978-0-12-382034-1

Cover image credit: Luminescent fungus, *Omphalotus nidiformis*, courtesy of Ray Kearney; Figure 7.8, Immuno-localisation of a highly expressed fungal effector-like protein in a *Populus trichocarpa–Laccaria bicolor* ectomycorrhizal root tip. A transverse cross section of a poplar root colonised by the symbiotic ectomycorrhizal fungus *L. bicolor*. The green signal is an immuno-localisation of the fungal effector protein MiSSP7 highly expressed in the hyphae of *L. bicolor* while staining with propidium iodide highlights the cell walls of the root cells, courtesy of Jonathan Plett and Francis Martin; Figure 1.21, Sporangium of the zygomycete *Phycomyces blakesleeanus*. The wall of the sporangium will blacken as it matures and then split open to release the sporangiospores, courtesy of Ron Wolfe Photography; and Figure 5.7, photon counting scintillation imaging of 14C-AIB in the mycelium, courtesy of Tlalka, Fricker and Watkinson.

Contents

7. Mutualistic Symbiosis Between Fungi and Autotrophs

SARAH C. WATKINSON

8. Pathogens of Autotrophs

LYNNE BODDY

9. Interactions with Humans and Other Animals

LYNNE BODDY

10. Interactions Between Fungi and Other Microbes

LYNNE BODDY

11. Fungi, Ecosystems, and Global Change

LYNNE BODDY

12. Fungi and Biotechnology

NICHOLAS P. MONEY

Preface

Life on Earth would look very different in the absence of Kingdom Fungi. Without wood-decomposing fungi, fallen timber would render forests impenetrable; without coprophilous fungi the landscape would be contoured by mountains of herbivore faeces, and without aquatic fungi, rivers and ponds would be clogged with plant debris. These dystopian fancies may be useful in pointing to the significance of fungi in the breakdown of biological debris, but they do not stand up to critical thinking for long. Without mycorrhizal fungi, there would be no forests in the first place, nor grasslands, herbivores, or herbivore faeces. Life on land has evolved with the participation of the fungi and would collapse without their continued activities. If fungi had not evolved, then their ecological roles might have been assumed by other groups of microorganisms, but this is the stuff of science fiction. Fungi did evolve and have diversified into a kingdom of more than 100,000 named species. The actual number of fungi is at least one order-of-magnitude higher. Fungi are a major component of the microbiome in almost every habitat.

A pair of familiar fungal species offers a study in contrasts that reflects the diversity of organisms within the kingdom: baker's yeast, *Saccharomyces cerevisiae*, and the button mushroom, *Agaricus bisporus*. Human civilisation would be inconceivable, or would require reconception, without our partnerships with *Saccharomyces*. The ease with which this yeast is cultured and its facility for metabolising glucose and producing carbon dioxide and alcohol has allowed humans to brew beer, make bread, and ferment wine for thousands of years. This single-celled fungus has been a model experimental organism for many decades, and the sequencing of its genome in 1996 was a landmark in modern biology. This was the first sequencing project for a eukaryote and revealed that the fungus housed 6000 genes on 16 chromosomes. More than one-fifth of these instructions match human genes, which is a persuasive reflection of the unity of all eukaryotes. Yeast cells grow by absorbing food and divide by creating buds on their surface. When sexually compatible strains of *Saccharomyces* mate, the resulting diploid cell divides by meiosis and produces four haploid ascospores.

It is difficult, at first glance, to equate *Agaricus bisporus* with yeast: one organism forms a gilled mushroom built from hundreds of thousands of filamentous cells, the other is a unicellular microbe. Like yeast, *Agaricus bisporus* is an organism that is partly a human invention. Wild versions of both species are different from cultivated strains. Untamed populations of the mushroom fruit beneath particular trees and shrubs, suggesting that they may be engaged in mutually beneficial symbioses with these plants. The farmed mushroom is grown as a saprotroph on compost. The fungus produces a colony of branching cells, called hyphae, which feed by secreting enzymes that decompose plant materials in the compost to release sugars and other small molecules that fuel their metabolism. Once the colony has accumulated sufficient biomass, and when the mushroom farmer manipulates the growth conditions, the colony undergoes the remarkable reorganisation that produces a flush of mushrooms.

Buttons develop from pinhead-sized groups of hyphae, and these inflate to form the familiar stem, cap, and gills of the fruit body or basidiome. Harvesting can occur early, before the gills are exposed, or later, when the mushroom has undergone a 1000-fold increase in volume. This hydraulic expansion process allows the wild mushroom to emerge from the ground and display its gills for the process of spore release. And what a marvel of natural engineering is the mushroom. Wild relatives of *Agaricusbisporus* can release an astonishing 31,000 spores per second, or 2.1 billion spores per day. The spores are shot from the gills by a catapult powered by the momentum of tiny droplets of fluid. Nothing like this happens in yeast.

Saccharomyces and *Agaricus* are members of the largest phyla of the fungi, respectively, the Ascomycota and the Basidiomycota, whose ancestral species began to diverge from one another 400 million years ago. These phyla are distinguished from one another by fundamental, or seemingly fundamental, differences in life cycles and developmental biology. If we adopt a broader view of fungal diversity, however, similarities between these great phyla are apparent, including commonalities in cell wall composition, trafficking of membranes within the cytoplasm, and features of metabolism and physiology. Once we recognise these characteristics, it becomes easier to embrace the fact that the single-celled yeast and the multicellular mushroom are different versions of the same kind of organism.

There is, of course, a great deal of subjectivity in organising the fungi, and the rest of life, into groups of taxonomic convenience. The scale of the inquiry is everything. After all, mushrooms and humans are different versions of the same kind of thing at the level of supergroupings of eukaryotes, because all animals and fungi are members of the Opisthokonta. This brief consideration of yeasts and mushrooms is useful because it indicates the breadth of morphological variation in the kingdom, but there is much more to the fungi besides yeasts and mushrooms. Fungal diversity and classification are introduced in Chapter 1.

Fungal cells are built from the same kinds of organelles as other eukaryotes, but possess many structures that are not encountered outside the kingdom. These include organelles involved in the hyphal mechanism of tip growth and plugs that protect wounded colonies from haemorrhaging cytoplasm. The mosaic of chitin and other polymers in the cell wall is another uniquely fungal attribute. Fungal cell biology and development are showcased in Chapter 2. Fungal developmental biology is a research arena that deserves greater attention. Despite tremendous advances in cell and molecular biology, we have very little information on the processes involved in the differentiation of root-like cords and rhizomorphs, resistant organs called sclerotia, and mushrooms. The puzzle of fungal multicellularity is one of the frontiers in mycology, and its solution requires the engagement of the brightest and most creative investigators.

Chapter 3 concerns the formation of microscopic spores, which is another unifying feature of the kingdom. Spores vary greatly in shape and size and help us to identify different groups of fungi, as well as individual species. They range from swimming cells, called zoospores, produced by the chytrids, to warty zygospores, multicelled conidia, and the beautiful spores of truffles patterned with delicate ridges. Within a single category of spores we see a range of dispersal mechanisms. Conidia of ascomycete fungi, for example, are dispersed passively by wind, rain, and insects, and by active mechanisms involving the explosive formation of gas bubbles and the elastic deformation of cell walls. Spore dispersal is a very important research area, because it affects fungal distribution and population biology, and the spread of plant and animal diseases (epidemiology). Airborne spores are a major cause of human allergy (Chapter 9), and there is growing evidence that the huge number of spores in the atmosphere influences cloud formation and rainfall patterns.

Genetic variation, sexuality, and evolution are discussed in Chapter 4. Researchers have turned a few fungi into model organisms for the study of genetics, including the aforementioned yeast, and the filamentous ascomycete, *Neurospora crassa*, which was instrumental in early research on gene expression. The life cycles of both species involve sexual reproduction in which pairs of compatible strains merge and form spores after sexual recombination. The notion of male and female is meaningless for fungi. Reproduction in some of the mushroom-forming basidiomycetes involves pairings between tens of thousands of different mating types. Fungal genetics is further complicated by difficulties in defining individual organisms. Should we regard two or more independent colonies that have separated from a parent mycelium as the same individual? Trickier still is the nature of the fungal species. Current work on these topics is informing wider questions on fungal variation, microevolution, gene flow, and other fundamental issues in evolutionary biology.

Fungal adaptation to the environment is the subject of Chapter 5. With the notable exception of aquatic species that form flagellate zoospores, fungi explore their environment through growth rather than motion. Growing fungi are at the mercy of the nutritional conditions in microscopic proximity to their cell surface. The expansion of a mycelium of interconnected hyphae allows the fungus to meet the challenges of localised nutrient depletion by transferring materials across the colony from regions where food is more plentiful. This adaptability allows fungi to travel through large volumes of soil and cope with exigencies of water and nutrient availability that exclude other microorganisms. Fungi are osmotrophs, absorbing nutrients from their surroundings using an array of secreted enzymes to decompose complex molecules and transport proteins to import the resulting harvest of small molecules through their cell membranes. The galaxy of proteins secreted by fungi is called the secretome, and its analysis is an exciting area of contemporary research. Fungal metabolism is another topic in Chapter 5. The primary metabolism of the fungi follows the same pathways that support other eukaryotes. Fungal secondary metabolism generates an incredible array of pharmacologically active compounds, mycotoxins and mushroom poisons, pigments, and volatile aromatic compounds.

Before the introduction of molecular methods to the study of fungal ecology, research on the roles of fungi in nutrient cycling and other ecological processes emphasised the importance of species made visible by their fruit bodies, or those that were amenable to pure culture. So much was missed. Molecular methods have changed our picture of microbial diversity in different habitats, and this work makes it clear that we have only begun to understand the ecological significance of the fungi. Metagenomic techniques and other methods used for environmental sampling and the analysis of ecosystem processes are introduced in Chapter 6.

The next four chapters (Chapters 7–10) cover interactions between fungi and other organisms. Fungi support plant productivity through mycorrhizal symbioses (Chapter 7) and damage and destroy plants through their activities as pathogens (Chapter 8). Mycorrhizal relationships include ectomycorrhizas, in which the colonies of mushroom-forming basidiomycetes, and a few ascomycetes, envelop the roots of trees and shrubs. These fungi expand into the surrounding soil, creating an absorptive network that supplies the plant host with water and dissolved minerals. The fungi benefit from these relationships by receiving carbohydrates from the plant. Arbuscular mycorrhizas are established by 200 species of Glomeromycota with 80% of the families of vascular plants. Fungi also engage in specialised kinds of mycorrhizal symbioses with orchids, species of parasitic plants, members of the Ericales, and bryophytes. Lichens and endophytic relationships are other examples of fungal mutualisms considered in Chapter 7. More than 18,000 ascomycete species and fewer than 50 basidiomycetes are lichenized; green algae and cyanobacteria are the photosynthetic partners in these intimate relationships. Pathogenic interactions with plants are introduced in Chapter 8. Plant pathogens include the rusts and smuts (Basidiomycota) and thousands of species of Ascomycota that infect every family of plants and cause billions of dollars of annual crop losses. Pathogens classified within the Oomycota (stramenopiles rather than fungi) have been studied by mycologists since the nineteenth century. *Phytophthora infestans* is the best known of these microorganisms, because it caused the potato famine in Ireland in the 1840s. The Oomycota, known as water moulds, are more closely related to brown algae and diatoms than they are to fungi. Functionally, however, they operate like fungi. Water moulds form branching colonies of tip-growing hyphae that penetrate their food sources, secrete digestive enzymes, and absorb small molecules to meet their nutritional needs. Similarities between the morphology, cellular organisation, and behaviour of the oomycetes and fungi offer a beautiful illustration of evolutionary convergence. Water moulds are included in this book, despite their lack of genealogical connection with the fungi.

Although fungal infections of humans are not as common as bacterial and viral diseases, human mycoses are widespread and are a significant cause of morbidity and mortality (Chapter 9). Most mycoses are opportunistic infections, stimulated by damage to the skin barrier, underlying metabolic problems, and compromised immune defenses. Fungal infections of the lung are established via the inhalation of infectious spores, and systemic infections can develop from normally harmless fungi that grow in the urogenital tract and gut, and on the skin. Fungal pathogens of other vertebrates and invertebrates are also addressed in this chapter. In recent decades a number of epidemic fungal diseases of animals have been recognised for the first time. These include: chytridiomycosis, caused by *Batrachochytrium dendrobatidis* (Chytridiomycota), which affects one-third of amphibian species; white-nose disease, caused by *Pseudogymnoascus destructans* (Ascomycota), which has killed 6 million bats in North America; and marine aspergillosis, caused by *Aspergillus sydowii* (Ascomycota), which threatens sea fan corals in the Caribbean. Mutualisms with animals include basidiomycetes farmed by ants and termites, and fungi that occupy the gut microbiome of vertebrates and invertebrates. Chapter 10 concerns interactions between fungi and other heterotrophic microorganisms. These include the competition for resources between different colonies of soil fungi, fungal parasitism of other fungi (mycoparasitism), and ecological relations between fungi and bacteria.

Mycological research has provided compelling illustrations of widespread biological responses to planetary warming. These investigations are the subject of Chapter 11. Changes in the prevalence of certain fungi can be tracked in the palynological record. A dramatic decrease in the prevalence of spores from dung fungi in the Pleistocene reflects the mass extinction of megaherbivores like the woolly mammoth and mastodon, resulting from human activity. In recent decades, alterations in the seasonality of mushroom fruiting have been measured in Europe that are linked to the extension of the growing season and delay in the first frost. Climate change is also affecting the distribution of lichens. Mathematical models predict an increase in the concentration of airborne allergenic spores and changes in the geographical distribution of plant diseases as warming proceeds.

Biotechnology is the subject of Chapter 12. The term biotechnology is often reserved for modern industrial processes involving genetically modified organisms, but a broader reading includes baking and brewing practices that originated in the ancient world. Mushroom cultivation on logs and horse dung are other examples of early biotechnology that has flavoured the omnivorous diet of our species. Methods for controlling fungal fermentations developed in parallel with progress in microbiology since the nineteenth century. The use of fungi to produce antibiotics and other pharmaceutical products is a more recent part of this endeavour and the introduction of molecular genetic manipulation of fungal strains has had a major impact on the business of biotechnology.

This book is written for undergraduates and graduate students, and will also be useful for professional biologists interested in familiarising themselves with specific topics in fungal biology. Few scientists identify themselves as fungal biologists or mycologists. This is a reflection of the ethos of modern biology that emphasises the study of questions about cell and molecular biology, genetics, ecological interactions, and so on, rather than encouraging societies of scientists to investigate all aspects of a particular group of organisms. In the last 50 years, academic departments have hired fewer and fewer mycologists, phycologists, entomologists, ornithologists, and so on. This is not necessarily a bad thing. There is a vibrant international community of cell biologists who study yeast, and none of these scientists call themselves mycologists. Filamentous fungi, including the ascomycete *Neurospora crassa*, have also attracted a good deal of attention from cell and molecular biologists, few of whom identify as mycologists. Ecological research is benefitting from more synthetic studies on interactions between groups of organisms, rather than the exclusive study of fungi or any other taxonomic category. Our growing recognition of the enormous breadth of biodiversity, the unifying molecular characteristics of all organisms, and the wealth and complexity of interactions between species may see the extinction of specialists in single groups of organisms. It will be interesting to see how the culture of biological research and education evolves.

The scope of mycological research has expanded in many areas in the 15 years since the publication of the second edition of this textbook. Molecular methods that seemed innovative in 2001 have been replaced with novel technologies, matched with increased computing power. For example, the laborious sequencing of short lengths of DNA, which was standard practice in 2001, has been supplanted by techniques that allow fast sequencing of genes and whole genomes. New techniques in light microscopy, including live-cell imaging, confocal microscopy, and high-speed video, have also had a huge impact on fungal biology. The resulting change of pace in many research areas presents a challenge for authors committed to covering the whole discipline of fungal biology. The variety of approaches necessitated by the diversity of topics has meant that the stylistic unity of previous editions is no longer feasible, and each of us has assumed responsibility for a subset of the chapters. Cross-referencing and a comprehensive index allow the reader to navigate between topics throughout the book.

Acknowledgements

The concept of a textbook encompassing the whole of mycology from a microbiological perspective was the brainchild of Michael Carlile. The first edition was published in 1994. The second edition in 2001 benefited from the co-authorship of the late Graham Gooday. The continued vigour of the book is largely due to the generous help we have received from mycological colleagues around the world.

We wish to renew our thanks to our friends who read or commented upon chapters or sections and provided illustrations for the first and second editions of this book. We now add our gratitude to those who have provided similar help in the preparation of the third edition.

Mary Preap of Elsevier provided the editorial stimulus for this project, and we thank here for her great patience and professional support throughout the development of this book. We thank the following scientific colleagues who looked at drafts of individual chapters: Susan Kaminskyj advised on the coverage in Chapter 2; Sara Branco, Gareth Griffith and Bruce McDonald reviewed Chapter 4, along with Clive Brasier, Nick Kent, Ursula Kües and Louise Glass who helped with specific sections; LB wrote some of Chapter 4 while holding a Miller Visiting Professorship at the University of California, Berkeley; Geoff Gadd provided feedback on Chapter 5, and Dan Eastwood on Chapter 6, with comments on specific sections from David Hibbett; Björn Lindahl appraised Chapter 7, with input on specific sections from Paola Bonfante and Peter Crittenden; Ali Ashby provided invaluable assistance with Chapter 8; Duur Aanen, Elaine Bignell, Matthew Fisher and Fernando Vega checked Chapter 9; Peter Jeffries and Sarah Johnston commented on Chapter 10; Don A'Bear and Anders Dahlberg assessed Chapter 11 and Dan Eastwood minded the gaps in Chapter 12.

Felix Bärlocher, Tom Bruns, Sarah Gurr, Frank Gleason, Neil Gow, Håvard Kauserud, Paul Kirk, Thorunn Helgason, Rosemarie Honegger, Tim James, Naresh Magan, Jan Stenlid and John Taylor answered additional questions in their areas of expertise.

Maribeth Hassett and Don A'Bear were very helpful in preparing composite images. Gordon Beakes, Dimitrios Floudas, David Hibbett, Jonathan Plett, Frances Martin and Mark Fischer prepared new figures. We also thank everyone who provided copies of published illustrations and gave permission for their use.

We apologise to everyone else who offered advice and encouragement, but whom we have failed to mention above, and again offer our thanks to them.

Finally, we thank members of our families, Anthony and Charles Watkinson, Ruth Knight, Diana Davis, Colin, Emma and Hannah Morris.

Note on Taxonomy

The taxonomy of the fungi is in continuous flux, and opinions differ about the value of competing schemes of classification. In light of the pace of change in phylogenetic research, we have opted to omit a detailed list of taxa that would, inevitably, become outdated during the life of this book.

The description of phyla provided in Chapter 1 serves as a guide to the broad taxonomic relationships among the organisms featured in this book. We refer to three phyla of zoosporic fungi, namely, the Blastocladiomycota, Chytridiomycota, and Neocallimastigomycota, rather than combining them in a single phylum as some of our colleagues advocate. A new phylum of zoosporic fungi, the Cryptomycota, is included in some classifications; this intriguing assemblage of microorganisms is discussed in Chapter 1. Mycologists have recognised the phylum Zygomycota for many decades, but this is not a monophyletic group, and we use the informal name zygomycete(s) in this edition of The Fungi.

For more information on the taxonomy of individual groups of fungi, readers should consult the wealth of online sources and untangle the terminology with the help of Kirk, P.M., et al., 2008. Dictionary of the Fungi, 10th edition. CAB International, Wallingford, United Kingdom.

CHAPTER

1

Fungal Diversity

Nicholas P. Money
Miami University, Oxford, OH, USA

EVOLUTIONARY ORIGINS OF THE FUNGI AND THEIR RELATIONSHIPS TO OTHER EUKARYOTES

The fungi originated as a distinctive group of unicellular eukaryotes in the Precambrian. Recent estimates of the origin of the fungal kingdom based on the analysis of molecular clocks range from 760 million years ago to 1.06 billion years ago (Figure 1.1). The reliability of date estimates based on molecular clocks is dependent upon calibration points derived from the fossil record, which is far from satisfying for the fungi. The oldest unambiguous fossils of fungi are described from Lower Devonian Rhynie chert (400 million years old). These fossils show spectacular preservation and include chytrid sporangia and zoospores, zygomycete sporangia, and ascomycete fruit bodies (Figure 1.2). Spores and arbuscules of Glomeromycota are also found in the roots of plant fossils in the Rhynie chert and even older spores of these fungi are reported from 460 million-year-old rocks. These structures provide evidence for early symbioses that are thought to have been essential for the evolution of land plants. The oldest fossils of Basidiomycota are 330 million-year-old hyphae with clamp connections, but it seems likely that this phylum arose much earlier.

Evidence of an ancient origin for the prevailing fungal phyla leaves us with considerable uncertainty about the precise timing of the development of the ancestral species. Based on the structural simplicity of the chytrids and absence of flagella among other fungi, it is inferred that the earliest fungi may have been simple, unicellular organisms that propelled themselves through water using a flagellum anchored to the posterior of the cell. The signature of the posterior flagellum is common to fungi and animals and both groups are combined in the supergrouping of eukaryotes called the Opisthokonta. The cells of other kinds of motile eukaryotes are equipped with flagella connected to the anterior of the cell. The choanoflagellates are one of the basal groups of eukaryotes related to animals and their structural similarity to the chytrids is one of many lines of evidence of common ancestry (Figure 1.3).

The Fungi
http://dx.doi.org/10.1016/B978-0-12-382034-1.00001-3

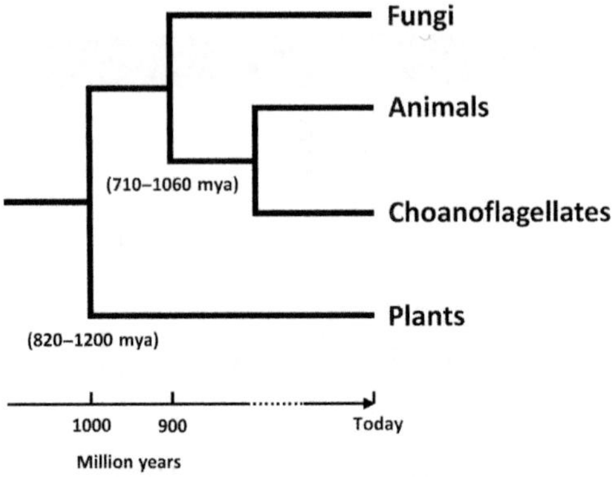

FIGURE 1.1 Evolutionary relationships between the fungi, animals, choanoflagellates, and plants with a timeline showing the emergence of the opisthokonts (fungi, animals, and choanoflagellates) approximately 1 billion years ago. The fungi are thought to have developed as a new kingdom of organisms between 710 and 1060 million years ago.

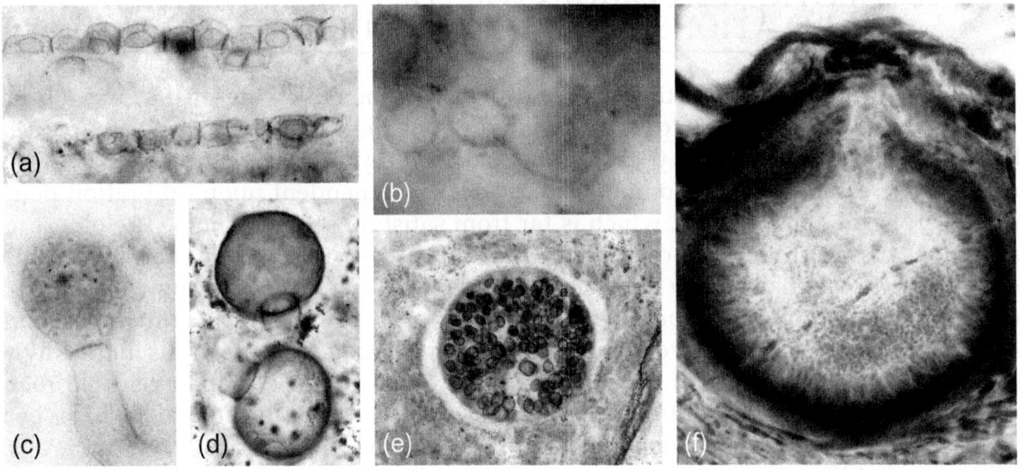

FIGURE 1.2 Fossil fungi in Lower Devonian Rhynie chert. (a) Zoosporangia of a chytrid inside host cells. Note exit tubes in two of the sporangia, through which zoospores were expelled. (b) Chytrid zoospore with (putative) single flagellum. (c) Zoosporangium of *Palaeoblastocladia milleri* (Blastocladiomycota). (d) Sporangia of a zygomycete fungus. (e) Sporocarp of species of arbuscular mycorrhizal fungus containing numerous spores. (f) Perithecium of *Palaeopyrenomycites devonicus* (Ascomycota). *Source: Thomas Taylor, University of Kansas.* (See the colour plate.)

THE CLASSIFICATION OF FUNGI

Pioneers of mycology sought to organise fungi into groups with shared characteristics and this morphological approach was successful in carving out the ascomycetes and the basidiomycetes and identifying sub-groups of closely related genera. In some cases, these pre-Darwinian schemes reflected evolutionary relatedness because some broad structural similarities among the fungi do reflect kinship. In many instances, however, these early investigations lumped unrelated organisms into groups defined by shared

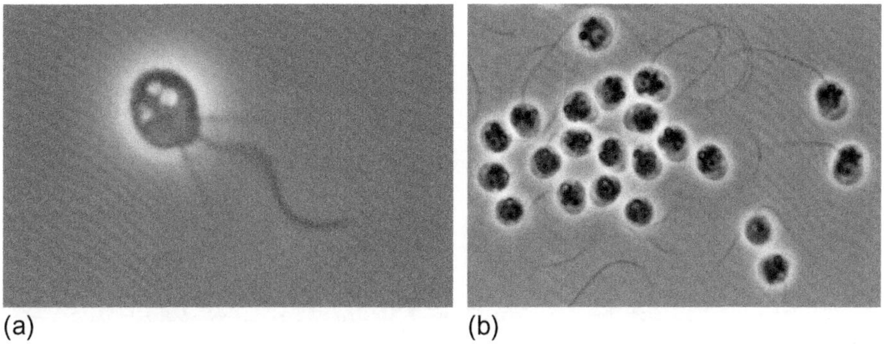

FIGURE 1.3 Similar morphology of single cells of a choanoflagellate and a chytrid fungus. (a) Marine choanoflagellate, *Monosiga brevicollis*. (b) Zoospore of freshwater chytrid, *Obelidium mucronatum*. *Source: (a) Stephen Fairclough (Creative Commons) and (b) Joyce Longcore, University of Maine.*

characteristics – such as spore colour – that held no evolutionary significance. Progress toward a natural classification of the fungi came with advances in microscopy and biochemistry in the twentieth century, but the advent of the polymerase chain reaction and other molecular genetic techniques in the 1980s transformed the study of fungal taxonomy. Since then, many of the traditional groups have been reorganised and renamed and the assignment of thousands of species remains in flux.

The study of fungal diversity spans the disciplines of taxonomy, classification, and systematics. Taxonomy concerns the identification, description, and naming of organisms; the assignment of these organisms to a hierarchy of groups – genus, family, order, and so on – is the task of classification; and systematics is the study of the evolutionary relationships between species and larger groupings of organisms. In theory, a fungal taxonomist might describe new species without considering their relationships to other fungi. She or he would proceed without any interest in systematics. In practice, modern taxonomists *are* concerned with evolutionary relationships between species, and fungal taxonomy has become synonymous with systematics. The goals of a field mycologist can be quite different. Classification is important to someone hunting for edible mushrooms: not recognising a death cap can be a fatal. The systematic position of the mushroom – its evolutionary relationship to other fungi – is immaterial to the collector. These differences in objectives have caused frustration among amateur mycologists who need reliable guides for identification (classification) but feel frustrated by frequent changes in scientific names. It probably makes sense for the amateur mycologist to ignore most academic research on systematics and concentrate on developing practical skills in identification. From the perspective of the biologist engaged in taxonomic research, however, the frequent name changes and adjustments to group assignments reflect the vibrancy of a field of inquiry that is benefiting from new technology and new ideas.

Thirty years ago, basidiomycetes including puffballs, earth-balls, earth-stars, stinkhorns, and bird's nest fungi, were placed in the Class Gasteromycetes. The only obvious shared characteristic among these fungi was their lack of the spore discharge mechanism found in the other mushroom-forming basidiomycetes and related rusts and smuts. In other words, this taxonomic entity was based upon the absence of a biomechanical character. Even then, however, mycologists who worked on these fungi were aware of many lines of evidence suggesting

FIGURE 1.4 Surprising relatives. (a, b) Field mushroom, *Agaricus arvensis* and puffball, *Lycoperdon perlatum*. (c, d) Bolete, *Boletus pinophilus* and earth-ball, *Scleroderma michiganense*. *Source: (a, c, d) Michael Kuo and (b) Pamela Kaminskyj.* (See the colour plate.)

that these organisms were not closely related. What we lacked, however, was proof of their relationships to other species. In the absence of this information, the Class Gasteromycetes was a convenient grouping for hundreds of distantly related basidiomycetes with an extraordinary range of fruit body types. With the introduction of molecular techniques for sequencing the genes of these fungi, and by comparing the similarity of these sequences with those of other basidiomycetes, the true relationships of the Gasteromycetes became clearer. The genetic data proves, for example, that puffballs are closely related to gilled mushrooms (Figure 1.4a, b). Additional information on the development of their spores suggests that puffballs evolved from ancestors with gills rather than vice versa. These findings are reflected in the modern classification that places the puffballs within the taxonomic Order Agaricales, which is the largest group of gilled mushrooms. Earth-balls produce fruit bodies that look similar to puffballs, but genetic comparisons show that these fungi are more closely related to mushrooms with tubes beneath their caps than to puffballs and their gilled relations. For this reason, the earth-balls are classified as members of the Order Boletales (Figure 1.4c, d).

The term gasteromycete remains a useful one for referring to the puffballs, earth-stars, and so on, but it is not a formal name of a taxonomic group today. The process of evolutionary convergence has often confused mycologists, as well as specialists in the taxonomy of other groups of organisms, leading them to guess at close phylogenetic affinities where none exist.

Mushrooms with gills offer another example of apparent kinship among fungi, whose fiction has been exposed by molecular phylogenetic studies. In addition to the Agaricales, fruit bodies with gills may have evolved independently within five other orders of basidiomycetes. There was no common gilled ancestor of gilled mushrooms. Molecular genetic analysis is the only objective method for sorting out who is related to whom.

Molecular Phylogenetic Analysis of the Fungi

Evolutionary relationships between fungi have been examined by comparing the sequences of ubiquitous eukaryote genes. Scrutiny of the ribosomal gene cluster is the most common choice for this research. This cluster encodes three subunits of ribosomal RNA (18S [**small subunit (SSU)**], and 5.8S and 28S [**large subunit (LSU)**] rRNA) and intervening regions, or **internal transcribed spacers**, identified as **ITS1** and **ITS2**. The two spacers plus the gene encoding the 5.8S subunit of the ribosome are called the **ITS region**. **External transcribed spacer sequences** (**ETS**) are located at the 5′ and 3′ ends of the cluster. ETS, ITS1, and ITS2 encode non-functional RNA molecules that are degraded when the functional rRNA is transcribed. Fungal genomes can contain more than 200 copies of the cluster organised in repeating sequences called **tandem repeats**. Individual copies are separated by a sequence called the **intergenic spacer region** (**IGS**) or **non-transcribed spacer region** (**NTS**). The high copy number of the gene cluster simplifies amplification using PCR even when samples of fungal DNA are very small.

An important attribute of the ITS sequences for molecular phylogenetic research is that they show significant variations between closely related fungi, and sometimes between populations within a single species. These variations are caused by insertions, deletions, and point mutations, which are conserved because ITS1 and ITS2 encode non-functional RNA molecules. Genes encoding functional products are subjected to stronger evolutionary pressure, which tends to dampen variation. In addition to its usefulness in phylogenetic research, portions of the ITS region are also amplified using diagnostic kits for rapid identification of fungi causing human infections and plant disease, and contaminating water-damaged buildings. Other genes used for phylogenetic research on fungi encode RNA subunits of the ribosome (LSU and SSU), and genes that encode proteins including translation factor 1-α, β-tubulin, actin, RNA polymerase II (*RPB1* and *RPB2*), and the minichromosome maintenance protein, *MCM7*.

Phylogenetic Trees

Evolutionary relationships between fungi are represented in the form of **phylogenetic trees**. The tips of the branches of these trees are occupied by living fungi and internal branch points, or nodes, represent their ancestors. Phylogenetic trees can be constructed using any measure of kinship, including morphological data, protein sequences, single genes, groups of genes, or, most reliably, by comparing whole genomes. Evolutionary relationships among fungi are analysed by comparing the ITS regions of multiple taxa. Trees derived from this information can be **rooted** or **unrooted**. Unrooted trees show the relatedness of organisms without indicating ancestry. Rooted trees converge upon a single **node** (the root) that represents a hypothetical common ancestor. Trees are often organised by including an **outgroup** that is selected as a plausibly distant relative that unites everything else in the tree. Trees that are displayed horizontally show close relatives one above the other on adjacent branches linked to a common ancestor. Connections between pairs of organisms, or groups of organisms, are discovered by following the branches back to shared branch points or internal nodes.

The passage of time is implicit in rooted trees, with the lengthiest pathways to an organism from the ancestor representing the longest evolutionary journeys. When fossils can be dated, actual time intervals can be estimated. Phylogenetic trees in which the lengths of the branches are scaled to evolutionary time are called **chronograms**. These are difficult to construct for fungi because the fossil record is so sparse. Nevertheless, the timing of the origins of some of the larger groupings has been estimated.

Once the necessary genetic sequences have been compiled and aligned with one another, phylogenetic trees can be constructed using a number of alternative methods. **Distance methods** organise sequences according to their overall similarity, computing the number of nucleotide substitutions between pairs of sequences. **Neighbour-joining** is the most popular of the distance methods and does not make any assumptions about evolutionary processes. **Character-based methods** are also used widely. These look at specific nucleotides and count insertions and deletions at each site. The advantage of these methods over neighbour-joining is that they can weigh the significance of changes in the nucleotide sequence differently. This is useful because some changes in a single sequence, relative to its comparison sequence, are uninformative. Neighbour-joining incorporates these changes into its trees; character-based methods can ignore these uninformative changes and focus on the more frequent and informative changes. **Maximum parsimony** is an example of a character-based method. This favours the simplest explanation for the similarities and differences between sequences. Parsimony looks at all possible trees and identifies the one that organises the sequences with the least evolutionary changes. **Maximum likelihood** is a more complicated character-based method that incorporates the lengths of branches into the tree that has the highest likelihood of being the correct representation of the phylogenetic relationships among the sequences. **Bayesian inference** is a third character-based method that generates a set of trees with roughly equal likelihoods.

When trees have been constructed, researchers can map various characteristics of the organisms represented in the tree onto its branches. This can inform us about the origin of particular features of cell biology, for example, or show how interactions between fungi and plant species evolved. Biogeographical patterns also emerge from these mapping exercises.

Operational Taxonomic Units and Environmental Sampling

Sequencing of fungal DNA allows researchers to extend the traditional taxonomic work of generations of scientists who described the morphology of fungi collected in the field and cultured in the laboratory. Unlike morphological characters, genetic markers provide an objective measure of relatedness and, for the first time, mycologists are able to organize different groups of fungi into a robust natural system of classification that reflects their evolutionary history. Molecular techniques are also used to study fungi that have not been cultured. By amplifying fungal genes from soil, water, or decaying wood, scientists have discovered new species and, more importantly, new larger groupings of fungi. Because we do not know how to grow these fungi, it is difficult to study them microscopically. While we can determine, for example, that a sample of estuarine mud contains the genes of fungi related to known species of chytrid, we do not know which of the cells seen with a microscope belong to these enigmatic microorganisms. In some instances we can do better than this and identify particular organisms in an environmental sample as the source of the amplified genes by hybridising genetic probes to the DNA inside their cells. These genetic probes are linked to a fluorescent dye so that the cells light up in the sample. This technique is called fluorescence in situ hybridization (**FISH**). Without this detailed investigation, we do not know whether genes that we have amplified have come from active fungi, from their inactive spores, or from remnants of the genome of damaged cells. Environmental sampling is a relatively new field of research for mycologists.

When we cannot determine whether a sequence has come from a distinctive species, we can describe the fungus as an **operational taxonomic unit** (**OTU**). An OTU can refer to any stipulated level of the taxonomic hierarchy, ranging from populations of organisms to single genetic sequences. Other names include **ENAS**, for environmental nucleic acid sequence, and **eMOTU**, for environmental molecular operational taxonomic unit.

Mycologists face a difficult challenge in reconciling traditional taxonomy with the wealth of new molecular data and with the revelation of an incredible variety of fungi that, until very recently, we had no idea existed. We know that similarities in morphology do not necessarily indicate shared heredity and that the malleability of fungal growth in response to changes in nutrient availability can mislead researchers into thinking that they are looking at distinct species. Another problem is encountered in the complex life cycles of some fungi, in which one organism is remodelled when it encounters a compatible mate for sexual reproduction and develops in a different way when it engages in clonal reproduction. One species can be mistaken for two or more species. By ignoring structure, molecular phylogenetic analysis has been very effective in assigning 'problematic' fungi to natural taxonomic groups. The use of these techniques has been less helpful in distinguishing between individual species because we do not have any independent measure of the degree of genetic difference to anticipate between fungal species. Some investigators suggest that a 3% difference in ITS sequence between two fungal cultures, or collections of fruit bodies, is indicative of separate species, but this judgement is not founded on sound experimental evidence. The percentage differences in ITS sequences between species are certain to vary among different groups of fungi. Resolution of these profound uncertainties is the great challenge of modern taxonomic research.

Implications of Molecular Phylogeny

Fungal classification is driven by the goal of producing a truly natural scheme that reflects evolutionary relationships between species. It is worth considering why this is viewed as an important objective. By recognising the evolutionary affinities of species, we immediately obtain useful information on the biology of a particular fungus by referring to what is known about its relatives. Interesting features of ecology, physiology, cell biology, and metabolism are often shared by related species, and so knowledge of a larger taxonomic group can inform research on a particular species. If, for example, two species in a fungal genus are capable of detoxifying oil-contaminated soil, it might be productive to study the biochemical attributes of other members of the genus. A second justification for investing time and money in this effort comes from the fundamental desire of biologists to understand nature. The pursuit of a natural classification of the fungi is revealing a wealth of information about the origins of these organisms, including estimates of the timing of the development of the major phyla. This is part of the story of life on Earth and there are few more interesting topics in science.

An important consideration in thinking about fungal classification is that the species with which we share our lives represent a tiny fraction of all of the fungi that have existed during the hundreds of millions of years of mycological history. This means that the closest relatives of many present-day species diverged from their common ancestor a very long time ago. In the intervening aeons of time, vast numbers of extinctions have pared away the majority of species, leaving huge gaps in the evolutionary tree. This makes it very difficult to unravel relationships between groups of fungi based on molecular analysis of living organisms. The scale of biological extinction is a huge challenge for phylogenetic studies on other groups of organisms, but the problem is alleviated when we look at the evolution of vertebrates, for example, by the availability of a rich fossil record. For these reasons, much of the evolutionary history of the fungi is obscure and this limits confidence in the details of the current natural, or evolutionary, classification of this diverse kingdom of microorganisms. With these shortcomings in mind, however, our picture of fungal diversity is incomparably richer than the textbook view of mycology even a generation ago. We turn now to the contemporary view of fungal classification.

FUNGAL PHYLA

Six phyla are recognised in the current taxonomic arrangement of the fungi:

Kingdom Fungi

Phylum Basidiomycota
Phylum Ascomycota
Phylum Glomeromycota
Phylum Blastocladiomycota
Phylum Chytridiomycota
Phylum Neocallimastigomycota

The evolutionary relationships between these phyla is shown in Figure 1.5. The name **Dikarya** has been proposed as a subkingdom that comprises the Basidiomycota and Ascomycota. The name refers to the formation of cells containing two nuclei during the process of sexual reproduction in these fungi (described later in this chapter). Genetic data indicate that some fungi do not fit neatly within any of the six phyla, but we do not have sufficient information to delineate additional phyla. These include 900 or more species of zygomycete, including *Mucor mucedo* that is used to illustrate simple sexual life cycles in laboratory classes. These organisms were identified as the Phylum Zygomycota in earlier classifications, but because these fungi do not seem to represent a cohesive phylogenetic assemblage, we will refer to them as zygomycetes in this book. Other fungi grouped in earlier classifications within the Zygomycota have been assigned to other phyla. In addition to the phyla listed above, there are more than 1000 species of animal parasite called microsporidians that many taxonomists regard as part of Kingdom Fungi. They are treated as a separate phylum called the Microspora in some classifications, but their evolutionary relationship to the other fungal phyla is not understood. In this portion of the book we will survey each of the major groups of fungi, describe their unifying characteristics, and provide examples of species.

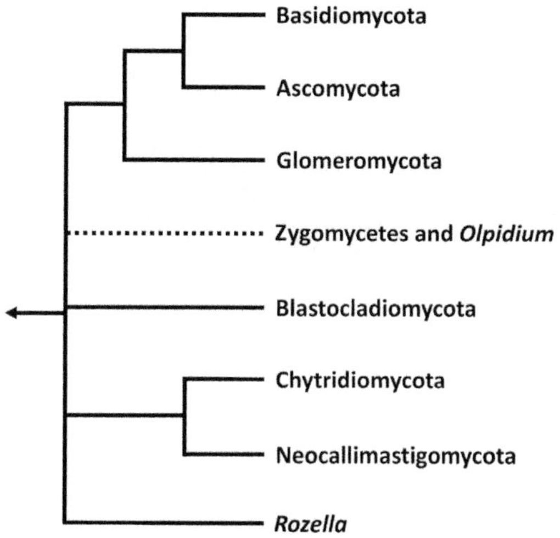

FIGURE 1.5 Phylogenetic tree showing relationships within Kingdom Fungi. *Rozella* is a genus of aquatic fungi that does not fit into existing phyla. It is a representative of a diverse group of fungi identified using molecular genetic methods that has been proposed as a seventh phylum called the Cryptomycota.

Phylum Basidiomycota

Overview and Unifying Characteristics

The 30,000 described species of Basidiomycota include mushroom-forming fungi, jelly fungi, yeasts, rusts, and smuts (Figure 1.6). These fungi play a multitude of ecological roles. Many mushroom-forming species support forest ecology through the formation of mycorrhizal symbioses with trees and shrubs and others by decomposing wood and leaf litter (Chapters 5 and 7). Basidiomycetes partner with termites and leaf-cutter ants in complex symbioses in which the fungi are farmed by the insects, and other basidiomycetes feed upon scale insects that they protect under a crust. Basidiomycetes also cause animal diseases, including potentially lethal infections of humans and the rusts and smuts are among the most important pathogens of plants (Chapters 8 and 9).

Molecular phylogenetic analyses are effective at separating the majority of the Basidiomycota from other fungi and this genetic distinction is reflected in a handful of shared structural and developmental characteristics. The Basidiomycota produce spores called **basidiospores**. Basidiospores form on the outside of cells called **basidia**. Nuclear fusion and meiosis occur within the basidia and the resulting tetrad of haploid nuclei is transmitted into buds that differentiate into the basidiospores. There are many variations upon this typical life cycle, including the formation of basidiospores on the surface of individual haploid yeast cells following mitotic division of the nucleus. A second common feature of the group is the **dolipore septum** that partitions successive compartments along the length of the basidiomycete hyphae. This septum is perforated by a central canal (pore) that is defined by a barrel-shaped swelling of the septum cell wall. Structural details are provided in Chapter 2. Nuclei cannot migrate through unmodified dolipore septa and their distribution within the hyphae that develop after the fusion of sexually compatible colonies involves the formation of **clamp connections** (see next section). Clamp connections are a third characteristic of this phylum. All

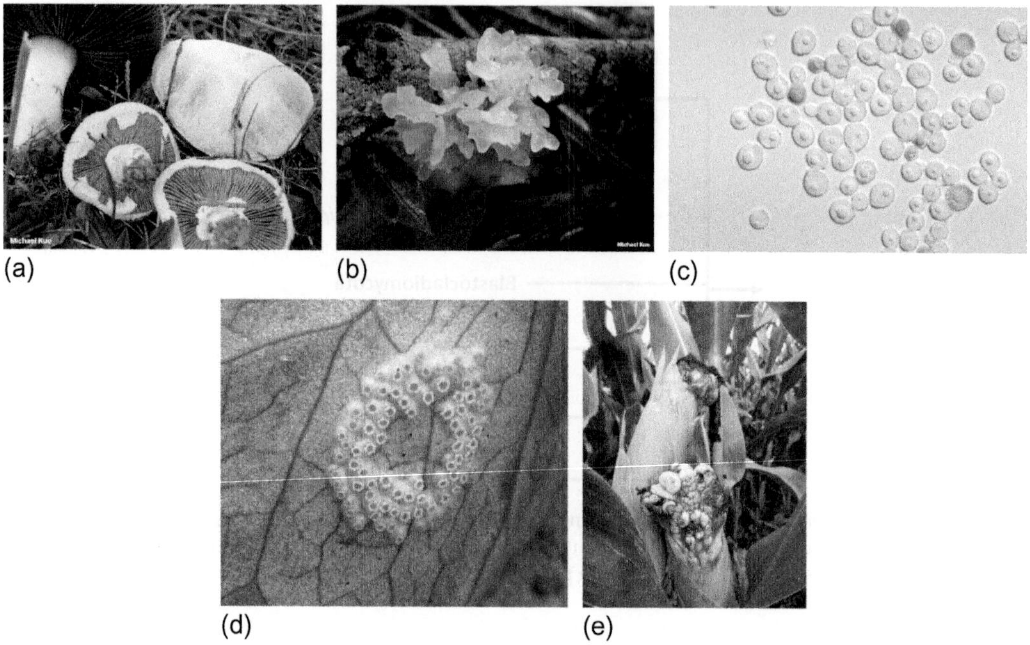

FIGURE 1.6 Diversity of the Basidiomycota. (a) *Agaricus campestris*, the field mushroom. (b) *Tremella fuciformis*, a jelly fungus. (c) *Malassezia globosa*, yeast associated with dandruff. (d) *Puccinia sessilis*, a rust, on *Arum maculatum*. (e) *Ustilago maydis*, corn smut. *Source: (a, b) Michael Kuo, (c) http://www.pfdb.net/photo/nishiyama_y/box20010917/wide/024.jpg, (d) http://upload.wikimedia.org/wikipedia/commons/f/f8/Puccinia_sessilis_0521.jpg, and (e) http://aktuell.ruhr-uni-bochum.de/mam/images/pi2012/begerow_maisbeulenbrand.jpg* (See the colour plate.)

basidiomycetes produce basidiospores, but basidiomycetes that grow exclusively as yeasts do not form them on basidia, nor do these fungi produce septa or clamp connections. But, any fungus with dolipore septa or clamp connections is a member of the Basidiomycota.

Basidiomycete Life Cycle

There are some important differences among the life cycles of the Basidiomycota, but the following description of the developmental processes typical of a mushroom-forming species serves as a template for understanding the way that other basidiomycetes operate (Figure 1.7). A single basidiospore germinates to form a branching colony of filamentous hyphae. This young mycelium is composed of multiple compartments, separated by dolipore septa, each containing a single haploid nucleus in most species. These nuclei are formed by mitosis, beginning with the first division of the single nucleus within the founding spore. The resulting mycelium is termed a homokaryon because all of its nuclei are genetically identical clones. It is also called a monokaryon because the compartments contain single nuclei (see page 103–106). Homokaryotic mycelia expand through soil and wood, or other food materials. Some species can produce fruit bodies without mating, but the homokaryons of most mushroom-forming fungi fuse with other homokaryons before fruiting. When a pair of sexually compatible homokayons merge they produce **heterokaryotic mycelia** in which each hyphal compartment contains a pair of nuclei, one derived from each mate. Compatibility is

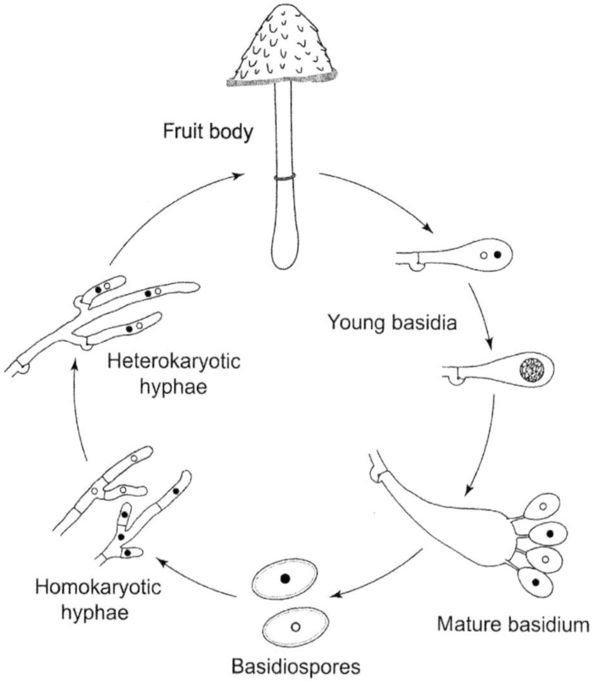

FIGURE 1.7 Life cycle of *Coprinus comatus*, the lawyer's wig. This is an example of a mushroom with a bipolar mating system that involves the fusion of homokaryotic colonies whose nuclei contain different alleles of a single mating type gene (open and closed circles). The resulting heterokaryon contains nuclei of both mating types. Fusion of these nuclei, followed by meiosis, occurs within basidia to produce four basidiospores.

determined by mating type genes (Chapter 4). The terms **dikaryon** and **dikaryotic mycelium** are also used to describe these colonies.

Clamp connections are crucial in the development of the heterokaryon (Figure 1.8). Consider an apical compartment of a heterokaryotic hypha: it contains two nuclei, one derived from each original mate. When these nuclei divide, a septum will form at a right angle across each mitotic spindle, creating three hyphal compartments; without the clamp connection, only one of the three compartments would be likely to contain nuclei of both mating types. Clamp connections are lateral branches that create bridges for nuclear movement between regions of the hypha that will become separate compartments. They allow the developing heterokaryon to populate each new compartment with nuclei of both mating types. This process of heterokaryon formation precedes the formation of fruit bodies in most Basidiomycota. It is regulated by interactions between the mating type genes within the cell that only permit clamp connections to form when the homokaryotic parents are sufficiently unrelated.

The elaboration of a mushroom from the colony of feeding filaments begins with the development of a knot of hyphae. As this pinhead-sized aggregate of cells enlarges, stem, cap, and gills become visible, priming the developing embryo, or button, for rapid expansion into the mature reproductive organ as soon as environmental conditions are conducive. **Basidia** form at the ends of hyphae whose tips stop growing at the surfaces of gills, spines, the interior of tubes, and other locations of spore production. These fertile tissue of a fruit

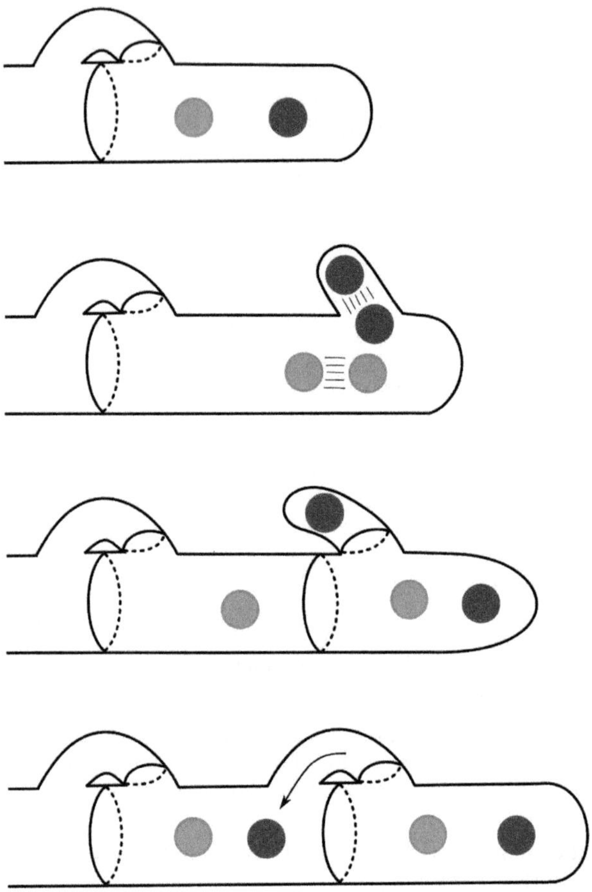

FIGURE 1.8 Formation of clamp connections on heterokaryotic basidiomycete hypha. *Source: Creative Commons.* (See the colour plate.)

body is referred to as the **hymenium**. Like the billions of other cell compartments within the tissue of the mushroom, the young basidia are heterokaryotic, containing copies of the nuclei derived from the parent homokaryotic colonies. These fuse and then undergo meiosis to produce four haploid nuclei, each of which is packaged into one of the four basidiospores formed from the basidium. (Four is the usual number of spores per basidium, but some species produce single spores from each basidium, and pairs and multiple spores are also quite common configurations.) The spores are discharged from the gill surface by a mechanism described as a surface-tension catapult that is detailed in Chapter 3. Relative to spore production over a flat surface, gills achieve a maximum 20-fold increase in surface area, and the output of spores from some mushrooms is astonishing. A single basidiome of *Agaricus campestris* can discharge 2.7 billion microscopic spores per day, or 31,000 spores per second; 2 million tubes on the underside of a bracket of the wood-rotting basidiomycete *Ganoderma applanatum* can shed 30 billion spores per day, or more than 5 trillion in the 6 months that the perennial bracket is active each year!

Taxonomic Groups Within the Basidiomycota

SUBPHYLUM: AGARICOMYCOTINA (MUSHROOMS, JELLY FUNGI, AND YEASTS)

The Agaricomycotina includes all of the Basidiomycota that form macroscopic fruit bodies, including mushrooms and jelly fungi, along with a variety of basidiomycete yeasts. There are three classes: **Agaricomycetes**, **Dacrymycetes**, and **Tremellomycetes**.

The Agaricomycetes contains the 16,000 species of identified mushroom-forming species. The variety of fruit body (**basidiome**) morphology in this group is remarkable: stalked umbrella-shaped mushrooms and brackets produce spores on the surface of gills or teeth, over flat or rippled surfaces, or over the inner surfaces of tubes; other species form branched basidiomes that resemble corals; basidiomes can form as gelatinous cushions or sticky, ear-shaped growths, and Agaricomycetes also produce vast numbers of basidiospores within closed fruit bodies, including puffballs and earth-stars (Figure 1.4). There are Agaricomycetes that develop underground, resembling ascomycete truffles, and others that form packets of spores in egg-like structures (peridioles) that are splashed from the cups of bird's nest fungi or propelled into the air by species of artillery fungus. As discussed in the 'The Classification of Fungi' section, the morphology of the basidiome provides little guidance in understanding evolutionary affinity among the Agaricomycetes. Some individual orders contain species with all manner of mushroom types. Examples of the orders will be introduced here.

Agaricales

This is the largest order within the Agaricomycotina, containing 8500 or more described species including the cultivated button mushroom, *Agaricus bisporus*, the ink caps (*Coprinus* and other genera), the lethal death cap, *Amanita phalloides*, hallucinogenic species of *Psilocybe*, and the world's largest organisms, *Armillaria gallica* and *Armillaria solidipes*. Most of these mushrooms are saprotrophs, feeding on plant debris in soils and rotting wood, or ectomycorrhizal fungi that absorb sugars from the roots of living trees and shrubs. A handful of species form mutualistic symbioses with ants and termites, and a few of the Agaricales are important plant pathogens. The first genome of a gilled mushroom, called *Laccaria bicolor*, was sequenced in 2008. This fungus forms mycorrhizas with pine, fir, birch, and poplar trees and is important in tree nurseries where it is added to soil to boost seedling growth. Its genome is quite large, containing 65 million base pairs (As and Ts, Gs and Cs), but is dwarfed by the 3 billion or so rungs on the ladder of the human genome. When we look at the number of genes that encode proteins, the comparison is more humbling: 20,000 for the mushroom, and between 20,000 and 25,000 for us. The fungus produces hundreds of enzymes that dissolve proteins, fats, and carbohydrates in the soil, but it lacks the usual catalysts for decomposing the cellulose and lignin of plant cell walls that are secreted by many other mushrooms. This is significant because it seems likely that *Laccaria* and other mycorrhizal fungi evolved from wood-rotting ancestors and must have lost these enzymes as they adopted their new lifestyle. In addition to fungi with umbrella-shaped mushrooms, the Agaricales encompasses species with clavarioid (cylindrical or club-shaped) fruit bodies, puffballs, and the bird's nest fungi with their extraordinary fluted fruit bodies from which spore-containing peridioles are ejected by raindrops.

Boletales

The Boletales include 300 species of *Boletus* whose spores are discharged from the surfaces of tubes beneath the mushroom cap. These fungi form ectomycorrhizas with the roots of forest trees. *Boletus edulis*, known as the king mushroom, cep or porcini, is a very important edible wild mushroom. It is harvested in Italy, Eastern Europe, China, Southern Africa, and North America. The dry rot fungus, *Serpula lacrymans*, is a saprotrophic member of the Boletales. This fungus is tremendously important as a cause of wood decomposition in buildings and is thought to have evolved from mycorrhizal ancestors. The cellar fungus, *Coniophora puteana*, is another member of the Boletales that damages buildings. There is some morphological diversity within the order, including the earth-balls (*Scleroderma* species, Figure 1.4d) that produce spherical fruit bodies superficially resembling puffballs, and the fruit bodies of *S. lacrymans* and *C. puteana* that take the form of crusts on the surface of the rotting wood.

Russulales

Species form all kinds of fruit bodies in this order (Figure 1.9), including mushrooms with gills (*Russula* and *Lactarius* [milk caps]), teeth (*Auriscalpium*), and tubes (*Bondarzewia*), others forming little lozenges and flattened crusts (*Peniophora* and *Aleurodiscus*), and still others producing intricate coral shapes (*Clavicorona*). There are mycorrhizal, saprotrophic, and a few parasitic Russulales.

Polyporales

Most of the 1800 or so species of Polyporales (shelf fungi) cause rot in standing trees and fallen logs. Their activity as wood decomposers is vital to the health of forest ecosystems. Many species are saprotrophs and grow exclusively on standing dead wood, logs, and other woody debris. Others, including species of *Ganoderma* and *Fomes*, attack living tissues and

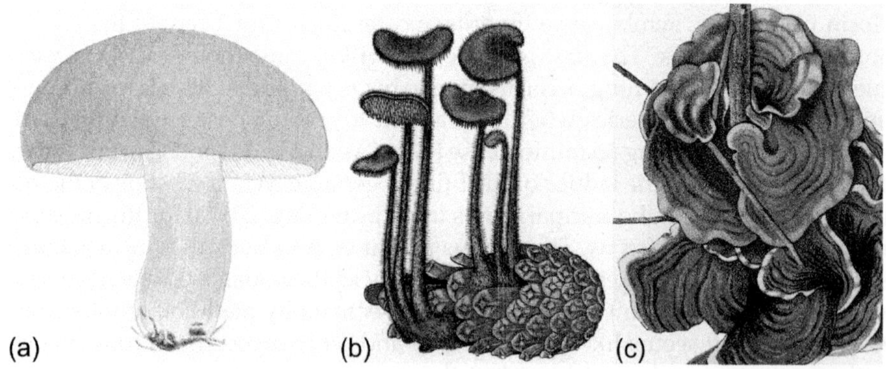

FIGURE 1.9 Diversity of fruit body morphology in the Russulales, Agaricomycotina. (a) Gilled mushroom, *Russula lepida*; (b) hydnoid or toothed fruit body of *Auriscalpium vulgare*; (c) flattened fruit body of *Stereum ostrea*, the false turkey tail. *Source: (a) Lange, J.E., 1940. Flora Agaricina Danica, vol. 5. Recato, Copenhagen; (b) Bulliard, P., 1791. Histoire des Champignons de la France. Chez L'auteur, Barrois, Belin, Croullebois, Bazan, Paris; (c) Inzenga, G., 1869. Funghi Siciliani Studii, vol. 2. Di Francesco Lao, Palermo.*

continue to decompose the wood of their dead hosts. *Ganoderma lucidum* (reishi) and other Polyporales have been used as natural medicines in traditional Chinese medicine. Purified cell wall polysaccharides from these fungi have a wide range of pharmacological activities. The majority of the bracket fungi that shed spores from tubes beneath their caps are Polyporales, but the order also includes hundreds of **corticioid** species whose fruit bodies are flattened crusts on the surface of logs.

Phallales

Stinkhorns and their relatives are among the minority of Basidiomycota that have lost the catapult mechanism of spore discharge (ballistospory) and interact with animals that disperse their spores. The common stinkhorn, *Phallus impudicus,* expands an epigeous (above ground) fruit body with a slime covered head in which the spores are embedded. A variety of volatile compounds diffuse from the slime (gleba) and attract carrion flies and other invertebrates. These animals act as vectors for stinkhorn dispersal when they carry the basidiospores on their bodies as well as consuming and defecating them some distance from the parent fruit body. The order includes other species whose foul-scented glebal slime is exposed on the surface of fruit body tissues that expand into cages (*Clathrus*) and star shapes (*Anthurus*).

Auriculariales

The edible Judas ear mushroom, *Auricularia auricula-judae,* is the best known fungus in this order. Its ear-shaped fruit bodies develop from mycelia on elder (*Sambucus*) and other woody plants. Basidiospores form on elongated basidia that are partitioned into four compartments by transverse septa. These form on the lower surface of the 'ears' and the spores are catapulted into the air beneath the fruit body. The fruit bodies have a rubbery texture and have been cultivated for centuries in Asia. *Exidia glandulosa,* witches' butter, is another species in this order that forms black fruit bodies on decaying wood. Its basidia are divided lengthwise into four compartments.

Jelly Fungi: Dacrymycetes and Tremellomycetes

Like the Auriculariales, jelly fungi in this pair of taxonomic classes form distinctive basidia. The basidium of the Dacrymycetales is shaped like a tuning fork and forms one spore at the tip of each of its two branches (epibasidia). In the Tremellomycetes, the basidium is divided into four separate compartments by septa that run through the cell longitudinally. Basidiospores develop at the tips of epibasidia that extend from each of the compartments. The fruit bodies of some species of both classes of jelly fungus are brightly coloured. In the Dacrymycetes, *Dacrymyces stillatus* forms tiny orange cushions on wet decomposing wood and *Calocera* species sprout orange spikes in similar locations. Colourful Tremellomycetes include the bright yellow or orange membranous fruit bodies of *Tremella mesenterica.* Its relative *Tremella fuciformis* is cultivated in China and is called the silver ear fungus. The life cycles of the Tremellomycetes involve switching between hyphal and yeast phases. This process occurs in *Cryptococcus neoformans* that adopts a budding unicellular morphology in the human nervous system where it causes life-threatening infections (Chapter 9). Species of *Tremella* are mycoparasites that feed on the colonies of other fungi in decomposing wood and some species form their gelatinous fruit bodies on the basidiomata of their hosts.

Subphylum: Ustilaginomycotina (Smuts)

The smuts are obligate pathogens of plants that cause devastating diseases of cereal crops including common bunt or stinking smut of wheat, caused by *Tilletia caries*, and corn smut (of maize), caused by *Ustilago maydis* (Chapter 8). The majority of the more than 1000 species of smut fungi infect flowering plants, but some cause diseases in conifers, ferns, and lycophytes. Related fungi within the Ustilaginomycotina include the yeast *Malassezia globosa* that is part of the natural microbial community on the human scalp and is associated with dandruff.

U. maydis has been studied in greater detail than other smuts. Genetic manipulation of this fungus, and the sequencing of its genome, has allowed researchers to use the smut as a model system for understanding interactions between pathogens and hosts plants. The fungus has also been used in cancer research. Disruption of a gene in the smut called *brh2*, which is related to the human tumour suppressor gene *BRCA2*, results in a deficiency in DNA repair mechanisms. This is consistent with a mechanistic link between mutations in the human gene and an increased risk of developing breast cancer.

The life cycle of *U. maydis* differs greatly from the mushroom life cycle in its details, but it shares the heterokaryotic phase with other members of the phylum (Figure 1.10). The smut alternates between a budding yeast phase and a filamentous mycelium; the yeast phase is saprotrophic and can be cultured in the lab, but the mycelium grows only within the tissues of the host plant. The yeast cells contain single haploid nuclei and mating, to form the heterokaryon, is controlled by two genetic loci, *a* and *b* (Chapter 4). The *a* locus contains genes that encode the precursors of sex hormones and receptors to these hormones. There are two versions of these loci, *a1* and *a2*, and the fungus can mate only with a strain of the opposite mating type; the *a1* pheromone docks with the *a2* receptor, and vice versa. When compatible strains are paired, they form conjugation tubes that grow toward one another and fuse at their tips. Fusion produces a dikaryon but this is stable and capable of infecting the host only when mating occurs between strains that differ in their *b* locus alleles. The *b* locus contains a pair of genes that encode transcription factors. There are more than 25 different *b* locus alleles, which means that there are hundreds of different combinations of compatible mating types that produce pathogenic heterokaryons.

As the heterokaryon penetrates the tissues of its host plant, it causes the development of tumours or galls. Tumour development within the flowers of *Zea mays* results in the formation of masses of swollen kernels. The mycelium within these galls is converted into blackened spores (the source of the name 'smut'), called **teliospores**, and nuclear fusion occurs during this process so that each spore contains a single diploid nucleus. The teliospores are dispersed by wind, but they can also lie dormant during winter months. Under appropriate environmental conditions, the diploid nucleus divides by meiosis and the spore germinates to produce a filamentous **promycelium** that forms haploid buds, or **sporidia**, on its surface. These sporidia are the source of the yeast phase that proliferates on the surface of the host plant before mating, dikaryon formation, and infection. The teliospore plays a comparable role to the basidium in the mushroom life cycle, as the site for nuclear fusion and subsequent meiosis. The spore-filled galls or tumours on maize are a Mexican delicacy called huitlacoche, which is used in soups, as a flavourful filling in tamales, in ice cream, and many other dishes.

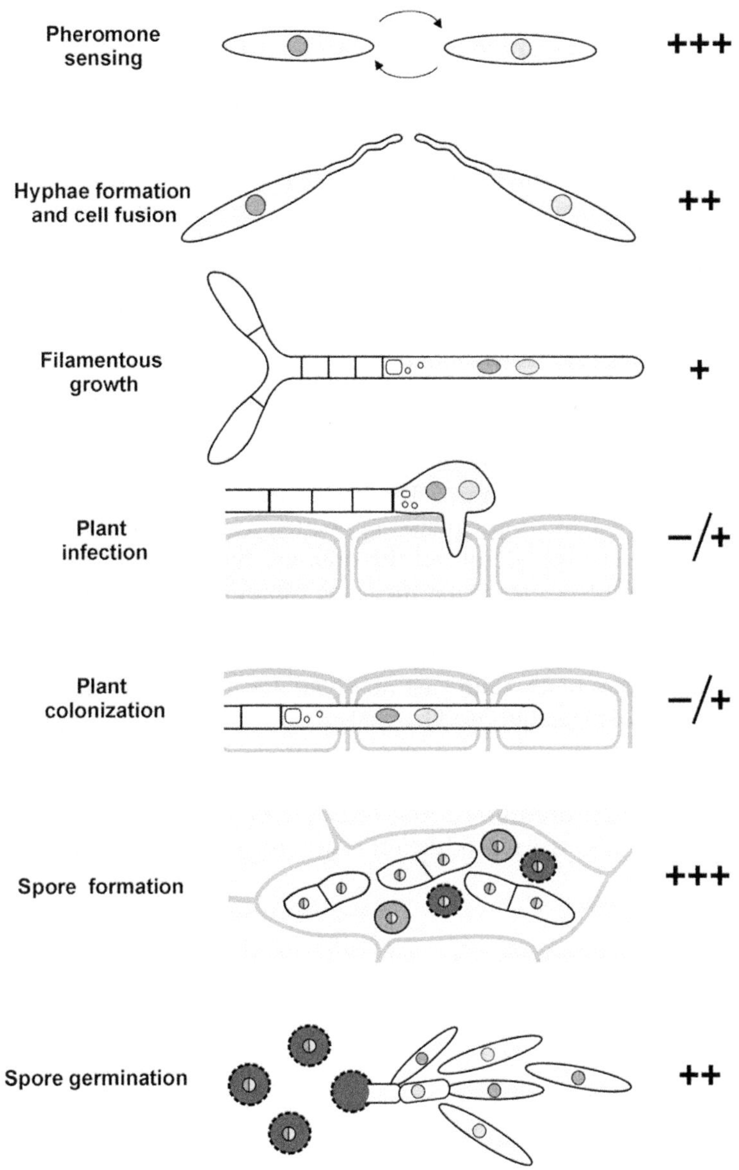

FIGURE 1.10 Life cycle of corn smut, *Ustilago maydis*. *Source: Fuchs, U. et al., 2006. Endocytosis is essential for pathogenic development in the corn smut fungus Ustilago maydis. Plant Cell 18, 2066–2081.*

The fungus has been eaten for centuries and was part of Aztec, Hopi, and Zuni cuisine long before the Spanish conquests in the sixteenth century.

The basidiomycete affinity of *U. maydis* is evident from its genome and is reflected in the dikaryotic phase of its life cycle. Like other basidiomycetes, the dikaryotic hyphae of the smut are partitioned by septa, but these are relatively simple structures compared with the dolipore septa of mushroom-forming species. *Ustilago* species also lack the ballistospore discharge mechanism that is emblematic of the phylum Basidiomycota. The mechanism does operate, however, in other smut fungi, including *T. caries*. There are many variations among the life cycles of the smut fungi. *Microbotryum violaceum* has a much simpler mating system than corn smut, with sexual compatibility controlled by a single genetic locus with a pair of alleles. This fungus infects the anthers of campion species (Caryophyllaceae), *Silene dioica* and *Silene alba*, that produce separate male and female plants (i.e. they are dioecious species). When the fungus infects the female flowers, it suppresses the formation of the ovaries and stimulates the production of stamens. The infected anthers fill with teliospores, completely subverting plant reproduction, and butterflies and other insect pollinators disperse the parasite's teliospores.

Subphylum Pucciniomycotina (Rusts and Allies)

Most of the more than 7000 species within this subphylum are rust fungi that are obligate parasites of plants. Septobasidiaceae form mutualistic symbioses with scale insects are closely related to the rusts. More distantly related Pucciniomycotina include plant pathogens that were classified originally as smuts. The basidiomycete affinities of the rusts are indicated by the formation of a dikaryotic phase in the life cycle that precedes nuclear fusion and meiosis. The rusts produce septa, but not the dolipore type of other basidiomycetes. They do, however, release ballistospores using the same catapult mechanism characteristic of the mushroom-forming species. The life cycle of many rusts is exceedingly complicated, involving as many as four distinct types of spore.

Puccinia graminis causes black stem rust of wheat and is often used to illustrate the processes of infection and dispersal utilised by rusts to move between two species of host plant and to recombine genes by sexual reproduction (Figure 1.11). Rusts often have alternate hosts and this cereal pathogen infects barberry, *Berberis vulgaris* and *Berberis canadensis*. *Puccinia* forms four types of spores: **urediniospores**, **teliospores**, and **basidiospores** on wheat, and **aeciospores** on barberry. Sexual reproduction is achieved by the transfer of gametes, called

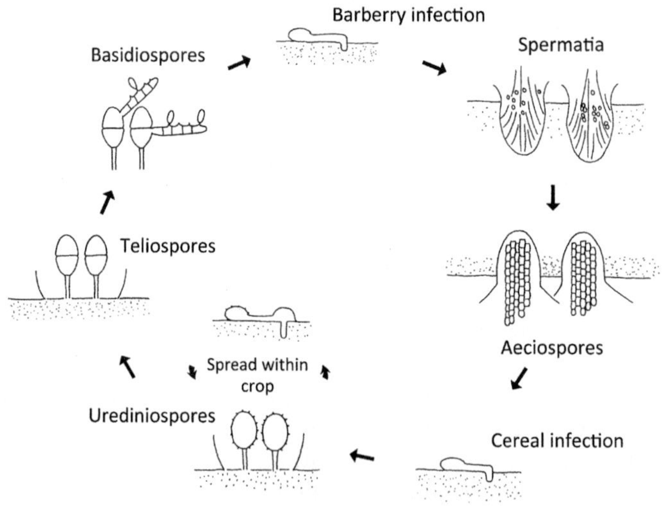

FIGURE 1.11 Life cycle of the rust, *Puccinia graminis*. *Source: Money, N.P., 2002. Mr. Bloomfield's Orchard. The Mysterious World of Mushrooms, Molds, and Mycologists. Oxford University Press, New York.*

spermatia (singular **spermatium**), which are transferred between pustules on barberry leaves by insects. The name rust refers to the reddish lesions that develop on wheat leaves and shed the reddish urediniospores. Urediniospores are dikaryotic, containing a pair of nuclei. They are dispersed by wind and germinate on leaf surfaces under humid conditions when the stomata are likely to be open. When the germ tube locates a stoma it inflates over the opening, forming an **appressorium**, and penetrates the leaf. The infectious mycelium develops between the leaf cells and produces branches called **haustoria** that penetrate the host cell walls. The haustoria form a tight connection with the host cell without destroying its plasma membrane, supporting sustained nutrient absorption from the living plant. This is characteristic of **biotrophic pathogens**. The rust can complete multiple rounds of urediniospore production, release, and cereal infection, allowing the disease to spread as an epidemic among susceptible hosts over a wide region. Toward the end of the growing season, uredial lesions are converted into black streaks producing teliospores. These telial lesions are responsible for the black stem reference in the common name for the disease. Teliospores of *P. graminis* have two cells supported on a stalk. Initially, each of the cells contains a pair of haploid nuclei, but these fuse to form a single diploid nuclei. In this condition, the thick-walled teliospores can survive within stubble during cold winter months. The teliospores germinate in the spring to produce a short septate hypha or promycelium from each cell. The single diploid nucleus in each cell of the teliospore divides by meiosis and the resulting haploid nuclei are packaged in four basidiospores. These spores are discharged into the air by the same catapult mechanism described in mushrooms and they infect barberry leaves.

Within the barberry leaf, the rust produces feeding haustoria again and forms flask-shaped structures on the upper leaf surface, called **spermagonia** that exude spermatia in a sugar-rich fluid. This nectar attracts flies and other insects that transfer spermatia between spermagonia on the same plant and to neighbouring plants. Each spermatium acts as a gamete, its single haploid nucleus fertilising a compatible spermagonium. There are two mating types (+) and (−) and cross fertilisation between these mating types converts the existing mycelium in the barberry leaf into a dikaryon. The final spore type, the aeciospore, is formed in pustules called **aecia** on the underside of the leaf. The dikaryotic aeciospores are spread by wind and infect susceptible wheat plants.

Other rusts have less complex life cycles missing one or more of the spore-producing stages described for *P. graminis*, or complete their life cycles on a single host species. *Gymnosporangium globosum*, which causes hawthorn rust, produces its spermatia and aeciospores on hawthorn (*Crataegus*) and its teliospores and basidiospores on evergreens. It lacks the urediniospore stage of the life cycle. Coffee rust is caused by *Hemileia vastatrix* that produces urediniospores, teliospores, and basidiospores on coffee (*Coffea*). The function of the basidiospores is unknown because a second host has not been identified. *Uromyces viciae-fabae*, which causes rust of faba bean (broad bean, *Vicia faba*), produces all four of the spore types and spermatia on the same species of host plant.

Phylum Ascomycota

Overview and Unifying Characteristics

The Ascomycota is the largest phylum of fungi encompassing more than 33,000 named species and a vast number of undescribed fungi. The phylum includes yeasts and filamentous fungi, fungi that partner with algae and cyanobacteria to form **lichen symbioses**, mycorrhizal species, saprotrophs, and pathogens of plants and animals. Ascomycetes are utilised in industrial applications, in food production and flavouring, and the fruit bodies of morels and truffles are prized edible fungi. Many species are known only as asexual fungi (**anamorphs**) that produce asexual spores (**conidia**) on stalks called **conidiophores** (Figure 1.12), but sexual phases (**teleomorphs**) have been identified in the life cycles of most ascomycetes that have been studied in detail. The sexual organs formed by ascomycetes are called **ascomata** (s. **ascoma**, Figure 1.13). Ascomata include open cup-shaped fruit bodies (**apothecia**), flask-shaped structures with a single vent for spore release (**perithecia**), and fruit bodies that develop as closed structures that open in a variety of ways to release spores (**cleistothecia**). Ascomata contain the characteristic spore-producing cells of the phylum called **asci** (s. **ascus**).

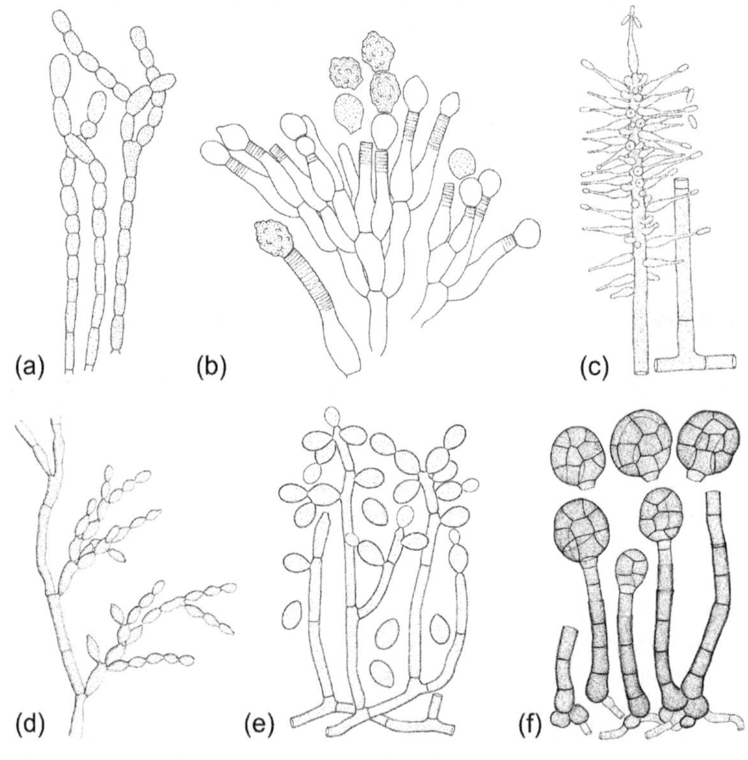

FIGURE 1.12 Selection of conidial stages, or anamorphs, of ascomycetes. (a) *Basipetospora variabilis*, a soil fungus. (b) *Scopulariopsis brevicaulis*, a saprotroph that grows in soil and causes opportunistic infections of humans. (c) *Harziella* (*Lepisticola*) *capitata*, which grows on fruit bodies of *Lepista nuda* (blewits). (d) *Tyrannosorus pinicola*, isolated from rotting wood. (e) *Haplotrichum chilense*, isolated from wood. (f) *Junewangia globulosa*, isolated from rotting plant stems. *Source: (a) Minter, D.W., Kirk, P.M., Sutton, B.C., 1983. Thallic phialides. Trans. Br. Mycol. Soc. 80, 39–66; (b) Minter, D.W., Kirk, P.M., Sutton, B.C., 1983. Holoblastic phialides. Trans. Br. Mycol. Soc. 79, 75–93; (c) Gams, W., Seifert, K.A., Morgan-Jones, G., 2009. New and validated hyphomycete taxa to resolve nomenclatural and taxonomic issues. Mycotaxon 110, 89–108; (d) Müller, E., et al., 1987. Taxonomy and anamorphs of the Herpotrichellaceae with notes on generic synonymy. Trans. Br. Mycol. Soc. 88, 63–74; (e) Partridge, E.C., Baker, W.A., Morgan-Jones, G., 2001. Notes on Hyphomycetes. LXXXII. A further contribution toward a monograph of the genus Haplotrichum. Mycotaxon 78, 127–160; (f) Baker, W.A., Partridge, E.C., Morgan-Jones, G., 2002. Notes on Hyphomycetes. LXXXV. Junewangia, a genus in which to classify four Acrodictys species and a new taxon. Mycotaxon 81, 293–319.*

FIGURE 1.13 Examples of fruit bodies, or ascomata, produced by Ascomycota in Subphylum Pezizomycotina. (a) Goblet shaped apothecia of *Urnula craterium*, the Devil's urn. (b) Flattened discoid apothecia on the thallus of a species of the lichen *Xanthoparmelia*. (c) Highly modified apothecium of the summer truffle, *Tuber aestivum*. (d) Perithecial stroma of *Cordyceps militaris* fruiting from parasitized caterpillar. *Source: (a) Michael Kuo, (b) http://en.wikipedia.org/wiki/Lichen#/media/File:Lichen_reproduction1.jpg, (c) http://upload.wikimedia.org/wikipedia/commons/8/89/Tuber_aestivum_Valnerina_018.jpg, (d) http://upload.wikimedia.org/wikipedia/commons/4/44/2008-12-14_Cordyceps_militaris_3107128906.jpg* (See the colour plate.)

The sexual spores of ascomycetes, called **ascospores**, form inside asci. This internal development of ascospores contrasts with the production of basidiospores on the outside of basidia in the Basidiomycota (compare Figure 1.14 with Figure 1.7). The hyphae of ascomycetes lack the dolipore septa and clamp connections of basidiomycetes; their septa have a single, central pore. Mobile organelles (microbodies) with dense protein cores, called **Woronin bodies**, plug the septal pores and isolate damaged hyphal compartments from the rest of the colony. These organelles are found in the largest subphylum, the Pezizomycotina, which contains 90% of the Ascomycota, but are absent from the other members of the phylum (whose groups are detailed below).

Ascomycete Life Cycle

Like the Basidiomycota, it is impossible to detail a single life cycle that is applicable to all species within the Ascomycota. The perithecial ascomycete, *Neurospora crassa*, is used to illustrate the life cycle here and a few of the variations in other species are discussed when the

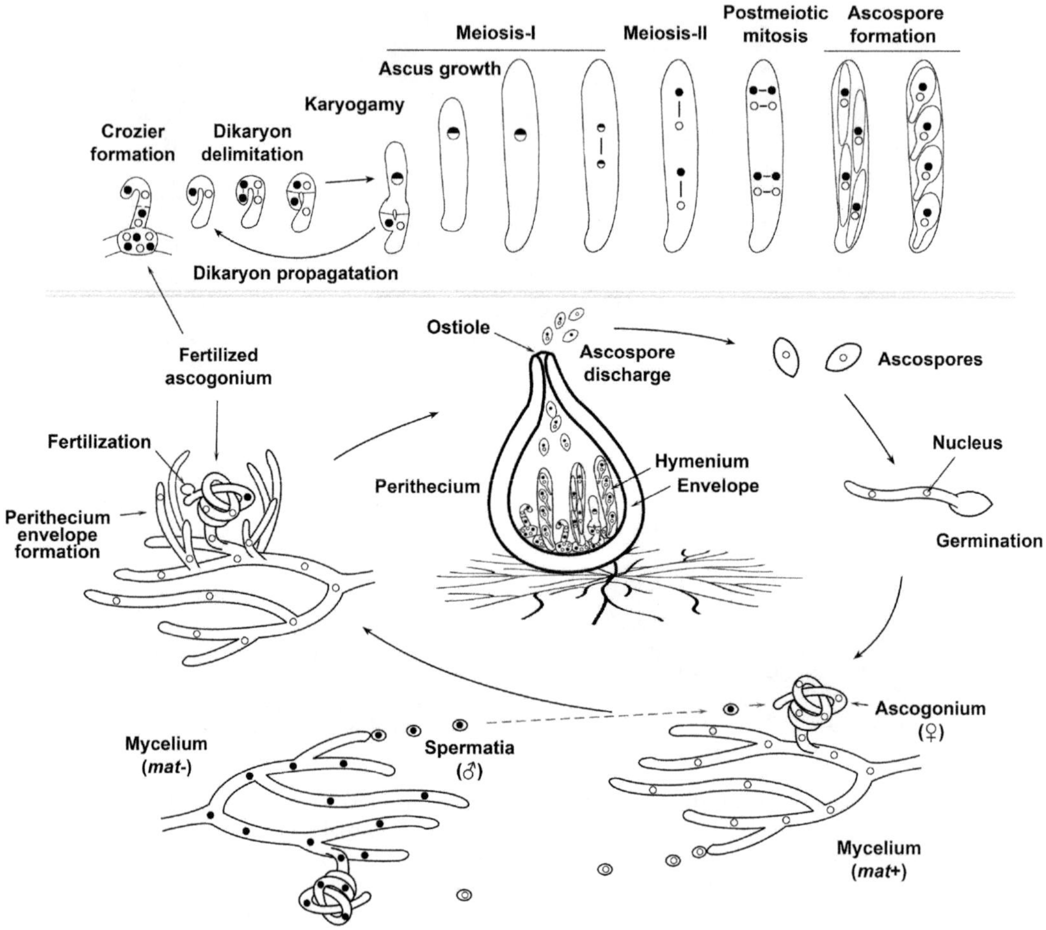

FIGURE 1.14 Sexual cycle of a filamentous ascomycete. *Source: Peraza-Reyes, L., Berteaux-Lecellier, V., 2013. Peroxisomes and sexual development in fungi. Front. Physiol. 4, 244.*

sub-phyla are introduced (Figure 1.14). Ascospores of *Neurospora* contain haploid nuclei that are replicated by mitosis when the spore germinates and distributed within the developing colony. The colony expands by hyphal extension and repeated branching, and septa form between multinucleate compartments. *N. crassa* is a heterothallic species with two mating types called *MAT A* and *MAT a*. All of the nuclei in a colony derived from a single ascospore are of the same mating type; the colony is homokaryotic. Sexual reproduction requires the formation of a heterokaryon that contains nuclei of both mating types. Unlike the basidiomycetes, ascomycetes do not form the extensive heterokaryotic colony or dikaryon. In *Neurospora*, a short-lived heterokaryon is formed within the developing ascoma. Colonies produce airborne **microconidia** (also called spermatia) that act as gametes, and these fuse with hyphae called **trichogynes** that project from the developing ascomata. The young ascoma is called the **protoperithecium**. Fusion between a microconidium and trichogyne of opposite mating type produces a heterokaryotic or ascogenous hypha. Different mating processes in other ascomycetes include (i) fusion of swollen male and female gametangia, and (ii) fusion of undifferentiated hyphae.

As the perithecium of *Neurospora* develops, the ascogenous hyphae form hook-shaped tips, called **croziers**. A pair of nuclei at the end of the crozier divides by mitosis and septum formation isolates two of the resulting four nuclei, one of each mating type, in the crook of the crozier. This cell extends to form the ascus. Its pair of nuclei fuse, and then divide by meiosis to produce four haploid nuclei. In *N. crassa*, each of these nuclei divides by mitosis and each of the resulting eight nuclei is packaged into a separate ascospore. A second mitotic division occurs within each developing ascospore so that it is equipped with an identical pair of nuclei. Additional divisions may occur so that the ascospore becomes multinucleate. After the formation of the ascus, the ascogenous hypha can form additional croziers and generate a cluster of asci within the fruit body.

The nuclear events that take place in the crozier are comparable to those within the young basidium of the Basidiomycota. (The process of spore formation in both groups involves the differentiation of cells containing pairs of nuclei, which is the characteristic that unites these phyla in the subkingdom Dikarya.) The subsequent development of the spores is very different in the pair of largest phyla in Kingdom Fungi and, currently, there is no evidence of evolutionary homology between the crozier and basidium. There are a lot of variations among the ascosporogenesis process in the Ascomycota, but in all cases nuclei along with portions of cytoplasm are separated within the ascus and packaged within individual cell walls. In most species, this is achieved by the formation of a cylinder of two membranes inside the ascus cytoplasm. This double membrane lies inside the plasma membrane that lines the inner surface of the ascus cell wall and is called the **ascospore-delimiting membrane**. The ascospore-delimiting membrane folds and fuses around the developing spores, and their cell walls are deposited in the sandwich formed by this structure. The cytoplasm that remains around the spores is called the epiplasm. This develops a mucilaginous consistency in some species and, in others, forms a watery fluid. The epiplasm is expelled with the spores during the discharge process (Chapter 3).

Taxonomic Groups Within the Ascomycota

SUBPHYLUM: PEZIZOMYCOTINA

Ten or more classes are designated within the Pezizomycotina. The subphylum includes saprotrophs that grow on woody and non-woody plant tissues, and on herbivore dung, parasites of plants and animals (particularly invertebrates), and partners in lichen and mycorrhizal symbioses. The variety of fruit bodies in the Pezizomycotina is remarkable. The cup fungi, with various types of apothecia, include the morels and truffles, many of the lichenized ascomycetes, species of *Rhytisma* whose apothecia are embedded in tar spots on leaves, and earth-tongues with club-shaped apothecia. The asci in these fungi open via lids in operculate species and by fracture of the ascus tip in inoperculate species. The asci are highly modified in the truffles and do not discharge spores into the air. Perithecial ascomycetes show comparable morphological diversity, ranging from species of *Neurospora*, *Sordaria*, and *Podospora* that form individual perithecia with a diameter of a few tenths of one millimetre, to *Xylaria*, *Daldinia*, and *Nectria* species whose multiple perithecia are embedded in the surface of larger fruit bodies called **perithecial stromata**. These fungi are ubiquitous in forest habitats where their colonies blacken decaying wood and their stromata develop as finger-like projections from the ground, and tiny antlers, single or multiple spheres, and crusts on the surfaces of logs. Perithecial stromata are also produced by *Cordyceps*, *Metarhizium*, and other arthropod pathogens, the cereal pathogen *Claviceps purpurea* (ergot), and *Epichloë* and related endophytes in grasses (Chapter 7). Tiny elongated perithecia formed by Laboulbeniales develop on the surface of insects and other arthropods, attached to the cuticle via a foot structure. These ascomycetes are regarded as parasites, but do not appear to incapacitate their hosts.

Cleistothecial ascomycetes include species of *Aspergillus* and *Penicillium* whose conidial phases are crucial in biotechnological applications including fermentation and food production, and the production of enzymes and antibiotics (Chapter 12). *Aspergillus* species are also significant as a source of carcinogenic aflatoxins in contaminated food and as pathogens of immune-compromised human patients where they cause **aspergillosis** (Chapter 9). Spore release from cleistothecia involves the release of asci by disintegration of the fruit body wall. In a different type of ascoma, called a **chasmothecium**, the asci are exposed to the air when the fruit body opens along a preformed line of weakness. *Erysiphe*, *Blumeria*, and other genera of powdery mildews form chasmothecia. Further structural diversity is seen in ascomycetes with **fissitunicate** asci in which spore release entails the splitting of the outer wall of the ascus, followed by inflation of the inner ascus wall before forcible discharge of the spores (Chapter 3). The fissitunicate asci develop inside distinctive types of ascomata called **pseudothecia** and **hysterothecia**. Ascomycetes with fissitunicate asci include many important plant pathogens including species of *Pleospora*, and the sexual stages of *Cochliobolus*, *Venturia*, and *Cladosporium*. *Alternaria* species are saprotrophs and plant pathogens, whose abundant airborne conidia are a significant contributors to asthma. Sexual stages are not known in these fungi, but molecular data show that *Alternaria* species are related to *Pleospora*. *Sporormiella* is another of the ascomycetes that produces fissitunicate asci. *Sporormiella* species grow on herbivore dung. Their spores are so prevalent in the palynological record that changes in their abundance identify the mass extinction of large herbivores in North America at the end of the Pleistocene.

SUBPHYLUM: SACCHAROMYCOTINA

Fungi in this group produce budding yeast cells. Morphological diversity is limited, but pseudohyphae, consisting of short bud-producing filaments, are formed under certain conditions, and some of the species produce tip-growing hyphae and elongated asci. In nature, the yeast *Saccharomyces cerevisiae* thrives on the sugar-rich diet available in ripened fruits. Its ability to metabolise glucose under conditions of low oxygen availability and produce carbon dioxide and alcohol has allowed humans to brew beer, make bread, and ferment wine and other alcoholic beverages for thousands of years. The yeast cell reproduces as a haploid cell by budding. When the single nucleus divides by mitosis, one of the daughter nuclei is shifted into the bud. A single cell may produce a series of buds and each leaves a scar on the surface of the mother cell upon separation. There are two mating types in *Saccharomyces*, designated a and α. Cells of opposite (compatible) mating type fuse to produce a single diploid cell and meiosis converts this cell into an ascus containing four haploid ascospores.

Candida is a large genus of saprotrophic yeast species that grow on many different foods including soil and rotting wood. *Candida* species have been isolated from seawater and fish guts, but the best known of these yeasts is *Candida albicans* that is a resident in 80% of human population where it grows in the gut. *C. albicans* can also cause infections ranging from proliferation of the fungus in the oral or vaginal mucosa (thrush), to life-threatening invasive growth of the hyphal phase of the fungus (**invasive candidiasis**) in patients with debilitated immune systems. Yeast cells of *C. albicans* are diploid and mating produces a tetraploid zygote. A parasexual programme of chromosome loss is responsible for restoring the normal diploid complement of chromosomes. *Dipodascus* is another genus within the Saccharomycotina. Species of *Dipodascus* switch between a yeast and hyphal phase, and their haploid cells conjugate to produce elongated asci that contain multiple spores.

SUBPHYLUM: TAPHRINOMYCOTINA

This subphylum contains five classes: Schizosaccharomycetes, Pneumocystidiomycetes, Neolectomycetes, Taphrinomycetes, and Archaeorhizomycetes. The Class Schizosaccharomycetes contains only three species of **fission yeast**, including *Schizosaccharomyces pombe*. Fission yeasts differ from budding yeasts in their mechanism of cell division: fission yeasts divide through the formation of a septum in the middle of the cell. This yeast is a saprotroph that grows on sugar-rich foods (e.g. fruits and honey). It is used as a model organism for research on the eukaryotic cell cycle and was the subject of investigations that were recognised by the award of the Nobel Prize for Medicine and Physiology in 2001 to Paul Nurse and colleagues. *Pneumocystis jirovecii* (Pneumocystidiomycetes) is an opportunistic pathogen of humans that causes pneumonia in patients with damaged immune systems. Species of *Neolecta* (Neolectomycetes) are the only members of this subphylum that produce a macroscopic fruit body: the other species are yeasts, although the plant parasitic Taphrinomycetes also form colonies of tip-growing hyphae within their host plants. The Archaeorhizomycetes were described in 2011 as an assemblage of hundreds of species of filamentous fungi whose rDNA was amplified from soil samples. These ascomycetes are associated with roots without forming typical mycorrhizal or endophytic structures. Although some of the Archaerhizomycetes have been cultured, little is known about their biology.

Phylum Glomeromycota

Ninety percent of all plants form mycorrhizal associations with fungi. The most widespread of these are arbuscular mycorrhizas produced by fewer than 200 species of fungi within the Phylum Glomeromycota. Genera include *Glomus*, *Acaulospora*, and *Gigaspora*. Nutrient transfer between the fungus and host plant occurs through intricately branched microscopic structures called **arbuscules** (meaning 'dwarf trees') that develop inside the living root cells (Figure 1.15). The mycorrhizal symbiosis is obligate for the fungi and they cannot be cultured in isolation from their hosts. Cellular characteristics that help identify these fungi include the development of non-septate hyphae and large spores (up to 800 μm in diameter) with multilayered walls. The spores are formed individually or in clusters. In some species, spore clusters are surrounded by a rind of hyphae forming a fruit body called a **sporocarp**. Each spore contains hundreds or thousands of nuclei. Sexual reproduction does not occur within these fungi but mixed populations of genetically distinct nuclei coexist within individual spores. It seems likely that these multi-genomic fungi have evolved through the accumulation of mutations without recombination (or with very infrequent recombination). New symbioses are formed from hyphal fragments in the soil, or when spores germinate close to the root surfaces, and fungi can also spread from a colonised plant to neighbouring hosts.

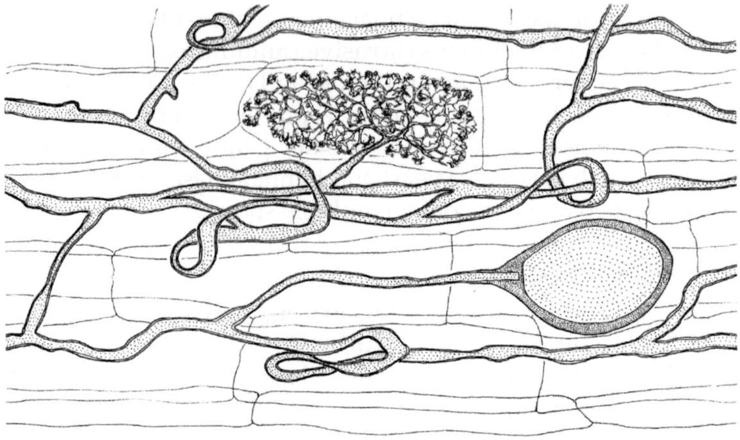

FIGURE 1.15 Arbuscular mycorrhizal fungus, Phylum Glomeromycota. Mycelium within plant root is connected to a single arbuscule and thick-walled spore. *Source: Roo Vandegrift.*

Geosiphon pyriforme is the only example of a non-mycorrhizal species within the Glomeromycota. It produces inflated bladder cells that contain endosymbiotic nitrogen-fixing cyanobacteria (*Nostoc punctiforme*). *Geosiphon* may be a member of an ancestral group within the phylum from which the mycorrhizal species evolved.

Phylum Blastocladiomycota

Blastocladiomycota, along with the Chytridiomycota and Neocallimastigomycota, are aquatic fungi that produce flagellate zoospores. It is important to underscore the absence of flagella in the majority of the fungi. No flagella are produced by the Basidiomycota, Ascomycota, Glomeromycota, and filamentous zygomycetes. Some molecular phylogenetic studies have concluded that the loss of flagella occurred once in the fungal lineage, suggesting that there was a single common ancestor for all of the non-flagellate groups. Blastocladiomycota live in freshwater habitats, mud, and soil where they operate as saprotrophs, decomposing plant and animal debris, or parasitize arthropods. Less than 200 species have been described. *Allomyces* species are saprotrophs that form separate haploid and diploid colonies with an unusual morphology. When *Allomyces* is grown on agar medium, it forms branched colonies of broad hyphae that lack septa. In liquid medium, or in samples of pond water, the hyphae are often stunted, producing short colonies attached to surfaces by a basal network of fine filaments referred to as **rhizoids**. Haploid and diploid colonies look the same, but when nutrients become limited, the hyphae stop extending and produce different types of reproductive structures at their tips (Figure 1.16). The diploid colony is called the **sporophyte** (drawing upon botanical nomenclature). This forms two different types of sporangia: **zoosporangia** and **meiosporangia**. The zoosporangia release diploid zoospores. Each spore has a single flagellum that pushes the spore through the water like a miniature tadpole. The single diploid nucleus in the spore contains a large nucleolus and is surrounded by a membrane-bound assemblage of ribosomes called the **nuclear cap**. This spore structure is one of the distinguishing features of the Blastocladiomycota. The

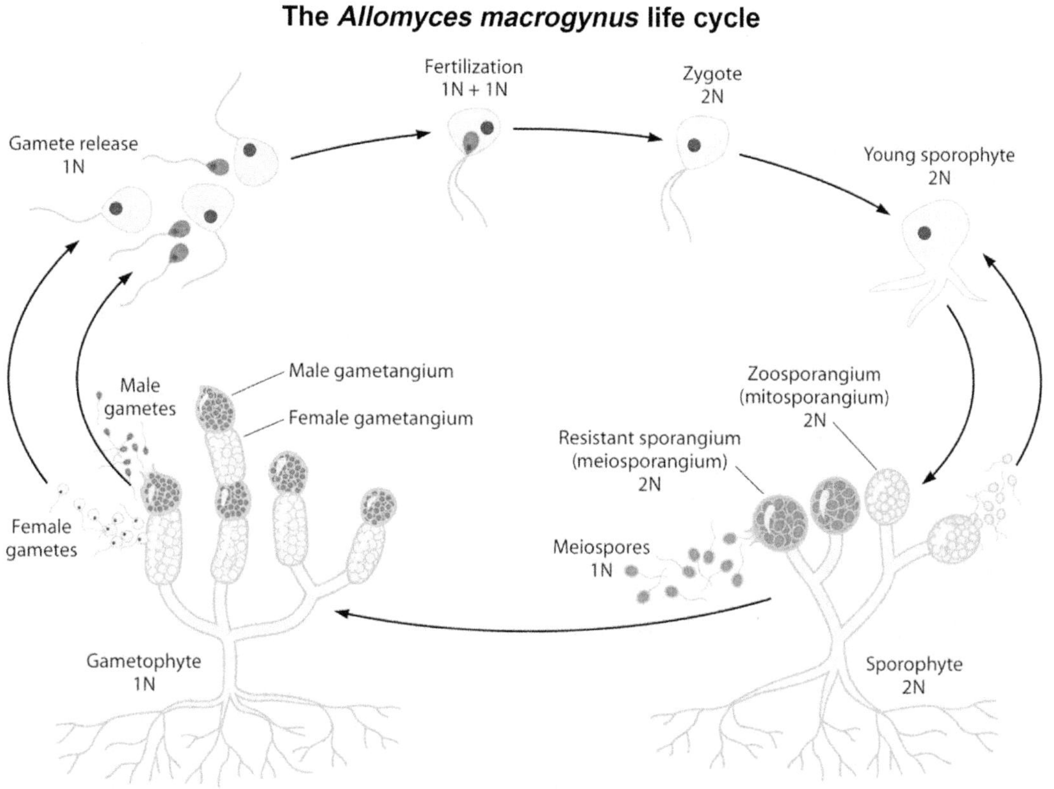

FIGURE 1.16 Life cycle of *Allomyces* (Blastocladiomycota). *Source: Lee, S.C., 2010. Microbiol. Mol. Biol. Rev. 74, 298–340.*

spores are chemotactic and direct their motion toward sources of dissolved amino acids. If they locate suitable food, the zoospores attach to the surface of the target, encyst, and form rhizoids that penetrate the underlying material. Branching hyphae of the new colony develop from the opposite side of the cyst and extend into the water. The importance of nutrient absorption by the rhizoids versus the hyphae is unclear, but may be determined by the relative concentrations of nutrients in the food base and within the surrounding water. Hyphal cultures on solid medium probably function like the cultured mycelia of other fungi, with most of the absorption of nutrients occurring at the hyphal apices as the colony periphery extends into fresh medium.

The second type of sporangium, the meiosporangium, also releases swimming spores, but these are formed by meiosis and give rise to haploid or **gametophyte** colonies. These colonies develop in the same fashion as the sporophytes, but produce terminal structures, which look like sporangia that release motile gametes rather than zoospores. The gamete-releasing structures are called **gametangia**. In *Allomyces macrogynus*, the male gametangia are formed at the ends of the hyphae, with the female gametangia directly behind them. The opposite arrangement occurs in *Allomyces arbusculus*. The male gametangia are coloured bright orange with gamma-carotene. The female gametangia and gametes release a sexual attractant, or pheromone, called **sirenin** to which the male gametes respond. After their release, male gametes swarm around the female gametangia and fuse with the emerging female gametes. The fused gametes produce a biflagellate zygote that swims through the water until it locates a suitable food source and encysts. Upon germination, the cyst produces a new sporophyte colony and the life cycle processes can be repeated.

The zoospores of *Blastocladiella emersonii* have a very similar structure to those of *Allomyces*, but this fungus produces an ovoid thallus rather than the more extended colony of branched hyphae characteristic of *Allomyces*. Nutrient limitation triggers the transformation of the thallus into a sporangium from which zoospores are discharged into the water. Species of a third genus in the Blastocladiomycota, *Coelomomyces*, are parasites of arthropods. *Coelomomyces psorophorae* has a complicated life cycle, reminiscent of the biology of some rusts, which involves the infection of mosquito larvae and copepods. Prospects for the development of *Coelomomyces* species as biocontrol agents against mosquito-borne infectious diseases seemed bright after the elucidation of its life cycle in the 1970s, but attempts to implement control methods have been unsuccessful.

Phylum Chytridiomycota

Chytridiomycota (true chytrids) are aerobic zoosporic fungi that operate as saprotrophs and pathogens in freshwater, brackish, and marine habitats, and are also abundant in soil. The host range of pathogenic chytrids includes other fungi, algae, plants, and amphibians. Pathogenic species include *Synchytrium endobioticum*, that causes wart disease in potatoes, and *Batrachochytrium dendrobatidis*, which causes amphibian **chytridiomycosis** that is implicated in the global decline in amphibian populations and extinction of multiple species. Like the Blastocladiomycota, the Chytridiomycota produce uniflagellate zoospores (Figure 1.3b). They do not form colonies of branched hyphae (mycelia) like non-flagellate fungi, but produce multinucleate, spheroidal bodies referred to as thalli (Figure 1.17). These are the

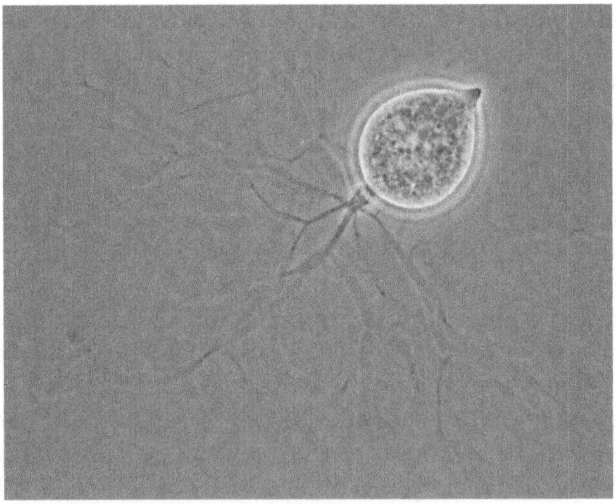

FIGURE 1.17 Thallus of the chytrid *Obelidium mucronatum* in process of differentiation into a sporangium from which zoospores will be released. Thin rhizoids spread from the base of the thallus. *Source: Joyce Longcore, University of Maine.*

feeding structures that absorb nutrients from their surroundings, and are transformed into sporangia that release zoospores when nutrients become limiting. Zoospore release occurs through discharge tubes that expand from the thalli (Chapter 3). The presence or absence of a lid at the tip of the discharge tube (**operculate** and **inoperculate** conditions) is a diagnostic feature for these fungi. The thalli of many chytrids are anchored to solid materials by finely branched rhizoids. The thalli of some parasitic species develop inside the cells of their hosts, others on the host surface, and multiple thalli of chytrids with the most complex morphology can be connected together in chains via a system of rhizoids to resemble the colonies of non-flagellate fungi. Chytrid zoospores are haploid and it is not clear whether sexual reproduction occurs in the life cycles of most chytrids. Sexual reproduction may involve fusion of haploid zoospores. If it does occur, sexual reproduction is followed by the development of a resting spore or sporangium and meiosis produces a new generation of recombinant haploid zoospores.

Phylum Neocallimastigomycota

The 20 species in this highly specialised phylum are anaerobic fungi that grow in the digestive tracts of herbivores. Their metabolism is fuelled by the decomposition of polysaccharides in the fibre of the host diet to produce sugars: the Neocallimastigomycota are **fibrolytic**. These fungi lack mitochondria. In the absence of oxygen, pyruvate oxidation is catalysed by pyruvate:ferredoxin oxidoreductase in hydrogenosomes with electron donation to a [Fe]-hydrogenase that produces hydrogen. Uniflagellate zoospores are produced by these fungi but multiflagellate spores are also described in some species.

Where Do *Olpidium* and *Rozella* Fit?

Two zoosporic genera that were originally assigned to the Chytridiomycota fall outside this phylum according to molecular phylogenetic analysis: *Olpidium* and *Rozella*. *Olpidium brassicae* infects the epidermal cells of cabbage roots, forming ovoid thalli that develop into zoosporangia (Figure 1.18). The zoospores are released into the surrounding soil water via exit tubes. They infect susceptible hosts by encysting on their root surface (epidermal cells/root hairs), penetrating the host cell wall, and transferring their cytoplasm from the cyst into the plant. The zoospores act as vectors for a number of different plant viruses. Other species of *Olpidium* infect nematodes and rotifers. *Olpidium* may be more closely related to the zygomycetes than other zoosporic species within the Chytridiomycota.

Rozella allomycis is an obligate parasite of *Allomyces* species (Blastocladiomycota) that infects both the haploid and diploid stages in its life cycle (Figure 1.19). Other *Rozella* species infect Oomycota (see the section 'Miscellaneous Microorganisms Studied by Mycologists'). The infection process resembles the mechanism of *Olpidium*: the fungus produces infectious zoospores that attach to the host colony, encyst, and penetrate the cell wall via a germ tube. The contents of the cyst pass through the germ tube and enter the host cell. Within its host, *Rozella* produces spherical thalli that absorb nutrients from the cytoplasm. These thalli are transformed into zoosporangia or thick-walled resting spores with a spiny surface. Molecular analysis shows that *Rozella* is a member of a large clade of fungi referred to as the **Cryptomycota**. Very little is known about the biology of species of Cryptomycota

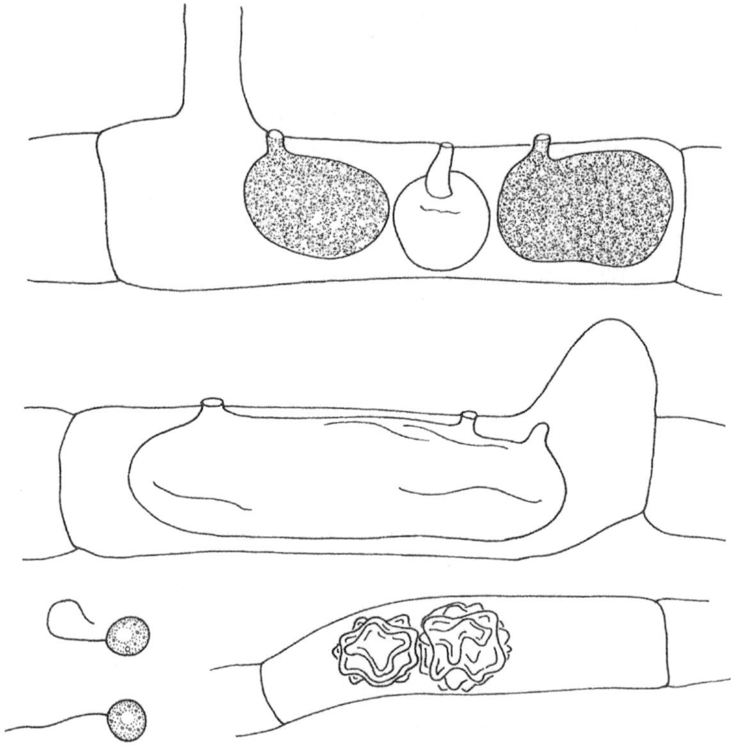

FIGURE 1.18 *Olpidium brassicae*, an aquatic fungus that infects cabbage roots using zoospores. *Source: Webster, J., Weber, R.W.S., 2007. Introduction to Fungi, third ed. Cambridge University Press.*

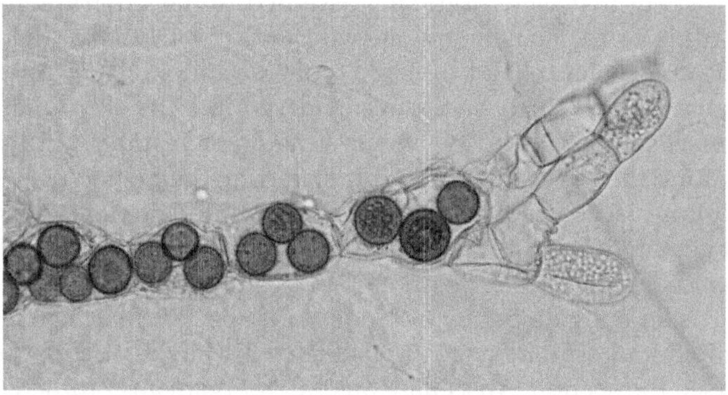

FIGURE 1.19 Brown resting sporangia of *Rozella allomycis* inside infected cells of *Allomyces* (Blastocladiomycota). *Source: Creative Commons.*

besides *Rozella*. They have not been cultured, but they have been detected, using molecular techniques, in soil samples, and in freshwater and marine ecosystems. Zoospores have been observed, along with cysts, and the cells of some species are attached to diatoms. Studies show that none of these cell types produce the cell wall chitin characteristic of other fungi. The diversity of these microorganisms may be as extensive as the genetic diversity of all of the other phyla of Kingdom Fungi. The Cryptomycota has been introduced as a new seventh phylum within the fungi.

Zygomycetes

Molecular phylogenetic studies have disrupted the assemblage of fungi that had been placed within the Phylum Zygomycota. These include *Mucor* and related genera (Order Mucorales) that assimilate sugars from diverse food sources, specialised insect pathogens (Entomophthorales), species that form asexual spores on elaborately branched stalks (Kickxellales, Figure 1.20), and predators and parasites of other fungi, amoebae, and soil invertebrates (Zoopagales). Species in another order, the Endogonales, produce coiled hyphae inside the cells of liverworts, hornworts, and at least one species of fern. It has been suggested that these fungi, rather than the

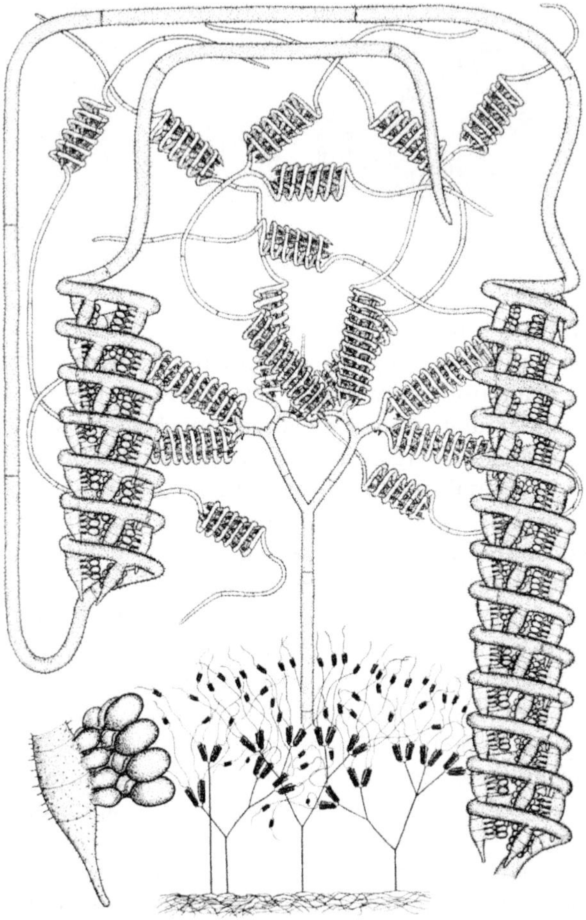

FIGURE 1.20 *Spirodactylon aureum,* a zygomycete fungus that grows on rodent dung and forms its spores on sporangiophores with a spectacular coiled structure. The fungus is shown in a variety of magnifications in this illustration. The spores are formed within the coils and are exposed in the drawing at bottom left. *Source:* Spirodactylon *is classified in the family Kixecellaceae. Benjamin, K., 1959. The merosporangiferous Mucorales. Aliso 4, 321–433.*

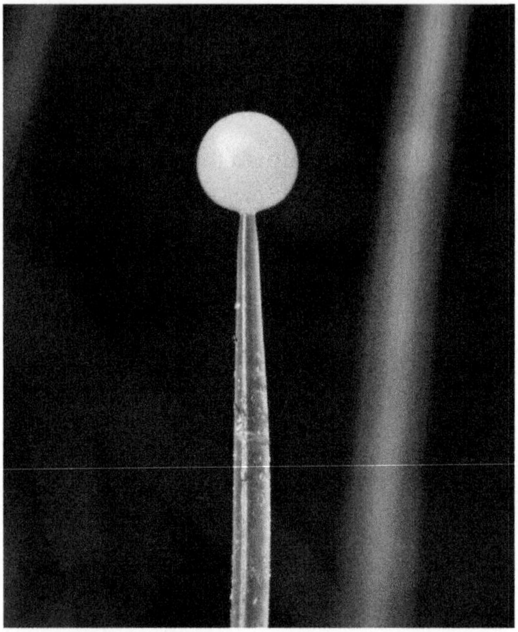

FIGURE 1.21 Sporangium of the zygomycete *Phycomyces blakesleeanus*. The wall of the sporangium will blacken as it matures and then split open to release the sporangiospores. *Source: Ron Wolf Photography.*

Glomeromycota, may have produced the earliest endomycorrhizal relationships that supported the evolution of land plants. Together, more than 900 species of these zygomycetes have been described but the relationship between the Mucorales and the other orders is unresolved. Nevertheless, many of these fungi share some important characteristics. The most familiar zygomycetes are food-spoilage microorganisms that proliferate on fruit and food scraps, producing fluffy growths of white aerial **sporangiophores** that develop bulbous tips called **sporangia** filled with spores (Figure 1.21). Sporangia of some species contain tens of thousands of these spores, called **sporangiospores**, which are dispersed in air currents. Zygomycetes are also common on less sugary matter including herbivore dung, forming the first phase of a succession of fungi that decomposes different substances within this abundant material. *Pilobolus* species are common examples of dung fungi and their mechanism of spore discharge is described in Chapter 3. Species of Mucorales are also encountered as opportunistic pathogens of humans and other animals (Chapter 9). Patients suffering from severe burns or uncontrolled diabetes are particularly susceptible to tissue invasion by colonies of these fungi. These rare infections are referred to as **mucomycoses** and can be exceedingly difficult to treat. The food-spoilage fungi and the coprophilous species have been called **primary saprotrophs** because they assimilate low-molecular-weight compounds, including sugars, rather than breaking down cellulose and other more complex macromolecules like the basidiomycetes. The metabolic properties of the Mucorales have been utilised in the production of Asian foods including the Indonesian (Javan) staple tempeh, in which *Rhizopus oligosporus* is used to ferment soybeans to produce a solid protein-rich cake. A number of other species of Mucorales are used to produce fermented bean curd from dried tofu in East Asian cuisine (Chapter 12).

Most of these fungi produce colonies of large, multinucleate, non-septate hyphae (the septate mycelia of the Kickxellales and some closely related fungi are an exception) and produce sexual spores called **zygospores** by a process called **conjugation** or **gametangial fusion** (Figure 1.22). Conjugation occurs between colonies of opposite mating type, designated plus and minus. There are a number of steps in this process that are controlled by pheromones. Colonies release precursor molecules that are converted by compatible colonies to trisporic acid. One strain (the minus strain) produces trisporol and the other strain (plus strain) produces 4-dehydrotrisporic acid. The minus strain produces the enzyme necessary to convert 4-dehydrotrisporic acid into trisporic acid, but lacks the enzyme needed to convert its own trisporol to trisporic acid. The plus strain carries out the complementary reaction, so that compatible colonies become suffused with trisporic acid, while the concentration of trisporic acid remains very low in interactions between colonies of the same mating type. This collaborative synthesis of trisporic acid causes the formation of aerial branches from the mycelia called **zygophores**. The zygophores grow toward one another, fuse at their tips, and swell. The swollen tips (**progametangia**) are separated from their mycelia by the formation of septa. At this stage, the pair of multinucleate swellings are termed **gametangia**; the contact region between these cells is then dissolved allowing nuclei of both mating types to mingle in a shared cytoplasm. The fused gametangia constitute the **zygosporangium**. A thick, heavily pigmented zygospore forms inside the zygosporangium; the available evidence suggests that nuclei of different mating types fuse to produce diploid nuclei in the zygospore. The process of zygospore germination is not observed easily under laboratory conditions, but the development of sporangiophores and sporangia (one per zygospore) has been reported in the literature. Meiosis occurs during the germination process and the **germ sporangia** release spores of a single mating type or of both mating types. This means that the post-meiotic abortion of nuclei of one of the mating types occurs in some species. The spores are dispersed in air currents and have the potential to establish haploid colonies on fresh food sources. There is considerable variation among the zygomycete life cycles, but the formation of some type of zygospore is a common feature of this unnatural assemblage of fungi.

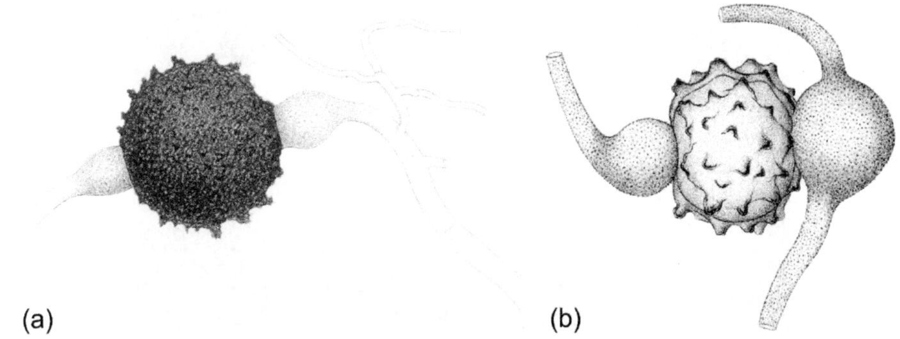

FIGURE 1.22 Mature zygospores formed between compatible strains of (a) *Mucor mucedo* and (b) *Chaetocladium jonesii*. *Source: Brefeld, O., 1872. Botanische Untersuchungen über Schimmelpilze, vol. 1. Verlag von Arthur Felix, Leipzig.*

MICROSPORIDIA

Molecular studies showing the consensus between microsporidian and fungal genes suggest that these obligate parasites are nested within Kingdom Fungi and may share a common ancestor with zygomycetes. Other work indicates that they are a closely related sister group to the fungi. Irrespective of their precise evolutionary relationships, the microsporidia are sufficiently close relatives of the fungi to be considered part of the purview of fungal biology. All of the 1300 described species parasitize animals, particularly insects, crustaceans, and fish, and some cause opportunistic infections in humans. Microsporidia are obligate parasites that grow and reproduce only inside host cells; they lack functional mitochondria and are dependent upon their hosts for their energetic needs. The genomes of some microsporidians are diminutive, smaller even than some prokaryote genomes, but others possess genomes as large as those of other fungi. But, in all cases, a relatively small subset of protein-encoding genes is encoded: 2000-3000 genes for microsporidia, compared with the 6000 genes of *S. cerevisiae* and 10,000 genes in *N. crassa*. The larger microsporidian genomes are filled with non-coding sequences. Some species lack many of the genes that control primary metabolism in other eukaryotes. Even the genes for glycolytic enzymes are missing from one species, *Enterocytozoon bieneusi*, suggesting that it is entirely dependent upon its mammalian hosts for ATP and NADPH. Within the animal, some microsporidians become surrounded by host mitochondria and the combined host–parasite cell is transformed into a giant spore-producing structure called a **xenoma** (Figure 1.23). The spores are highly resistant and can remain infectious for years. The spore contains a harpoon-like infection apparatus consisting of a coiled polar filament that unravels and penetrates the host plasma membrane. The filament provides a conduit for the rapid migration of the spore cytoplasm into the host cell.

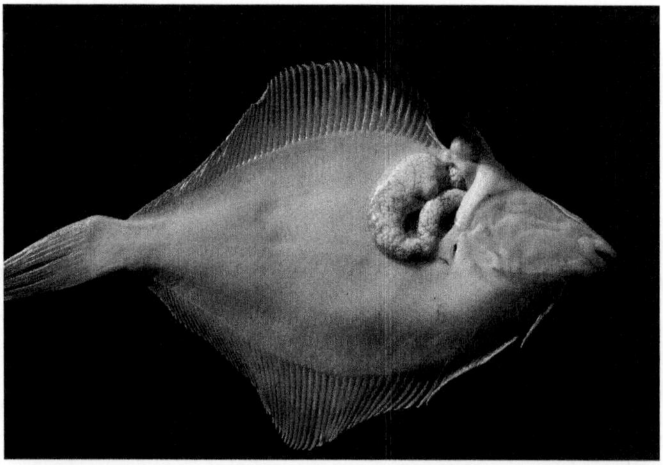

FIGURE 1.23 Xenoma produced by microsporidian (*Glugea stephani*) infection of dab (*Limanda limanda*). Fish caught at the Vlakte van de Raan on the Belgian continental shelf (southern North Sea). *Photograph by Hans Hillewart (Source: Creative Commons).*

MISCELLANEOUS MICROORGANISMS STUDIED BY MYCOLOGISTS

Water moulds within the **Oomycota** (a phylum in the eukaryote supergrouping called the Stramenopila) have been studied by mycologists since the nineteenth century. These microorganisms are more closely related to diatoms and brown algae than they are to fungi, but they produce branched colonies of tip-growing hyphae, reproduce by spore formation, and operate as saprotrophs and pathogens (Figure 1.24). In other words, they are fungi in operational terms, but do not share close evolutionary affinity with species within Kingdom Fungi. Because mycologists have studied these microorganisms for so long, a significant part of our understanding of the way that hyphal organisms work has come from experiments on species within the Order **Saprolegniales** (e.g. *Saprolegnia ferax*, *Achlya bisexualis*) and Order **Pythiales** (e.g. species of *Phytophthora* and *Pythium*). These organisms will be discussed elsewhere in the book.

Mycologists have also studied slime moulds (Figure 1.25). The **Acrasiomycetes** (acrasid cellular slime moulds) and the **Dictyosteliomycetes** (dictyostelid cellular slime moulds) are groups of protists with an amoeboid feeding phase. *Dictyostelium discoideum* is viewed as a model for understanding cell–cell communication and the evolutionary origins of multicellularity. **Myxomycetes**, or 'true slime moulds', are the familiar slime moulds whose multinucleate plasmodia flow over the surface of rotting wood, before releasing spores from a variety of beautiful fruiting structures. The mechanism of cytoplasmic streaming (shuttle streaming) within the plasmodia of *Physarum polycephalum* has been the subject of intensive cell biological research. **Protosteliomycetes** are related to the Myxomycetes but form phagocytic amoebae and, in some species, small plasmodia. Finally, the **Plasmodiophoromycota** is a group of parasitic microorganisms that are related to the Myxomycota. The group includes *Plasmodiophora brassicae* that causes club root of cabbage. Finally, the **Labyrinthulomycota** are aquatic microorganisms that grow as networks of branched 'slime tubes' (slime nets) containing motile cells (Labyrinthulales), or as zoospore-producing thalli (Thraustochytriales) that

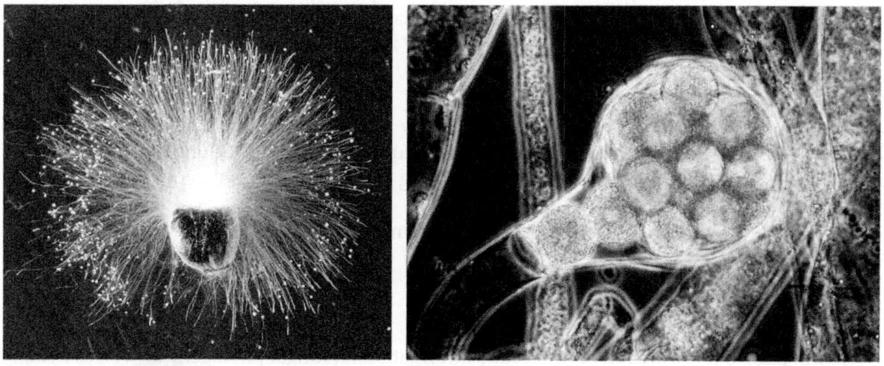

(a) (b)

FIGURE 1.24 Oomycete water moulds, which are stramenopiles rather than fungi. (a) Filamentous hyphae of an *Achlya* species growing from a hemp seed immersed in water. (b) Sexual oospores of a *Saprolegnia* species within a structure called an oogonium. *Source: George Barron, University of Guelph https://dspace.lib.uoguelph.ca/xmlui/handle/10214/3955*

FIGURE 1.25 Plasmodium of a slime mould *Leucocarpus fragilis. Source: http://curbstonevalley.com/wp-content/uploads/2010/02/LfragilisNet.jpg*

resemble the thalli of chytrids. The Labyrinthulomycota are stramenopiles, distantly related to the Oomycota. The biology of these diverse groups of slime moulds is not featured in this third edition of 'The Fungi'.

Further Reading

Berbee, M., Taylor, J.W., 2010. Dating the molecular clock in fungi – how close are we? Fungal Biol. Rev. 24, 1–16.

Bidartondo, M.I., Read, D.J., Trappe, J.M., Merckx, V., Ligrone, R., Duckett, J.G., 2011. The dawn of symbiosis between plants and fungi. Biol. Lett. 7, 574–577.

Ebersberger, I., de Matos Simoes, R., Kupczok, A., Gube, M., Kothe, E., Voigt, K., von Haeseler, A., 2012. A consistent phylogenetic backbone for the Fungi. Mol. Biol. Evol. 29, 1319–1334.

Hibbett, D.S., et al., 2007. A higher-level phylogenetic classification of the Fungi. Mycol. Res. 111, 509–547.

James, T.J., et al., 2006. Reconstructing the early evolution of Fungi using a six-gene phylogeny. Nature 443, 818–822.

Jones, M.D.M., Forn, I., Gadelha, C., Egan, M.J., Bass, D., Massana, R., Richards, T.A., 2011. Discovery of novel intermediate forms redefines the fungal tree of life. Nature 474, 200–203.

Lücking, R., Huhndorf, S., Pfister, D.H., Plata, E.R., Lumbsch, H.T., 2009. Fungi evolved right on track. Mycologia 101, 810–822.

Rosling, A., Cox, F., Cruz-Martinez, K., Ihrmark, K., Grelet, G.-A., Lindahl, B.D., Menkis, A., James, T.Y., 2011. Archaeorhizomycetes: unearthing an ancient class of ubiquitous soil fungi. Science 333, 876–879.

Schoch, C.L., et al., 2014. Finding needles in haystacks: linking scientific names, reference specimens and molecular data for Fungi. *Database* bau061.

Taylor, T.N., Krings, M., Taylor, E.L., 2015. Fossil Fungi. Academic Press, Amsterdam.

Weblinks

http://tolweb.org/tree/ offers a wealth of up-to-date information on fungal phylogeny as well as detailed descriptions of the major groupings of the fungi.

http://eol.org/pages/5559/overview

http://genome.jgi.doe.gov/programs/fungi/1000fungalgenomes.jsf

CHAPTER

2

Fungal Cell Biology and Development

Nicholas P. Money

Miami University, Oxford, OH, USA

ORGANELLES, CELLS, ORGANS

Fungi are eukaryotes and much of their cell biology is shared with animals, plants, and protists. Fungal cells are built from the same kinds of organelles as other eukaryotes. They have a plasma membrane, nuclei, and complicated endomembrane system. Most species have mitochondria. A few organelles are not found in other kingdoms. These include a dense assembly of secretory vesicles called the Spitzenkörper that is located in hyphal tips, and the Woronin body of Ascomycota that serves as an intracellular plugging device that stops cytoplasmic leakage. The chitinous cell wall is another distinctive feature of fungal cells. This chapter will provide an overview of the structure of fungal cells, discuss how they grow and multiply to form yeast colonies or branching mycelia, and describe the developmental processes that result in the formation of complex, multicellular organs including cords, sclerotia, and mushrooms (Figure 2.1).

CELL STRUCTURE

The Cell Wall

The cytoplasm of growing yeasts and mycelia contains a higher concentration of salts and sugars than the surrounding fluid. This osmotic differential drives the net influx of water through the plasma membrane and this causes cell expansion. Unlike many protists, fungi do not limit expansion by exporting water using contractile vacuoles, but do so by constructing a cell wall on the surface of the plasma membrane. As water enters the cell, the plasma membrane is pressed against the inner surface of the wall, resulting in the development of hydrostatic pressure or **turgor**. The increase in internal pressure allows the cell to approach a condition of homeostasis in which water influx matches the increase in cell volume that occurs during growth.

The wall is a highly dynamic structure, resisting expansion over much of its surface, but extending in specific regions including hyphal tips and yeast buds. The adaptive significance of the wall is controversial. It is important to avoid the chicken-and-egg trap of

http://dx.doi.org/10.1016/B978-0-12-382034-1.00002-5

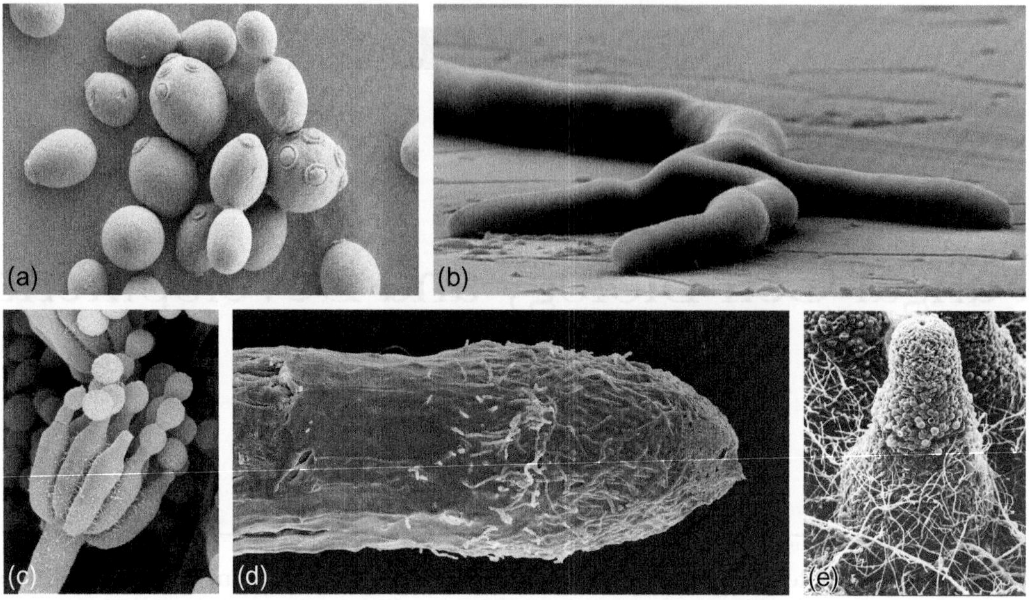

FIGURE 2.1 Fungal cells and multicellular structures imaged using scanning electron microscopy. (a) Yeast cells of *Saccharomyces cerevisiae*. The older cells have faint birth scars, stretched since separation from their mother cells, and bud scars produced by the formation of daughter cells. (b) Hypha of *Aspergillus niger*. (c) Conidiophore of a *Penicillium* species. (d) Rhizomorph of *Armillaria gallica*. (e) Perithecium of *Sordaria humana*. *Source: (a) Kathryn Cross, Institute of Food Research, Norwich. (b) Geoffrey Gadd, University of Dundee. (c) Richard Edelmann, Miami University, Ohio. (d) Levi Yafetto, University of Cape Coast, Ghana. (e) Nick Read, University of Manchester.*

suggesting that the cell wall functions to resist the explosive effects of turgor, because the cell would not be pressurised without the resistive behaviour of its wall. The cell wall allows the cell to generate turgor pressure, so perhaps it is more fruitful to think about why turgor might be useful. We will come back to this issue later in the chapter.

The fungal cell wall is a porous macromolecular composite assembled at the surface of the plasma membrane (Figure 2.2). It contains stress-bearing microfibrils of **chitin**, linear polymers of glucose, or **glucans**, and a variety of **cell wall proteins** (CWP). The chitin polymer is built from β-1→4-linked monomers of the amino sugar, *N*-acetyl-D-glucosamine (Figure 2.3). Adjacent chitin chains assemble into hydrogen-bonded antiparallel arrays, producing microfibrils that can reach lengths of more than 1 μm. Chitin microfibrils have tremendous tensile strength; when chitin is disrupted, the cell loses its osmotic stability and may rupture. **Chitosan**, or β-1→4-glucosamine, is a polymer of the deacetylated sugar that is produced by many fungi in addition to chitin. **β-1→3-glucan** is often the most abundant wall polymer (Figure 2.4). The β-1→3-glycosidic linkage in glucans twists the polymer and three glucan chains form a triple helix that is held together by hydrogen bonds. β-1→3-glucans are connected with β-1→6-glucans in the mature wall structure to produce a highly branched elastic network of polymers. The structural proteins in the cell wall are **glycoproteins** with *N*- and *O*-linked carbohydrates. These include **mannoproteins**, which are glycosylated with mannose-rich chains, and other glycoproteins with both mannose and

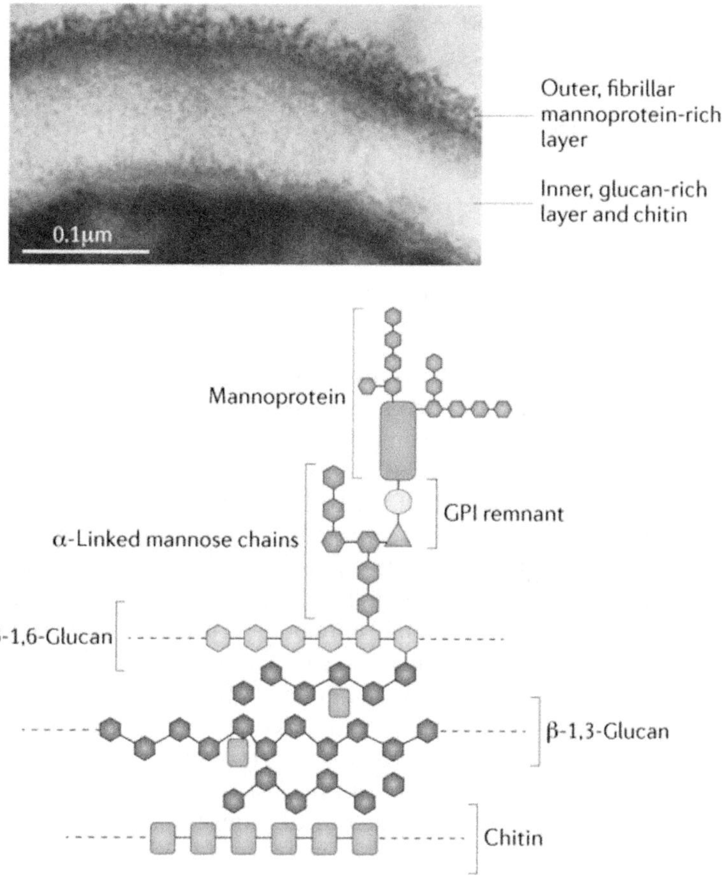

FIGURE 2.2 Structure of the fungal cell wall. Transmission electron micrograph of *Candida albicans* cell wall with diagram showing chemical composition. *Source: Cassone, A., 2013. Development of vaccines for Candida albicans: fighting a skilled transformer. Nature Rev. Microbiol. 11, 884–891.*

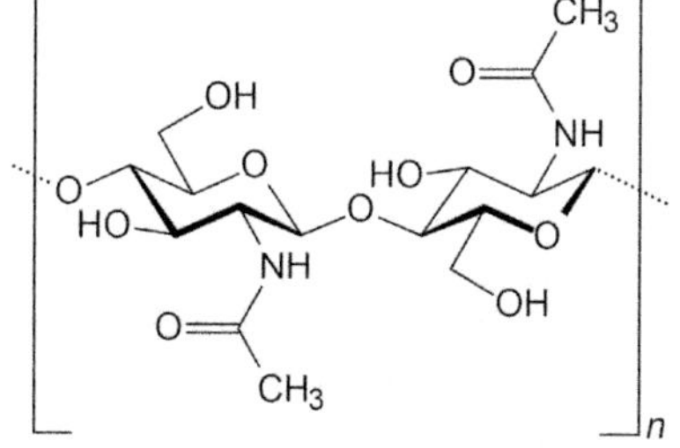

FIGURE 2.3 Chemical structure of the cell wall polymer chitin.

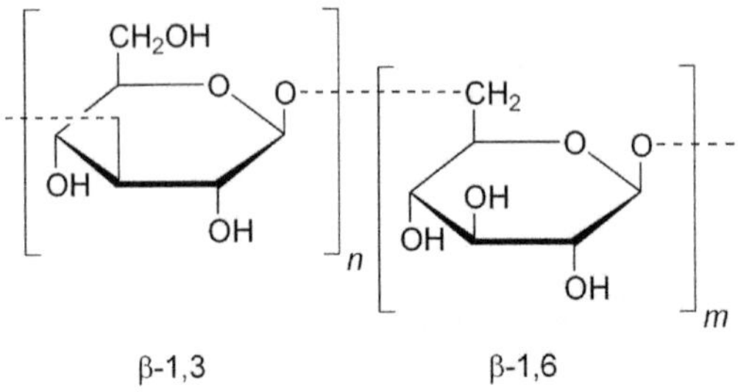

FIGURE 2.4 Chemical structure of the cell wall glucan.

galactose residues. Cell wall glycoproteins are connected to the plasma membrane by a glycophosphatidylinositol (GPI) anchor and cross-linked to the chitin microfibrils and glucans.

Although chitin is a stress-bearing component within the extracellular matrix of most fungi, the maxim that 'fungi have chitinous walls' can be misleading. β-1 → 3-glucan is often the dominant wall polymer and up to half the wall of some fungi is proteinaceous. The relative proportions of chitin, glucans, and glycoproteins vary according to fungal species. Chitin is a minor constituent of the cell wall of the yeast, *Saccharomyces cerevisiae*, and is synthesised when the mother cell produces a bud. The bulk of the yeast cell wall is a scaffold of β-1 → 3-glucans capped by a layer of mannoproteins. The β-1 → 3-glucans are cross-linked to β-1 → 6-glucans and to a variety of CWPs. The cell wall of fission yeast, *Shizosaccharomyces pombe*, is very different, containing glucans with varied glycosidic linkages (α-1 → 3-, β-1 → 3-, and α-1 → 3-glucan) but completely lacking chitin. The same polymers are found in mycelial fungi, but chitin is more important in these species, accounting for 10% or more of the dry weight of the wall. In *Neurospora crassa*, for example, β-1 → 3-glucans and chitin form an inner layer, which is covered by a protein–polysaccharide complex. There is no β-1 → 6-glucan in the cell wall of this ascomycete.

Chitin synthesis is catalysed by **chitin synthase** which is an integral membrane protein. Chitin synthase extrudes chitin chains through the plasma membrane and the new polymers hydrogen bond with one another and crystallise into microfibrils. *Saccharomyces cerevisiae* has three chitin synthases (Chs1p, Chs2p, and Chs3p); the filamentous ascomycete *Aspergillus fumigatus* has seven chitin synthases, and four have been identified in *Neurospora crassa*. Glucan synthases operate in a similar fashion to the chitin synthases. They are integral membrane proteins and form long chains of glucans by the sequential addition of glucose residues. Crosslinks between adjacent branches within glucan molecules and linkages between the glucans, chitin, and glycoproteins are vital for maintaining wall strength. Glucan synthesis is a potential target for antifungal agents. A family of drugs called **echinocandins** that bind to **β-1 → 3-glucan synthase** show promise in the treatment of aspergillosis and candidiasis (Chapter 9).

Most of the wall proteins are glycosylated during transit through the secretory pathway. Delivered to the plasma membrane by exocytosis and tethered by GPI anchors, their sugar residues form covalent links with other wall polymers and help to maintain cell shape. Glycoproteins also serve signalling and transport functions, participate in fusion with other cells (e.g. agglutinins involved in cell–cell recognition during mating reactions), and function in adhesion to surfaces, biofilm formation, and pathogenesis. They also mediate the absorption of compounds from the surrounding environment and protect the cell from noxious substances. The wall is also rich in enzymes. Some of these enzymes are involved in the synthesis of other wall components and are critical for the continuous remodelling of the wall during growth. The function of many cell wall enzymes is unknown.

The electron microscope reveals that the walls of some fungi are organised into discrete layers, but this does not mean that the wall has a plywood-like structure that separates each of the different kinds of molecule. The wall is more like a fibre reinforced polymer (e.g. fibreglass), with multiple interwoven components within each layer. Another important feature of the wall is its dynamic nature. Even the thick cell wall of a long-dormant spore is reconfigured and rendered more fluid as the cell germinates. Although there are many features of cell wall structure and function that are unique to the fungi, there are numerous similarities between this structure and the extracellular matrix of other eukaryotes. The plant cell wall contains stress-bearing microfibrils of cellulose (β-1→4-glucan), but it also contains glucans with β-1→3 linkages. The extracellular matrix of animal cells is constructed from collagens intermeshed with proteoglycans, including glycoproteins. Fungal cell walls are particular types of extracellular matrix.

Plasma Membrane

The sterol molecule, **ergosterol** is a distinguishing component of the fungal plasma membrane (Figure 2.5a). Ergosterol performs the same function as cholesterol in animal membranes, namely, modulating membrane fluidity and permeability through its interactions with phospholipids and other membrane constituents. The absence of ergosterol in animals allows treatment of a broad range of fungal infections (mycoses) using antifungal agents that inhibit its synthesis (Chapter 9). **Terbinafine hydrochloride** (Lamisil), which is used to treat athlete's foot and other superficial infections caused by dermatophytes, inhibits an

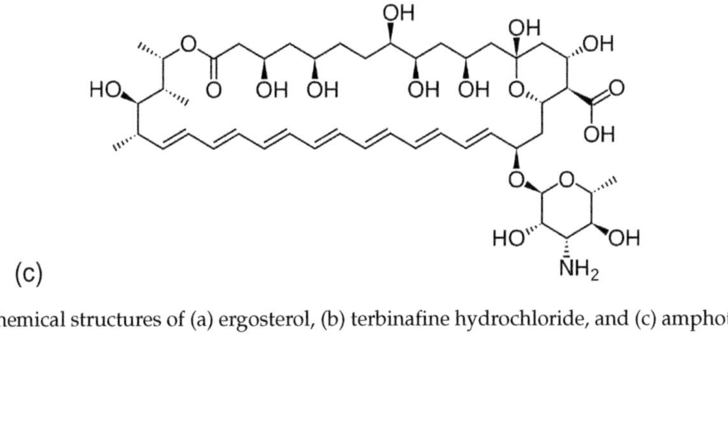

FIGURE 2.5 Chemical structures of (a) ergosterol, (b) terbinafine hydrochloride, and (c) amphotericin B. *(Source: Creative Commons).*

enzyme involved in ergosterol synthesis called squalene epoxidase (Figure 2.5b). Disruption of sterol biosynthesis causes lysis of the target cells. **Amphotericin B** is an antifungal agent that is used to treat more serious infections, including cryptococcal meningitis (Figure 2.5c). Amphotericin B binds to ergosterol and, like terbinafine hydrochloride, is thought to interfere with membrane integrity.

The lipid bilayer of the fungal plasma membrane contains proteins that function in solute transport, signal transduction, cell wall synthesis, and as anchors for the underlying cytoskeleton. **Lipid-anchored proteins**, including G proteins, are linked covalently to the lipid bilayer via lipidated amino acid residues (or by the GPI anchor described in the previous section). Peripheral membrane proteins are associated with the membrane by electrostatic forces and other kinds of non-covalent interactions. **Integral membrane proteins** are permanent residents of the phospholipid bilayer. These include **transmembrane proteins**, or integral polytopic proteins, that function in transporting ions and molecules through the plasma membrane. **Integral monotopic proteins** are embedded in only one side of the membrane (rather than threading through the entire membrane).

Ion transport functions catalysed by transmembrane proteins are critical for understanding fungal physiology (Figure 2.6). The plasma membrane acts as a semipermeable barrier to the diffusion of many ions and small molecules. For example, protons cannot diffuse freely through the membrane. They are extruded from the cytoplasm by an enzyme, or **ion pump,** called the proton **(H^+)-ATPase**. Proton extrusion is an example of **primary active transport.** This creates an electrochemical gradient with a reduction in pH at the surface of the fungal cell and a negative voltage inside the cell (as much as −250 mV). This voltage, or membrane potential, can be measured with a microelectrode inserted through the cell wall and underlying membrane (Figure 2.7).

The electrochemical gradient established by proton ATPase activity is vital to the absorptive feeding mechanism characteristic of the fungi, because it powers the import of small molecules including sugars and amino acids. This cellular physiological mechanism is very elegant. The proton ATPase moves protons against their concentration gradient, so that protons will flow into the cytoplasm as soon as a diffusion path is opened. As a hypha digests protein-rich food, a localised pool of amino acids accumulates around the cell. Because biomolecules, including amino acids, are at far higher concentration inside the cytoplasm than

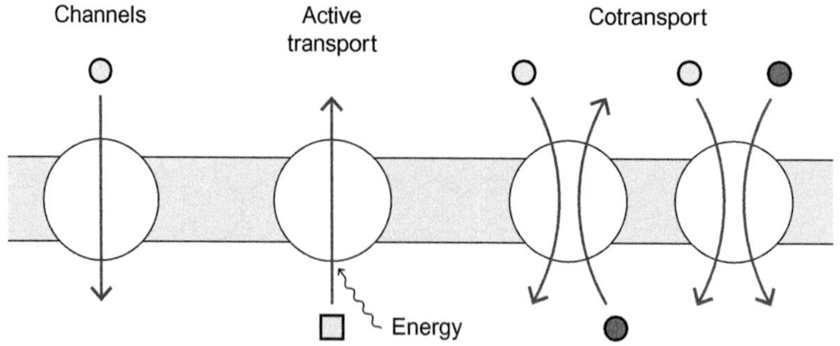

FIGURE 2.6 Transmembrane proteins that transport ions and molecules between fungal cells and the surrounding environment. *Source: Mark Fischer, Mount St. Joseph University, Cincinnati.*

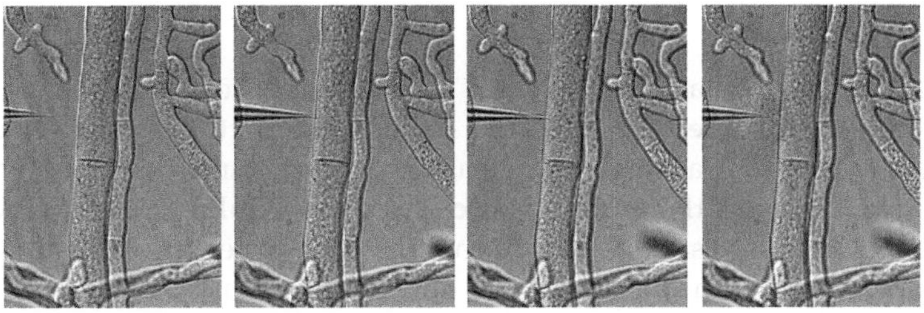

FIGURE 2.7 Microelectrode piercing hyphal cell wall and plasma membrane during a recording of the membrane potential. *Source: Roger Lew, York University, Ontario.*

the surrounding environment, their import will not occur passively by diffusion, even if physical pathways through the membrane are available. The concentration gradient is in the wrong direction. Transmembrane proteins called **carrier proteins** solve this problem by harnessing the influx of protons to the movement of amino acids. Carrier proteins function by undergoing a specific conformational change in response to the flux of each proton that shuffles an amino acid molecule from the exterior of the cell into its cytoplasm. The amino acid is captured and then released by the carrier protein. This is an example of **secondary active transport**.

The carrier protein that conducts amino acid import is an example of a **symporter**. Other symport proteins couple proton influx to the import of potassium ions and sugars. Antiporters couple the passive influx of a proton, or another ion, to the export of a different ion or a molecule. Na^+/H^+ antiporters have been characterised in yeast and filamentous fungi. These proteins export one sodium ion for each proton that passes into the cell and are important in maintaining ionic homeostasis and determine salt tolerance in saline environments. Carrier proteins can also provide aqueous pathways for the **facilitated diffusion** of ions and molecules that are more diluted within the cell, but cannot diffuse through the hydrophobic interior of the plasma membrane. **Ion channels** are another subset of integral membrane proteins that control ion import and export. Some ion channels act as gates that open and close in response to changes in membrane potential or to mechanical signals including stretching of the membrane. Calcium channels regulate the concentration of calcium ions in the cytoplasm. Calcium is maintained at very low levels in the cell, but bursts in concentration caused by channel opening may serve as important signals regulating cellular development. The concerted action of all these transport proteins determines the chemical composition of the cytoplasm, the food supply for the mitochondria, the export of waste metabolic materials, and exclusion of environmental toxins.

In the 1980s, it was thought that the electrical activity of cells caused by all of these ion movements reflected important developmental cues. The unequal distribution of ion pumps along fungal hyphae, for example, produces a halo of ionic current with the net influx of positive charge at the growing hyphal tip. More recent studies suggest that these patterns reflect the feeding activity of the hypha and have no deeper meaning for developmental biologists searching for signals that determine how fast these cells extend and where branches appear.

Endomembrane System

The cytoplasm of hyphae and yeast cells is packed with the membrane-bound compartments of the **endomembrane system**. These organelles function in the secretory (export) and endocytotic (import) pathways that sustain growth and development. The endomembrane system includes the **endoplasmic reticulum, Golgi apparatus, vacuoles**, and **vesicles** (Figure 2.8). The system is very dynamic, both in terms of the movement of the different components within the cytoplasm and the biochemical activities within them. Analysis of the endomembrane system has been revolutionised by the use of **fluorescent protein tagging** methods to pinpoint the distribution of specific molecules within the cell (Figure 2.9). Researchers can track the locations and movement of these fluorescent probes using laser scanning **confocal microscopy**. Major advances in fungal cell biology have been made using these techniques in **live cell imaging** in conjunction with modern methods of molecular genetics and **comparative genomics**. The computer-based tools of comparative genomics have highlighted genes in the fungus that share sequences with genes with proven activity in the endomembrane system of humans. The actual function of these genes in the fungus can then be studied by comparing the growth and development of a mutant strain in which the genes have been deleted, with wild-type strains expressing the functional genes. Live cell imaging is used to track the location of proteins in the wild-type

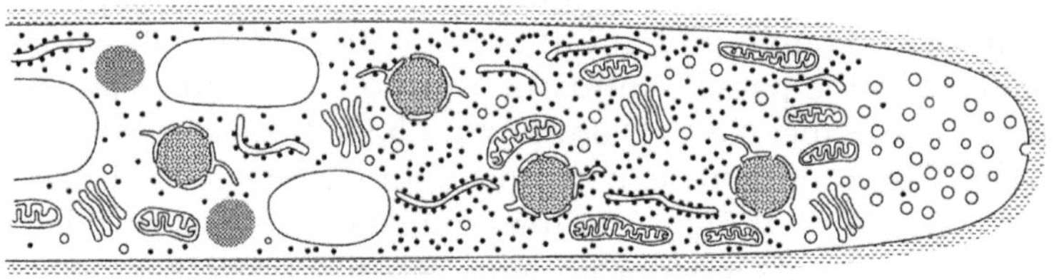

FIGURE 2.8 Diagram of the endomembrane system within a fungal hypha. *Source: www.cronodon.com*

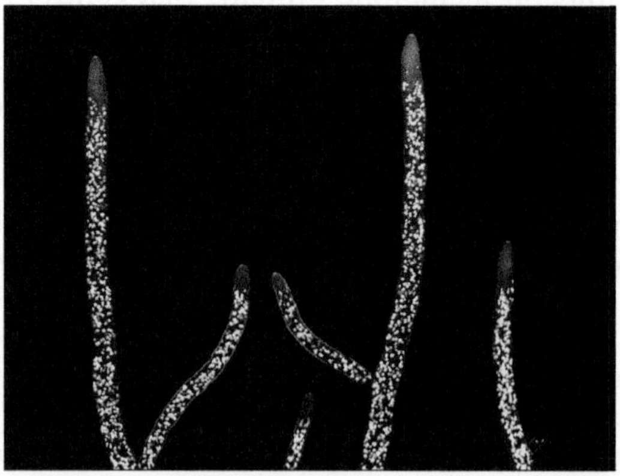

FIGURE 2.9 Confocal image showing tip-growing and branched, multinucleate hyphae of *Neurospora crassa*. The nuclei are shown in green (nuclear-targeted GFP) and membranes, especially the plasma membrane and secretory vesicles, are shown in red (stained with FM4-64). *Source: www.fungalcell.org* (See the colour plate.)

strains and protein–protein interactions can be visualised when two or more fluorescent probes are introduced into the same cells.

The **endoplasmic reticulum** is an interconnected system of membranes from which most other membranes originate. In *Saccharomyces cerevisiae*, the endoplasmic reticulum is organised as an array of contiguous sheets and tubules that link to the plasma membrane. In filamentous fungi, the endoplasmic reticulum is organised as stacks of membrane-bound sacs or **cisternae**. Most of these cisternae are studded with groups of ribosomes called polyribosomes, constituting **rough endoplasmic reticulum**. Other portions of the endoplasmic reticulum, called **smooth endoplasmic reticulum**, are not decorated with ribosomes. The whole assembly contains mixtures of rough and smooth endoplasmic reticulum and transitional regions from which the cisternae of the **Golgi apparatus** are generated. Studies with fluorescent dyes show that the endoplasmic reticulum is a mobile organelle rather than a static platform. The Golgi in fungi is a very different structure from the organelles in other eukaryotes. The stacks of cisternae, called dictyosomes, found in animals and plants are replaced by single cisternae perforated with holes and tubular extensions that are dispersed throughout the fungal cytoplasm. The term **Golgi equivalent** is often used to describe this fungal organelle, but Golgi apparatus has the virtue of being less cumbersome. The Golgi develops by the coalescence of vesicles that bud from the endoplasmic reticulum and it functions in the modification of proteins for delivery to the plasma membrane by exocytosis. In preparation for these functions, the proteins are cleaved, folded into their tertiary structures, glycosylated and phosphorylated in the Golgi. Molecular labels or **signal sequences** are also added to proteins that specify their destination after release from the Golgi. These proteins include the integral membrane proteins, CWPs, and secreted enzymes that catalyse the breakdown of polymers and fuel the fungus with absorbable nutrients. Cell wall polysaccharides are also generated in the Golgi and reach the cell surface by exocytosis.

Vacuoles form the largest endomembrane compartments in fungal hyphae (Figure 2.10). These are highly mobile organelles, capable of extension, retraction, and peristaltic shape changes. These structures also occur in yeast where their function in the endocytotic pathway for sorting and recycling proteins has been studied in most detail. In filamentous fungi, the vacuoles form an interconnected system of elongated tubules behind the tip and rounded vacuoles in older portions of the cell. Dissolved substances are free to diffuse throughout these compartments, which is consistent with a passive transport function for these organelles. In addition to their endocytotic function, vacuoles may also serve as a repository for waste products from metabolism and for heavy metals and other toxins absorbed from the environment.

Cytoskeleton

The shape of yeast cells and filamentous hyphae is determined by physical interactions between the cell wall, whose composition and mechanical properties vary at different points on the cell surface, the pressurised cytoplasm, which tends to inflate the cell, and the cytoskeleton. The importance of the cytoskeleton in this shaping process has been demonstrated by live cell imaging of its behaviour, experiments with inhibitors that disrupt specific cytoskeletal components, and work with mutants with various defects in cytoskeletal function (Figure 2.11). The fungal cytoskeleton comprises three polymers: actin microfilaments

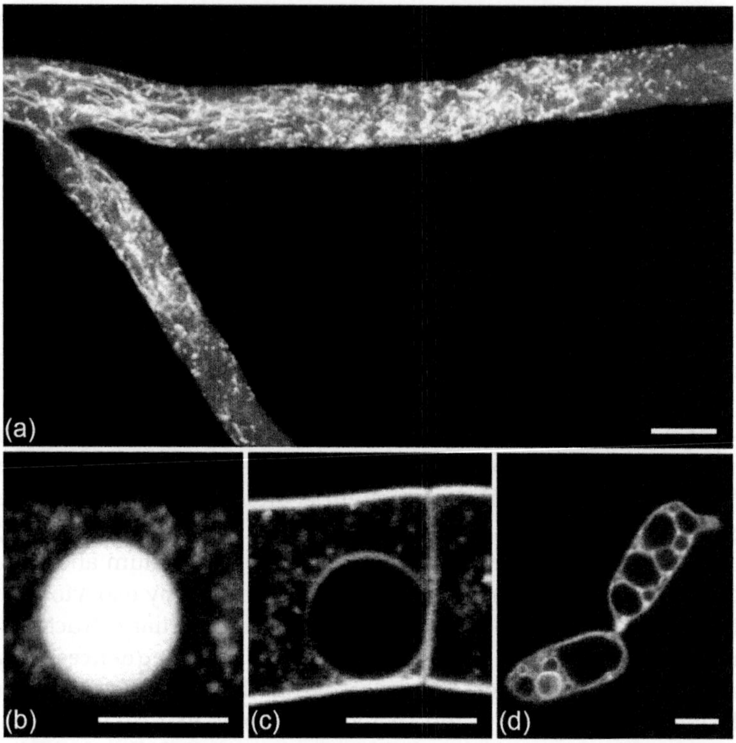

FIGURE 2.10 Vacuoles in (a–c) *Neurospora crassa* and (d) *Colletotrichum lindemuthianum*. (a) Tubular vacuolar network in apical hyphal compartment and branch stained with carboxy-DFFDA. (b) Large and small spherical vacuoles in subapical hyphal compartment stained with carboxy-DFFDA. (c) Vacuole membrane of large spherical vacuole in subapical hyphal compartment stained with FM4-64. (d) Vacuole membranes in conidia stained with MDY-64. Note that the two conidia are fusing via conidial anastomosis tubes. All scale bars = 10 μm. *Source: www.fungalcell.org*

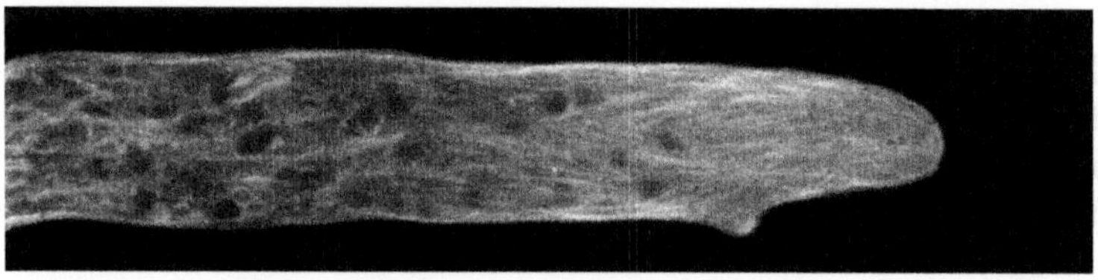

FIGURE 2.11 Confocal image of a growing hypha expressing ß-tubulin-GFP, localised in microtubules (green), and co-labelled with FM4-64 to show distribution of membranes (red). Microtubules extend into the negatively stained Spitzenkörper at the tip. A subapical swelling that will become a hyphal branch is highlighted with a concentration of vesicles that will become a separate Spitzenkörper. *Source: Patrick Hickey.* (See the colour plate.)

(F-actin), microtubules, and septins. These form an intracellular scaffold that probably plays a role in maintaining cell shape as well as directing the traffic of organelles. The structural polymers regulate cell shape by serving as directional pathways for the movement of vesicles. The trafficking functions are dependent upon molecular motors, including myosin, dynein, and kinesin, which glide along the actin filaments and microtubules. Interactions between

F-actin and septins are important in controlling the direction of trafficking, but septins do not interact directly with the motor proteins.

Actin microfilaments are assembled from globular actin monomers (G-actin) that polymerize as pairs of intertwined helices. Each filament is polarised, with the addition of new G-actin monomers to one end (called the plus, or barbed, end) and depolymerization at the other end (the minus or pointed end). The plus ends of the microfilaments tend to be oriented toward sites of growth and relative rates of polymerization and depolymerization determine the length of the filaments. Interactions between F-actin and actin binding proteins organise the microfilaments into cables, patches, and rings. Actin cables function in exocytosis, forming pathways for the movement of vesicles powered by myosin motors. Patch structures are involved in endocytosis and actin rings function in cell division and the formation of septa (see next section).

As their name suggests, microtubules have a tubular form. They are assembled from 'protofilaments' of paired α- and β-tubulin monomers (forming heterodimers), whose parallel arrangement creates a tube with an outer diameter of 25 nm. Like the microfilaments, the microtubules have a plus and a minus end and this polarity dictates the direction of organelle movement along the filament surface: organelles equipped with kinesin motors move toward the plus end of the microtubules, and dynein sends them in the reverse direction. In common with the role of microtubules in other eukaryotes, fungal microtubules also position nuclei, form the mitotic spindle, and drive chromosome separation.

Less is known about the third type of cytoskeletal polymer, called septins. Septins are types of GTPases that can be configured in a variety of ways to produce rods, longer filaments, and sheets. GTPases hydrolyse the purine nucleotide guanosine triphosphate (GTP) into guanosine diphosphate and inorganic phosphate, and perform critical functions in signal transduction, protein synthesis, cell differentiation, and vesicle transport. Septins become concentrated at the periphery of fungal cells and play important roles in tip growth and the control of cell shape.

GROWTH AND CELL DIVISION

The quest for an integrated model of tip growth in filamentous fungi is a continuing endeavour for experimental mycologists. Rather more is known about the mechanism of bud formation in yeast and we will look at this first.

Growth and Cell Division in Budding Yeast and Fission Yeast

Bud formation in *Saccharomyces cerevisiae* is regulated throughout the cell cycle, producing one daughter cell at each mitotic division (Figure 2.12). The position of the bud on the cell surface is determined during the G1 phase of the cell cycle. In haploid cells, the new bud will develop next to the scar left by the separation of the preceding daughter cell. Buds form at opposite poles of a diploid mother cell, alternating from one end of the ovoid cell to the other from division to division. The new bud site is designated by specific marker proteins in the cytoplasm during the previous cell cycle. These are recognised during G1 by a GTPase and associated proteins constituting the Rsr1p GTPase module; these proteins interact with another protein complex, the polarity establishment Cdc42p GTPase module, which directs

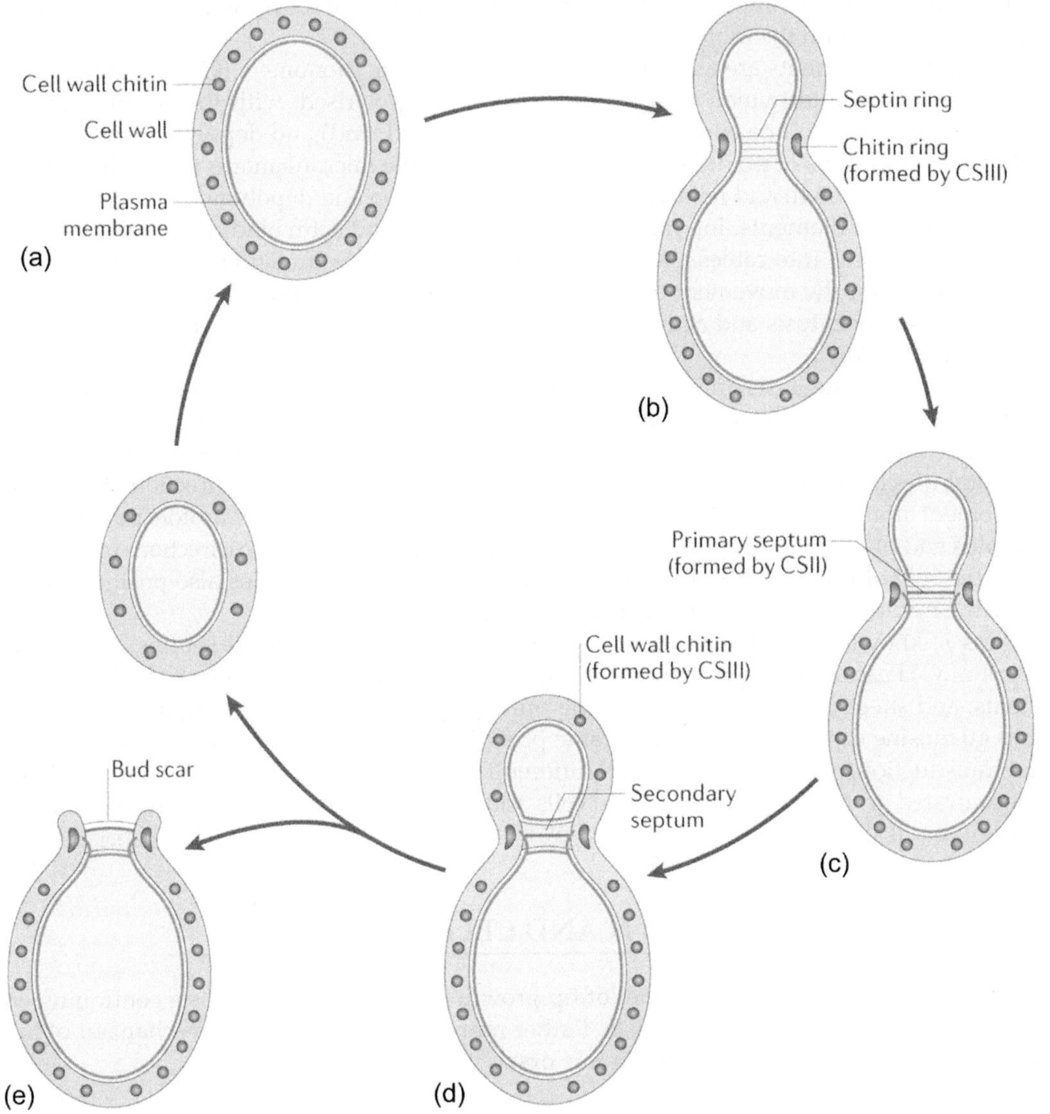

FIGURE 2.12 Cell division in *Saccharomyces cerevisiae*. *Source: Cabib, E., Arroyo, J., 2013. How carbohydrates sculpt cells: chemical control of morphogenesis in the yeast cell wall. Nat. Rev. Microbiol. 11, 648–665.*

the construction of the new cell. Toward the end of G1, and throughout S and S/G2, septins, myosin, actin, and associated molecules are organised at the neck of the bud. The septins form a ring at the neck that seems to function as a template, or scaffold, upon which a second ring of actomyosin is constructed as mitosis takes place. The septin ring is also thought to be important in positioning the mitotic spindle along the mother–daughter axis. Exocytosis provides new membrane for the growing bud and chitin synthesis is localised to the bud neck, directed by the septin ring. This produces a septum that divides the mother and daughter cell at the completion of mitosis. The contractile actomyosin ring guides membrane formation during septum formation. Unlike the rest of the yeast cell wall, the septum is rich in chitin. Chitin synthases are localised at the neck of the bud in anaphase and create the septum when mitosis is completed. The coordination between the intricate processes of bud formation and the cell cycle is an active area of research. A signalling cascade called the mitotic exit network, or MEN, plays an important role in this biochemical control.

Fission yeast, *Schizosaccharomyces pombe*, produces cylindrical cells that divide by the formation of septa. The division site is determined by the position of nucleus that is, in turn, fixed in the centre of the cell by microtubules. The position of the nucleus is communicated by the export of a protein, Mid1p, that marks the cell cortex. Proteins related to Mid1p are involved in cell division in *Drosophila* and humans, suggesting conservation of the mechanism of division site selection in multicellular organisms. Cell division in fission yeast shows a number of similarities to the process in budding yeast, including the assembly of an actomyosin ring that directs the synthesis of new plasma membrane and cell wall at the division site.

Nuclear Division

The nuclear envelope of animal cells disassembles in prometaphase and is absorbed by the endoplasmic reticulum. The envelope of paired membranes reassembles around each daughter nucleus at the end of anaphase. This type of nuclear division is described as an 'open mitosis'. 'Closed mitosis', in which the intact nuclear envelope encloses the spindle and chromosomes, is characteristic of most fungi. This has been studied in greatest detail in *Saccharomyces cerevisiae*, and a similar process takes place in *Neurospora crassa*. There are significant variations in the mitotic process among other ascomycetes, including the partial breakdown of the nuclear pore complex in *Aspergillus nidulans*. Open mitosis may be widespread among the Basidiomycota. Unlike the fragmentation of the nuclear envelope that occurs in animal cells, the entire envelope is stripped away from the dividing nucleus in the basidiomycete *Ustilago maydis* and recycled in telophase. The structures that organise the microtubules of the mitotic spindle in fungi are called **spindle pole bodies.** These **microtubule organising centres** perform the same function as the centrosomes of animal cells, but they are given a distinct name because they do not contain centrioles. The spindle pole body is attached to the nuclear envelope and is duplicated through a series of discrete steps that proceed throughout interphase. The behaviour of the nuclear envelope during mitosis and the structure of the spindle pole body suggest that the mitotic mechanism in the fungi may have had an independent evolutionary origin from the process in other eukaryotes.

Tip Growth in Filamentous Fungi

Tip growth in hyphae is a continuous process of extension that is not coupled to nuclear division in the same manner that bud formation and mitosis are connected in yeast. The colony, or mycelium, of the ascomycete *Neurospora crassa* can expand radially at a rate of a few millimetres per hour with the repetitive formation of lateral branches. As each hypha extends, its population of nuclei increases through asynchronous mitosis, but the formation of new cells, or hyphal compartments, does not seem tightly coordinated to nuclear division. Mitosis is described as 'parasynchronous' in *Aspergillus nidulans*, and other ascomycetes, because waves of nuclear division progress along the hyphae so that adjacent nuclei are engaged in mitosis at the same time. As the hyphae elongate, septa form across (at a right angle to) the long axis of the mitotic spindles of dividing nuclei. But because most nuclei divide without directing septum formation, each compartment of a hypha can contain hundreds of nuclei. Hyphal structure is different in other phyla. Mycelia produced by most basidiomycetes have uninucleate (homokaryotic) hyphal compartments before mating, and binucleate (heterokaryotic) compartments following the fusion of sexually compatible colonies. Nuclear division is coupled to cell division via the formation of septa in the basidiomycetes. Zygomycetes form non-septate multinucleate hyphae.

The apex of a growing hypha is a site of concentrated vesicle traffic. Vesicles arriving at the tip create new membrane surface through the process of exocytosis and deliver new cell wall materials. A dense nugget of vesicles in the hyphal tip called the **Spitzenkörper** plays an important role in the growth process (Figure 2.13). It contains macrovesicles, also known as apical vesicles, and microvesicles, plus ribosomes and cytoskeletal components. In the ascomycete *Neurospora crassa*, the microvesicles are packed into the core of the Spitzenkörper, which is surrounded by macrovesicles. The microvesicles are called **chitosomes**, referring to their function in chitin biosynthesis. A protein involved in glucan synthesis has been identified in the same region of the cell as the macrovesicles, suggesting that these larger vesicles are involved in the synthesis of other cell wall polymers. The organisation of these components varies between species, but the structure, biochemistry, and behaviour of the Spitzenkörper offer compelling evidence for its role as an exocytotic apparatus (or vesicle supply centre) that controls cell wall synthesis at the hyphal tip. The activity of the Spitzenkörper is regulated by a pair of protein complexes called the **exocyst** that forms a bridge between the vesicles and the cell membrane,

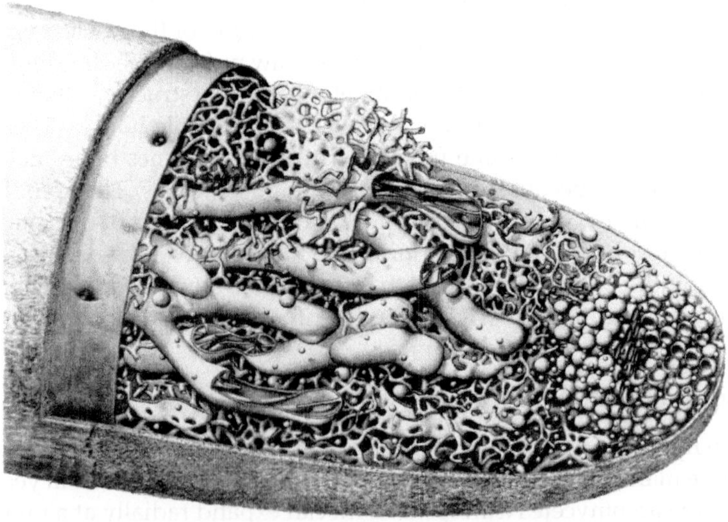

FIGURE 2.13 Diagram showing the inside of a hyphal tip packed with vesicles that form the Spitzenkörper whose position directs the direction of growth. *Source: Girbardt, M., 1969. Die Ultrastruktur der Apikal region von Pilzhyphen. Protoplasma 67, 413–441.*

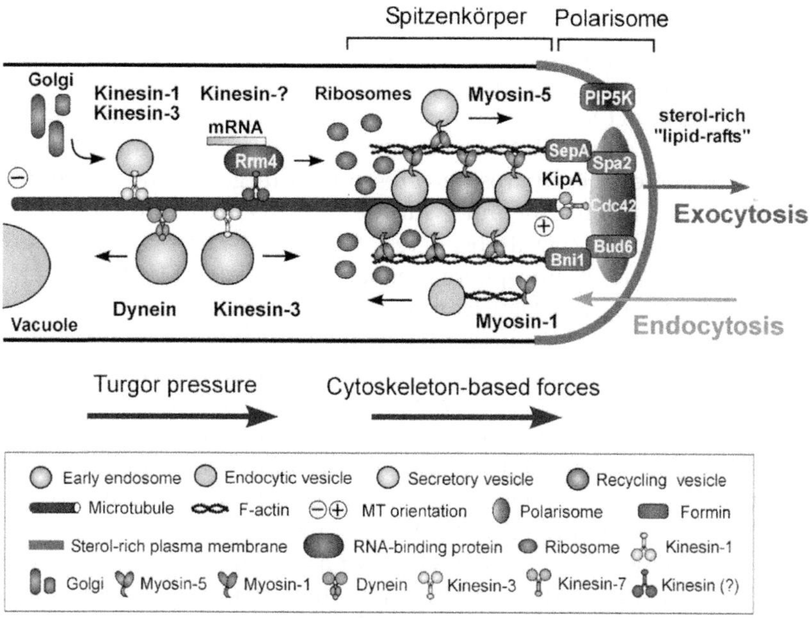

FIGURE 2.14 Model of tip growth showing some of the molecular components in the extending hypha. *Source: Steinberg, G., 2007. Hyphal growth: a tale of motors, lipids, and the Spitzenkörper. Euk. Cell 6, 351–360.* (See the colour plate.)

and the **polarisome** that controls the position of the Spitzenkörper (Figure 2.14). The activity and position of the Spitzenkörper are absolutely critical for hyphal growth, branching, and mycelial development. Shifts in the position of the organelle result in changes in the direction of hyphal extension. If the cluster of vesicles remains in the centre of the tip, the hypha will grow straight; if it shifts to one side, the hypha will reorient its growth axis in this direction. Small Spitzenkörpers develop at sites along hyphae where new branches emerge and occupy the tips of new growth axes. Computer models have been effective in 'growing' virtual hyphae, generating hyphal tips of different shapes, and forming branches by moving and duplicating a vesicle supply centre whose behaviour mimics the Spitzenkörper.

The Spitzenkörper functions in the export of materials packaged in vesicles, but the process of membrane (vesicle) import, or endocytosis, may be equally important in fungal growth. Much of this is hypothetical. Endocytosis is thought to occur behind the growing hyphal tip where patches of actin are concentrated in a subapical collar. As the tip extends, the enzymes that control cell wall synthesis, and molecular complexes that maintain the position of the Spitzenkörper, are swept back. Endocytosis in the subapical region of the hypha would serve to recycle these cell components and enable their repositioning at the tip. According to this **apical recycling model**, endocytosis assumes a pivotal role in controlling cell shape. Put simply, the balance between exocytosis and endocytosis will determine the increase in surface area of the hyphal tip and this may control a range of developmental processes including the formation of spores at the tips of aerial hyphae (Chapter 3). Researchers have developed a detailed picture of the molecular mechanisms of endocytosis in yeast, and there is an intensive effort to understand these processes in filamentous fungi.

Tip growth depends upon water influx to support the increase in cell volume, and exocytosis to provide the new plasma membrane and cell wall materials of the expanding surface. Osmosis causes water influx when the cytoplasm contains a higher concentration of dissolved ions and molecules than the surrounding fluid. Net water influx occurs until the chemical potential of water (water potential) of the cell and its surroundings equilibrate (Chapter 5). This is achieved by an increase in hydrostatic pressure, or turgor, in the cytoplasm as the plasma membrane is pressed against the inner surface of the cell wall. In most instances, growing hyphae are pressurised to a few atmospheres of hydrostatic pressure. (The pascal, symbol Pa, is the SI unit for pressure; 1 atmosphere is equal to 100 kPa or 0.1 MPa.)

Turgor pressure is an important feature of growing hyphae, in the sense that the cell is always at risk of rupturing if its surface is damaged (see Woronin bodies below), and plasma membrane and cell wall synthesis must be regulated carefully to allow controlled expansion. Our understanding of the mechanisms that permit cell wall polymers to slip past one another as new materials are incorporated into the wall is very patchy. The study of these biomechanical processes is a potentially fruitful research topic for new investigators. We have learned a great deal about the different components in the fungal cell wall, but know very little about their interactions during hyphal growth. Turgor pressure acts to smooth the cell surface and, presumably, pushes the cell wall polymers apart as the tip expands. But turgor does not operate as any kind of special 'driving force', any more than water influx drives growth. The walled cells of plants are also pressurised, but the cells of animals and many protists maintain ionic homeostasis with their surroundings or export water using contractile vacuoles and do not generate any significant turgor pressure. In other words, if many eukaryotes can dispense with turgor pressure, why is this intracellular pressure essential for hyphae? The answer may lie in the special feeding mechanism of filamentous fungi.

Hyphae are microscopic mining devices that envelop and permeate dead plant tissues and other sources of nutrients. Many of the foods used by fungi have low concentrations of soluble sugars and other readily digested molecules and are rich in polymers, including complex polysaccharides, proteins, and lipids. Hyphae are superbly adapted for solubilizing these materials. Rather than digesting solids from the outside and working inward, colonies of branched hyphae penetrate their food and digestion is achieved over a large area of cell surface. This is called **invasive growth**. The mechanism involves the interlinked processes of enzymatic digestion and pressure-driven penetration. Hyphae secrete enzymes from their growing tips. These enzymes hydrolyse polymers, releasing low molecular weight molecules that are absorbed by the cell (Chapter 5). This digestive mechanism renders the food more fluid, or, at least, less of a physical obstacle to growth. By loosening the cell wall polymers at the hyphal tip, the cell allows a proportion of its internal turgor pressure to press upon the surrounding material. These invasive pressures have been measured using miniature strain gauges and range from a few tenths of one atmosphere to two atmospheres (up to 200 kPa). The combination of enzymatic digestion and pressure-driven penetration allows fungi to penetrate an astonishing range of solid substances.

Enzymes secreted from hyphae may diffuse into the environment and catalyse the digestion of macromolecules some distance from the cell surface. Some enzymes seem to be less mobile, remaining in the cell wall or close to its surface. If these act as nutritional enzymes, molecules will be solubilized much closer to the cell, with the invasive hypha mining food from the immediate vicinity of its cell wall. Fungal nutrition will be considered in greater detail in Chapter 5.

Aerial Hyphae and Conidiophores

Fungal colonies cultured on agar medium often cover themselves with a forest of aerial hyphae. This tangle of filaments can fill the airspace above the agar and squash itself by extending against the lid. Extension of these aerial hyphae is driven by the same mechanisms described in the previous section, but the cells encounter a different set of environmental challenges. Hyphae-penetrating agar must exert substantial force to push through the gel matrix and do so by loosening the cross-links between polymers in the apical cell wall, thereby applying a proportion of their internal turgor pressure against their surroundings. The requirement for force generation is greatly reduced for a cell extending into the air, but internal pressure is necessary to support the vertical orientation of the cell. This is obvious, based on the observed collapse of aerial hyphae when they are exposed to dry air. Before hyphae can reach the air above the colony, they must overcome the surface tension of the air–water interface. This problem seems to be addressed by a combination of the continued exertion of turgor-derived force and the secretion of hydrophobic proteins called **hydrophobins** that reduce the surface tension at the interface.

Hydrophobins are small, cysteine-rich, water-repellant proteins that are secreted on the surface of aerial hyphae and fruit bodies. They have been studied in the mushroom-forming basidiomycete, *Schizophyllum commune*. Targeted deletion of one of the hydrophobin genes in this fungus, *SC3*, results in the resulting mutant's inability to produce aerial hyphae. The protein, SC3, operates as a surfactant, self-assembling as a monolayer at an air–water interface and reducing the surface tension of the water. The reduction in surface tension may allow hyphae of the fungus to escape a fluid environment and grow into the air. Secretion of the protein continues with the extension of the aerial hyphae, coating them with a hydrophobic layer. Hydrophobin secretion is also important in facilitating the attachment of hyphae to hydrophobic surfaces. In many species, aerial hyphae differentiate into conidiophores (spore stalks) that produce asexual spores, or conidia. We will return to the process of conidium formation, or conidiogenesis, in the next chapter. The secretion of hydrophobins onto the surface of developing mushrooms (basidiomata) is important in cementing hyphae together, waterproofing the surface of the reproductive organ, and supporting gas exchange by preventing tissues from becoming saturated with water.

Septa, Woronin Bodies, and the Septal Pore Complex

Septal structure is quite different in the Ascomycota and Basidiomycota. In filamentous ascomycetes, septa are perforated by a single, centrally located pore. Open pores allow the transmission of organelles, including nuclei, between compartments. The movement of organelles is clearly visible in active hyphae viewed with a conventional light microscope. Organelles move toward the growing tip, and in the opposite direction toward older compartments. Some organelles move along relatively straight lines: these structures are carried along actin microfilaments, or microtubules, powered by motor proteins. More obvious bulk flow of cytoplasm is also a feature of active hyphae, with pulses of tip-directed flow, interrupted by retrograde motion. These movements are probably driven by tiny differences in turgor pressure along the hypha. The septate hyphae of the subphylum Pezizomycotina (Ascomycota) contain organelles called Woronin bodies

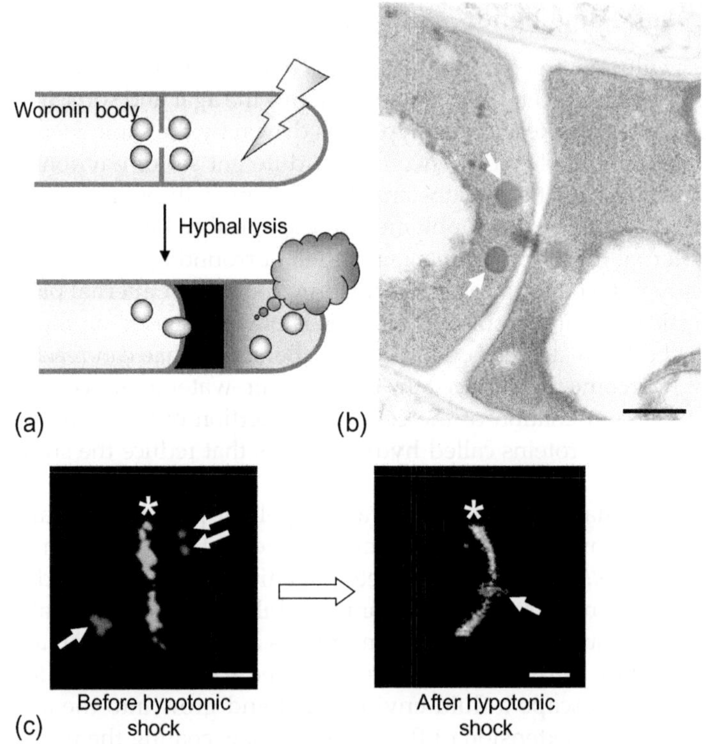

FIGURE 2.15 Morphology and function of the Woronin body. (a) Schematic of Woronin body function. (b) Transmission electron micrograph showing Woronin bodies (arrows) in *Aspergillus oryzae*. Scale bar: 500 nm. (c) Confocal images of Woronin bodies (red arrows) and septa (green asterisks) stained with fluorescent labels before (left) and after (right) hyphal tip bursting. Scale bar = 2 μm. *Source: Maruyama, J., Kitamoto, K., 2013. Expanding functional repertoires of fungal peroxisomes: contribution to growth and survival processes. Front. Physiol. 4, 177.* (See the colour plate.)

that protect the mycelium from catastrophic injury following damage to one or more cell compartments (Figure 2.15). Rupture of the cell wall of a hypha leads to loss of the pressurised cytoplasm and leakage would continue unabated in the absence of some type of sealing mechanism. Woronin bodies isolate the damaged portion of a colony by plugging septal pores on either side of a wound, allowing the rest of the colony to continue growth. Woronin bodies range in size from 100 nm to more than 1 μm and can be seen with a light microscope in some species. The organelle is a type of **peroxisome**. The membrane-bound structure contains a dense core that develops as a self-assembling hexagonal crystal of a single protein called HEX-1. *hex-1* mutants are prone to 'bleeding', and show many developmental defects.

Basidiomycetes produce **dolipore septa** (Figure 2.16). The central canal (pore) of this structure is surrounded by a barrel-shaped swelling of the septum cell wall. Nuclei cannot migrate through unmodified dolipore septa and their distribution within the hyphae that develop after the fusion of sexually compatible colonies involves the formation of **clamp connections**

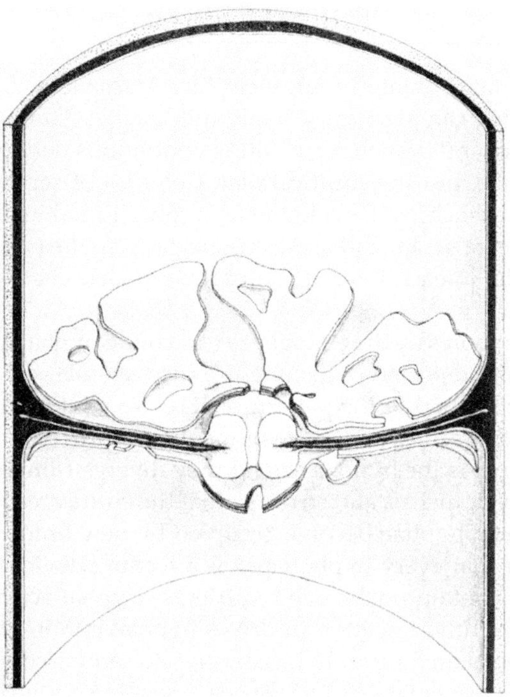

FIGURE 2.16 Dolipore septum of *Rhizoctonia solani. Source: Bracker, C.E., Butler, E.E., 1963. The ultrastructure and development of septa in hyphae of Rhizoctonia solani. Mycologia 55, 35–58.*

(Chapter 1). Both open ends of the dolipore swelling are surrounded by membranes of the **septal pore cap** derived from the endoplasmic reticulum. When the hypha is injured, these membranes collapse and seal the open ends of the septal swelling, performing the same function as the Woronin bodies.

The evolution of complex multicellular structures, from elaborately branched and interconnected colonies to mushrooms and other kinds of fruit bodies, is associated with the development of mechanisms for protecting hyphae from uncontrolled leakage of cytoplasm. This makes sense from an economic point of view. A mushroom constructed from millions, or hundreds of millions of hyphae represents a tremendous investment on the part of the fungal colony. The probability of injury from abrasion or from invertebrates during fruit body expansion seems high. Without a mechanism for isolating damaged hyphal compartments, the operation of the entire organ would be compromised. This process also allows differentiation of the mycelium into compartments that assume specialised functions including survival and reproduction. A simple illustration of this is seen in zygomycetes where the formation of a septum at the base of a sporangium precedes the development of spores; the rest of the colony is non-septate. It seems possible that the sealing mechanisms in the Ascomycota and the Basidiomycota evolved independently, allowing species in both phyla to elaborate complex kinds of feeding colonies and multicellular organs.

THE MYCELIUM

Hyphal extension in non-septate fungi, including *Mucor* and its relatives (the zygomycetes), produces tubes that can extend for many millimetres. Three-dimensional colonies of these fungi develop by lateral branching, creating continuous networks of cytoplasm. These colonies are multinucleate, but not multicellular. Colonies of septate fungi also proliferate through tip growth and branching. Development is complicated in basidiomycetes and ascomycetes by the formation of septa, and clamp connections in the basidiomycetes (Chapter 1), and the number of nuclei in each hyphal compartment varies between different taxonomic groups.

Germination of single spores by the emergence of a young hypha, or germ tube, followed by continuous elongation and repeated branching produces circular colonies, or mycelia, whose superficial form may be likened to the pattern of spokes radiating from the hub of a bicycle wheel (Figure 2.17). There are, however, developmental features of the mycelium that limit the usefulness of this analogy. As the hyphae extend they diverge from one another and primary branches form secondary branches, and so on, so that the entire area becomes occupied by the fungus. Any gaps between hyphae become occupied by new branches. This infilling makes sense when we consider that every hyphal apex is a feeding device and that any substantial gaps in the colony may contain unabsorbed nutrients. Hyphal activity on exposed surfaces is only part of the picture. Invasive growth drives hyphae into its food producing a mature colony with a three-dimensional form. In basidiomycetes and ascomycetes, hyphal branches fuse with one another to create highly interconnected networks or webs. Connections between hyphae, or **anastomoses,** provide pathways for the bulk flow of cytoplasm and regulated movement of organelles over the cytoskeleton (Figure 2.18). Anastomoses can form between

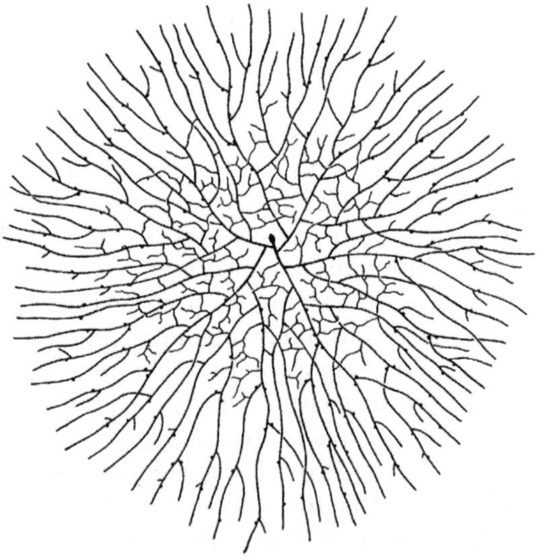

FIGURE 2.17 Young colony or mycelium of a fungus that has grown outwards from a single spore at the center of the drawing. *Source: Buller, A.H.R., 1931. Researches on Fungi, vol. 4. Longmans, Green, and Co., London.*

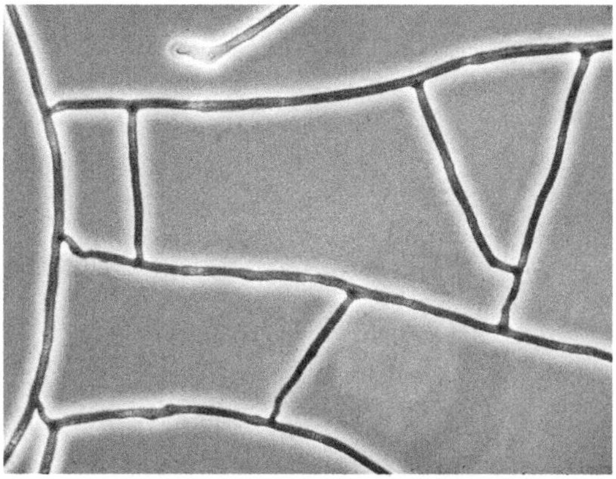

FIGURE 2.18 Anastomoses between hyphae of *Sordaria*. *Source: George Barron, University of Guelph.*

germlings directly after spore germination in some ascomycetes, as well as later in the development of the mycelium. Viewed under the light microscope, the growing colony is revealed as a complex network of tubes supporting the continuous motion of fluid. The interconnected architecture of the colony is important because it allows resources to be shuttled from one location to another tangentially as well as radially. The significance of this resource distribution is apparent when we consider the behaviour of basidiomycetes that digest wood.

Living trees and decaying wood in forests are connected by populations of mycelia formed by mycorrhizal fungi and saprotrophs. Mycelia can span large territories in these ecosystems, disbursing nutrients from portions of the colony embedded in a rotting log, for example, to other parts of the colony spanning out in search of fresh nutrient sources. The architecture of mycelia varies according to species and to environmental conditions. Young colonies can be very dense, forming a thick weft of filaments close to the point of origin. Others remain more diffuse, extending rapidly rather than concentrating in a single area. When one part of the colony encounters a promising resource, the form of the entire colony can change. It is worth underscoring that the exploratory part of the mycelium is pursuing genetically determined algorithms of colonial behaviour rather than expressing any intent. This probably seems obvious, but it can be easy to convey impressions of fungal intelligence when we think sloppily about the complex signalling that controls development. The redistribution of resources between distant parts of the mycelium has been examined in beautiful experiments on cultured wood-decay basidiomycetes. The formation of multicellular 'organs', called cords, is a very important part of this process, and we will examine their structure in the next section.

Hyphal size and shape show considerable variation among the fungi. Hyphae range in diameter from a few micrometres in many ascomycetes and basidiomycetes to much larger cells, with a diameter of more than 20 μm in zygomycete fungi. Tip shapes range from perfect hemispheres to more pointed forms and some variations are seen within individual cultures. Differentiation of hyphae occurs within mycelia producing hyphal swellings and cells with thickened walls. The function of these features is often obscure. More complex morphological changes are common, too, and these can be matched to particular functions. **Appressoria** are inflated cells produced by plant pathogens on the leaf surfaces of their hosts (Figure 2.19). These are initiated as hooks and swellings on the host and expand into domed cells that adhere tightly to the plant cuticle.

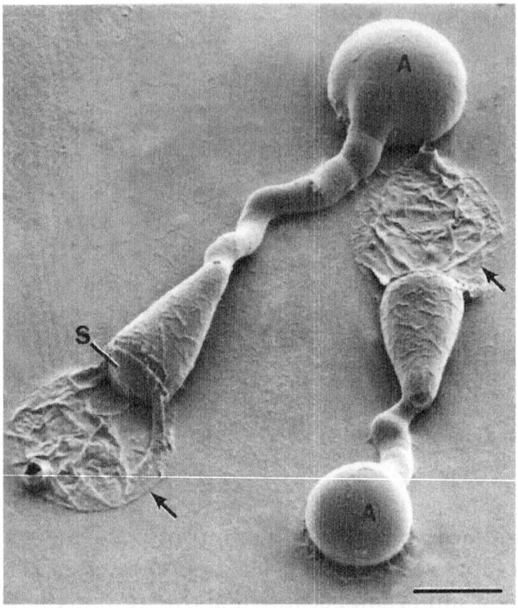

FIGURE 2.19 Young appressoria of the rice blast fungus, *Magnaporthe grisea*, connected by germ tubes to their conidia. The conidia deflate (arrows) as their contents are transferred into the appressoria (A). A septum (S) within one of the conidia is evident between the collapsed portion of the spore and the adjacent cell that has not emptied its cytoplasm. *Source: Money, N.P., Howard, R.J., 1996. Confirmation of a link between fungal pigmentation, turgor pressure, and pathogenicity using a new method of turgor measurement. Fungal Genet. Biol. 20, 217–227.*

Slender hyphae, or infection pegs, grow from the base of the appressoria and penetrate the leaf. Appressoria of some species form over stomata and penetration occurs when the stomata are open. In other diseases, penetration occurs directly through the intact cuticle and underlying cell wall and is a mechanical process that seems to be dependent upon the exertion of force derived from cytoplasmic turgor pressure (Chapter 8). Following penetration of the host, hyphae of biotrophic pathogens produce **haustoria** that absorb nutrients from the plant without destroying the infected cell. Arbuscular mycorrhizal fungi classified as Glomeromycota produce analogous structures called **arbuscules** that form the interface between the fungus and its host. Arbuscules are highly branched cells that control nutrient transfer from plant to fungus and vice versa. Nematophagous or nematode-trapping fungi produce a variety of hyphal structures ranging from adhesive knobs to constricting rings to snare their prey. Distinctive hyphal shapes are often seen in cultures, but the functions of these modifications are not understood.

MULTICELLULAR ORGANS

Cords and Rhizomorphs

The use of the term 'organ' to describe the multicellular structures of fungi is subjective. Animal organs are differentiated structures that perform specific functions. A mycelium may be viewed as an organ by this definition, but the degree of differentiation among its hyphae is very limited. A higher degree of differentiation is achieved through the formation of cords by mycelia of wood-decay and ectomycorrhizal fungi and it seems justifiable to describe the more complex of these structures as organs. We will return to this issue of terminology when we consider the structure and function of fruit bodies later in the chapter.

Strands, **cords**, and **rhizomorphs** vary in complexity from bundles of hyphae, whose cell walls adhere to one another to produce slender cylinders of unpigmented cells on the surface of a culture dish, to fat pipes with a diameter of a few centimetres that can extend for hundreds of metres. The larger pipes are formed from hundreds of thousands of hyphae, develop a complex internal anatomy, and are sometimes tipped with a rounded, mucilaginous cap. Variations in the internal structure of these organs make it difficult to discriminate between strands, cords, and rhizomorphs. The term rhizomorph may be useful to designate the larger of these invasive organs that have an identifiable tip that pushes through the soil. Cord is the preferred term for other linear organs without an organised tip.

Rhizomorphs facilitate the spread of *Armillaria* species between host plants and rotting wood, and support the coverage of vast territories by colonies of these mushroom-forming fungi. The surface of the rhizomorph is covered with a peripheral layer of thin hyphae that encloses a cortex of radially oriented hyphae (Figure 2.20). Hyphae of various sizes run along the length of the organ beneath the cortex. These longitudinally oriented cells constitute the medulla. The largest of these medullary hyphae, toward the centre of the organ, are dead and devoid of cytoplasm. A gas-filled lumen occupies the centre of the rhizomorph. Within the medulla, it seems likely that the smallest hyphae (just beneath the cortex) are the youngest and most active cells. As the rhizomorph elongates, these cells are pushed toward the centre of the organ by new hyphae growing directly beneath the cortex. The displaced medullary cells enlarge, eventually becoming inactive and line the empty lumen. Physiological experiments suggest that some of the medulla cells act as conduits for fluid transport and the term 'vessel hypha' has been applied to the largest hyphae that have this presumed function. Beautiful experiments have been performed on fluid translocation within cords and rhizomorphs using radioactive tracers, but there are many unresolved questions about the physiology of rhizomorphs. The gas-filled lumen aids oxygenation of rhizomorphs that bury themselves within soil and rotting wood.

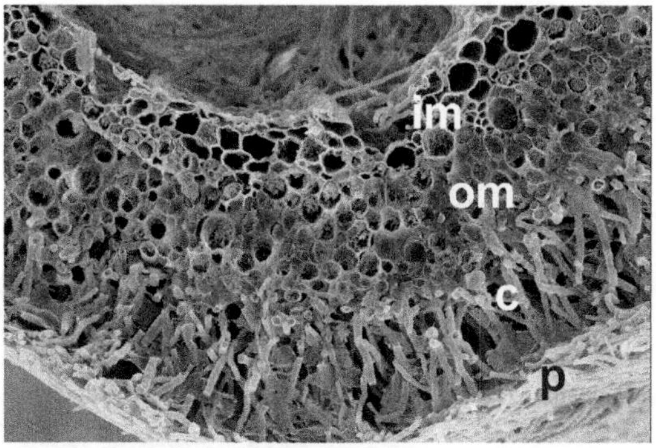

FIGURE 2.20 Anatomy of the rhizomorph of *Armillaria gallica* revealed by fracturing the frozen specimen and viewing in a scanning electron microscope. (p) peripheral layer of hyphae, (c) cortex, (om) outer medulla, and (im) inner medulla. The inner medulla surrounds an open central cavity. *Source: Yafetto, L., Davis, D.J., Money, N.P., 2009. Biomechanics of invasive growth by Armillaria rhizomorphs. Fungal Genet. Biol. 46, 688–694.*

Experiments on rhizomorphs of *Armillaria gallica* reveal that the organs extend at faster rates than individual hyphae growing in their normal unbundled form. This accelerated extension is driven by a combination of apical growth of cells at the tip of the rhizomorph and intercalary growth of hyphae behind the apex. Like hyphae growing individually, the cells of the rhizomorph exert a pressure of up to one atmosphere (100 kPa), which provides the organ with mechanical force to overcome physical obstacles in its path.

Cords vary from loose bundles of hyphae to more complex structures with interior vessels enclosed in a thick outer rind. Rather than forming an organised tip like a rhizomorph, hyphae aggregate into cylindrical cords behind tip-growing hyphae that spread into a fan. Cords often develop between separate woody resources colonised by a single mycelium.

Cords and rhizomorphs allow fungi to mobilise food resources and water from one location and transport them over long distances to support other parts of an extended mycelium. They greatly extend the area that may be explored by a single mycelium. These organs allow dry-rot fungi that specialise in the destruction of timber in buildings to bridge nutrient deserts of brick and concrete and colonise dry wood by translocating nutrients and water from wetter locations. Dry-rot fungi include *Serpula lacrymans*, which causes tremendous damage to buildings in Europe, and an equally destructive basidiomycete, *Meruliporia incrassata*, which destroys homes in North America.

Sclerotia

Sclerotia are hardened masses of hyphae that serve as survival structures for ascomycetes and basidiomycetes (Figure 2.21). Sclerotia can be rounded, flattened, or elongated. Their sizes range from the 0.1-mm-diameter microsclerotia of the plant pathogen *Macrophomina phaseolina*, to the 30 cm sclerotia of the edible Australian fungus *Laccocephalum mylittae* that weigh several kilogrammes. Sclerotial development sometimes occurs when nutrients are running out, but many are formed in active cultures showing that there are other stimuli for the growth of these structures. Their development involves the repeated branching of hyphae and formation of closely spaced septa. Differentiation of hyphae within the sclerotium produces a central medulla of thin-walled hyphae rich in lipid and glycogen reserves. These cells are surrounded by a cortex of hyphae with thicker walls and an outer layer, or rind, of

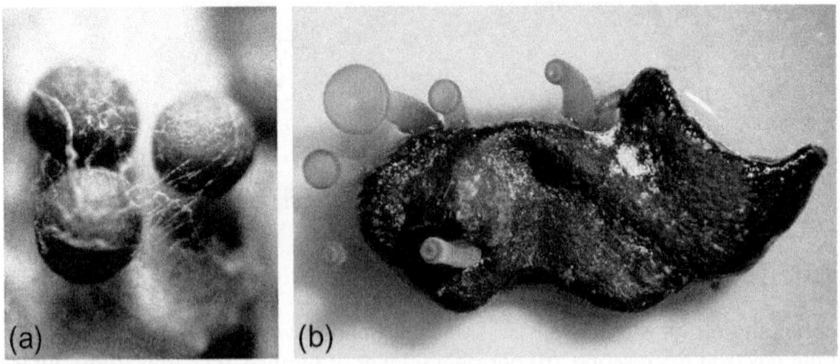

FIGURE 2.21 Sclerotia of *Sclerotinia sclerotiorum*. (a) Dormant and (b) germinating to produce apothecia. *Source: www.sclerotia.org*

cells that are highly melanized in some species. Some fungi form a feeding mycelium when their sclerotia germinate; others produce fruit bodies. Elongated sclerotia known as ergots are produced by species of *Claviceps*, including *Claviceps purpurea* that causes ergot of rye. Ergots develop in autumn, survive the winter, and germinate in the spring to produce a stalked fruit body from which infectious ascospores are discharged into the air. Ergots of *Claviceps purpurea* contain toxic alkaloids, including ergotamine that causes vasoconstriction. Ingestion of this toxin in the form of bread baked from contaminated flour has caused outbreaks of gangrene, loss of limbs, and resulted in many deaths. *Sclerotium sclerotiorum*, that causes white mould of flowers and vegetables, produces lumpy sclerotia as over-wintering structures. Germination of these sclerotia produces stalked fruit bodies tipped with cup-shaped apothecia. Sclerotium production by plant pathogens has been studied in the greatest detail, but these structures are also formed by ectomycorrhizal and ericoid mycorrhizal fungi, saprotrophs, and fungi that have adopted a variety of other lifestyles. The distribution of these species in diverse taxonomic groups shows that sclerotial development is an example of evolutionary convergence.

Pseudosclerotia and Pseudosclerotial Plates

Pseudosclerotial development promotes the survival of a fungal colony by incorporating the material on which it is growing. Species of *Ophiocordyceps* mummify the bodies of the invertebrates that they have killed within pseudosclerotia. Stalked fruit bodies bearing perithecia develop from the pseudosclerotia of these ascomycetes. Pathogenic species of *Monilia* (*Sclerotinia*) produce similar structures around the infected fruits of their host plants. The polypore *Laccocephalum basilapiloides* is the Australian 'stone-making fungus', whose 8 cm-diameter pseudosclerotium binds sand grains, root fragments and other debris, and supports a single fruit body.

Pseudosclerotial plates are thin sheets of mycelium that incorporate wood and other organic matter. The hyphae of these structures are highly branched and pigmented with melanin, forming a barrier that resists penetration by water and the hyphae of other fungi. Pseudosclerotial plates are often visible as black zone lines in rotting wood. *Armillaria* species confine 'decay columns' of wet wood rot within pseudosclerotial plates. Conversely, the ascomycete *Xylaria hypoxylon* maintains the wood beneath its pseudosclerotial plates in a dry condition.

Fruit Bodies: Ascomata and Basidiomata

The formation of multicellular fruit bodies is the most complex and least understood developmental process in the fungi. Fungi are regarded traditionally as microorganisms because many species (e.g. zoosporic fungi, yeasts) are visible only using a microscope, and others exist in microscopic form for much of their life cycles. This distinction can seem irrational, however, when we consider macroscopic fruit bodies (mushrooms) including the ascomata of morels (Ascomycota), or basidiomata of the Agaricomycotina (Basidiomycota). Indeed, the fruit bodies of many of the basidiomycetes can be difficult to ignore. The caps of *Termitomyces titanicus* mushrooms that grow from abandoned termite nests in West Africa can expand to 1 m in diameter. Even larger fruit bodies are produced by wood-decay fungi. A metre-long crust produced by a white rot fungus called *Phellinus ellipsoideus* on the underside of an oak log was reported in 2010 on Hainan Island in China. It weighed 500 kg and shed an estimated one trillion spores per day. In this respect, fungi are unlike other microorganisms, including bacteria and many protists, which are invisible to the unaided eye. Nevertheless, the fungi are considered part of the purview of microbiologists.

Limited differentiation of tissues is an important feature of fruit body development. Many kinds of cells are visible when thin sections of a young plant stem are studied with a microscope: guard cells form stomata in the epidermis, cortical cells lie beneath the epidermis, and phloem, cambium, and xylem cells are organised in vascular bundles. The same exercise performed on the stem (or stipe) of a mushroom reveals a considerably simpler anatomy. Mushroom stems and caps are built from thin-walled hyphae that differ only in diameter, the spacing of septa, and frequency of branching. This anatomical plainness is a little disconcerting. At least 16,000 different species of mushroom-forming basidiomycete have been described and each has a distinctive basidiome. Besides the conventional umbrella-shaped mushroom, these fungi form brackets, coral-shaped fruit bodies, little spindles, discs, ruffled crusts, puffballs, and phallic mushrooms tipped with stinking slime. All of these beautiful organs are formed by filamentous hyphae that grow from the feeding mycelium. Different types of hyphae are recognised in some fruit bodies, including thin-walled generative hyphae, thick-walled skeletal hyphae, and elaborately branched binding hyphae. This limited differentiation affects the texture of the fruit body, with a high proportion of skeletal and binding hyphae, for example, producing the harder and less flexible basidiomata of certain bracket mushrooms.

Sections of fruit bodies viewed with the transmission electron microscope show tightly packed cells that resemble the parenchyma of plants (Figure 2.22). This structure is also apparent when the surfaces of a perithecium or other types of ascomata are viewed with a scanning electron microscope (Figure 2.1e). Investigators looking at the anatomy of rhizomorphs in the

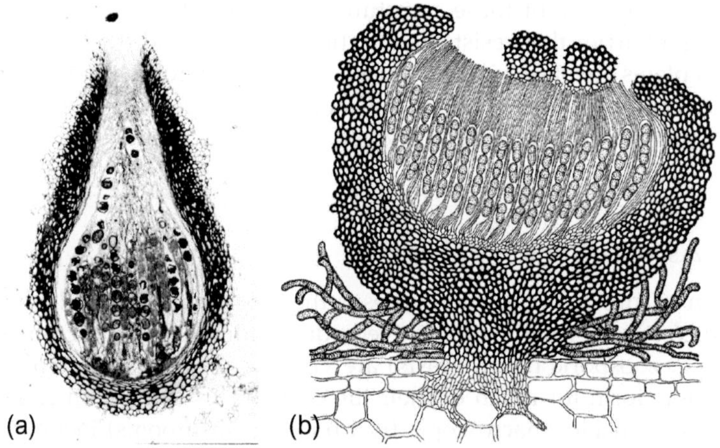

FIGURE 2.22 Pseudoparenchyma tissue formed by branched and interwoven hyphae in the walls of ascomata. (a) Transmission electron micrograph of perithecium of *Sordaria humana*. (b) Ascoma (pseudothecium) of *Pseudoparodia pseudopeziza*. *Source: (a) Read, N.D., Beckett, A., 1985. The anatomy of the mature perithecium in Sordaria humana and its significance for fungal multicellular development. Can. J. Bot. 63, 281–296 and (b) Müller E., von Arx, J.A., 1962. Beit. Kryptogamen. Schweiz 11 (2), 922 p.*

1970s believed that they had identified a group of tiny isodiametric cells in the tip of the organ that might function as a meristem. These observations produced a confusing picture of fungal development, until careful microscopic analysis demonstrated that fruit bodies are formed by the intermingling and branching of hyphae and frequent formation of septa to produce short hyphal compartments. The resulting tissue is called **pseudoparenchyma**, for its superficial resemblance to the parenchyma tissue of plants. Recognition of the hyphal nature of every fungal structure is critical for understanding fungal development because it highlights the importance of 'invisible' differentiation at the molecular level in sculpting mushrooms and other fungal organs. The complex form of ascomata and basidiomata derives from intricately choreographed interactions between hyphae that reflect their positions within the fruit body.

Environmental and genetic factors that control fruit body development are treated later in this book. The focus here is on the developmental processes that take place once fruiting is initiated. The tissue that forms ascomata is composed of hyphae that contain haploid nuclei. Fusion of sexually compatible hyphae occurs within the fruit body and produces the cells from which the asci develop. This delayed mating distinguishes the ascomycetes from the basidiomycetes, as described in Chapter 1. Basidiomata are produced by heterokaryotic hyphae that contain nuclei of two compatible mating types. Information on the genetic control of the development of ascomata is limited. Genetic studies have concentrated on a handful of 'model' ascomycetes including *Aspergillus nidulans*, *Neurospora crassa*, and *Podospora anserina*. Several genes involved in primary metabolism affect sexual development. Sexual development is disrupted when these genes are mutated, but vegetative growth of hyphae is unaltered. This indicates, as one would anticipate, that fruit body development involves distinctive molecular processes that are not activated as long as the fungal colony is feeding.

The transition between the two phases in the life cycle is driven by a number of environmental factors, but nutrient availability is the primary stimulus. The production of fruit bodies represents a tremendous investment in resources for the colony and demands a high level of metabolic activity. This is the reason that mutants defective in mitochondrial activity can be sterile (unable to form fruit bodies). Increased respiration during sexual differentiation produces more reactive oxygen species (ROS) and oxidative damage to DNA and proteins is deleterious for developing ascomata. Not surprisingly, genes encoding protective superoxide dismutases, oxidases, and peroxidases that control the production and degradation of ROS are also crucial for fruit body formation. Signal transduction pathways (cascades) involved in fruiting have also been examined and a catalogue of G proteins, G protein receptors, cAMP-dependent protein kinases, GTP-binding proteins, and other signalling molecules and transcription factors have been characterised in the model ascomycetes. Once the sexual cycle has been initiated, genes that control the positioning of nuclei and the cell cycle are essential for fruit body differentiation, and tubulin and other components of the cytoskeleton are implicated in these mechanisms. Genes involved in the synthesis of cell wall components are also critical, but researchers are a long way from understanding ascoma development at the level of molecular genetics. This is another of those topics in fungal biology with the potential for huge advances by future researchers.

Research on basidiome development in basidiomycetes is similarly limited. Formation of a mushroom requires the transfer of cytoplasm from a large volume of supporting mycelium. Experiments on this process on Petri dish cultures of *Coprinopsis lagopus* indicate that 50% of the biomass in the mycelium may be shuttled into developing basidiomata in response to the appropriate environmental cues. Fruiting begins (or becomes visible) with the formation of aggregates of branched hyphae called knots. These knots expand to form the embryonic fruit body, or **primordium**. Differentiation of cells occurs in the centre of the primordium demarking regions that will become the stipe, cap, and gills of the mushroom. Because mushroom development has evolved multiple times, it is not surprising that it occurs in different ways. In **gymnocarpic** development, the cap enlarges from the top of the elongating stipe and the hymenium is exposed for most of the expansion process. Gymnocarpic development is characteristic of *Boletus, Clitocybe, Lactarius,* and *Russula*. **Angiocarpic** development refers to protection of the immature hymenium. In some mushrooms, the hymenium differentiates within an internal cavity in the primordium and is exposed when the fruit body expands. In others, tissues derived from the surface of the primordium surround the hymenium. A mantle of hyphae, called the **universal veil**, wraps around the entire primordium of the paddy straw mushroom, *Volvariella volvacea*, and species of *Amanita* (Figure 2.23). This is torn apart when the mushroom expands and remains as a basal cup, or **volva**, and as scales on the surface of the cap. Another sector of tissue within some primordia is stretched into a sheet that covers the underside of the gills. This **partial veil** is pulled away from the outer rim of the expanding cap and remains as a ring, or **annulus**, around the stipe in species of *Amanita* and as a thin, cobweb-like drape called the **cortina** in *Cortinarius*. Volva and ring structures are important features for identifying many mushrooms. The **pseudorhiza** is another diagnostic structure associated with some basidiomycetes. This is a root-like extension of the stipe that connects the fruit body to its buried source of nutrients. *Xerula radicata* and *Collybia fusipes* that decompose woody roots are examples of fungi that form pseudorhizas. Colonies of *Termitomyces* are cultivated by termites and produce long pseudorhizas to raise their fruit bodies above the termite mounds.

The mechanism of expansion differs among the basidiomycetes. In some species, expansion is due to the inflation of preexisting hyphae with very limited formation of new hyphal branches. In other cases, stipe elongation involves a new programme of branching and septation. Some mushrooms expand through a combination of inflation and the development of new hyphal compartments, with the balance between these processes differing according to

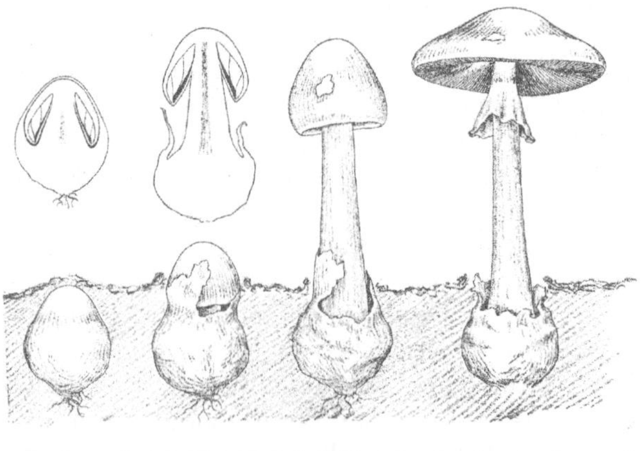

FIGURE 2.23 Mushroom expansion in *Amanita phalloides*, the death cap. *Source: Longyear, B.O., 1915. Some Colorado Mushrooms. Agricultural Experiment Station of the Agricultural College of Colorado, Fort Collins, CO.*

location within the fruit body. Intercalary extension of hyphae is another important process in mushroom expansion with evidence of the synthesis of new cell wall components along the length of the hyphae. The remarkable feature of this phase of development is its speed: mushrooms can show a 1000-fold increase in volume in a few hours! The mechanism is hydraulic, meaning that the inflation is driven by the uptake of water from the environment. This water absorption allows the expanding tissue to exert a hydrostatic pressure of one atmosphere or more against any surrounding obstacles, which allows mushrooms to emerge from rotting wood, hard-packed soil and even to burst through asphalt paving.

The sectors of tissue within the primordium that prefigure the different regions of the mature basidiome differ in the density of the hyphae (how tightly they are packed together), the hyphal diameter, and the frequency of septation. Surprisingly, small pieces of tissue from anywhere within the primordium will generate a feeding colony if they are transplanted onto nutrient agar. This simple experiment illustrates the totipotency of fruit body tissues. Irrespective of their position within the primordium, hyphae are not committed, irreversibly, to a particular developmental fate. This observation marks a sharp contrast to the cells in animal embryos in which differentiation of the cells into various tissues is an irreversible process. Animal stem cells are the important exception to this feature of animal development. Part of the reason for the flexibility in the developmental fate of cells in the basidiome is that most of them perform a purely mechanical function by supporting the spore-producing tissues. Totipotency is lost in the hyphal compartments at the tips of the cells that emerge on the gill surface that become basidia. The basidia are the specialised sites for nuclear fusion, meiosis, and spore formation and these cells show irreversible commitment to these processes.

Knowledge that differentiation of tissues within the basidiome is limited does not help us to understand how the distinctive shapes and sizes and colours of mushrooms are formed. Analysis of fungal genomes has failed to find genes that are homologous to key genes that are universal players in animal development with names like *Hedgehog* and *Notch*. Computer simulations are effective at generating virtual mushrooms from groups of filaments whose behaviour is governed by a handful of rules. These rules include the degree to which neighbouring filaments attract or repel one another as they extend, the frequency of branching, the angles at which those branches grow, and the gravitational response of the growing tips. The power of these models lies in the fact that so few rules can specify a mushroom. The computer simulations also hint at the reason that the kinds of developmental genes ubiquitous among animals are absent in the fungi. According to the models, a developmental clock dictating the expression of successive waves of cell attraction and repulsion might be sufficient to shape everything from a mushroom with delicate gills to a fat bracket sticking out of a dying tree. Having advanced this possibility, however, investigators still need to identify the cell biological mechanisms that enable hyphae to sense the position of their neighbours, control branching, and perceive gravity.

Most of the ongoing molecular studies on fruit body development are directed at an ink cap, *Coprinopsis cinerea*, and the split gill mushroom, *Schizophyllum commune* (Figure 2.24). Experiments on these fungi have singled out a few of essential regulatory genes. These include the *THN* gene in *Schizophyllum* that is involved in the formation of aerial hyphae and fruit bodies. This is interesting given the shared challenges for hyphae and multicellular fruit bodies in 'escaping' the submerged mycelium. Genes that encode hydrophobins and are regulated by mating type-genes are essential participants in the fruiting process. Other

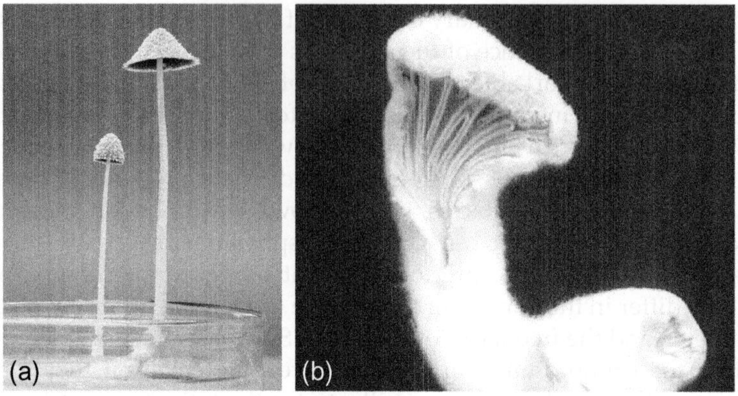

FIGURE 2.24 Cultured fruit bodies of (a) *Coprinopsis cinerea* and (b) *Schizophyllum commune*. *Source: (a) Hajime Muraguchi, Akita Prefectural University, Japan and (b) www.mycology.adelaide.edu.au*

participants include genes encoding lectins (carbohydrate-binding proteins), and, in common with the process of ascomata development, oxidative enzymes and enzymes involved in carbohydrate metabolism. Despite significant advances in research on fungal multicellularity, however, it is important to recognize that researchers are nowhere close to pinpointing the genes that distinguish the fruit bodies of the 16,000 species of mushroom from one another.

Further Reading

Berepiki, A., Lichius, A., Read, N.D., 2011. Actin organization and dynamics in filamentous fungi. Nat. Rev. Microbiol. 9, 876–887.

Howard, R.J., Gow, N.A.R. (Eds.), 2007. The Mycota, Volume 8, Biology of the Fungal Cell. second ed. Springer Verlag, New York.

Jedd, G., 2011. Fungal evo-devo: organelles and multicellular complexity. Trends Cell Biol. 21, 12–19.

Lew, R.R., 2011. How does a hypha grow? The biophysics of pressurized growth in fungi. Nat. Rev. Microbiol. 9, 509–518.

Read, N.D., Goryachev, A.B., Lichius, A., 2012. The mechanistic basis of self-fusion between conidial anastomosis tubes during fungal colony initiation. Fungal Biol. Rev. 26, 1–11.

Richards, A., Veses, V., Gow, N.A.R., 2010. Vacuole dynamics in fungi. Fungal Biol. Rev. 24, 93–105.

Steinberg, G., 2007. Hyphal growth: a tale of motors, lipids, and the Spitzenkörper. Euk. Cell 6, 351–360.

Steinberg, G., Martin, S., 2011. The dynamic fungal cell. Fungal Biol. Rev. 25, 14–37.

Taylor, J.W., Ellison, C.E., 2010. Mushrooms: morphological complexity in the fungi. Proc. Natl. Acad. Sci. USA 107, 11655–11656.

Voisey, C.R., 2010. Intercalary growth in hyphae of filamentous fungi. Fungal Biol. Rev. 24, 123–131.

Weblink

http://www.gerosteinberg.com/introduction.php

CHAPTER

3

Spore Production, Discharge, and Dispersal

Nicholas P. Money
Miami University, Oxford, OH, USA

DIVERSITY, DEVELOPMENT, AND FUNCTIONS OF SPORES

It is impossible to grasp the diversity of species in Kingdom Fungi by studying their vegetative colonies. Zoosporic fungi have the least complicated feeding structures, with many chytrids, for example, producing nothing more complex than a spherical thallus that absorbs food from the single plant cell in which it develops. The mycelia of non-zoosporic phyla are immeasurably more complicated in terms of their structure and development, but it is rarely possible to identify a fungus beyond its phylum by looking at its hyphae. The presence of dolipore septa is diagnostic of a basidiomycete, but few, if any, fungal biologists could say whether hyphae viewed on a microscope slide belong to a giant puffball or a jelly fungus. When we examine spores, and the structures producing the spores, we have a much better prospect of more detailed identification. Spore morphology is one of the most important features used for visual identification of fungi and the astonishing range of spore types is a powerful reflection of their evolutionary diversity (Figure 3.1). Without assessing differences between spores and fruit bodies, efforts to develop a taxonomy of the fungi would have been futile. The study of spores is a study of evolution.

Asexual Spores

Nuclei within asexual spores are produced by mitotic division so that the spores are clones of the parent mycelium. The simplest mechanism of spore formation involves the differentiation of preformed mycelium. Spores generated in this fashion are called **thallospores**. There are two categories of thallospore: **arthrospores** are produced by the fragmentation of hyphae into compartments separated by septa, and thickening of the cell wall of a hyphal compartment forms a **chlamydospore**. **Sporangiospores** are asexual spores formed inside a walled sporangium. Sporangiospores include the spores of zygomycetes, which are exposed to air by splitting of the mature sporangial wall, and motile zoospores of chytrids expelled into water from their zoosporangia. Asexual spores produced on stalks, or **conidiophores**, are called **conidia** (singular **conidium**).

http://dx.doi.org/10.1016/B978-0-12-382034-1.00003-7

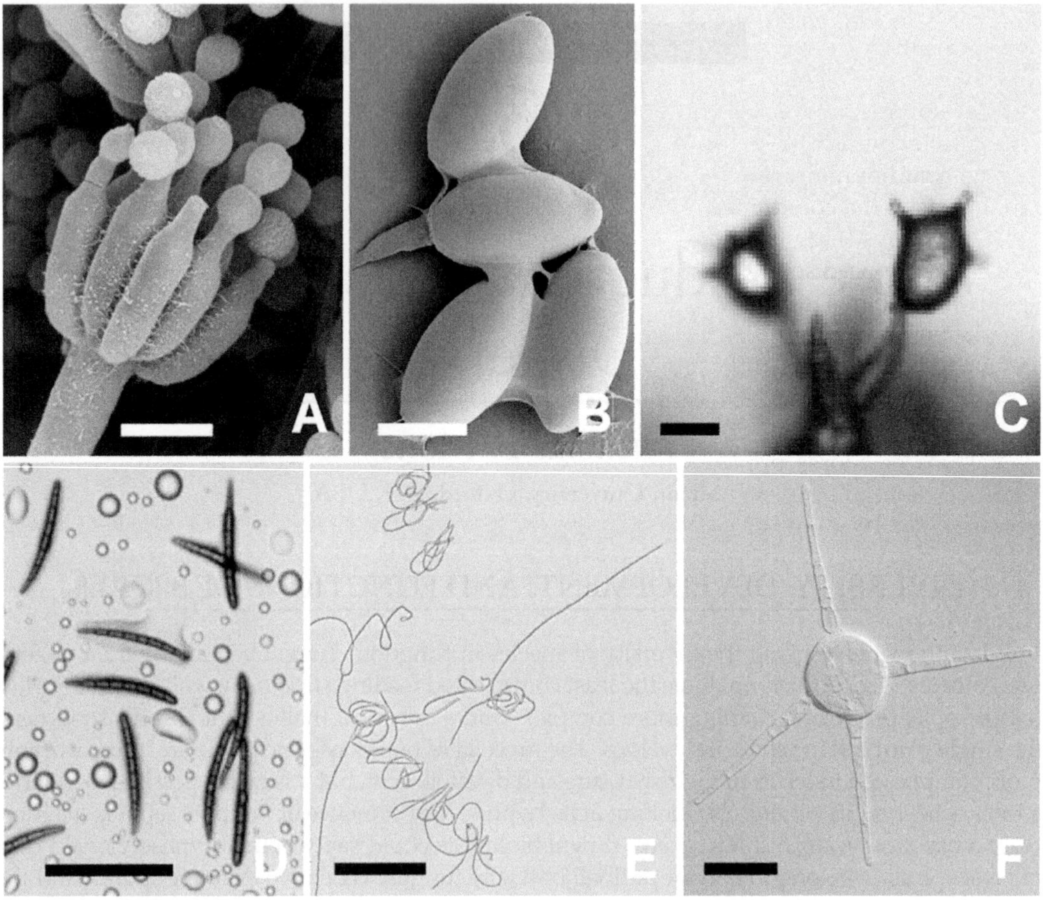

FIGURE 3.1 Sampling of the morphological diversity of fungal spores. (a) Spherical conidia of *Penicillium* species produced by phialides. (b) Ellipsoidal ascospres of *Podospora anserina*. (c) Polyhedral basidiospores of *Aleurodiscus oakesii*. (d) Fusiform ascospores of *Geoglossum nigritum*. (e) Filamentous ascospores of *Cordyceps militaris*. (f) Star-shaped aquatic conidium of *Brachiosphaera tropicalis*. Scale bars (a–c) 10 μm, (d, e) 100 μm, (f) 20 μm. *Source: Fischer et al., 2010.*

More than 100 terms have been used to discriminate between different types of conidium. These include annellospore, botryo-aleuriospore, closterospore, polarlocularspore, and stalagmospore. Some of these nouns refer to the shape of the spore, others to a mechanism of development (**conidiogenesis**). To anyone other than an expert, the distinctions between many of these spore types are quite opaque. The 'Dictionary of Fungi' definition of an annellospore (or annelloconidium), for example, refers to 'holoblastic conidiogenesis in which the conidiogenous cell by repeated enteroblastic percurrent proliferation produces a basipetal sequence of conidia.' Deconstruction of this description would require a separate chapter on taxonomic research on the formation of conidia (asexual spores) and we have opted, instead, to refer the reader to this specialized literature. The variety of conidial shapes and developmental processes has little taxonomic value beyond its use in describing species. Many unrelated genera share the same manner of spore formation and closely related species can show entirely different mechanisms of conidiogenesis.

Given the importance of conidiogenesis in the reproduction of fungi of clinical and agricultural significance, it is surprising that this has received so little attention from researchers in recent decades. Electron microscopic studies on conidial fungi have documented many types of developmental mechanisms (more than 40 have been described), but their cellular and molecular controls have not been analyzed in any detail. Most conidia are formed on stalks called conidiophores. They develop at the tips of the conidiophore, or on branches from the main axis of the conidiophore, as single spores, or in chains. Chains of spores are formed in different ways (Figure 3.2). Segmentation of the conidiophore by the development of multiple closely spaced septa creates spores from the existing structure of the conidiophore. Other fungi form chains from the free ends of conidiophores and conidiophore branches. Spores can inflate before they are separated from other cells by the formation of a septum (described as blastic development), or separate without prior expansion (thallic). Chains can develop with the youngest spores at the base (basipetal arrangement) or youngest at the tip (acropetal). The variation can seem limitless.

Chains of spores produced from cells called phialides are very common among ascomycetes, including *Aspergillus* and *Penicillium* (Figure 3.3). Phialides are vase-shaped cells

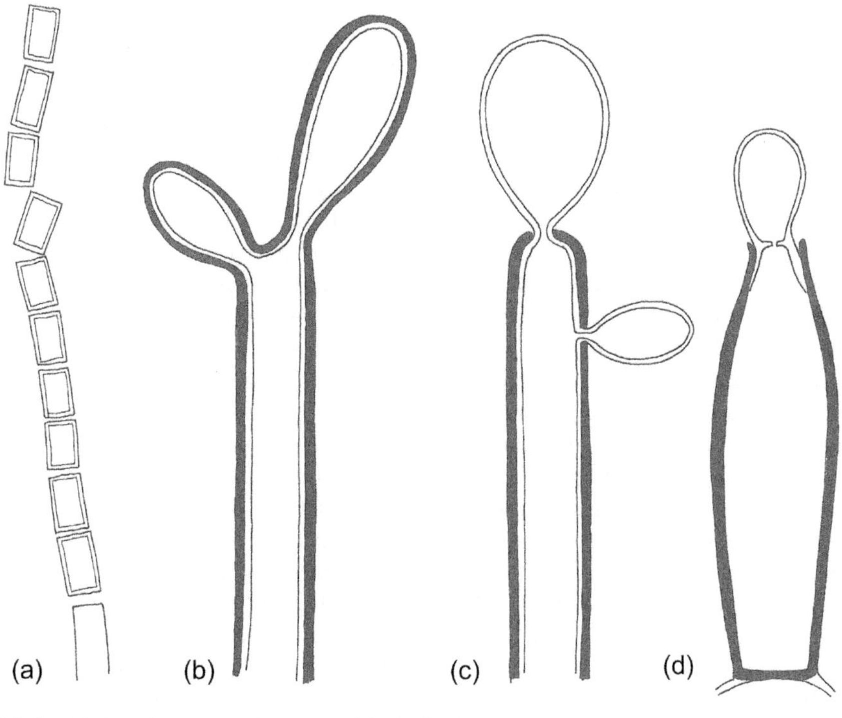

FIGURE 3.2 Modes of conidium formation. (a) Thallic development: the conidial initial does not enlarge before it separates from the conidiophore. (b) Holoblastic development: all cell wall layers expand during conidium formation. (c) Enteroblastic development: only the inner layer of the cell wall expands through an aperture in the outer wall of the conidiophore. (d) Phialidic development: conidium is formed by the synthesis of new cell wall material within the neck of the phialide. *Source: Webster, J., Weber, R.W.S., 2007. Introduction to Fungi, third edition. Cambridge University Press.*

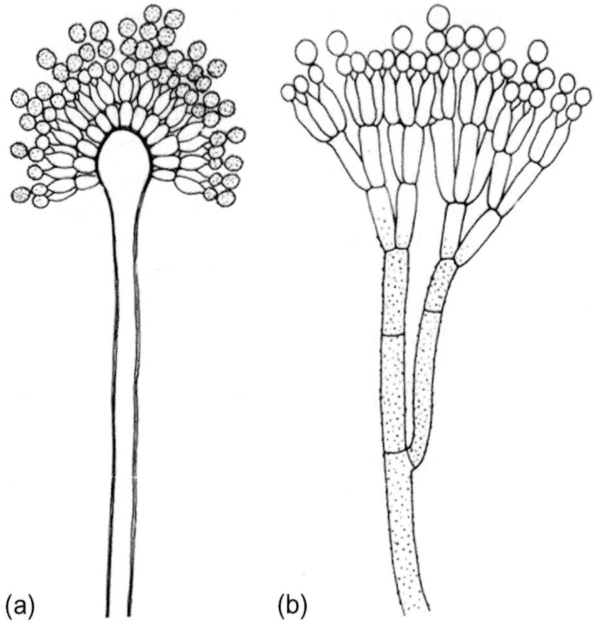

FIGURE 3.3 Conidiophores of (a) *Aspergillus* and (b) *Penicillium* bearing clusters of phialides that generate chains of spores.

(shaped, more explicitly, like a Greek amphora), that develop on conidiophores. They produce chains of conidia by extruding new cell walls through their open necks (think of pulling a turtleneck sweater over your head) and pinching-off uninucleate portions of cytoplasm (Figure 3.2d). The formation of each conidium is coordinated with mitosis in the phialide such that the conidium is separated from the phialide by the formation of a septum through the axis of the mitotic spindle. Phialides produce basipetal chains of spores. In *Aspergillus*, multiple phialides develop over the dome of the swollen tip of the conidiophore called the ampulla. As each phialide generates a chain of conidia, the single conidiophores can support hundreds of spherical spores. The conidiophore of *Penicillium* has a branched apex with one or more phialides produced at the tip of each branch. Like *Aspergillus*, the activity of multiple phialides allows a single conidiophore to support a mass of conidia. These fungi produce millions of dry spores per square centimetre of colony surface.

Conidiogenesis requires hyphae to switch from programs of indeterminate elongation to more determinate growth processes that shape and detach spores. The alterations in cellular form during sporulation have been studied using electron microscopy but very little is known about the underlying molecular genetic controls. The formation of a single phialide must involve a period of limited elongation of a hyphal branch, cessation of extension, and swelling of the cell behind the apex to produce the ampulla shape. Subsequent events include the formation of a septum at the base of the phialide, perforated by a pore to maintain cytoplasmic continuity with the subtending conidiophore, followed by successive programs of tip swelling, differentiation of conidium initials, and separation from the phialide tip by septation. The enzymes involved in conidiogenesis include catalysts for cell wall synthesis and reorganization of the cytoskeleton. Again, however, this hypothetical laundry list is not supported, nor embellished, by critical studies on the genetics of spore formation. Mutant analysis has identified genes involved in the signalling pathways involved in conidiogenesis in *Aspergillus*. These include genes that exert broad regulatory control over conidiogenesis, genes that encode transmembrane receptors and G-proteins, MAP kinase genes, and multiple transcription factors. However, we remain a long way from solving how each conidial fungus exerts precise control over the shape and size of its spores. This is another example of a fertile area for researchers that would produce significant advances with the application of existing molecular technology coupled with creative experimental design.

Sexual Spores

Fungi produce three types of sexual spore: basidiospores (Basidiomycota), ascospores (Ascomycota), and zygospore (zygomycetes). Because these spores are among the defining features of the major phyla, their development is described in Chapter 1 in the context of fungal life cycles.

Spore Functions

Spores serve as vehicles for transmitting fungi through space and time. This sentence is a bit melodramatic, but captures the fundamental function of spores. Asexual spores carry identical copies of the genes of the parent colony. Sexual spores transmit versions of genes recombined from parental strains. Spores are often formed in response to nutrient limitation. When a colony is in contact with an abundance of nutrients it invests in hyphal growth and branching, which has the effect of capturing energy from an expanding field of activity. When the food is exhausted, it makes more sense to escape from the coming famine and seek a new source of nutrients. This can be done by moving away from the parent mycelium, or by staying put and waiting for a fresh pulse of nutrients to become available in the same spot. To these ends, many spores are adapted for dispersal, for carrying the genome of individual strains through air or water. Following dispersal, deposition of spores some distance from the mycelium may provide access to an environment suitable for the development of a new colony. Other kinds of spores are survival capsules, adapted for allowing the individual to ride out hostile environmental conditions with the potential for future germination. Speaking broadly, the first category of spores may be relatively fragile and poorly provisioned for long-term survival. The second category – resting spores – tend to have thicker cell walls and are packed with nutrient reserves that support longer-term survival. Nutrient limitation is not the only stimulus for spore formation. Mushroom formation follows seasonal patterns for many species and the timing of fruiting may bear little relation to the availability of food.

Airborne spores are subject to rapid dehydration and conidia and basidiospores of plant pathogenic fungi often lose viability within a few hours. Spores of ectomycorrhizal basidiomycetes can survive for at least 4 years under various storage conditions. In the soil, these spores create a 'spore bank' that is primed for establishing symbioses with new tree seedlings. Rust teliospores will not germinate until they have endured a 6-month period of dormancy and, beyond the required dormancy, teliospores can survive in the soil for at least 4 years.

Under controlled environmental conditions in the laboratory, some rust teliospores survive for more than 14 years. Experiments on teliospores of *Tilletia indica*, which causes Karnal bunt of wheat, show that only 3–4% of the spores remain viable after 7 years. Regression analysis based on these investigations indicates zero viability of spores within 13–18 years under field conditions and 38 years in the lab.

SPORE SIZE AND SHAPE

Spores vary greatly in size, from 3 μm-long basidiospores of certain bracket fungi, to the 'giant' spores of lichenized Ascomycota that measure up to 300×100 μm. Corresponding estimates of spore mass, based on density measurements between 1.0×10^3 and 1.3×10^3 kg m^{-3}, range from 1 pg to 2 μg. Most variations in spore shape are quite puzzling from an adaptive perspective. Our familiarity with the physical behaviour of large objects like ourselves can lead to fuzzy thinking about the aerodynamics of spores. Spores with appendages may look like they would remain airborne for longer than simple spherical spores, but the mass of the microscopic spore has a far greater influence upon its sedimentation rate. (This is true of spores in aquatic environments too and they are addressed later in this chapter.) We know this because sedimentation rates estimated for non-spherical spores using figures for the size of spheres of equal mass provide a good match to experimental data. Sedimentation rates for spherical spores with a diameter of 5–10 μm range from 1 to 4 mm s^{-1}, meaning that they take 4–17 min to fall 1 m through still air. Many mushrooms produce spores within this size range. The slow settling speed prolongs their exposure to air currents beneath the mushroom cap that can sweep them away from the fruit body. Some of the variations in spore shape may be related to developmental constraints and to mechanisms of spore discharge. The different shapes and sizes of ascospores are fitted to the asci in which they develop (and vice versa) and spore and ascus shape affect the launch process by controlling whether spores are discharged one at a time with a pause between shots, in a stream with one spores following the next, or as a single projectile of connected spores. In basidiomycetes, spore shape affects discharge distance by controlling the size of the fluid droplet (Buller's drop) whose motion catapults the spore into the air (pp. 86–89).

SPORE DISCHARGE

In this book, we use the term **spore discharge** to refer to the separation of fungal spores from their parent colonies and fruit bodies, and **spore dispersal** for their subsequent movement. Discharge often launches spores over a short distance, whereas dispersal can involve travel over vast distances through the atmosphere. The spores of many fungi are displaced from their parent colonies by physical disturbance resulting from airflow, raindrops, vibration of the surface supporting the colony, or by the activities of animals. These are referred to as **passive discharge mechanisms**. **Active discharge mechanisms** are powered by hydrostatic pressure, fast movements induced by cytoplasmic dehydration, and by the utilization of surface tension force.

Physical Obstacles to Fungal Motion: Air Viscosity and Boundary Layers

The physical challenges encountered by fungi and other microorganisms are very different from those experienced by large animals. Gazelles and exceptional humans achieve horizontal leaps of 9 m. The arc of their flight paths is dominated by inertial forces and neither animal is slowed by the viscosity of the air through which it moves. Things are very different for ascomycetes that shoot their spores from cup-shaped apothecia and for basidiomycete yeasts that propel their ballistospores from the surface of infected flower petals. This is because air represents a viscous obstacle to fungal movement and remarkable launch speeds are necessary to propel spores even for short distances. The ratio of inertial to viscous forces is described by the non-dimensional term Reynolds number (R_e): big animals experience high R_e; spore movement is a low R_e process. The same scaling principles make greater intuitive sense in the more viscous medium of water. A dolphin can glide many metres through the ocean after a single stroke of its flukes (high R_e, inertia trumps viscosity), whereas a chytrid zoospore stops dead the instant it stops lashing its posterior flagellum (low R_e, viscosity rules).

The horizontal movement of actively discharged fungal spores ceases within a fraction of a second after launch and the spore falls toward the ground. This is the case with coprophilous (dung) fungi whose spores (or spore clusters or sporangia) are shot onto vegetation surrounding their colonies. For the majority of fungi, however, discharge mechanisms get spores airborne and longer-distance dispersal is driven by wind. Wind dispersal cannot occur, obviously, unless the fungus reaches mobile air currents. This poses a problem for a microscopic particle that develops within the boundary layer of slow-moving air close to the colony surface. When air flows around any solid object, drag creates a mantle of slow-moving air close to the object's surface; at the interface itself, the air is stationary. There is an inverse relationship between airspeed and boundary layer thickness, and this is another situation in which the R_e term is useful. At low airspeed, air movement is dominated by viscosity, the boundary layer is thickest, and R_e is low. Inertia dominates airflow patterns at higher airspeeds and under these conditions of elevated R_e, eddies and vortices develop. The boundary layer surrounding leaves of terrestrial plants varies from approximately 0.1 to 9.0 mm in thickness depending upon leaf size and wind speed. The same considerations apply to the motion of water around solids, but boundary layers are much thicker, and R_es lower, in the aquatic environment.

Discharge by Airflow and Drying, Electrostatics, and Cavitation

Airflow and Drying

The physical disturbance of fungal colonies by airflow is thought to serve as the main stimulus for separating conidia from their conidiophores. Airflow also causes the release of other kinds of spores including the sporangiospores of zygomycetes, and uredospores and teliospores in rusts. Despite the importance of this passive spore release mechanism, it has not been examined in detail in many species. Interesting experiments using miniature wind tunnels have shown that low airspeeds are effective at releasing uredospores from infected cereal leaves. In some instances, clumps of spores are liberated into the airflow rather than single spores. Other experiments have reached a surprising conclusion. Dense clouds of spores are ejected into the airflow when colonies are exposed to wind, but this is followed by a sharp decline in spore release even when airflow is continuous. The number of spores released at any airspeed is increased at low levels of relative humidity, but the pattern of an initial burst of spore release followed by a greatly reduced aerosolisation seems to be a common phenomenon. Studies on conidial ascomycetes show that 98% or more of the spores remain attached to the surface of a colony at low airspeeds (Figure 3.4). Conidia of many fungi seem to be firmly attached to their conidiophores, remaining in chains and clumps despite considerable buffeting. Physical calculations and experiments employing miniature strain gauges show that shear forces produced by wind speeds of 10 m per second, or 36 km per hour, are necessary to separate the spores of some fungi from their colonies. The apparent strong adhesion between spores and their conidiophores is counterintuitive. One possibility is that passive spore release works by allowing colonies to maintain a slow trickle of conidia into the air over a long period of time. A more likely explanation is that the culture conditions and design of experiments fail to replicate the behaviour of fungi in nature, but the source of this discrepancy is not clear. The effectiveness of passive spore release seems evident from the ubiquity of fungi that utilize this dispersal mechanism and the high densities of their spores in air samples.

Electrostatics

The surface of fungal spores becomes electrically charged in dry air and these charges have a significant effect upon spore motion. It has been proposed that electrostatic charges are involved in spore release, but calculations show that forces produced by charge separation are too small to detach conidia from their conidiophores.

Cavitation (Explosive Bubble Formation)

Cavitation occurs in the walled cells of fungi and plants whose contents are subjected to negative pressures sufficient to exceed the tensile strength of water in the cytoplasm. Gas bubble formation inside air-dried ascospores of Sordariomycetes is the most familiar example

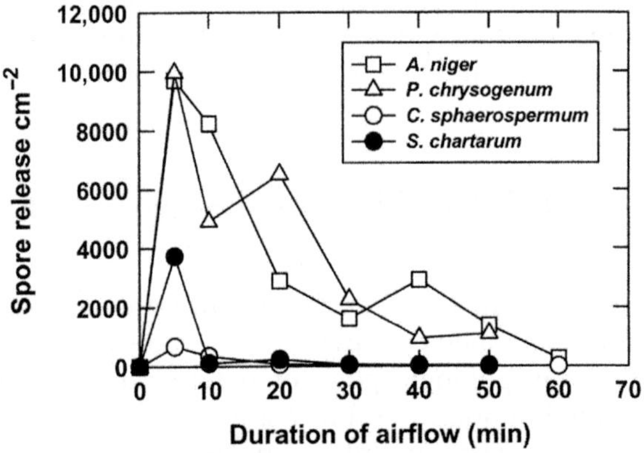

FIGURE 3.4 Plot showing conidial resistance to release from surfaces exposed to slow airflow (1.6 m per second). Spore release shows initial burst followed by sustained period of low-frequency dispersion. Species in this plot are *Aspergillus niger*, *Cladosporium sphaerospermum*, *Penicillium chrysogenum*, and *Stachybotrys chartarum*. *Source: Tucker, K., Stolze, J.L., Kennedy, A.H., Money, N.P. 2007. Biomechanics of conidial dispersal in the toxic mold Stachybotrys chartarum. Fungal Genet. Biol. 44, 641–647.*

of the cavitation process for mycologists (Figure 3.5). The analogous cavitation-powered sporangium of ferns is well known among botanists. As water evaporates from the cytoplasm of the spore, or other type of cell, its walls are placed under increasing tension. If the cell wall is pliable it will crumple, allowing the dried cytoplasm to occupy a reduced volume; if the cell wall is rigid it resists compression and, if drying continues, negative pressure causes water in the cytoplasm to fracture creating a vapour-filled bubble. This occurs in the conidia and conidiophores of certain fungi in which it is linked to spore discharge. The mechanism has been studied in the fruit pathogen *Deightoniella torulosa*. The apical cell of the conidiophore of this fungus has a bulbous tip whose sidewalls are thickened (Figure 3.6). Cavitation causes the thin-walled tip of the bulb to flex outward and this motion propels the spore into the air at speeds of up to 0.6 m per second, propelling the spore over a distance of almost 0.5 mm. Cavitation has also been described in a handful of other conidial fungi, raising the possibility that this discharge mechanism may be widespread.

Raindrops and Vibration

Dry spores and wet spores are puffed and splashed from the surface of their colonies by raindrops. Spores can be splashed over short distances by this mechanism or can be carried over longer distances by wind either as free spores or associated with water droplets. The impact of raindrops exerts much larger forces on colonies than wind disturbance. A raindrop with a diameter of 0.5 mm is one million times heavier than a spore; scaling up to human dimensions, the 'raindrop' would weigh 100 kilotons and splatter the target with the force of 200 kg of high explosive! Raindrops can discharge spores after falling freely and reaching their terminal velocity or when travelling at slower speeds as secondary droplets shed from vegetation. Raindrops can also discharge spores indirectly by causing vibration. Finally, tiny droplets of water in mist capture and disperse spores as they move through vegetation colonized by fungi.

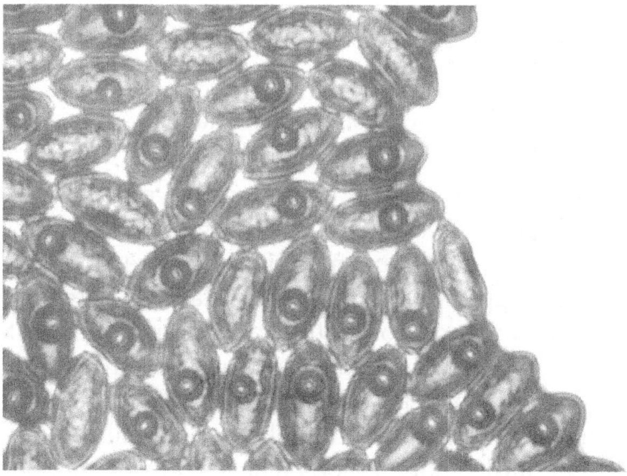

FIGURE 3.5 Ascospores of *Neurospora tetrasperma* exposed to dry air, containing single cavitation bubbles in their cytoplasm.

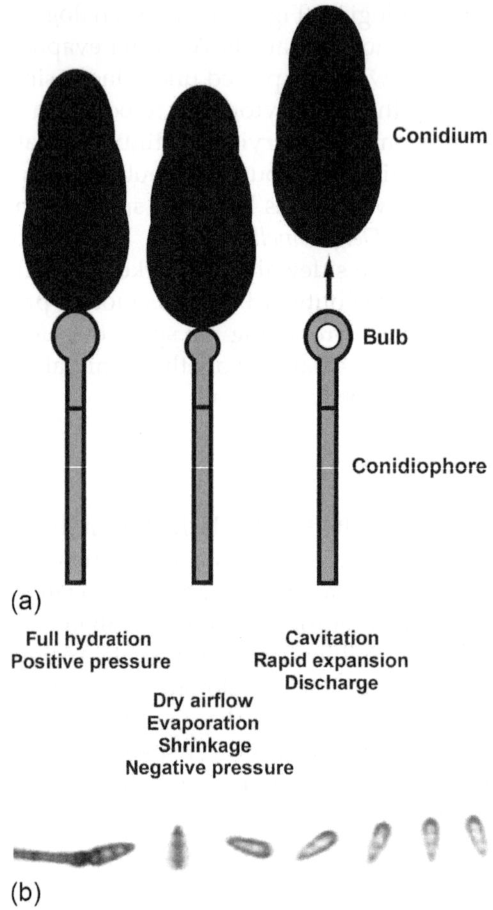

FIGURE 3.6 Conidial discharge in *Deightoniella torulosa*. (a) Diagram showing mechanism of conidial discharge involving shrinkage of conidiophore bulb, followed by explosive expansion driven by the formation of a cavitation bubble. (b) Composite of individual frames from high-speed video recording of conidial discharge showing launch and rotation of spore as it moves from left to right of the microscope field of view. The mean launch speed of *Deightoniella* conidia is 0.24 m per second. Video captured at 50,000 frames per second.

Raindrops compress the loosely packed spore bags of puffballs (Agaricales) and earthballs (Boletales) forcing clouds of basidiospores into the air through irregular tears in the fruit body jacket, or peridium, or through a more regular pore or nozzle on its upper surface. The same mechanism discharges spores from earth-stars (Geastrales), which often have a more intricate nozzle formed by a cone of tooth-shaped flaps. The most complex mechanism of splash discharge has evolved in the bird's nest fungi (Agaricales). The basidiospores of these fungi are enclosed within packets called peridioles. The peridioles are exposed in the mature fruit body that creates a structure that looks, superficially, like an egg-filled nest. The fruit bodies vary from irregularly shaped blobs whose walls break open and ooze the peridioles in a mucilaginous matrix (*Mycocalia* and some species of *Nidularia*), to wide-mouthed cups (*Crucibulum* and *Nidula*), and more angular conical cups (*Cyathus*). Peridiole discharge in *Cyathus* is a spectacular mechanism (Figure 3.7). Each peridiole is tethered to the interior of the fruit body by a cord that is coiled inside a purse. When a raindrop hits the lip of the fruit body it plunges to the bottom of the cone, dislodging the peridiole as water is ejected in a high arc into the air. The peridiole is carried with the ejected water and can fly at 5 m per second (18 km/h) over a horizontal distance of more than one metre. The cord remains coiled during the flight of the peridiole, unravelling if its free end brushes against a leaf or twig (Figure 3.8). The free end of the cord (the hapteron) is coated with a powerful adhesive that sticks to vegetation; when this happens, the momentum of the peridiole carries it beyond the obstacle and its cord unravels, acting as a brake. The cord performs an analogous function to the elastic tether used in bungee jumping. Peridioles of *Cyathus* are often found with their cords coiled around twigs or leaf petioles close to the parent fruit body. This complex mechanism may be an adaptation associated with the coprophilous nature of some bird's nest fungi, situating the discharged peridiole away from the dung in a perfect position for unintentional ingestion by an herbivore (see section on Dispersal of Spores by Animals).

FIGURE 3.7 Splash discharge of peridioles in the bird's nest fungus, *Cyathus olla*. Selected frames from high-speed video obtained at a frame rate of 3000 frames per second. Time in milliseconds from beginning of sequence is shown in the bottom right of each frame. The peridiole is ejected by the upward displacement of water from the interior of the basidiome when water drop hits the rim of the peridium. Scale bar = 3 mm. *Source: Hassett, M.O., et al., 2013. Splash and grab: biomechanics of peridiole ejection and function of the funicular cord in bird's nest fungi. Fungal Biol. 117, 708–714.*

FIGURE 3.8 Diagram showing splash discharge of peridioles of the bird's nest fungus and their mechanism of attachment to vegetation. The funicular cord is packed within a purse before discharge. The force of the raindrop fractures the purse leaving the sticky end of the cord exposed during the flight of the peridiole. Deployment of the funicular cord occurs when the hapteron contacts an obstacle. The process is completed in less than 200 ms. *Source: Hassett, M.O., et al., 2013. Splash and grab: biomechanics of peridiole ejection and function of the funicular cord in bird's nest fungi. Fungal Biol. 117, 708–714.* (See the colour plate.)

Turgor Pressure

Ascospore Discharge

Explosive discharge of asci drives some of the fastest movements in nature with launch speeds exceeding 30 m per second, or 100 km per hours clocked using ultra-high-speed video cameras. The ecological and agricultural importance of the mechanism is apparent from the observation that the Ascomycota is the largest fungal phylum and the most numerous group of plant pathogens. The considerable diversity in ascus structure, ascospore development, and spore morphology among the Ascomycota has significant effects on the dynamics of the discharge process (Figure 3.9). The simplest mechanism operates in pathogens like species of *Taphrina* that belong to the basal lineage of the Ascomycota called the Taphrinomycotina. For example, in *Taphrina deformans*, which causes peach leaf curl, the ascus is exposed on the infected leaf surface, splits open at its tip, and expels a cloud of infectious spores. In the Saccharomycotina, most of the yeast species engage in passive distribution of ascospores that relies upon digestion of the ascus wall. Exceptions to this include the discharge of needle-shaped spores from *Eremothecium* and *Metschnikowia*, and the slow extrusion of ascospores in *Dipodascus*. In the Pezizomycotina, asci are formed within multicellular fruit bodies or ascomata. The chasmothecia of the Erysiphales (powdery mildews, e.g. *Phyllactinia*) crack open to expose their explosive asci; cleistothecia of the Eurotiales (e.g., *Eurotium*) contain non-explosive asci that spill from the ascomata when its wall fragments. Apothecial ascomycetes with operculate asci include *Ascobolus immersus*, a coprophilous (dung-inhabiting) fungus that has been used for research on the mechanism of ascospore discharge. The 'enormous' asci of this species (up to 1 mm long), project from the ascoma, open via an operculum, and expel clusters of eight spores embedded in mucilage at a velocity of up to 18 m per second. Other species of *Ascobolus*, and ascomycetes that form larger apothecia, release plumes of ascospores from the discharge of multiple asci. This is called 'puffing' and creates an updraft of air that serves to drag volleys of spores to greater heights than spores discharged when single asci fire (Figure 3.10). Puffing is also seen in the stalked apothecia of *Sclerotinia*, which is an inoperculate apothecial fungus.

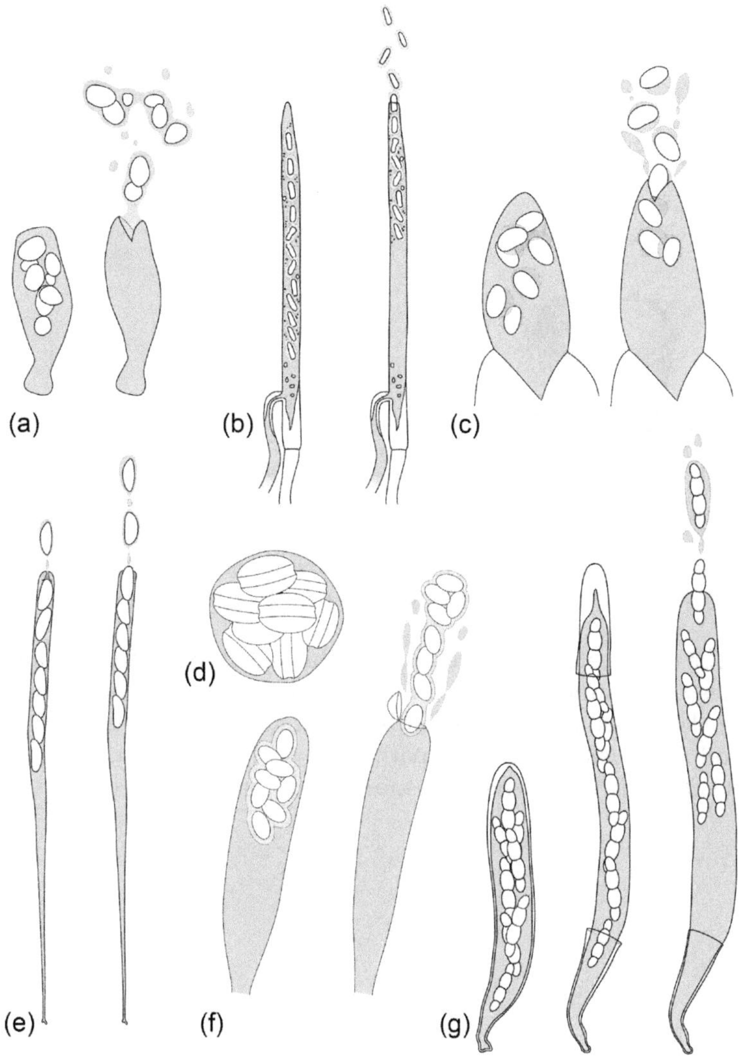

FIGURE 3.9 Diversity of ascus morphology and ascospore discharge mechanisms. (a) *Taphrina deformans* (Taphrinomycotina). Asci of this plant pathogen are exposed on the surface of peach leaves. They split open at the tip and discharge multiple spores in a single shot. (b) *Dipodascus macrosporus* (Saccharomycotina) produces ascospores with a mucilage coating. The spores are extruded slowly through the torn apex of the ascus. This fungus forms exposed asci. (c) Single asci of the powdery mildew *Podosphaera pannosa* (Pezizomycotina, Erysiphales) form in chasmothecia (one ascus per ascoma) and open via a slit at the apex. (d) Multiple non-explosive asci of *Emericella nidulans* (Pezizomycotina, Eurotiales) form inside cleistothecia. The ascospores of this fungus are ornamented with a double flange. (e) Asci of *Xylaria hypoxylon* (Pezizomycotina, Xylariales) discharge spores through constricting ring (apical apparatus). Ascomata of this fungus are perithecia. (f) Operculate asci of *Ascobolus immersus* (Pezizomycotina, Pezizales) expel spores when lid or operculum flips open. Ascomata of this species are apothecia. (g) Ascospore discharge from fissitunicate asci of *Pleospora herbarum* (Pezizomycotina, Pleosporales) occurs after outer wall of ascus ruptures to allow expansion of the inner wall. The fruit body of this species is a pseudothecium. *Source: Mark Fischer, Mount St. Joseph University, Cincinnati.* (See the colour plate.)

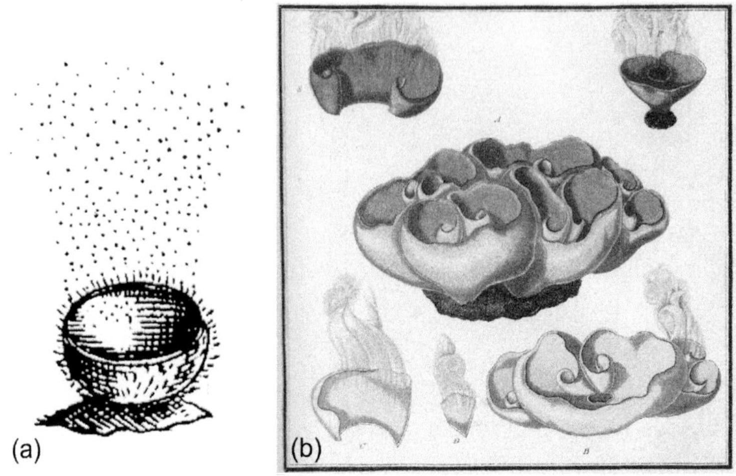

FIGURE 3.10 Classical illustrations of mass discharge or puffing of ascospores from apothecia. (a) First published illustration by Micheli, P.A. 1729. Nova Plantarum Genera. Florence, Bernardi Paperinii. (b) *Source: Bulliard, P. 1791. Histoire des champignons de la France, ou, Traité él émentaire renfermant dans un ordre méthodique les descriptions et les figures des champignons qui croissent naturellement en France. Paris. Chez L'auteur, Barrois, Belin, Croullebois, Bazan.*

In perithecial ascomycetes, the ascus apex is thickened in the form of an apical apparatus that operates as a sphincter, maintaining ascus pressure and separating spores as they are discharged. The discharge process in these fungi has been studied in greatest detail in *Podospora* and *Sordaria* (Figure 3.11); pathogens with this type of ascus include species of *Nectria* and *Gibberella*. Explosive ascospore discharge is not a universal feature of the

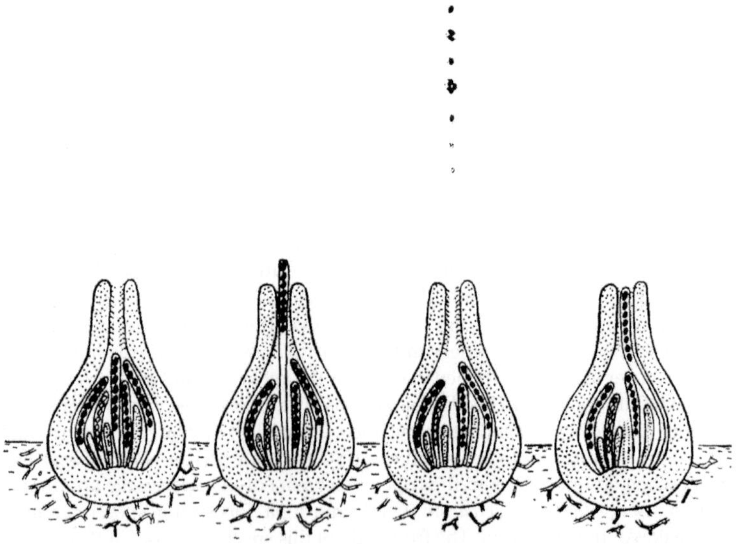

FIGURE 3.11 Ascospore discharge in *Sordaria fimicola*. Asci elongate through the neck of the perithecium, one at a time, discharge their spores, and retract, providing space for expansion of the next ascus. *Source: Ingold, C.T., 1971. Fungal Spores: Their Liberation and Dispersal. Clarendon Press, Oxford.*

perithecial ascomycetes. Many species exude masses of spores and sticky sap to form cirrhi that protrude from their perithecia. Spores of these fungi are dispersed, secondarily, by insects. *Ophiostoma* species work in this fashion, their ascus walls dissolving within the perithecium to liberate an ooze of spores. The distribution of the slow-exudation process among groups of perithecial ascomycetes that exhibit explosive spore discharge is consistent with the loss of this 'violent' mechanism in multiple lineages.

Finally, many ascomycetes with bitunicate asci engage in a two-stage discharge mechanism in which the outer wall (ectotunica) of the ascus ruptures, allowing the inner wall (endotunica) to elongate, and discharge the spores when its apex ruptures. The bitunicate ascus, also referred to as fissitunicate, is formed in fruit bodies called **pseudothecia** that resemble perithecia or other types of fruit body. Important pathogens that form bitunicate asci include species of *Cochliobolus*, *Mycosphaerella*, *Pleospora*, and *Venturia*.

The explosive action of the individual ascus has been studied for more than a century, and in most species we know that the ascus functions as a pressurized spore gun whose dehiscence spurts spores into the air along with a stream of sap. Progress in understanding the biomechanics of ascus function has come from a combination of high-speed video microscopy, spectroscopic analysis of ascus fluid, and mathematical modelling (Figure 3.12). Asci of *Neurospora tetrasperma* discharge four ascospores that may remain connected by mucilage during flight, or separate from one another. Before discharge, the spores are bathed in fluid containing inorganic ions (potassium and chloride) and sugar alcohols that generate an internal turgor pressure of a few atmospheres. When the tip of the ascus bursts open, the spores and surrounding fluid are expelled at an average speed of 16 m per second. The spores and fluid droplets travel over a horizontal distance of up to 20 mm. A similar mechanism has been demonstrated in *Gibberella zeae*, which causes head blight of cereals.

Discharge distance varies from a few millimetres to tens of centimetres among the ascomycetes. Species of *Podospora* are among the record holders, with *Podospora dicipiens* firing its spores for more than 0.5 m. These spores are shot over a distance of 16,000-times the length of the spore. A comparably impressive cannon would propel a human over a distance of 30 km. In reality, gravity brings a human fired from a circus cannon to the ground just a few metres from the muzzle. Gravity certainly affects the range of the ascus, but drag from the air acts as a far greater brake. Spores decelerate during flight at a rate that depends on this drag force and the mass (or inertia) of the spore according to Newton's Second Law, force = mass × acceleration. As the size of the spore increases, both the drag and mass increase. The mass, however, growing as the cube of the spore volume, becomes increasingly significant so that larger spores experience less deceleration than smaller ones and traverse greater distances. For the same ascus turgor pressure and launch speed, larger projectiles travel farther than smaller ones. This principle probably accounts for the evolution of mucilage coats and appendages that connect the spores and increase projectile mass in coprophilous ascomycetes like *Ascobolus* and *Podospora*, whose asci exhibit the greatest range.

Active Conidial Discharge

A number of turgor-driven mechanisms of conidial discharge have been described, but there has been little recent work on their operation. It seems likely that these are quite widespread and may be important in determining the spread of crop diseases. Conidia of the rice blast fungus, *Magnaporthe grisea* (Pezizomycotina), are propelled over a distance of

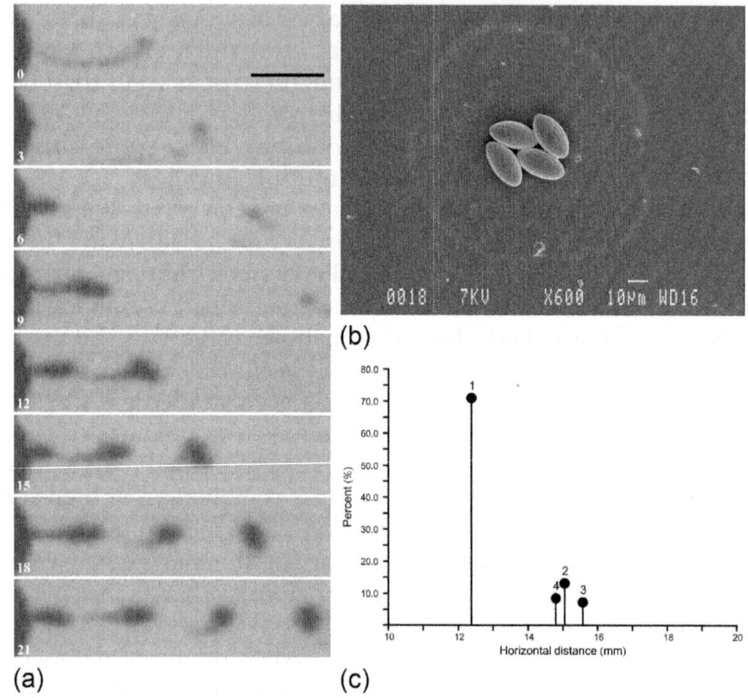

FIGURE 3.12 Ascospore discharge in *Neurospora tetrasperma*. (a) Launch process captured using high-speed video camera running at one million frames per second. Composite shows every third frame from 21 µs of footage. In the first two frames, a plug of cell wall material is discharged from the tip of the ascus. This is followed by the release of the four spores inside the ascus. The maximum launch speed in this fungus is 32 m per second or 115 km per hour. The spores tumble during flight. Scale bar = 20 µm. (b) Scanning electron micrograph of spores shot from single ascus onto a glass slide. The spores are surrounded by a dried deposit of ascus sap and the plug of material discharged from the tip of the ascus rests below the spores in this image. (c) Distribution of spore discharge distances measured from spore deposits: 70% of the spores travelled singly and covered a mean distance of 12 mm; spores that travelled in groups of 2, 3, or 4 spores connected by strings of ascus sap moved farther. This illustrates the relationship between projectile mass and range.

up to 0.5 mm under humid conditions. Details of the mechanism are unclear, but it seems to involve the rupture of pressurized stalks that connect the spores to their conidiophores. A clearer case of turgor-driven conidial discharge is seen in species of *Nigrospora*. This genus includes *Nigrospora oryzae* that causes ear rot of corn. The globose conidium of *Nigrospora* is subtended by a supporting cell and connecting ampulliform cell. Squirting of fluid from the ampulliform cell through a nozzle in the supporting cell propels the 20 µm diameter conidium over a distance of 1–2 cm. Conidial discharge may also be driven by supporting cells whose walls spring outward under pressure but do not burst. This mechanism has been described in *Epicoccum nigrum*, *Arthrinium cuspidatum*, and *Xylosphaera furcata*. The large conidia of entomopathogenic species of *Conidiobolus*, *Erynia*, *Entomophthora* (secondary conidia), and *Furia* are discharged by a similarly rapid pressure-driven eversion of the two-ply septum between the spore and conidiophore apex. The range of this mechanism can reach several centimetres. A related process is also involved in discharging rust aeciospores. Rather than

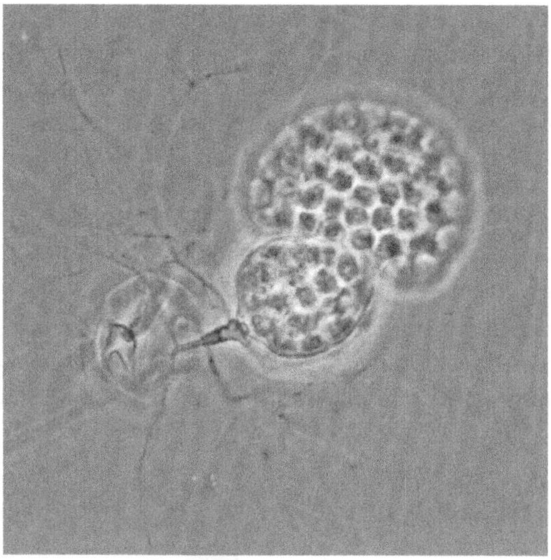

FIGURE 3.13 Sporangium of the chytrid *Obelidium mucronatum* expelling mass of zoospores. The wall of the sporangium is elastic and shrinks as the contents are discharged. *Source: Joyce Longcore, University of Maine.*

septal eversion, pressure boosted by high humidity results in the sudden rounding-off and mutual repulsion of aeciospores, but the mechanical details have not been examined. The efficacy of the mechanism is obvious, however, from the shower of spores shot to distances of up to 1.0 cm from mature aecia.

Zoospore Discharge

There is considerable variation in the process of sporangial emptying in the Chytridiomycota (Figure 3.13). In the simplest cases, there is no obvious pressure-driven discharge: the papilla opens, sometimes at the tip of an elongated discharge tube, and the spores swim around in the chamber of the sporangium until they find the exit. In some species the papilla is capped by an operculum that is shed to initiate discharge and in others the spores pour into a vesicle. Hydrostatic pressure seems to be more important in these cases. Comparable mechanisms are found in zoosporic water moulds (Oomycota, Straminipila). In these protists, the spores are shot into the water when a papilla at the tip of the sporangium breaks, or its material is stretched into a vesicle, releasing pressure from the sporangium.

Pilobolus (Zygomycete), Basidiobolus, and Entomophthora

The *Pilobolus* 'squirt gun' is probably the best known device for spore discharge. Turgor pressure is generated osmotically within a fluid-filled translucent sporangiophore that develops from the herbivore dung on which *Pilobolus* thrives. In the commonest species, *Pilobolus kleinii*, a black-pigmented sporangium filled with 30,000–90,000 spores forms at the tip of the sporangiophore. The sporangiophore is usually 2–5 mm in height and its sporangium has a diameter of 0.5 mm. The region of the sporangiophore beneath the sporangium swells to form a bulb that functions as a lens to direct the phototropic bending of the sporangiophore towards sunlight. The fluid within the sporangiophore contains dissolved salts and sugar alcohols and is pressurized to 5 atmospheres (500 kPa). Discharge of the sporangium occurs when a ring of cell wall at the tip of the vesicle fractures, propelling the sporangium and jet of sporangiophore fluid over a horizontal distance of up to 2.5 m (Figure 3.14). The average launch speed is 9 m per second or 32 km per hour.

There are a handful of similar mechanisms among the fungi, including the processes of conidial discharge in *Basidiobolus* and *Entomophothora*. In *Basidiobolus*, the apical part of the conidiophore is discharged along with the spore, so that the projectile resembles a microscopic rocket; the spore and this subtending conidiophore tip may separate after launch by the septal eversion process that launches some conidial fungi (see section on Active Conidial Discharge). In *Entomophthora*, the primary conidium is discharged along with a stream of sap

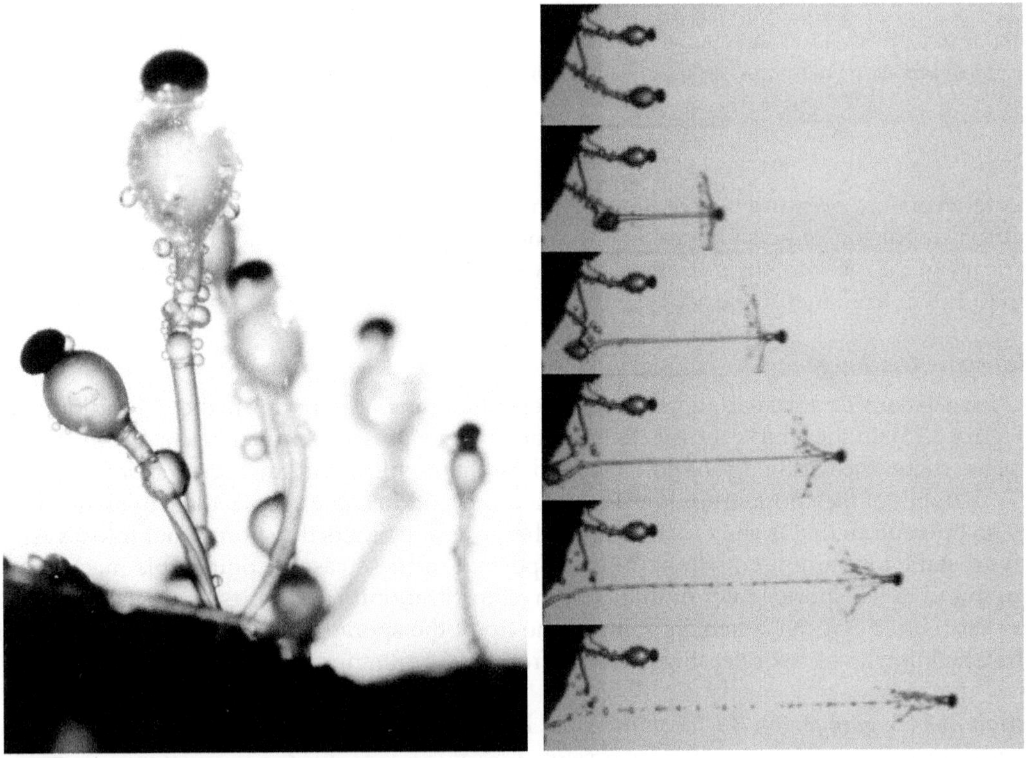

(a) (b)

FIGURE 3.14 Sporangia and sporangial discharge in *Pilobolus kleinii*. (a) The translucent stalk is a few millimetres tall and filled with pressurized fluid. The swelling at the top serves as a lens that focuses sunlight on pigments that allow the fungus to point toward the sun. (b) Discharge of the black sporangium occurs when it separates from the stalk and is blasted up to 2.5 m through the air at a speed of 32 km per hour. This sequence of images was edited from a high-speed video recording captured at 50,000 frames per second.

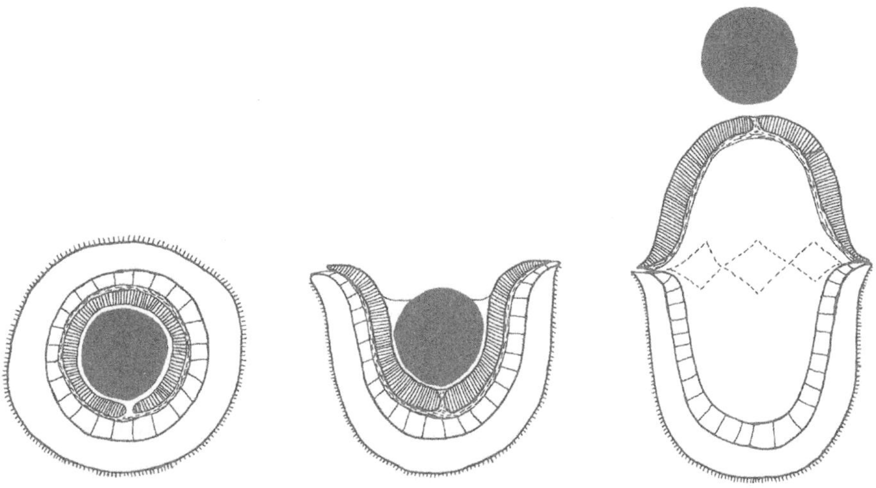

FIGURE 3.15 The artillery fungus, *Sphaerobolus stellatus* sliced through the centre to show its multiple tissue layers. (Left) unopened fruit body; (middle) open fruit body with black capsule bathed in fluid within cup, and (right) capsule (gleba) jettisoned from triggered cup. *Source: Burnett, J.H., 1976. Fundamentals of Mycology, second edition. Edward Arnold, London.*

from its conidiophore; the septal eversion mechanism is responsible for discharging the secondary conidium of *Entomophthora* species (p. 82), illustrating how different mechanisms of spore liberation can operate in a single fungal species.

Sphaerobolus (Basidiomycota)

Sphaerobolus propels a 1 mm-diameter spherical spore-filled projectile called the gleba over distances of up to 6 m (Figure 3.15). The anatomy of the basidiome of *Sphaerobolus* is unusually complex for a fungal organ. It is spherical, no more than 2 mm in diameter, and is composed of six layers of interwoven hyphae surrounding the central gleba. At maturity, the surface of the organ fractures and opens outward to form a star-shape, with the exposed gleba sitting in its centre. In this form, the mature basidiome is best described as a cup within a cup. A slightly flaccid tennis ball, pushed inward on one side to form an unstable dimple, serves as an excellent model for the ripe fruit body. The gleba is bathed in fluid derived from the innermost tissue layer and is supported by the underlying pair of tissue layers that form an elastic 'membrane' called the peridium. The two layers of the peridium consist of a palisade of elongated, radially oriented cells whose long axes point inward, and a backing layer of tangentially oriented thin hyphae. The palisade cells solubilise sugars from glycogen reserves and become pressurized by osmosis. When the peridium is untriggered in its concave form, the exposed ends of the palisade cells are more compressed than their bases. The resulting strain within the walls of these cells is relieved when the peridium flips outward, propelling the gleba into the air at a velocity of 9 m per second (the same as *Pilobolus*). This is associated with an audible 'pop'. Buller (1933) estimated that the power required for glebal discharge is about 1×10^{-4} horsepower or 0.1 W.

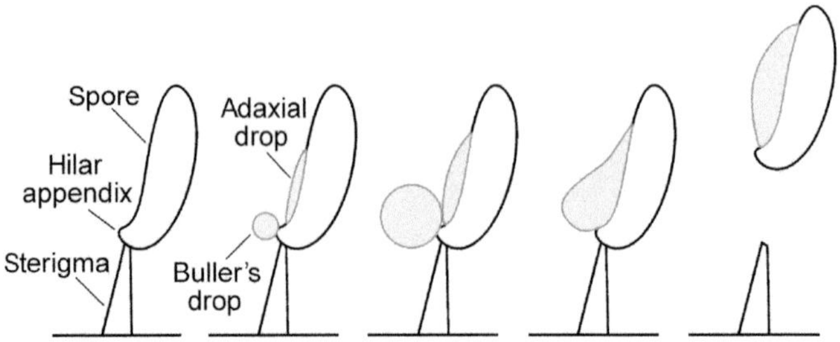

FIGURE 3.16 Mechanism of ballistospore discharge. *Source: Mark Fischer, Mount St. Joseph University, Cincinnati.*

Surface Tension

Most Basidiomycota utilize the process of ballistospore discharge (ballistospory), which is powered by the rapid motion of a fluid droplet over the spore surface (Figure 3.16). This mechanism is responsible for launching basidiospores from the gills, spines, and tube surfaces of mushroom-forming fungi, and also features in the life cycles of basidiomycete yeasts and the phytopathogenic rusts and smuts. The wide distribution of ballistospory and the similarity of the mechanism throughout the phylum suggest that it had a single origin in an ancestral group. Ballistospores have an asymmetric shape, with a prominent bulge at their base, called the hilar appendix, adjacent to the region of contact with the pointed stalk or sterigma. A few seconds before discharge, fluid begins to condense on the spore surface in two locations: (i) as a prominent droplet on the hilar appendix (called Buller's drop), and (ii) on the adjacent spore surface (called the adaxial drop). Once initiated, Buller's drop expands for a few seconds, and then, the spore and drop are discharged from the sterigma. Discharge occurs when Buller's drop reaches a critical diameter and contacts and coalesces with the adaxial drop on the adjacent spore surface. When Buller's drop moves, mass is redistributed from the hilar appendix in the direction of the free end of the spore. This imparts momentum to the spore, and the spore springs from its sterigma. Fluid movement is driven by the reduction in free energy (surface tension) when the two drops fuse, so we refer to the mechanism as a surface tension catapult.

The fluid that accumulates on the hilar appendix (Buller's drop), and on the adjacent spore surface (adaxial drop), occurs by condensation of water from the surrounding air. Condensation follows the release of mannitol and other osmotically active compounds onto the spore surface that cause a localised reduction in the chemical activity of water (or water potential) and vapour pressure. It is not clear how the osmolytes are delivered to the spore surface in sufficient concentration to act as nuclei for condensation of water, but electron microscopic studies suggest that these compounds may be prepackaged in the form of a discrete organelle in the cytoplasm of the maturing spore.

The basidiomycete yeast, *Itersonilia perplexans*, causes petal blight of *Chrysanthemum* and also causes a destructive disease called black streak on edible burdock (*Arctium lappa*). It forms large ballistospores in pure culture and has been used for biomechanical research on the discharge mechanism for many years. The launch of the spore is too fast for conventional video analysis, but has been captured using ultra-high-speed video cameras. The average initial velocity of the spore is $0.7\,m\,s^{-1}$, which is slow relative to explosive discharge of ascospores. In terms of acceleration, however, this is an impressive launch: motion of Buller's

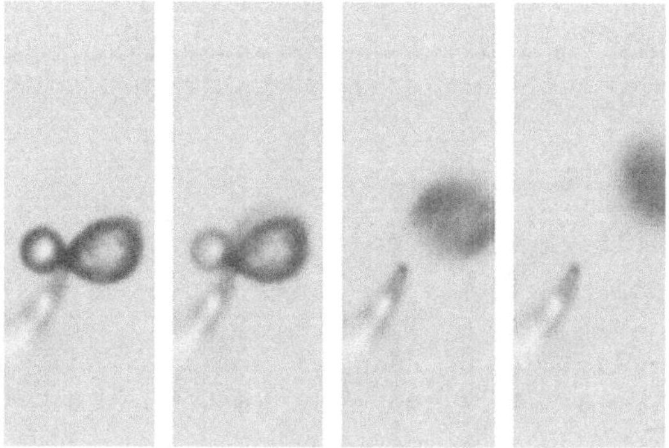

FIGURE 3.17 Basidiospore discharge shown in successive images from high-speed video recording captured at 100,000 frames per second. The fluid droplet at the base of the spore in the first frame coalesces with fluid on the adjacent spore surface in the second frame, which makes the spore jump into the air at an acceleration of 10,000 g. *Source: Pringle, A., et al., 2005. The captured launch of a ballistospore. Mycologia 97, 866–871.*

drop and separation of the spore from the sterigma is completed in less than 10 μs, implying an acceleration of 10,000 *g*. (Figure 3.17). A variety of other basidiomycetes have been studied using high speed video and maximum launch speeds approach 2 m per second. This is interesting in light of C. T. Ingold's maxim, that 'the basidium is a spore-gun of precise range'. The ballistospore discharge mechanism is capable of propelling spores over distances of no more than 1–2 mm. The maximum range for *Itersonilia* is estimated at 1.2 mm, but many spores are propelled over distances of only a few tenths of one millimetre. This must be sufficient to clear the boundary layer on the surface of the host plant. The boundary layer varies from 0.1 to 9.0 mm in thickness depending upon wind speed. If the fungus is positioned so that its spores will fall free from the plant surface after discharge, the boundary layer does not present a problem. Spore wastage may be significant for the colony on the upper surface of a petal, however, because they will fall back on the plant after a brief flight. Similar considerations apply to other ballistosporic yeasts, jelly fungi, rusts, and smuts. For mushroom-forming species, the limited range of the discharge mechanism must be tightly controlled to ensure release into the surrounding air. Spores formed on gills, for example, must be propelled over a limited distance so that they do not hit the opposite gill. The control of discharge distance is a complex issue that involves modifications in spore morphology that affect the size of Buller's drop (Figure 3.18). These variations have been studied in many species of Basidiomycota using high-speed video.

Models of the effect of viscous drag upon spores of different sizes show good agreement between the launch speeds measured from video data and the range of the mechanism. For example, *Itersonilia* spores will travel almost 0.4 mm with an initial velocity of 0.7 m per second. The rapid acceleration of a stationary spore into air is an impressive feat of microengineering, but it is important to recognize that spore motion is also an example of impressive deceleration: as soon as the spore is discharged, the viscosity of the air begins to drag it to a halt. At any lower initial velocities, the microscopic particle would make little headway at all through air. The direction of the launch has very little effect upon the distance that the spore is propelled. In other words, the same initial velocity will shoot a spore upwards or downwards over the same distance. The influence of gravity is imperceptible until the spore is braked by air viscosity. For spores that are shot horizontally, the typical flight path may be

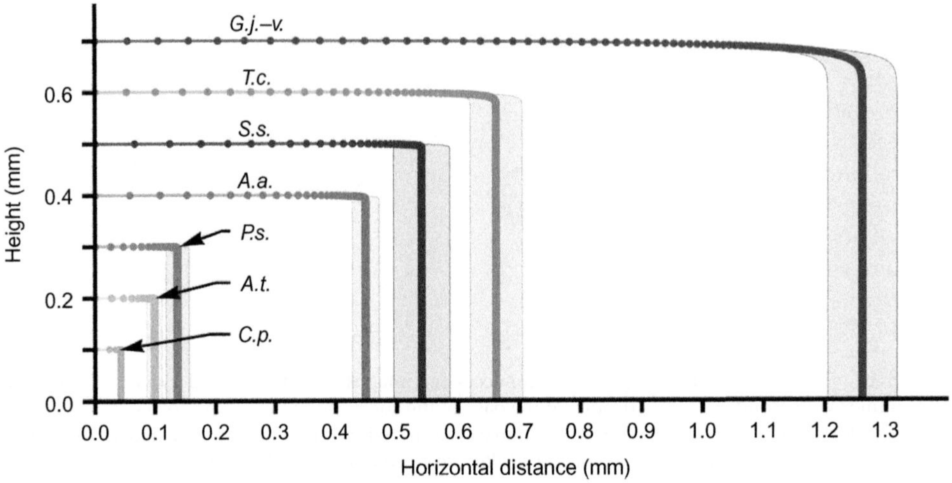

FIGURE 3.18 Model spore trajectories for seven basidiomycetes species based on measurements of spore size and launch speed and using Stokes model of viscous drag. To aid visualization, spores were launched horizontally from arbitrary heights. Positions of spores at 50 μs intervals indicated by dots. The variation in horizontal range predicted from the measured variation in launch speeds (± standard error; Table 1) for each species is represented by the shaded region around each trajectory. Species initials: *G.j.-v., Gymnosporangium juniperi-virginianae; T.c., Tilletia caries; S.s., Sporobolomyces salmonicolor; A.a., Auricularia auricula; P.s., Polyporus squamosus; A.t., Armillaria tabescens,* and *C.p., Clavicorona pyxidata. Source: Stolze-Rybczynski, J. L., et al., 2009. Adaptation of the spore discharge mechanism in the Basidiomycota.* PLoS ONE *4(1): e4163.*

aptly described as a 'Wile E. Coyote Trajectory', recalling the tragic canine featured in Warner Brothers cartoons. This trajectory is critical to the effectiveness of the mechanism in mushrooms whose spores are propelled from the surface of gills, spines, and tubes.

Calculations show that the hypothetical link between drop movement and spore discharge makes sense energetically: minuscule droplets of fluid certainly possess sufficient energy (in the form of surface free energy or surface tension) to discharge spores the appropriate distance. Second, it is clear from mathematical models that any change in spore and drop size will have dramatic effects on the launch speed and discharge distance. For constant spore size, an increase in the radius of Buller's drop will tend to catapult the spore over a greater distance. Similarly, any decrease in drop size will reduce the launch speed and range. Calculations show that the relative size of the spore and drop are the crucial variables. The hydrophobic nature of the spore surface surrounding the hilar appendix causes the water condensing in this spot to remain as a discrete drop until the moment of discharge. Because Buller's drop forms on the end of the hilar appendix, longer appendices will tend to support larger drops because the drop must expand further before it will contact and run over the adjacent spore surface. This is one example of the kind of microscopic morphological detail that may control the dynamics of the launch process. The remarkable variation in spore morphology among the Basidiomycota may reflect the control of discharge speed and distance necessitated by the evolution of diverse fruit body forms.

Experiments in wind tunnels show that mushroom shapes may enhance spore release by interrupting airflow, reducing airspeed directly beneath the cap in a way that protects falling spores from being blown back onto the gills. Evaporation of water from mushrooms has additional effects on dispersal. Evaporative cooling of the spore-producing tissues has been measured with thermocouples and this temperature drop is thought to promote condensation of water on the spore surface. Evaporation of water from mushrooms has the added effect of creating local airflow patterns that may sweep spores away from the fruit body.

The reliance of the ballistospore discharge mechanism upon condensation limits spore discharge to wet environments. Discharge of the ballistosporic stages of pathogenic yeasts, rusts, and smuts can occur only under conditions of high ambient humidity. Mushroom-forming fungi have some degree of control over the humidity of the air between their gills, but even these organisms are restricted to moist habitats and fruit after rainfall. One mushroom, discovered in Oregon, produces submerged fruit bodies in rivers and maintains air pockets between its gills in the form of trapped bubbles. Ballistospore discharge proceeds in the usual fashion and the spores accumulate in rafts at the bottom of the gills and drift downstream in water currents. The surface tension catapult mechanism is not found in other fungi, but an analogous form of fluid movement discharges the spores of the protostelid slime mould *Schizoplasmodium cavostelioides* and related species.

DISPERSAL OF SPORES AND AEROMYCOLOGY

Plant Disease Epidemiology

The concentration of airborne fungal spores decreases in proportion to the distance from their point of origin. For spores dispersed from an infected tree, for example, the number of spores per cubic metre of air is highest close to the plant surface and decreases in the direction of the airflow. Dispersal patterns are of obvious significance to plant pathologists trying to predict the spread of an epidemic disease and the modelling of spore clouds is part of the specialized research field called plant disease epidemiology. Models of dispersal and disease transmission tend to be very complicated because a large number of variables affect spore movement. The concentration of spores at a particular distance from the source is determined by multiple factors including: (i) the number of spores released per minute, (ii) windspeed, (iii) air turbulence, (iv) convection currents, and (v) the sedimentation rate of the spores. When this physical puzzle is scaled up from a single source to an infected stand of trees, or to fields of infected crops, the complexities of disease forecasting become evident. Epidemiologists must also consider genetic variations in plant disease resistance, weather conditions, and other factors in forecasting the severity of the disease once the spores are deposited on the plant surface.

Mushroom Spores and Atmospheric Chemistry

Experiments on mushrooms have shown that most spores released from the cap are deposited quite close to the basidiome (Figure 3.19). In a study of ectomycorrhizal mushrooms, 95% of the spores fell within 1 m of the basidiome. This seems exceedingly wasteful, given the likely premium upon longer distance dispersal from the parent colony, and may speak to the reasons that fungi produce such large numbers of spores. It has been suggested that

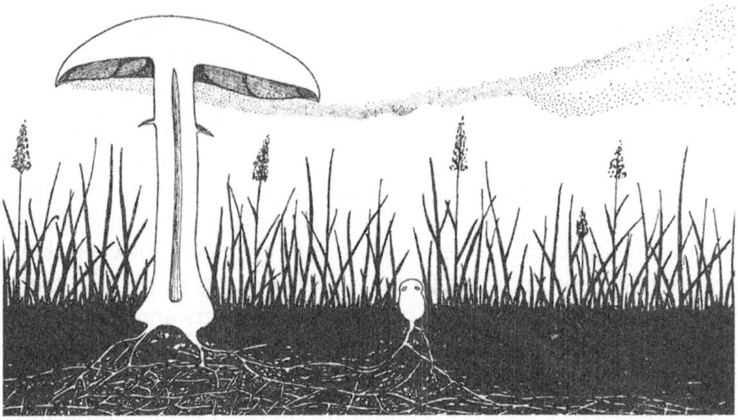

FIGURE 3.19 Cloud of spores dispersed from gills of *Agaricus arvensis*, the horse mushroom. *Source: Buller, A.H.R., 1909. Researches on Fungi, vol. 1. Longmans, Green, and Co., London.*

sampling limitations may have led investigators to conclude that such a high proportion of the total spore output settles close to the fruit body. Nevertheless, dense spore deposits do accumulate beneath mushrooms producing a visible carpet on the surrounding vegetation. Spore deposits on the surface of the caps of bracket fungi reveal the additional wastage of spores deposited from air currents swirling around the fruit body.

Irrespective of the proportion of spores that land close to their source, the fecundity of many species ensures that large numbers of spores are carried longer distances from the parent colony. If only 5% of the spores released from a mushroom travel more than 1 m, at least 135 million spores per day will escape the immediate neighbourhood of a single basidiome of *Agaricus campestris* (that discharges an estimated 2.7 billion spores), and more than one billion spores per day will disperse from *Ganoderma applanatum* (that discharges 5 trillion spores during 6 months of annual activity). Fungi invest huge resources in sporulation and natural selection has 'calculated' the likely losses and small probability of long-distance transmission and survival. The fate of the spores that do land close to the parent mycelium is unknown.

Despite the impediments to becoming airborne, an estimated 50 million tons of fungal spores are dispersed in the atmosphere every year, corresponding to more than 10^{23} spores. This approximation comes from studies by atmospheric chemists who have measured the concentration of mannitol in the air above rainforests. Sugar alcohols are a good proxy for spore numbers because they are carried on basidiospores and ascospores: mannitol, which is concentrated in Buller's drops and in ascus sap, clings to the spore surface after discharge. Spore clouds above rainforests are particularly rich in basidiospores from saprotrophic and mycorrhizal species. It has been suggested that these mushroom spores may act as nuclei for cloud formation, supporting the heavy rainfall that sustains these ecosystems.

Sampling Methods

Until the introduction of molecular methods for identifying fungi, there was no substitute for the painstaking microscopic examination of air samples to determine the numbers of airborne spores and their identity. A very simple form of air sampling is carried out by exposing microscope slides coated with adhesive. This was used in the nineteenth century to collect spores of plant pathogens moving through crops and, early in the twentieth century, for analyzing air samples using aircraft and balloons. The spores on these slides can be identified, or, at least, assigned to a genus or larger taxonomic grouping. They can be counted too, but it is impossible to relate these numbers to the concentration of spores in the air unless the volume of air passing over the slides is known. This limitation was addressed with the development of a series of ingenious instruments for air sampling.

The earliest volumetric sampling devices were tested in the nineteenth century and relied on trapping spores in cotton plugs. Impactor traps that deposited spores on sticky slides were introduced in the 1940s and refined as the Burkard spore trap (based on the original Hirst trap) that remains in use today. This device employs a vacuum pump to suck air through a slit at a defined rate and particles are collected on sticky slides or adhesive tape. By moving the slide or tape beneath the slit at a controlled speed using a clockwork mechanism, the Burkard trap is used to record spore identity and concentration for up to 1 week. Rotorod traps offer a competing design in which spores are collected on the surface of adhesive tape attached to U-shaped arms that are rotated through the air by a motor. Both types of trap are also used for pollen counting. Other kinds of impactor traps expose Petri dishes containing culture medium to the air. These allow the investigator to estimate the concentration of viable spores from the number of colonies that develop in the Petri dish after a short incubation.

Filtration devices that employ disposable cassettes have become very popular instruments for studying airborne spores. After filtering a measured volume of air using a vacuum pump, the filters are removed from the cassettes and transferred to a microscope slide for identifying and counting of spores (Figure 3.20). The standard unit for spore concentration is number per cubic metre. Spore concentrations in outdoor air vary greatly, of course, according to geographical location, seasonal patterns of plant growth and decomposition, and weather conditions. Average spore concentrations in cities range from a few hundred to a few thousand spores per cubic metre. A study in San Diego showed averages of 200 spores m^{-3} during calm wind conditions rising to 80,000 spores m^{-3} during high wind. Similar numbers have been recorded in many cities in North America and Europe, with exceptional readings in excess of 100,000 spores per cubic metre.

Most samples of outdoor air are populated by spores of ascomycetes, with conidia of *Alternaria*, *Aspergillus*, *Cladosporium*, and *Penicillium* being the most frequent types. Other fungi are more common where particular crops are cultivated. The frequency of mushroom spores rises in heavily forested locations. Record spore concentrations have been associated with farming practices. A British study of farm buildings found a maximum concentration of 68 million spores m^{-3} when mouldy hay was shaken. Spores of *Aspergillus* and *Mucor* dominated these extraordinary samples. Higher spore numbers quoted in various sources may have conflated spores produced by actinobacteria (called actinomycetes in older literature) with the somewhat lower frequencies of fungal spores.

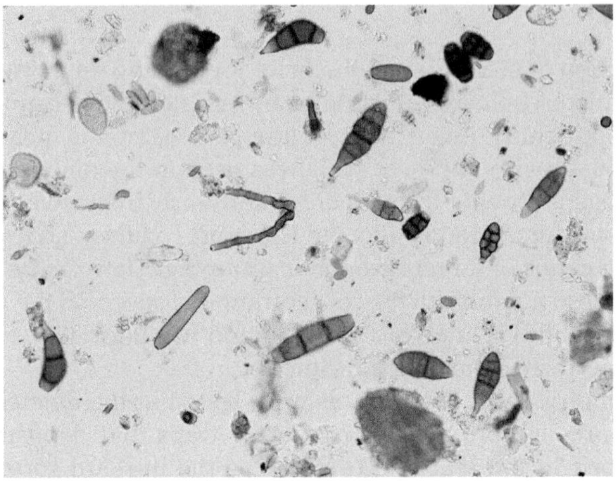

FIGURE 3.20 Air sample with spores and spore fragments captured by filtration. *Source: Estelle Levetin, The University of Tulsa.*

Besides their significance in forecasting allergy symptoms, information on the prevalence of fungal spores is also of interest to patients suffering from a variety of illnesses including chronic obstructive pulmonary disease and cystic fibrosis. Both conditions are exacerbated by lung infections caused by fungi, including *Pneumocystis jiroveci, Scedosporium prolificans*, and species of *Aspergillus*. Inhalation is the route for many fungal infections, including histoplasmosis, coccidioidomycosis, and deep-seated mycoses of the central nervous system caused by *Cryptococcus* species. Information on the concentrations of spores of these opportunistic pathogens is useful for understanding geographic hotspots of infection by different fungi.

The study of aeromycology is benefitting from the introduction of molecular methods and metagenomic research is identifying broad patterns in the diversity of fungi. Molecular analysis of air samples taken from locations in the middle of continents, coastal ecosystems, and above the ocean showed that mushroom spores dominated the air in continental locations. The proportion of ascomycete spores increased over coastal habitats and exceeded the numbers of mushroom spores in the air samples collected at sea. Related studies have tracked seasonal changes and shown, as one would predict, increases in mushroom spores during peak periods of fruiting. Molecular studies have also revealed a surprising diversity of fungi in buildings and found a greater diversity of fungi in temperate compared with tropical locations. Surprisingly, the latitude of the sampled building is a better indicator of fungal diversity than its construction materials. This is interesting in light of the observed correlation between the incidence of childhood asthma and distance from the equator.

DISPERSAL OF SPORES BY ANIMALS

Animals are important vectors for the dispersal of fungal spores. Given the great sweep of fungal diversity and abundance of invertebrates in the habitats in which fungi grow, we are probably aware of only a small fraction of these dispersal mechanisms. The discharge mechanisms that have evolved in many coprophilous fungi, including bird's nest fungi, diverse ascomycetes, and the zygomycete *Pilobolus*, deposit spores on vegetation that is eaten by herbivores. After passage through the digestive system of the animal, the spores are released at varying distances from the parent mycelium when the animal defecates. The vector does not gain any benefit from these interactions. The fungus is simply exploiting an available source of mobility in its reproductive strategy. In at least one case the vector is harmed by its interaction with a coprophilous fungus. Parasitic nematodes can hitch a ride on the *Pilobolus* sporangium when it is squirted into the air and cause lung disease in mammals.

Besides the carriage of spores inside animals (endozoochory), fungi are dispersed on the surface of their vectors (ectozoochory). Ascospores that ooze from their asci to form sticky cirrhi that are smeared on the bodies of insects. In some cases, the ascomycete may secrete a specific chemical attractant; in others, contact between fungus and vector is non-specific. Stinkhorns and other members of the Phallales attract invertebrate vectors to their fruit bodies using volatile chemicals in a way that resembles the relationships between flowers and their animal pollinators (Chapter 1). A comparable strategy is used by ascomycete truffles and basidiomycete false truffles that form subterranean or **hypogeous** fruit bodies. Ascomycete truffles are members of the Pezizales and evolved from cup fungi that discharge their ascospores into the air. The black or Périgord truffle, *Tuber melanosporum*, and the white truffle, *Tuber magnatum*, are the best-known species, but there are hundreds of other truffles. Dispersal of *Tuber* species by mammalian vectors is stimulated by fungal synthesis of a potent mammalian pheromome called alpha-androstenol. Boars are among the animals that produce this steroid in their saliva and this accounts for the attraction of sows to truffles and their traditional use in truffle hunting.

False truffles have evolved from gilled mushrooms in the Russulales and poroid mushrooms in the Boletales. Evolution of the hypogeous morphology is thought to involve an intermediate stage in which the spore-producing tissues remain covered as the fruit body develops. Fruit bodies of this kind are called **secotioid**. The secotioid morphology may reduce water loss by evaporation from the fertile tissues, but limits, or completely eliminates, wind dispersal of spores. The evolution of the secotioid and hypogeous fruit bodies is accompanied by the loss of the spore discharge mechanism using Buller's drop and adaptations that favour consumption by animals. The loss of ballistospory in false truffles is matched by a loss of active ascospore discharge in ascomycete truffles. This process is thought to have occurred during the evolution of *Rhizopogon*, a genus of ectomycorrhizal false truffles related to species of *Suillus* (Boletales) that form poroid mushrooms. *Rhizopogon* spores are spread by numerous species of small mammal in coniferous forests.

Provocative experiments on a bioluminescent mushroom species, *Neonothopanus gardneri*, suggest that the emission of green light according to a circadian rhythm acts as a lure for insects during darkness. Investigators found that acrylic models of fruit bodies illuminated with green light-emitting diodes were more effective at attracting insects than non-luminescent control models. The role of insects as vectors for spore dispersal in *Neonothopanus gardneri* is unproven, but this species grows in Brazilian forests at the base of palm trees where the fruit bodies are shielded from wind. In sheltered habitats of this kind, insect dispersal may be an important adjunct to wind dispersal for luminescent and nonluminescent mushrooms.

AQUATIC SPORES

Non-Motile Aquatic Spores

Spores of aquatic fungi engaged in leaf decomposition are plentiful in calcareous streams. These are conidia of ascomycetes, and a smaller number of basidiomycetes, called Ingoldian fungi after C. T. Ingold (1905–2010), who discovered them in the 1930s. It has been suggested that some of the Ingoldian fungi are endophytes and that their colonies are dispersed by leaf abscission. Ingoldian spores are striking for their elaborate shapes, including stars with four limbs connected to a central hub (tetraradiate conidia), crescents, sigmoids, commas, and miniature cloves (Figure 3.21). They form at the tips of conidiophores that develop at the surface of leaves and become concentrated in bubbles of white foam that accumulate around rocks and fallen logs obstructing water flow. Collection of foam samples provides a convenient way to study these fungi and pure cultures can be obtained by harvesting spores from leaves after a short incubation in the lab.

Spore morphology is no more important in determining sedimentation rate in water than it is for spores dispersed in air. The high viscosity of water relative to air slows the sedimentation rate of spores from millimetres *per second* to millimetres *per minute*, but most experiments show that conidia with appendages fall through the water column at the same speed as more compact spores. Indeed, one experiment showed that intact spores of marine fungi settled faster than spores whose appendages had been disrupted by sonication. The unusual shapes of aquatic spores require an alternative explanation.

The most compelling answer is that the broader span of spores with unusual shapes increases the probability that they will collide with submerged plant materials. The largest

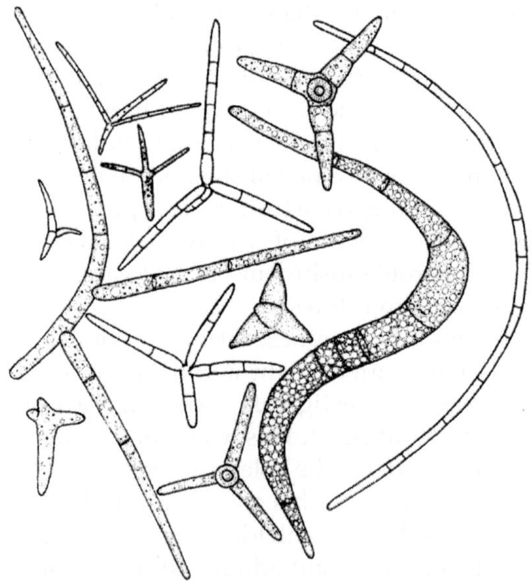

FIGURE 3.21 Variety of conidia produced by Ingoldian fungi in freshwater habitats. *Source: Webster, J., Weber, R.W.S., 2007. Introduction to Fungi, third edition. Cambridge University Press.*

conidia produced by the Ingoldian *Brachiosphaera tropicalis* have an effective diameter of 0.4 mm (Figure 3.1f). A spherical spore of this diameter would weigh approximately 40 μg; the tetraradiate spore with a central hub and slender arms is 400 times lighter, producing a similar probability of hitting a leaf fragment yet saving a considerable investment in cytoplasm. This calculation is a little simplistic because the spherical spore might reduce its volume of active cytoplasm by expanding a large fluid-filled vacuole. Also, the surface area of the tetraradiate spore is only 30 times less than the surface of the sphere, which means that the economy in cell wall production is more modest than the potential reduction in cytoplasm. Nevertheless, the concept of the spore as a search vehicle probably explains the significance of the beautiful spore shapes in these fungi. The utility of the spore morphology with multiple appendages is evident from its convergent development in basidiospores of the marine wood-rotting basidiomycete *Nia vibrissa*.

The extended shapes of these aquatic spores may confer other advantages. Experiments on tetraradiate conidia show that when one arm of a spore strikes a target, the spore pivots around this point of attachment allowing the fungus to make a stable three-point landing. Leaf colonization begins when the tips of the arms cement themselves to the surface and slender hyphae grow from the triangle of contacts. Enhanced dispersal in surface films is another possible benefit of this spore morphology and helps explain the concentration of spores in foam. It has been suggested that spores trapped in foam may become airborne as the bubbles collapse. This would explain how some of these aquatic fungi establish themselves as endophytes in plants growing above the water.

A different adaptation is observed in aeroaquatic conidia that form at the air–water interface in stagnant ponds. These spores develop by helical growth of hyphae to form barrels with an air bubble trapped in the centre. Dispersal occurs by floating on the surface of the water and these fungi colonize leaves and decaying wood. Many other fungi that grow on plant debris in aquatic environments do not show any obvious morphological adaptations to their habitats.

Motile Aquatic Spores

Motile fungal spores called zoospores have a single posterior flagellum that pushes them head first through the water (Figure 3.22). This is the structural signature of the Opisthokonta supergrouping that encompasses the fungi and animals. Zoospores are produced by Blastocladiomycota, Chytridiomycota, Neocallimastigomycota, and diverse zoosporic fungi of uncertain taxonomic assignment included in the Cryptomycota (Chapter 1). Exceptions to the uniflagellate structure of fungal zoospores are found in some of the anaerobic gut fungi in the Neocallimastigomycota that produce spores with multiple flagella. The majority of zoospores function in dispersal and allow the fungus to locate new sources of nutrients. Motile uniflagellate cells also serve as gametes in the sexual cycles of Blastocladiomycota.

The fungal zoospore is a spherical or ovoid cell that lacks a cell wall. The absence of a wall means that the cell must regulate water influx without developing turgor pressure: unregulated osmosis would burst the naked zoospore. Contractile vacuoles have been observed in some chytrid zoospores and additional control of water influx is achieved via active ion exchange through the spore membrane. Zoospore propulsion is driven by high-frequency undulation of the flagellum from base to the tail and a velocity of 100 μm per second (20-times

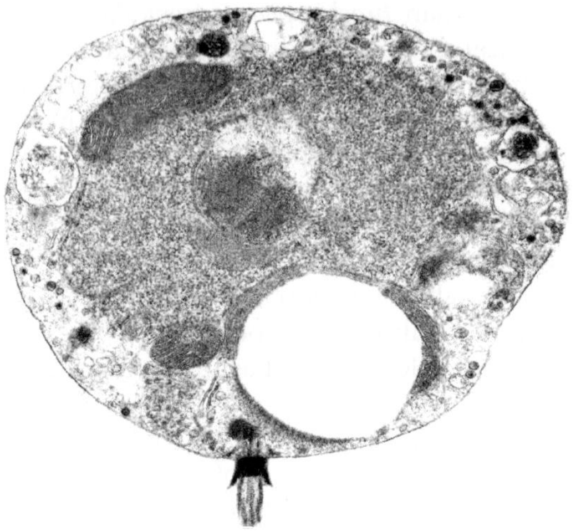

FIGURE 3.22 Transmission electron micrograph of a single zoospore of the chytrid *Chytridium lagenaria*. The base of the single flagellum is visible at the bottom of the spore. The large circular structure is a contractile vacuole. *Source: Peter Letcher, University of Alabama.*

cell length per second) is typical for a chytrid. Zoospores can swim for many hours in a culture dish or glass microscope chamber. They spend most of the time swimming in straight lines or following circular paths. Unlike the flagellate cells of many other eukaryotic microorganisms, chytrid spores swim without rotation of the cell. This smooth gliding motion is interrupted by momentary flicks, jerks, and changes in direction. Changes in direction are controlled by bending of the flagellum toward its base so that it acts as a rudder. Zoospores stop swimming periodically too, even though the flagellum keeps lashing from side-to-side and curling around the stationary cell. Chytrid zoospores show amoeboid motion over surfaces for relatively short distances and can switch repeatedly between this behaviour and swimming freely in the water.

Unlike airborne spores, which have no need to draw upon nutrient reserves until germination, zoospores are powered by the continuous oxidation of lipids and other stored fuels. This limits their period of activity. A zoospore swimming at an average speed of 25 μm per second (allowing for frequent stops) for 5 h would travel 0.5 m. Observations on the erratic swimming patterns of zoospores in the lab suggest that a journey over this distance in a straight line is unlikely. Flagellar movement and amoeboid locomotion are probably effective over quite short distances and allow the spores to explore limited zone in which they can detect chemical gradients that provide cues to nutrient availability. Experiments have shown that dissolved amino acids and sugars attract chytrid zoospores. It is likely that more distinctive compounds released from host cells are also used for chemotaxis by species that infect plants and animals. Adhesion to host surfaces is accompanied by retraction of the flagellum into the cell and the formation of a cell wall to create a cyst. Penetration of the host cell occurs via the growth of a penetration hypha from the cyst. Motile zoospores and cysts that are unattached to surfaces may be dispersed passively over long distances in water trickling through soils and carried by water movement in aquatic habitats.

Much more is known about mechanisms of zoospore dispersal in plant pathogenic oomycetes (Stramenopila) including species of *Phytophthora* and *Pythium*. Zoospores of these microorganisms have paired flagella that emerge from the side of the kidney-shaped cell. One flagellum points ahead of the swimming zoospore and is covered with fine filaments called mastgonemes, and the other lashes behind the cell. Both flagella undulate from base to tip. The presence of the mastigonemes on the anterior flagellum redirects its thrust so that it pulls the spore through the water. The posterior flagellum acts as a rudder and does not generate much propulsion. Oomycete zoospores rotate around the long axis of the cell and follow a wider helical path as they swim. Like the zoospores of fungi, swimming zoospores of oomycetes show frequent changes in direction and are adapted for nutrient detection over distances of a few centimetres.

Further Reading

Fischer, M.W.F., Stolze-Rybczynski, J.L., Davis, D.J., Cui, Y., Money, N.P., 2010. Solving the aerodynamics of fungal flight: How air viscosity slows spore motion. Fungal Biol. 114, 943–948.

Ingold, C.T., 1971. Fungal Spores: Their Liberation and Dispersal. Clarendon Press, Oxford.

Kirk, P.M., Cannon, P.F., Minter, D.W., Stalpers, J.A. (Eds.), 2008. Dictionary of the Fungi. 10th ed. CAB International, Wallingford, United Kingdom.

Lacy, M.E., West, J.S., 2007. The Air Spora: A Manual for Catching and Identifying Airborne Biological Particles. Springer-Verlag, Berlin, Heidelberg, New York.

Money, N.P., Fischer, M.W.F., 2009. Biomechanics of spore discharge in phytopathogens. In: Deising, H. (Ed.), The Mycota, second ed. In: Plant Relationships, Vol. 5. Springer Verlag, Berlin, Heidelberg, New York, pp. 115–133.

Webster, J., 1987. Convergent evolution and the functional significance of spore shape in aquatic and semi-aquatic fungi. In: Rayner, A.D.M., Brasier, C., Moore, D. (Eds.), Evolutionary Biology of the Fungi. Cambridge University Press, Cambridge, pp. 191–201.

CHAPTER

4

Genetics – Variation, Sexuality, and Evolution

Lynne Boddy
Cardiff University, Cardiff, UK

Genetics deals with variation and inheritance, and as such, forms the basis for understanding why fungi behave as they do. Ultimately, understanding for example, why one fungal species or even strain is pathogenic to perhaps only one species or variety of plant, how one fungus recognises another, why some fungi grow faster or decompose organic matter more rapidly than others, how the characteristics of fungi and the relationships between them and other organisms change with time, all comes down to understanding their genetics. Fungal ecologists, conservation biologists, plant and invertebrate pathologists, medical mycologists, biodeterioration specialists and those concerned with the exploitation of fungi in horticulture, and in the pharmaceutical, food and chemical industries, all need to understand basic fungal genetic concepts. Further, many fungi have traits that make them ideal for genetic and evolutionary biology research: many can be grown in pure culture under controlled laboratory conditions, minimising environmental variability that might conceal or confound genetic differences; many are haploid for much of their lives, allowing mutant alleles of genes to be readily detected phenotypically; many have high growth rates, short lifecycles, uninucleate haploid cells at some stage of the life cycle, large numbers of progeny from each cross, low chromosome numbers, and small geonomes. The ascomycetes *Aspergillus nidulans*, *Neurospora crassa*, and *Saccharomyces cerevisiae* have many of these features, and are among the genetically best understood organisms.

In this chapter we define first what constitutes a fungal individual and population, and consider fungal species concepts. The rest of the chapter is divided into two main sections, the first concerned with the different lifecycles and sexual processes found within the diverse Kingdom Fungi. Then we review fungal variation, sources of variability, evolution, and formation of new species.

DEFINING INDIVIDUALS, POPULATIONS, AND SPECIES

There are three concepts essential for population genetics, ecology, and conservation – the individual, the population, and the species.

The Fungi
http://dx.doi.org/10.1016/B978-0-12-382034-1.00004-9

What Is an Individual? Definition, Recognition, and Consequences

Defining Individuals

For multicellular animals and plants, identification of what is an individual is often easy – one lion is an individual. Unicellular fungi and those with very limited mycelial growth can reasonably be regarded as individuals, but what constitutes an individual with larger mycelial fungi is harder to define, as with plants that form stolons or suckers and with clonal animals such as coral. The difficulty arises because individual compartments of a mycelium are capable of independent existence and production of new colonies, and colonies do not have a predetermined body shape and size (i.e. they have indeterminate growth). If large mycelia become fragmented, or asexual spores are produced, colonies are formed which are physically separated but genetically identical to each other. This occurs commonly in nature. Organisms related in this way are termed clones. Following the terminology of plant biologists there are thus two aspects to defining individuals: a whole group of genetically identical individuals (clones) may be defined as a **genet** and each asexually originating part of the genet can be termed a **ramet**.

Not only can mycelia fragment, but they can also join. If hyphae of two or more members of a clone come into contact with one another then they may fuse (pp. 102–104) in the same way that anastomoses are formed within a single colony (p. 56), resulting in coalescence of mycelia to form an individual functional unit. In the past it was thought that coalescence would occur whenever mycelium of the same species, irrespective of genetic origin, encountered each other; the resulting mycelium was thought to contain many different nuclei derived from merging of different individuals which could be 'switched' on and off according to environmental circumstances. The occurrence of somatic incompatibility (pp. 102–104), however, means that such mosaics are likely to be rare. Glomeromycota are an exception, with individual mycelia containing many different nuclei.

Genets and ramets vary considerably in their size and age. In the field, the sizes achieved depend on, among other factors, lifestyle and ecological strategies, age, size of resources colonised, and interaction with other organisms. Many fungi are restricted to the resources (e.g. leaf, twig, or branch) which they colonise and hence the size they can reach cannot be larger than the resource. Typically they would be much smaller because of the presence of other individuals of the same or different species. Some fungi tend to form mycelia that do not grow much more than a few cm radius, however large the resource or time given (e.g. *Penicillium* spp.). In contrast, some basidiomycetes can form extensive long-lived genets and ramets in soil and leaf litter. For most ectomycorrhizal (pp. 205–228) and saprotrophic leaf litter decaying species, genets are usually less than 10 m diameter, but species that form cords (pp. 59–60), rhizomorphs (pp. 59–60), and fairy rings can be larger. The largest organisms on the planet are probably *Armillaria* species with genets occupying up to 8 or so hectares on the floors of North American forest, with estimated ages of a thousand or more years. Little is known about the dynamics of ramets in the field, but their size, shape, distribution, and number will be constantly changing, especially during times of the year when growth is rapid (Figure 4.1).

Definition of an individual is important because it forms a baseline upon which numerical comparisons are often made, though use of biomass might be a less ambiguous metric. The actual definition of an individual depends on circumstances. The IUCN (International Union

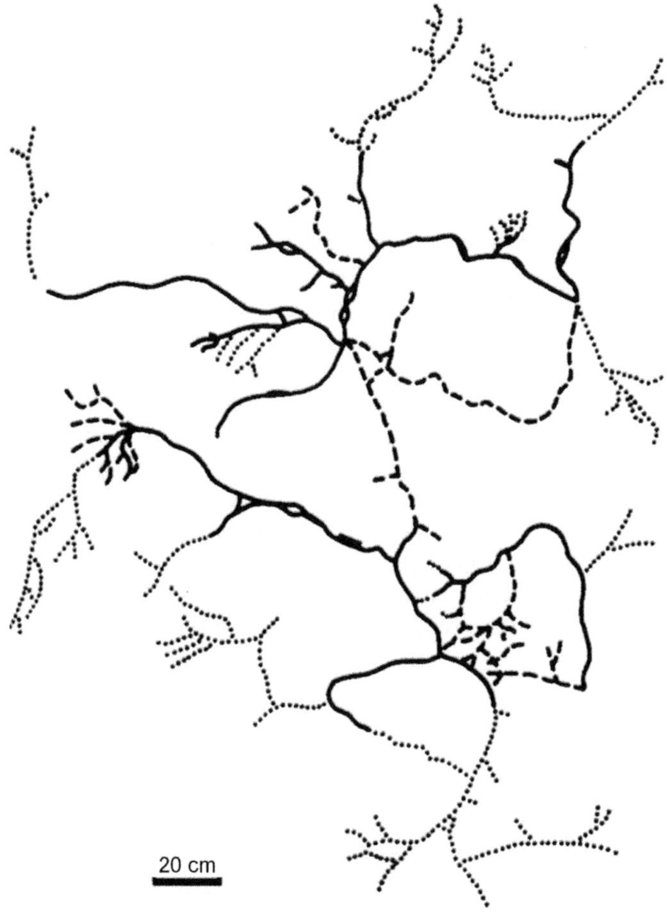

FIGURE 4.1 A map of a mycelial cord system of *Phanerochaete velutina* at the soil–litter interface in a temperate deciduous woodland, excavated by carefully removing the surrounding leaf litter, and then recovered with litter. The continuous and dashed lines indicate the system when it was first excavated. The dotted lines indicate new mycelial cords 13 months later, while the dashed lines indicate cords which were no longer present. Loss of one of the cords represented by a dashed line caused the original genet to be split into two ramets. *Source: Redrawn from Thompson, W., Rayner, A.D.M., 1993. Extent, development and functioning of mycelia cord systems in soil. Trans. Br. Mycol. Soc. 81, 333–345.*

of Conservation) defines an individual as the smallest entity capable of independent survival and reproduction (i.e. a ramet, not a genet). However, the cost and complex logistics of gathering genetic data on fungi over extensive geographical areas has led fungal conservationists to use the concept of functional individuals, rather than genets or ramets, for red-listing (pp. 392–394). In genetic studies the genet would usually be used to define an individual. Whichever appropriate definition is being used, it is essential that like is compared to like, when any quantitative comparisons of numbers of individuals are being made.

Roles and Consequences of Hyphal Fusion and Somatic Incompatibility

There are advantages of fusion: within a colony, fusion between hyphae can result in the reconnection of two streams that had previously branched (termed anastomosis). This allows the formation of networks (p. 56) and the ability to transport water, nutrients, and signals tangentially rather than just radially within a colony. Fusion of genetically identical colonies can allow pooling of resources and cooperation between different regions, be it following fusion of large mycelia of saprotrophic cord-forming basidiomycetes, or ectomycorrhizal mycelium, or of small colonies developing from asexually produced conidia (p. 103). Heterokaryon formation, when two genetically different hyphae of the same ascomycete species fuse, provides potential benefits of functional diploidy (p. 109) and the ability to generate mitotic genetic exchange in the absence of sexual reproduction during the parasexual cycle (pp. 122–123).

Fusion with non-self can, however, expose an individual and its genome to risks such as harmful nuclei, mitochondria, plasmids, viruses, retrotransposons, and other selfish genetic elements (pp. 132–134). Examples of these include: the presence of a gene in *Neurospora crassa* that enables nuclei containing it to replace other nuclei; in *Neurospora* and other fungi some mitochondria have genomes that render them defective in respiratory function but capable of normal replication; some strains of *Podospora anserina* have a plasmid that causes senescence; and there are viruses that can affect growth and fruit body development (pp. 357–358). Somatic incompatibility provides a defence against the spread of these.

Somatic Incompatibility

The ability to distinguish self from non-self is ubiquitous among all organisms, and is fundamental to distinguishing one individual from another. Filamentous fungi recognise self from non-self by incompatibility systems termed vegetative incompatibility, mycelial incompatibility, or somatic incompatibility (particularly in basidiomycetes). Within a species, recognition of non-self typically occurs following fusion of hyphae, and incompatibility usually results in death of the fusion cell. In basidiomycetes, recognition has a genetic basis and is regulated by one to three or, possibly more genes with multiallelic loci (termed vegetative compatibility (VC), *vic* or *het* loci). If two mycelia have different alleles at one or more *het* loci recognition as incompatible occurs. VC does not necessarily imply genetic uniqueness, and the relationship between the two depends on the number of loci and alleles involved. With basidiomycetes, most mycelia (usually dikaryons, p. 113) isolated from the natural environment tend to be incompatible when paired (Figure 4.2), implying that vegetative incompatibility groups usually correspond to genetic individuals, though this is not true for all species. With ascomycetes, VC types are commonly, but not always, associated with genetic individuals. Sometimes dominant VC types can spread over large distances, especially when a pathogen is invasive and a founder effect (p. 130) or selection of a more fitted VC type occurs, as in the spread of the Dutch elm disease fungus *Ophiostoma novo-ulmi*. With the basidiomycete *Serpula lacrymans*, low genetic variation in the founder populations (p. 130) has led to breakdown in the correlation between genetic uniqueness and VC groups.

More is known about hyphal recognition systems in ascomycetes than in basidiomycetes. In ascomycetes two types of genetic systems regulate vegetative incompatibility, termed allelic, and non-allelic, the former being most common. In both types, incompatibility occurs if there are differences at any of the *vic* or *het* loci. In allelic systems, incompatibility occurs

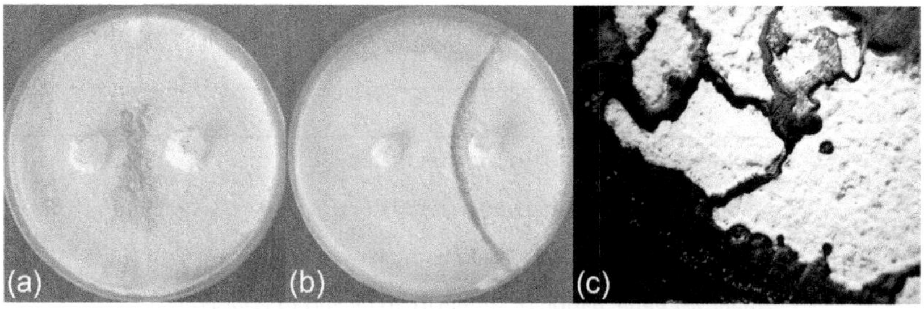

FIGURE 4.2 Vegetative incompatibility in wood decay fungi. When self-pairings are made in artificial culture, hyphae fuse and the two mycelia function as one because they are somatically compatible (a), but when mycelia from non-self pairings meet there is a clear rejection response (b), as seen here with *Trametes versicolor*. (c) Mycelia of *S. hirsutum* have grown out from individual decay columns in wood demarcated by interaction zone lines and are clearly separated from each other. *Source: (c) Rayner A.D.M., Todd, N.K., 1979. Population and community structure and dynamics of fungi in decaying wood. Adv. Bot. Res. 7, 333–420.*

if there are different alleles at the same loci, whereas in non-allelic systems, incompatibility occurs as a result of interaction between two genes at different loci. *Aspergillus nidulans* and *Neurospora crassa* only have allelic systems, but *Podospora anserina* has both allelic and non-allelic systems. The number of *het* loci identified and characterised varies between 7 and 11 depending on species (e.g. *Aspergillus nidulans, Ophiostoma novo-ulmi, Cryphonectria parasitica, Neurospora crassa,* and *Podospora anserina* have at least 8, 7, 7, 11, and 9 loci, respectively) (Table 4.1). The vegetative incompatibility systems of basidiomycetes probably have some of the same genetic characteristics but differ in others (e.g. mating genes do not appear to be involved in the basidiomycetes but sometimes they are in the ascomycetes) (Table 4.1).

With ascomycetes, as with basidiomycetes, recognition follows fusion. Prior to fusion, hyphal tip growth stops after contact with another hypha of the same species, the cell wall is broken down by hydrolytic enzymes at the point of contact, and a bridge is made between the two. Plasma membranes then fuse and cytoplasmic contents of the two compartments mix. Fusion between fungi identical at all *het* loci results in compatibility, and is frequently associated with changes in cytoplasmic flow (Figure 4.3). The mechanistic basis of self-fusion has been studied in most detail for fusion between conidial anastomosis tubes (CATs) during initiation of colonies of *Neurospora crassa*. CATs, produced by most ascomycetes, are short, specialised hyphae or cell protrusions that grow out from and link conidia or germlings to form interconnected networks (Figure 4.4). They have also been seen linking rust uredospores. The process of CAT fusion occurs in a continuum of six developmental stages: (1) induction, (2) chemotropism, (3) adhesion of cells, (4) cell wall breakdown and remodelling, (5) plasma membrane fusion, and (6) achievement of cytoplasmic continuity (Figure 4.4).

When there are differences at one or more *het* loci, rapid compartmentalization, death of the fusion cell, and often adjacent cells occurs (Figure 4.4). Within a few minutes of fusion, granules form within the cytoplasm of the fused compartment, septal pores (pp. 53–55) connecting with adjacent cells are blocked, cytoplasm becomes vacuolized, and these vacuoles burst, releasing proteases and other degradative enzymes. Destruction of the fusion cell is often complete within 30 min. Different species probably share the same cell death machinery as microscopic processes are similar, despite being mediated by different *het* loci.

TABLE 4.1 Characteristics of Vegetative Incompatibility and Related Genes in *Neurospora crassa* and *Podospora anserina*

NEUROSPORA CRASSA	
matA-1	Mating type gene transcription regulator
mata-1	Mating type gene transcription regulator
het-c	Allelic *het* gene; signal peptide with glycine-rich repeats
het-6	Allelic *het* gene; region of similarity to *tol* and to *P. anserina het-e*
un-24	Allelic *het* gene; large subunit of ribonucleotide reductase
tol	Suppressor; has a coiled-coil, leucine-rich repeat; some sequence similar to *het-6* and *P. anserina het-e*
vib-1	Suppressor; nuclear localisation sequence
PODOSPORA ANSERINA	
het-c	Non-allelic *het* gene against *het-d*; glycolipid transfer protein (glycolipids are involved in interactions between cells
het-d	Non-allelic *het* gene against *het-c*; GTP-binding domain; similarity to *het-e* and to *N. crassa tol* and *het-6*
het-e	Non-allelic *het* gene against *het-c*; GTP-binding domain; similarity to *het-d* and to *N. crassa tol* and *het-6*
het-s	Allelic *het* gene; prion-like protein
idi-1, idi-3	Vegetative incompatibility (VI)-related genes; signal peptide, induced by non-allelic (*het-c/e* and *het-r/v*) incompatibility
idi-2	VI-related gene; signal peptide, induced by *het-r/v*
mod-A	VI-related gene; modifier of *het-c/e, c/d*, and *r/v* incompatibility; SH3-binding domain, involved in signal transduction, cytoskeletal proteins, protein–protein interactions
mod-D	VI-related gene; modifier of *het-c/e* incompatibility; subunit of G protein, involved in signal transduction
mod-E	VI-related gene; modifier of *het-r/v* incompatibility; heat-shock protein
pspA	VI-related gene; vacuolar serine protease; induced by non-allelic (*het-c/e* and *het-r/v*) incompatibility

Note that names have been assigned independently in different fungi so, for example, het-c *in* N. crassa *is not the same as in* P. anserina*.*
Adapted from Moore et al., 2002; Glass and Kaneko, 2003

What Is a Population?

A population comprises an assemblage of individuals of a species. Delimitation of the assemblage depends on the researcher and the questions being asked. The boundaries can range to include all of the individuals in a single organic resource, or in a forest, or in a geographic region, the ultimate outer boundary depending on the species limit set by hosts, tolerance of climate, etc. The boundaries set by a researcher may not be the borders for gene flow (pp. 128–130), as airborne spores can sometimes disperse over long distances. Another

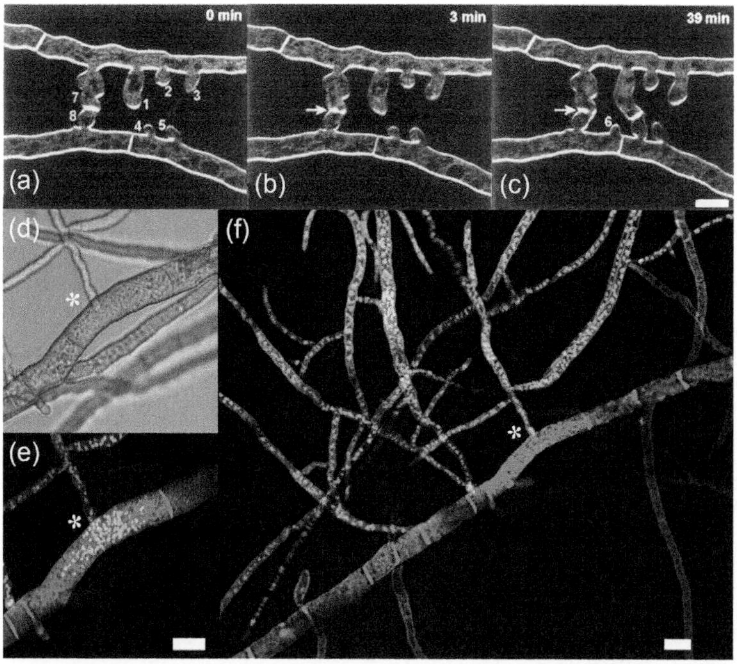

FIGURE 4.3 Vegetative compatibility and incompatibility. Stages of the hyphal fusion process (a–c) and non-self rejection responses (d–f) in *Neurospora crassa* viewed with confocal microscopy. (a–c) Show induction, homing, and fusion of hyphae in a mature colony. (a) Three branches (labelled 1, 2, and 3) from one hypha have started growing towards two short branches (4 and 5) on the opposite hypha. Two branches (7 and 8) have already fused. (b) A fusion pore has started to open between the two fused hyphal branches (arrowed), and in (c) it is completely open, allowing cytoplasmic continuity. (c) Branches 1 and 4 have fused, and another side branch (6) is developing. Bar = 10 μm. (d–f) Fusion and subsequent rejection of non-self hyphae. (d) A brightfield image showing a thin hypha of one strain fusing with the underside of a wide hypha of another strain (*). In (e) and (f) the same interacting hyphae have been labelled with a membrane selective red (grey in the print version) dye and one of the strains has been labelled with a nuclei selective green (light grey in the print version) H1-GFP. (e) A confocal image of (d) showing that nuclei (fluorescing green (light grey in the print version)) have migrated from the narrow hypha into the wider one. (f) One hour later incompatibility is evident. The compartment where fusion has occurred is stained intensely red (grey in the print version) due to increased permeability of the plasma membrane, and the nuclei have broken down, as evidenced by lack of green (light grey in the print version) fluorescence. *Source: (a–c) Hickey, P.C., Jacobson, D.J., Read, N.D., Glass, N.L., 2002. Live-cell imaging of vegetative hyphal fusion in Neurospora crassa. Fungal Genet. Biol. 37, 109–119. (d–f) Read and Roca, 2006.* (See the colour plate.)

concept is that of the metapopulation, which comprises local populations each with its own probability of going extinct. Uncolonised regions can be colonised from other populations within the metapopulation. So each log in a forest will contain its own population of a species, and when a new branch falls it will be colonised from the other populations in the forest; long-term survival of the species occurs at the metapopulation level.

Important characteristics of populations include their size and whether this is changing, their age structure, and their genetic variation. Fungal individuals are characterised by their **genotype**, and how this is actually expressed (i.e. how the traits are manifested) – **phenotype**, and populations comprise individuals having different genotypes and phenotypes. The

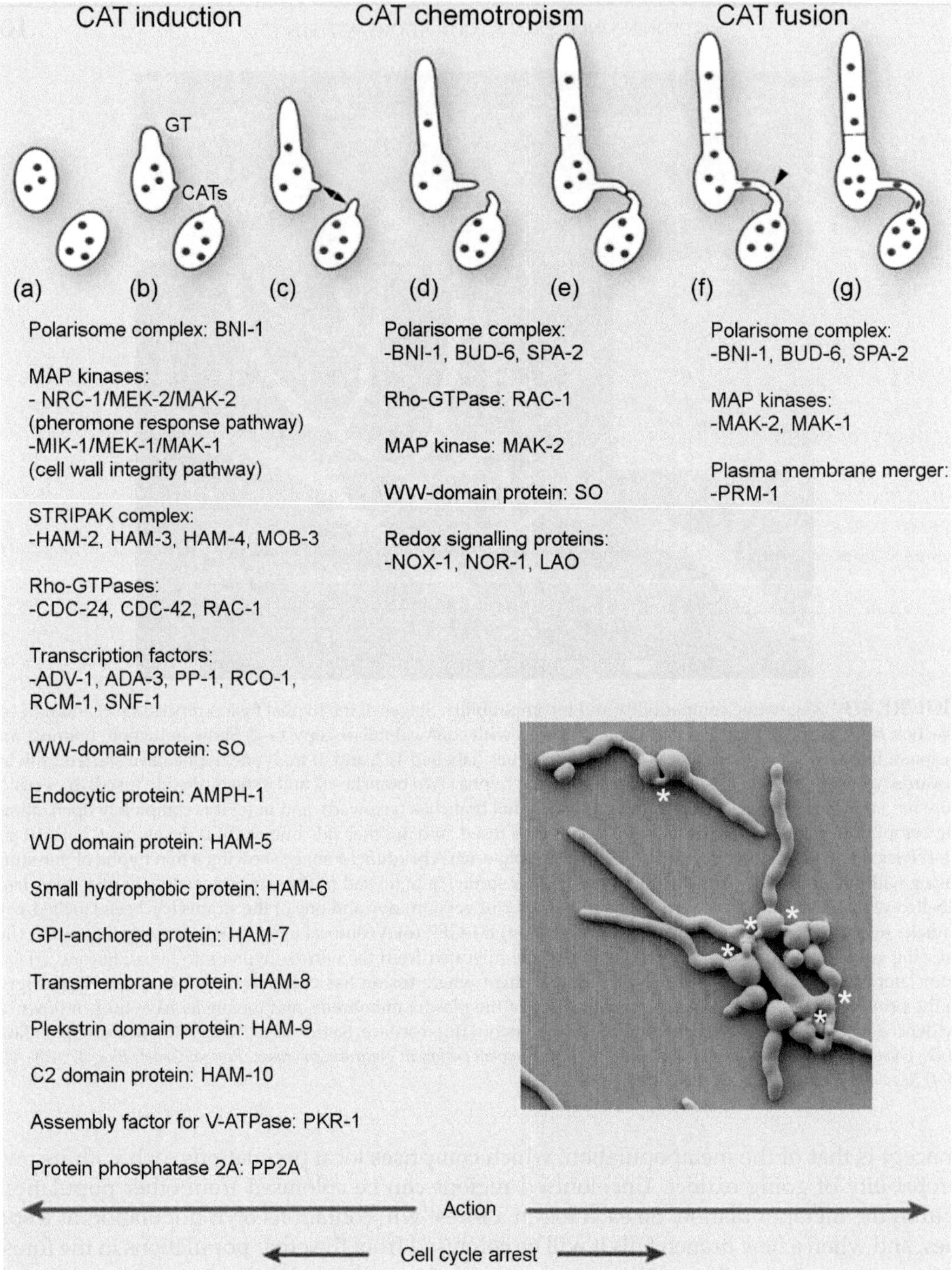

FIGURE 4.4 Self-recognition and the formation of conidial anastomosis tubes (CATs) in ascomycetes. The continuum of six developmental stages in CAT fusion is shown diagrammatically, together with the signalling networks and proteins involved at different stages. (a) Ungerminated macroconidia which contain 3–6 nuclei (circles). (b) Germ tubes (GT) grow out, and CATs form when conidia are close together, indicating the possibility that quorum sensing has a role in CAT induction. Actin cables accumulate within the cell at the sites

biological processes that affect populations are the focus of population biology, including the forces that shape the genetic composition of populations, such as mutation, recombination, drift and selection, and these are explained in the section 'Microevolution' (pp. 123–137).

What Is a Species?

The **species** is the fundamental unit of biological classification, but there are different ways of defining a species and practical difficulties in defining and delimiting species. Fungal species can be defined and recognised based on phenotype, reproductive isolation, and genetic isolation. Historically, as with plants and animals, fungi were first categorised by Linnaeus, based on morphological similarities (i.e. phenotype), – the **morphological species** concept. In the mid-1800s, Elias Fries laid the foundations for fungal classification based on reproductive morphological features, such as spore size, shape, and colour, and whether macroscopic basidiomycete fruit bodies had pores, gills, crusts, or the spores were enclosed. The main difficulty is in finding characters that define the boundaries of a species: the morphology of a fungus can be very variable depending on physiological state and environmental conditions; closely related species can have very different characters; and unrelated organisms can have evolved similar morphologies by different routes (convergent evolution). Within the Agaricomycetes, for example, the order Russulales contains species with seven different sexual reproductive fruit body types – gills, pores, teeth, clubs, crusts, epigeous gasteromycete, and hypogeous gasteromycete; hydnoid fruit bodies (i.e. with teeth) are also found in the Polyporales, Thelephorales, Hymenochaetales, Gomphales, and Cantharellales (Figure 1.6). Other phenotypic characters have been used to augment morphological characters for fungi with simple morphology and those that have industrial importance. These include substrate utilisation in yeasts, and temperature and water potential for growth of *Penicillium*.

The **biological species** concept is most commonly used by macrobiologists, and of particular use to the geneticist. It is based on reproductive isolation, and defines the normal limits of genetic exchange. A biological species consists of all the populations that are able to mate successfully to produce viable offspring. Delimiting the species requires considerable study, since isolates of a fungus from different parts of the world must be brought together and mating tests carried out. Mating tests sometimes reveal that what was thought on morphological grounds to be one species is, in fact, two or sometimes more. However, while such mating tests may yield an unambiguous demarcation, sometimes matings between different isolates

---

FIGURE 4.4—CONT'D from which CATs subsequently emerge, and an apical cap of actin is present in the CAT tip. CATs and germ tubes both lack the Spitzenkörper (pp. 50–51) that is important in tip growth of other vegetative hyphae. (c) Genetically identical cells communicate with each other by releasing a chemoattractant from their tips (arrowed). (d) CATs become orientated along the gradient of chemoattractant resulting in growth towards each other. This is achieved by what has been termed the ping-pong mechanism: two CAT tips homing in on each other rapidly and repeatedly switch between two different signal proteins – MAK-2 and SO – at their tips. When SO is transiently present at the tip of one CAT, MAK-2 is present at the tip of the other and vice versa, acting alternately as a signal sender and signal receiver. (e) When tips make contact, they stop growing and adhere to each other. The cell walls are probably remodelled in the vicinity of the site of contact to prevent leakage when a pore forms between the two CATs. (f) A fusion pore is formed (arrowed) by localised cell wall breakdown and remodelling, and merger of the two plasma membranes. (g) The cytoplasm from both CATs, including all organelles, can then move between the two. Actin, but not microtubules, is required for the process of CAT fusion. The inset is a scanning EM showing a network of germinated conidia formed by fusion of CATs (position indicated by asterisks). *Source: Read et al., 2012.*

are only partially successful, yielding few progeny or ones of reduced vigour. Then it may be difficult to decide whether two isolates belong to the same biological species. A further difficulty is that this definition of a species is based on reproductive isolation, but reproductive isolation is only one step towards speciation (pp. 135–137). Intersterility is the stage at which speciation becomes irreversible, but it can occur at different times ranging between early and very late stages of speciation.

The biological species concept is difficult to employ with fungi that appear to have an entirely asexual lifecycle and with those with homothallic mating systems (pp. 119–120). Molecular methods, however, now make it possible to delimit species by determining the extent of genetic isolation or to put it another way, the limits within which genetic recombination has occurred. This **phylogenetic species** concept is based on analysis of the congruence of genealogies constructed from DNA sequences of appropriately polymorphic loci (e.g. the various parts of the ribosomal RNA operon (SSU, LSU, and 5.8S), as well as several protein coding genes (rpb1, rpb2, efla, and tef1), or of whole genomes. This approach is applicable to fungi that have no obvious mating as well as to those that do. Phylogenetic analysis often reveals three or four species (Figure 4.5) where mating reveals perhaps two

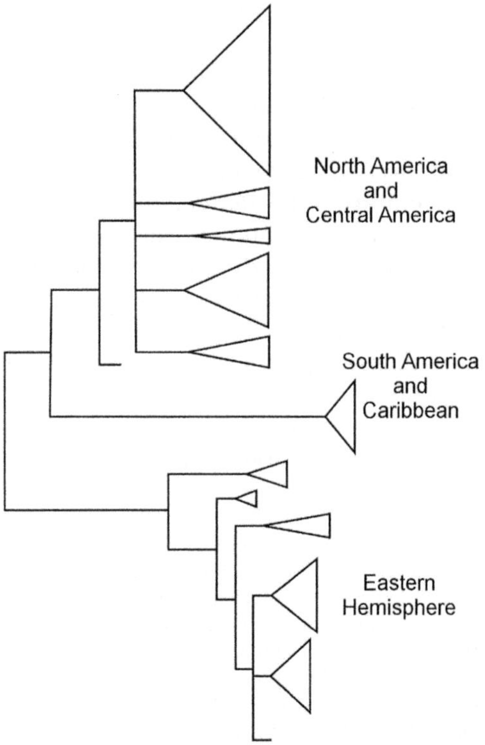

FIGURE 4.5 Phylogenetic species recognition in *Schizophyllum commune*. The intergenic spacer (IGS) of the nuclear ribosomal repeat was sequenced in 195 cultures of individuals from South, Central and North America, the Caribbean, Europe, Asia, and Australasia. Three clades (phylogenetic species) were revealed, one containing isolates from North and Central America, a second containing isolates from South America and the Caribbean, and a third with isolates from the Eastern Hemisphere. Based on morphology and mating, it had previously been thought that this was a single species. *Source: Taylor et al., 2000, based on information from James et al., 2001.*

or three, and morphology, only a single species. In a study on *Neurospora* in which two species – *Neurospora crassa* and *Neurospora discreta* – were distinguished on morphological grounds, seven were found based on ability to mate, and eight based on phylogenetic analysis. This is because with fungi, genetic isolation precedes reproductive isolation; both of these precede morphological divergence, and morphological differences are few – at least relative to most macroorganisms.

About 120,000 fungal species have been formally described so far, but there are likely to be over 5 million (Chapter 1), and new species are being continuously discovered. A newly discovered fungus is only accepted as a new species after it has been described and named in accordance with the International Code of Nomenclature for algae, fungi, and plants. A Latin binomial has to be provided, consisting of a generic name followed by a specific epithet. Often a new species is sufficiently similar to one already known to be assigned to an existing genus, but sometimes it is necessary to create a new genus to accommodate it. A description in English or Latin has to be provided, a type specimen has to be selected, and the specimen deposited in a fungarium (previously, herbarium). Thereafter, all subsequent specimens that sufficiently resemble the type specimen are considered to belong to the same species.

Taxonomists can differ greatly in what they regard as a sufficient resemblance. Some are 'splitters', assigning specimens with small differences to separate species, and others are 'lumpers', combining specimens with considerable differences in the same species. In the genus *Fusarium*, for example, some authorities have recognised scores of species and others less than a dozen. As knowledge increases, especially with the use of molecular approaches, species may be split and new names applied or even combined where the oldest name may be applied. This has to be done in accordance with the rules of the International Code for the new names to be correct. With the rise in sequencing of DNA from environmental samples, many species are being discovered for which no living material is available to fulfil the traditional rules for describing and naming species. New rules were recently agreed to accommodate these.

LIFE CYCLES AND THE SEXUAL PROCESS

The Roles and Consequences of Sex, Outcrossing, and Non-outcrossing

The role of sex is often regarded as the promotion of genetic variability through outcrossing. A species with a lot of genetic variability will be more likely to produce genotypes able to cope with changed environmental conditions than other species with little variability. A consequence of sex may, hence, be an increased capacity, through variation, to survive challenging environments or rapidly evolving parasites or competitors. Natural selection, however; favours features that increase the ecological fitness of an individual and its immediate progeny, not those that might be of value to the rest of the species, or remote posterity. So, the value of sex relates to the immediate products of sexual fusion or of meiosis. Sex has several different roles indicated below, and these may vary between species that have different life cycles, and as fungi are exposed to different environments.

The vast majority of mutations are deleterious rather than beneficial. Different cell lineages, however, will accumulate different unfavourable mutations. Outcrossing will hence result in a dikaryon or diploid in which the effect of mutations, provided that they are recessive, will be masked by the complementary, correctly functioning alleles, and effects known as **complementation**. Also, the recombination that occurs when the haploid state is restored by meiosis will result in some progeny with fewer deleterious alleles than either of the haploid parental strains, counteracting a gradual tendency to accumulate unfavourable mutations.

Outcrossing may result in **heterozygote advantage**, in which the heterozygote performs better than either corresponding homozygote. This is well known in plants and animals, and has also been shown in yeast. Sex will also produce new variants, some potentially less susceptible to parasites than the parent strains. In fungi, the most common parasites are viruses (pp. 357–358), usually transmitted by vegetative hyphal fusion. The sexual process generates, through outcrossing and recombination, progeny with new VC genotypes which will not fuse with and acquire the viruses of the parental VC genotypes. In summary, sexual reproduction allows fungi to cope with changes in the environment they inhabit and threats posed by instability of their own genome.

Despite the many benefits of outcrossing there are also benefits to restricting it. In particular, restricted outcrossing keeps fit combinations of genes together, rather than being separated during recombination. In diploids, lack of outcrossing cleanses genomes of recessive mutations, some of which may have been deleterious; and it allows expression of phenotypes that are recessive but advantageous. Self-fertility provides reproductive assurance since sexual reproduction can occur in the absence of a mate. This is very advantageous when population sizes are small, for example, when there is dispersal limitation and very few migrants colonise a new habitat. So sex (i.e. nuclear fusion followed by meiosis) has a role in fungi, even when outbreeding will not result. Another possible role is in the repair of damage affecting both strands of the DNA molecule, for which a template from a homologous chromosome is needed. Meiosis would then allow copying from the template to occur during crossing over (gene conversion), as happens in the *Saccharomyces* species, amongst others.

While most fungi can reproduce sexually, some do not appear to do so, or at least only mate infrequently. Though there can sometimes be benefits of not outcrossing, as indicated in the previous paragraph, usually the benefits of genetic recombination, sex and outcrossing, are considerable. So how do fungal species with no known sexual processes survive? Molecular phylogeny (pp. 5–8) shows that mitosporic fungi belong to Ascomycota and Basidiomycota. This rather limited divergence from sexual forms indicates that they have not had a long evolutionary history. It seems likely that the loss of sexuality commonly results in short-term advantages but perhaps ultimate extinction. However, although many populations are clonal, molecular variation of mitosporic species shows that some genetic recombination occurs. Sexual reproduction of mitosporic fungi may sometimes occur in nature, as recently shown with *Aspergillus fumigatus* (p. 114), and another mechanism that may generate recombination is the **parasexual cycle** (pp. 122–123).

Glomeromycota, with the sexual process appearing to be completely lacking, is an outstanding exception to the idea that loss of sexuality may eventually lead to extinction. These fungi, which form arbuscular mycorrhizas (pp. 206–212), as well as being abundant today, occur in fossil material of the earliest land plants, and diverged from other fungal groups over 400 million years ago. The multinucleate spores of these fungi contain a population of genetically different nuclei. Live three-dimensional imaging and mathematical modelling showed that when spores of *Claroideoglomus etunicatum* are formed, rather than a single founder nucleus dividing, a stream of nuclei enter from nearby hyphae. This perhaps provides an alternative to sex with natural selection operating to promote the survival of individuals with an optimally balanced population of genetically different nuclei. Also, these fungi, as valuable partners for their plant hosts, may receive some protection from the effects of hostile environments, but selection is most likely to operate on traits of the fungi in soil.

Types and Phases of Life Cycles

The sexual process in fungi, as in other eukaryotes, has three key steps: (1) **cell fusion** (plasmogamy) between two haploid cells, which are uninucleate in many fungi and genetically different, resulting in a cell with two different haploid nuclei; (2) **nuclear fusion** (karyogamy) of the two (typically) haploid nuclei giving a cell with a single (typically) diploid nucleus; and (3) **meiosis**, which results in four haploid cells. There is considerable variation amongst phyla in terms of which structures fuse during plasmogamy, and in which structures karyogamy and meiosis occur (Table 4.2). The timing of these events during the lives of fungi, and the intervals between these events, also varies considerably, resulting in a diversity of life cycles which can be categorised into five basic types (Figure 4.6), and are described below after considering the different types of plasmogamy.

TABLE 4.2 Distinguishing Features of the Lifecycles of Fungi and Fungus-Like Oomycetes that Reproduce Sexually

	Type of plasmogamy[a]	Site of karyogamy	Site of meiosis	Type of sexual spore
Fungus-like oomycetes	Gametangial copulation	Oospores	Gametangia	Oospore
Chytridiomycota	Gametes fuse	Zygote	Sporangium	Oospore
Zygomycetes	Gametangia fuse	Zygospore	Zygospore	Zygospore
Ascomycota	Spermatisation (e.g. *Neurospora*)	Ascus initial[b]	Ascus[b]	Ascospore
	Gametangial copulation (e.g. *Arachnotis*)			
	Somatogamy (e.g. Free-living *Saccharomyces*)			
Basidiomycota	Somatogamy in the majority	Basidium initial[c]	Basidium[c]	Basidiospore
	Spermatisation between pycniospores and hyphae in Uredinales	Teliospores	Basidium	Basidiospore (haploid). Teliospore (dikaryotic then diploid); pycniospores/ spermatia (haploid) (see Table 8.6)

[a]*See text for explanation of types of plasmogamy.*
[b]*See Figure 1.14.*
[c]*See Figure 1.7.*

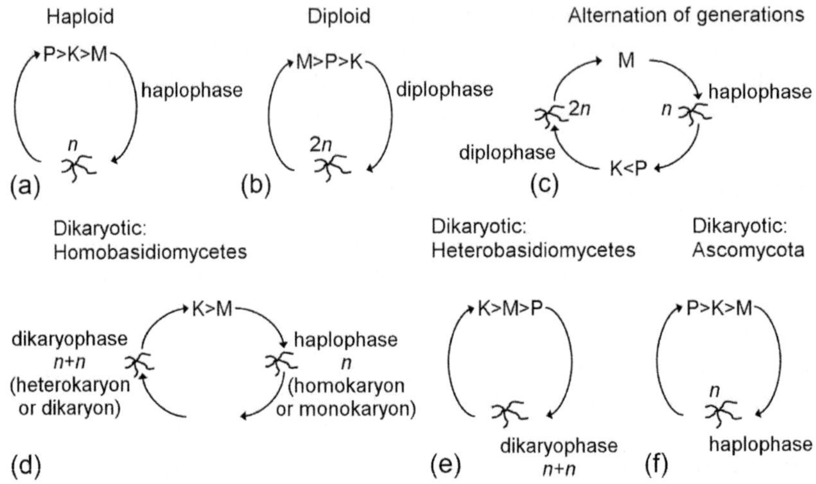

FIGURE 4.6 Fungal life cycles. There are five basic types, four of which are shown here: (a–c) haploid, diploid, and haploid/diploid respectively. The fifth is a completely asexual life cycle where nuclear fusion and recombination does not occur. The dikaryotic type could be considered as three subtypes (d, e, f). The life cycles differ depending on the extent of time between plasmogamy (P, cell fusion), karyogamy (K, fusion of nuclei) and meiosis (M).

There are five main types of plasmogamy, most commonly between cells which do not differ in morphology (isogamy), but sometimes where the cells are of different size (anisogamy) – as in fungus-like oomycetes, the larger being referred to as female and the smaller as male: (1) **Gametangial copulation** occurs in the fungus-like oomycetes, in which small male antheridia produce fertilisation tubes which grow chemotropically towards, form branches around, and then fuse with the larger female oogonia, allowing nuclei to pass through a fine penetration peg and fuse with oospheres. In some Ascomycota, a thin tube (trichogyne) grows from the gametangium (termed an ascogonium) to the antheridium, and a nucleus passes along the trichogyne to the ascogonium. (2) **Fusion of gametes** – unicells, in which at least one, usually both, are motile. Commonly they are the same size (isogamous zoospores), but occasionally one is larger than the other (anisogamous), and in the chytrid genus *Monoblepharis*, the motile male gametes are released from an antheridium and penetrate a large static female oogonium (heterogamy or oogamy) to fertilise the oospheres. (3) **Fusion of gametangia** in which first, typically, morphologically identical zygomycete zygophores grow together, swell to form progametangia, differentiate gametangia, and subsequently fuse to form a zygote, which develops into a thick-walled zygospore (Figure 1.22). (4) **Spermatisation** involves fusion between uninucleate non-motile cells and a 'female' gametangium. In *Neurospora*, for example, the trichogyne of the female curves around a conidium and the spore nucleus travels along the trichogyne into the ascogonium. In rusts (Basidiomycota), pycniospores are transferred to receptive hyphae. (5) **Somatogamy** involves fusion of structures which are morphologically no different from other vegetative parts, including between hyphae, and between yeast cells. All types of plasmogamy are under tight hormonal control.

There are five main types of lifecycle: haploid, diploid, haploid–diploid, dikaryotic, and asexual, characterised by the number of nuclei and ploidy of the nuclei and whether or not mating occurs (Figure 4.6):

Haploid lifecycle. For the majority of the life cycle the vegetative mycelium is haploid. When plasmogomy occurs, it is quickly followed by karyogamy and usually followed quickly by meiosis (e.g. in zygomycetes). In zygomycetes, meiosis occurs in zygospores and if they lie dormant, the spore remains a diploid, but mycelia that arise following germination are haploid.

Diploid lifecycle. The vegetative thallus is diploid. Following meiosis, plasmogamy, and karyogamy soon occur. For example, *Saccharomyces cerevisiae* can exist as a haploid, diploid, or polyploid, but in nature is mostly diploid. Though some *Candida* spp. are haploid (e.g. *Candida lusitaniae* and *Candida guilliermondii*), the human pathogen *Candida albicans* (pp. 306–307) is usually diploid in its vegetative state (but see *Asexual life cycle*, p. 114).

Haploid–diploid lifecycle. Vegetative growth occurs both in haploid and diploid phases, the transition from haploid to diploid occurring by plasmogomy, and the transition from diploid to haploid resulting from meiosis. The haploid and diploid phases are morphologically similar, but with diploid cells often larger than haploid ones (e.g. in the yeast cells of *Saccharomyces cerevisiae*).

Dikaryotic lifecycle. Dikaryotic (literally two nuclei) cells are characterised by the presence of two haploid nuclei, and this situation is characteristic of Basidiomycota and, in a sense, of many Ascomycota. Following plasmogamy, the two nuclei do not fuse immediately giving rise to a dikaryon. In Ascomycota only the ascogenous hyphae and crozier cells are dikaryotic (Figure 1.14). In most Basidiomycota, somatic cells are probably dikaryotic for much of the life cycle (Figure 1.17). When basidiospores (haploid) germinate, the mycelium typically has one nucleus per compartment. This situation continues until plasmogamy occurs with a sexually compatible (pp. 114–118) mycelium, and then each compartment has two nuclei.

While the majority of basidiomycetes have a single nucleus per compartment prior to mating, some have several or many (e.g. in the genera *Coniophora*, *Stereum*, and *Phanerochaete*). These are obviously not monokaryons (mono means one), and are referred to as coenocytic homokaryons (i.e. many nuclei of the same origin, which are not necessarily identical). Following plasmogamy, each compartment contains nuclei from both mycelia, and these mycelia are referred to as heterokaryons (different nuclei). Monokaryons can also be referred to as homokaryons, and dikaryons as heterokaryons.

Dikaryons (or heterokaryons) are also formed in the so-called **Buller phenomenon**: dikaryotic (or heterokaryotic) or diploid mycelia can provide one nucleus (or more in species with many nuclei per compartment) to unmated monokaryotic (or homokaryotic) or haploid strains, termed di–mon (or he–ho) mating. For example, heterokaryon–homokaryon (he–ho) matings occur in *Stereum hirsutum*, dikaryon–homokaryon (di–mon) matings in *Schizophyllum commune*, and diploid and haploid strains of *Armillaria gallica*. Sometimes homokaryotic sectors arise in a heterokaryotic mycelium, allowing the possibility of remating with an appropriate homokaryon. Reassortment of nuclei in this way has been seen in *Heterobasidion annosum* in somatic incompatibility (pp. 102–104) interaction zones in the lab, and in dense populations in wood in the field.

How long mycelia remain as homokaryons has rarely been studied; some basidiomycetes may only have haploid nuclei for hours or days before successful mating/plasmogomy transforms them to dikaryons. Rare basidiomycetes may not encounter a mating compatible mycelium for many weeks or years, and may even live their whole life as a monokaryon. Even common basidiomycetes, such as *Trametes versicolor* can sometimes persist as homokaryons in nature for several years.

When a dikaryon has formed, this state (i.e. one nucleus from each of the two original mycelia in each cell/compartment) is maintained at cell division by a mechanism involving formation of clamp connections (Figure 1.8). Eventually, the mycelium may form sexual fruit bodies, the two nuclei fusing to form a diploid in the cells (basidium initials) that quickly become basidia (Figure 1.7; p. 11). So the diploid is confined to a single cell type in the life cycle. Meiosis then occurs in the basidia and each basidiospore produced contains a hapoid nucleus, which subsequently germinates to give a haploid mycelium.

The rusts (Uredinales) and the smuts (Ustilaginales) are obligate, biotrophic plant pathogenic basidiomycetes with complex life-cycles involving several spore types (Table 8.6). The sexual stage is pathogenic, and mating typically occurs in association with the host. The smut fungus, *Ustilago maydis*, for example, exists as a non-pathogenic budding yeast. When two mating type compatible cells meet and fuse, dikaryotic hyphae are produced that are able to infect the host plant, maize (pp. 16–18).

Asexual lifecycle. In some fungi the sexual process has not been seen. In Ascomycota and in Basidiomycota, species which appear only to reproduce asexually are termed **mitosporic** fungi. However, lack of evidence for sexual reproduction in the lab does not necessarily mean that it does not occur in nature. *Aspergillus fumigatus*, for example, was thought for a long time to be asexual, but is now known to have a sexual cycle – under certain nutrient and environmental conditions (sealed plates of oatmeal agar, incubated for 6 months in the dark at 30 °C) it produced cleistothecia (p. 20) containing ascospores. For several *Aspergillus* species hitherto considered to be asexual, whole genome sequences have revealed the presence of suites of genes associated with parts of the sexual cycle in other ascomycetes, including for mating and pheromone response, meiosis, and development of fruit bodies. Population genetic studies have also shown evidence of genetic recombination (linkage disequilibrium) which indicates past sexual activity.

Many plant and animal pathogens are considered to be asexual, but this is likely due to the fact that only one clone has the set of genes needed to infect the appropriate plant or animal. Although *Magnaporthe oryzae* is, for example, considered to have an asexual life cycle, there appears to be a sexual population in India, and the global distribution of clones and clonal lineages probably reflects rare 'escapes' from the sexual population. The human pathogen *Candida albicans* was, until recently, thought to be completely asexual, always being diploid, however, mating between cells with opposite mating types (see below) does occur. However, as well as this heterothallism (see below), homothallism also occurs.

Mating Systems

Fungi have mating systems or breeding systems that determine whether or not individuals of the same species can mate. Some fungi are self-fertile but many fungi have genetic systems that prevent mating between very genetically similar individuals, (i.e. self-sterile) so that genetic diversity will be increased. With basidiomycetes, since somatic (vegetative) incompatibility (pp. 102–104) often prevents non-self mycelium from fusing to form a stable connection containing nuclei from both, this rejection mechanism has to be overcome before successful mating can occur. With ascomycetes, somatic (vegetative) incompatibility is suppressed during mating, provided that mating occurs between female reproductive structures (protoperithecia) and a male cell, and does not require hyphal fusion (heterokaryon formation) prior to perithecia formation. Compatible matings are determined by mating-type (MAT) factors. The *MAT* loci have a complex genetic structure. In *Coprinus cinereus*, for example, there are four sites each having two closely linked loci with multiple alleles. However, in population genetics, these complex loci can usually be treated as if they were simply multiple alleles at two loci, A and B, i.e. $A_1 \ldots A_n$ and $B_1 \ldots B_n$, or in the case of some basidiomycetes one locus $A_1 \ldots A_n$.

Successful mating is achieved if alleles at the mating type loci differ. Since mating occurs between different genetic strains, the fungi that operate these systems are termed **heterothallic** (Greek, *hetro* = different, *thallos* = young shoot), and the controlling systems are termed **homogenic incompatibility** (Greek, *homo* = same) or equally appropriately **heterogenic compatibility**, since mating fails if strains are of identical **mating type**; the sexual process only occurs if two strains which differ in mating type interact.

Several kinds of mating systems, as described below, have been recognised in the fungi. Many fungi have two mating types, effectively two sexes. Morphological differences are often not evident and they are referred to as, for example, + and −, A and B, A and a, or A and α. Some fungi or fungus-like eukaryotes produce visibly different structures involved in plasmogamy, as described above, and these are often referred to as male and female structures. However, it is inappropriate to equate two sexes with male and female as mycelia arising from a single haploid spore are often able to produce both male and female structures, but fertilisation will only occur as a result of an encounter with the opposite mating type. For example, a haploid mycelium of *Neurospora crassa* can produce female protoperithecia and conidia which are potentially male gametes. It is only in the fungus-like oomycetes that sexual differentiation is involved in the promotion of outcrossing. The mating systems of some fungi are more complex, with more sexes, and these are normally referred to with letters and numerical subscripts.

Population biologists studying diploid plants and animals use terms such as parents, progeny, outbreeding, self-fertilisation, etc., but these terms have slightly different meanings when applied to fungi, because in the vast majority, the vegetative thallus is haploid, at least for a while. So the vegetative thallus can be treated as 'self' rather than as a gamete. A major distinction is between fungi that operate **outcrossing** strategies, that promote genetic exchange, and **non-outcrossing** strategies where genetic exchange is promoted less (Figure 4.7). The term outcrossing encompasses situations where mating occurs between different haploid genotypes irrespective of whether they derived from the same spore source (e.g. the same basidiocarp or ascocarp) or from a different spore source. On the other hand, the term **inbreeding** does denote origin from the same spore source, and outbreeding denotes origin from different spore sources.

Mating Systems that Promote Outcrossing

Breeding Systems with Two Mating Types: Dimixis

In those fungi studied to date, the two mating types differ with respect to the allele present at a single mating type locus. Mating is only successful between haploid cells or mycelium that differs at the mating type locus. The system ensures that mating does not take place between the genetically identical progeny of a single haploid cell. A diploid cell that arises from mating will carry both mating types, and when meiosis occurs the resulting haploid progeny will be half of one mating type and half of the other. So there is 50% chance that the encounter

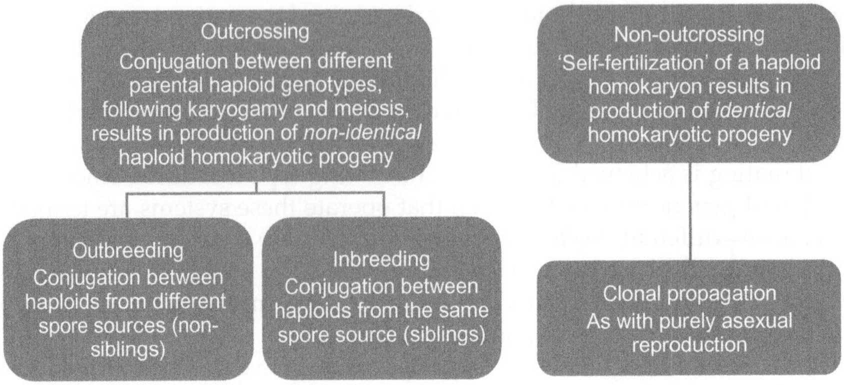

FIGURE 4.7 Terminology for describing breeding strategies in fungi with a haploid phase (i.e. a yeast/mycelium with nuclei derived from a single haploid nucleus). *Source: Rayner, A.D.M., Boddy, L., 1988. Fungal Decomposition of Wood: Its Biology and Ecology. John Wiley, Chichester.*

between two individuals derived from a single diploid will result in mating (the **inbreeding potential**). Since species with this breeding system only have two mating types in the whole population, there is also a 50% chance of an encounter between two unrelated individuals resulting in mating (**outbreeding potential**) (Table 4.3). Thus, although selfing is prevented, the likelihood of mating between close relatives is not reduced.

Breeding Systems with Many Mating Types (Diaphoromixis): Unifactorial (Bipolar) Incompatibility

Some Basidiomycota have a single mating type locus (and are hence 'unifactorial') but within the whole population have a large number of mating type alleles. Mating occurs between any haploid mycelia that differ in mating types (i.e. have different alleles of the mating type locus). So the diploid basidium nucleus is heterozygous for mating type (e.g. A_1A_2). Hence, two of the haploid spores borne on a basidium will be of one mating type (e.g. A_1) and two of the other (e.g. A_2). This is why this type of system is sometimes termed **bipolar**. Since any diploid strain yields two mating types, the inbreeding potential, as with breeding systems having only two mating types, is 50%. That is, if the mycelia from a large number of basidiospores from the same fruit body are paired in all combinations, 50% of the pairings will mate successfully. Since, however, a large number of mating types can occur in a population, almost all encounters with non-sibling strains can be compatible and the outbreeding potential can approach 100% (Table 4.3), so there is a bias towards outbreeding.

Breeding Systems with Many Mating Types (Diaphoromixis): Bifactorial (Tetrapolar) Incompatibility Systems

Many Basidiomycota have mating systems with two unlinked mating type factors, designated *AB*. Alleles at both the *A* and the *B* loci must differ for successful mating. The *A* and *B* genes control different parts of the mating process. In *Coprinopsis cinerea* and *Schizophyllum commune*, the *A* genes control the development of clamp connections

TABLE 4.3 Characteristics and Examples of Fungi Having Different Mating Systems that Promote Outcrossing

System	Examples	Number of mating type loci	Number of different alleles at each locus in the whole population	Designation of mating types	Inbreeding potential (%)[a]	Outbreeding potential (%)[b]
Two mating types	Most zygomycetes; *Neurospora crassa* (Ascomycota); *Saccharomyces cerevisiae* and *Schizosaccharomyces pombe* (Ascomycota, yeast)	1	1	±, *Aa*, *aα*	50	50
Unifactorial (bipolar) incompatibility	*Coprinus disseminatus* and *Stereum gausapatum* (Basidiomycota)	1	Many	$A_1 \ldots A_n$	50	Nearly 100
Bifactorial (tetrapolar) incompatibility	*Coprinopsis cinerea* and *Schizophyllum commune* (Basidiomycota)	2	Many	$A_1 B_1 \ldots A_n B_n$	25	Nearly 100
Modified tetrapolar incompatibility	*Ustilago maydis* (Basidiomycota, smut)	2	2 at locus *a*, many at locus *b*	$A_1 B_1 \ldots A_2 B_n$	25	50

[a]*The probability that a randomly encountered sibling (product of the same diploid parent) will be compatible.*
[b]*The probability that a randomly encountered unrelated individual will be compatible.*

(Figure 1.8), which maintain nuclei from both parents in each compartment. The *B* genes regulate exchange of nuclei between both partners, migration of nuclei from one mycelium through the other and vice versa, and encode for pheromone and receptor systems. Differences at both loci are necessary for successful mating; differences at only one of the loci result in semicompatibility.

If one of the nuclei in a fungus has mating factors A_1B_1 and the other A_2B_2, basidiospores can possess nuclei that have the same mating factors as either of the parents (i.e. A_1B_1 or A_2B_2), and A_1B_2 and A_2B_1. Such systems with four possible combinations of mating factors are termed **tetrapolar**. When haploid mycelia derived from basidiospores from the same fruit body are paired in all combinations, only 25% of crosses will mate successfully because only 25% of the crosses will have different alleles at both mating loci (Table 4.4), so there is a greater outbreeding bias than with bipolar systems. In a population that contains many *A* and *B* factors, the outbreeding potential will approach 100%. Recombination between the subunits of the A or the B factors can occur during meiosis, yielding a new mating type compatible with the parent types. The frequency of such recombination between mating factors varies greatly between species and strains, and is also influenced by environmental conditions. Where it is high, new mating types will arise with high frequency, so the inbreeding potential will greatly increase.

TABLE 4.4 Mating among the Haploid Progeny from the Fruit Body of a Basidiomycete with Tetrapolar Incompatibility

Mating type alleles	A_1B_1	A_1B_2	A_2B_1	A_2B_2
A_1B_1	–	–	–	+
A_1B_2	–	–	+	–
A_2B_1	–	+	–	–
A_2B_2	+	–	–	–

The basidium nuclei have the mating genotype A_1B_1, A_2B_2. Meiosis gives rise to haploid progeny with four possible mating type genotypes – A_1B_1, A_1B_2, A_2B_1, A_2B_2. Mating is only completely successfully if encounters are between mycelia that differ with respect to both mating type factors. The chequerboard shows that only 25% of the possible encounters satisfy this condition. The inbreeding potential is thus 25%.

Modified Tetrapolar Incompatibility

In some Basidiomycota there are many *B* factors but only two *A* factors. This resembles typical tetrapolar incompatibility in giving an inbreeding potential of 25%, but resembles incompatibility systems with two mating types in having an outbreeding potential of 50% (Table 4.3).

The Evolution of Mating Type Genes and Their Function in Fungal Life Cycles

Mating type genes are central to the operation of the fungal life cycles described above. While emphasis has traditionally been placed on differences in breeding systems between phyla, more recent analysis of these sex-determining regions of the genome in zygomycetes, ascomycetes, and basidiomycetes has revealed some underlying homologies of the genes involved. Mating type loci of fungi operate like sex chromosomes in more complex eukaryotes, to control compatibility between sexual partners, enable sexually compatible haploid cells to recognise and attract each other, and to prepare the cell for sexual development after fertilisation.

Genetically controlled mating compatibility – heterothallism – was first discovered in the basal lineage zygomycetes, in *Phycomyces blakesleeanus*, named after Albert Blakeslee, the American discoverer of fungal heterothallism. The mating type locus encodes a type of transcription factor belonging to the High Mobility Group (HMG). The plus and minus strains of *Phycomyces blakesleeanus* each carry a different HMG. The function of these regions in controlling sexual development was shown when sexual spores were found in a mutant haploid strain with a diploid mating type region. The two HMG's segregate with mating type in crosses. One of them consists of a longer DNA sequence than the other; the sequenced genome revealed it to have been expanded by repetitive elements of DNA. Such expansions are known to suppress recombination at the regions of the DNA where they occur, and might explain why this mating type region has persisted intact through evolution of more complex systems. Such suppression of recombination by repetitive elements in mating type loci may drive the expansion of sex-determining regions, not only in fungi but in other eukaryote lineages as well, even perhaps leading to the evolution of sex chromosomes like our own.

In heterothallic ascomycetes there is typically one mating type locus with two forms, and the study of these in mating of the model yeast *Saccharomyces cerevisiae* provides one of the most complete analyses of developmental gene expression in eukaryotes. The so-called *MAT* loci determine haploid and diploid cell identity. They function to ensure outbreeding by encoding sex-specific expression of many genes required in mating. In both *MATa* and *MATα* haploid cells, sex pheromones, and their receptors are expressed under the control of transcription factors encoded at the *MAT* loci. The transcription factors involved in mating of *Saccharomyces cerevisiae* include two alternative homologous sequences ('homeodomains') HD1 and HD2, one in each mating type. *MATα* cells express a DNA-binding protein (alpha-box protein) that activates transcription of its alpha sex pheromone, as well as receptors to sense the pheromone secreted by compatible cells of the *MATa* mating type. The *MATa* cells have a different transcription factor, encoded by an HMG domain, which activates transcription of a-pheromone and α-pheromone receptors, setting up reciprocal pheromone attraction between neighbouring compatible haploid cells. Pheromone binding at mating then triggers the expression of all the genes involved in mating, via an intracellular mitogen activated protein (MAP)-kinase cascade. After mating, the mating type genes help to establish and maintain the characteristics of the diploid cell.

In basidiomycetes, some of these genes are conserved. The A locus contains genes for both HD1 and HD2 homeodomain transcription factors, and the *B* locus contains genes for pheromones and pheromone receptors. The heterodimer of HD1 and HD2 protein can be formed by any pair of protein products of allelic gene pairs, provided these are not identical. At the *B* locus, pheromone and receptor genes have been gathered together. Basidiomycota are the only lineage known to have evolved tetrapolar outbreeding systems. Within the phylum, a variety of derived sexual cycles has evolved from tetrapolar ancestry. In some species, bipolarity is secondarily generated, either by fusion of the two loci or by loss of *B* mating type function (although not by loss of the *B* genes, because these still function in the regulation of development, albeit in mating-type independent fashion). Anamorphic fungi in some ascomycetes and basidiomycetes with no known sexual phase in their life cycles contain sex-encoding regions, with the practical implication that reactivating this latent potential for sexual recombination might provide a useful breeding strategy for industrial strain development. Simplification of mating systems to reduce or obviate the need for mating seems also to have resulted from adaptation to niches including, for example, *Cryptococcus* the human pathogen (Chapter 9), and *Ustilago* the smut pathogen of plants (Chapter 8). However, tetrapolar mating remains widespread in other lineages including Agaricomycetes.

Systems Restricting Outcrossing

Self-fertility

Some fungi are self-fertile, that is the sexual process can occur between genetically identical cells. Self-fertility is also referred to as **homothallism** (Greek, *homo* = same). Homothallism is termed **primary** if there is no evidence of a heterothallic ancestor. If it is clear that an earlier heterothallism has been circumvented, the term **secondary homothallism** is used. For example, *Neurospora tetrasperma* (like *Neurospora crassa*) has two mating type alleles, but instead of an ascus containing eight uninucleate ascospores, only four are produced, each with two nuclei, one of each mating type. The resulting vegetative mycelia and protoperithecia also contain nuclei of both mating types, so fertilisation by conidia of a different strain is not required for ascus production. Basidia of the cultivated mushroom, *Agaricus bisporus*, have two binucleate spores per basidium instead of the four uninucleate spores characteristic of most Basidiomycota. A unifactorial incompatibility system occurs but the two nuclei in a basidiospore are commonly of different mating type, hence, the mycelium that arises from a single basidiospore is usually able to produce fertile fruit bodies. This could be an advantage for fungi in situations where encountering a compatible mate is unlikely, for example following long-distance spread and colonisation of new habitats. Occasionally, a spore of *Agaricus bisporus* only has one nucleus. In *Moniliophthora perniciosa* (the witches' broom pathogen of cocoa (*Theobroma cacao*)), there is usually one nucleus per spore on each basidium, but in the same basidiocarp, up to about 8% of spores can be binucleate, and occasionally trinucleate. So, in Ascomycota and Basidiomycota, there is probably a continuum from full outcrossing to secondary homothallism.

While some species are completely self-fertile, other species have some populations that are non-outcrossing while others are outcrossing. For example, the wood-decay basidiomycete *Stereum sanguinolentum* has many (>30) distinct clonal subpopulations in Northwestern Europe, but outcrossing populations in Austria and North America. As well as non-outcrossing sexual systems, clonal subpopulations can be generated by the production of asexual spores, especially by Ascomycota, and extensive clones can develop even in populations with sexual outcrossing mechanisms.

Strains of the yeast *Saccharomyces cerevisiae* have two mating types, ***a*** and *α*. Some strains appear to be homothallic, with mating occurring among the progeny of a single haploid cell. The apparent homothallism is the result of a switch from ***a*** to *α* or of *α* to ***a*** mating type which can occur at cell division (Figure 4.8a). The molecular basis of mating type switching is the replacement of the genetic factor of the mating type (*MAT*) locus by a copy of the alternative factor resident at a silenced locus (i.e. a locus at which the gene is not expressed), there being one silenced locus for each mating type (*HML* homologous to *MAT* α and *HMR* homologous to *MAT**a***). Recombination between *MATα* and *HMR*(***a***) or between *MAT**a*** and *HML*(*α*) results in a switch in mating type.

Homothallic and heterothallic strains differ in genotype at a locus which initiates mating type switching, possessing the alleles *HO* or mutant *ho*, respectively. *HO* encodes an endonuclease that induces a DNA double strand break within the *MAT* locus, which provides a substrate to initiate the recombination event with the opposite silent locus. At each mitotic division of a free living haploid yeast cell, *HO* induction occurs within the original mother cell allowing mating with the adjacent, genetically identical daughter cell (Figure 4.8b). Outbreeding is also possible between cells that are of opposite mating type from a meiotic event or if it encounters a cell having a different mating type in the environment (Figure 4.8b).

Mating type switching also occurs in the homothallic fission yeast *Schizosaccharomyces pombe*. *Schizosaccharomyces pombe* is distantly related to *Saccharomyces cerevisiae*, and mating type switching has apparently evolved independently in the two species. It may also occur in apparently homothallic filamentous fungi, but it would be less easy to detect.

Self-fertility does not prevent outcrossing but can reduce the likelihood of it occurring, as encounters are more likely to take place between cells of common parentage than between unrelated cells. The absence of the requirement for encountering mycelia/cells with a complementary mating type may have short-term advantages, especially if no other spore type is produced, or if meiospores (spores produced following meiosis) have some special role. It is not clear how far homothallism restricts recombination, as molecular variation in nature has been studied only occasionally. Recombination has been found in populations of secondarily homothallic species and clonal populations (e.g. *Neurospora crassa*).

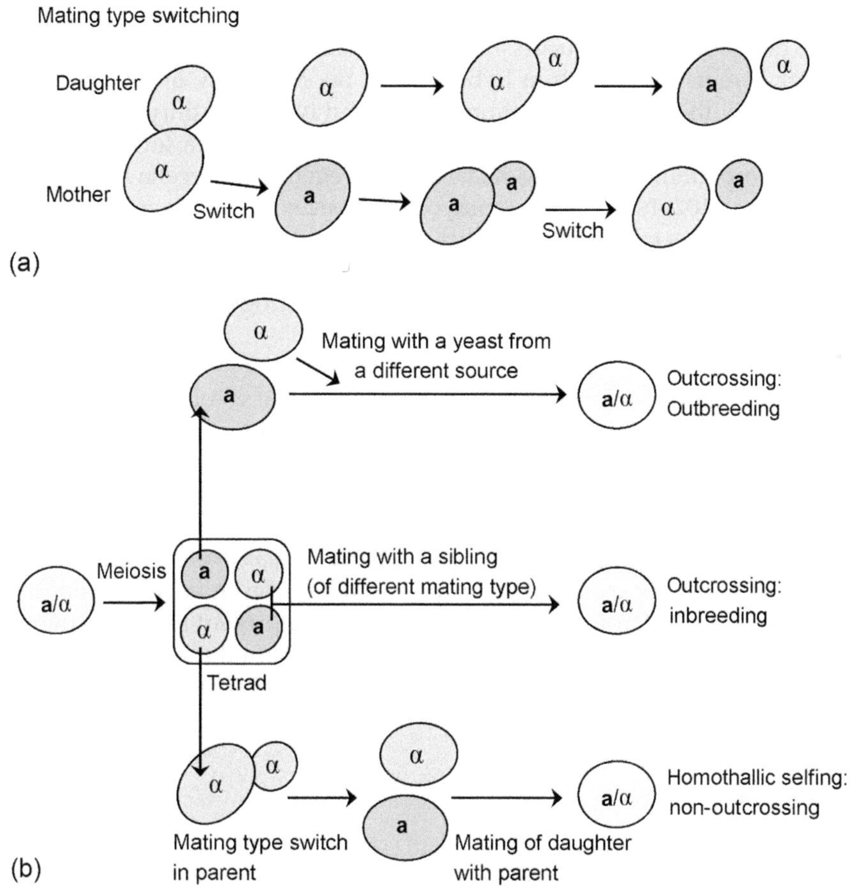

FIGURE 4.8 Mating and mating type switching in ascomycete yeast. The complexity allows multiple options for inbreeding and outbreeding. (a) Mating type switching occurs in haploid *Saccharomyces cerevisiae*. The *MAT* locus switches form (i.e. from **a** and α mating type and vice versa, in the G1 phase of every cell cycle, allowing self-mating (b). (b) Outcrossing is also possible between cells that are of opposite mating type, either from a meiotic event (inbreeding from mating with siblings) or if a yeast encounters a cell having a different mating type in the environment (outbreeding). Intratetrad mating leads to a slower loss of heterozygosity, in subsequent diploid generations, than homothallic selfing, which fixes loci instantly.

Fertility Barriers

Though, as discussed above (p. 114), many fungi are self-sterile and have systems that promote outcrossing, fertility barriers can limit the extent of outcrossing. Two main types are genetic disharmony and heterogenic incompatibility. If, through geographical separation, there is little gene flow between two populations for a long time, genetic differences between the populations may arise (i.e. **genetic disharmony**). Such differences could lead to the inefficient performance or failure of some step in the complex interactions between the participants that occur in mating. As a result, mating could fail completely or only a few progeny may be produced. Furthermore, efficient metabolism and normal development are dependent not only on the action of individual genes but on harmonious interaction throughout the genome. Hence, even if mating is successful, an unsatisfactory interaction between the genes from dissimilar parents may result in hybrids with poor viability or which are sterile, and unable to compete with either parent strain. The second type of fertility barrier – **heterogenic incompatibility** – occurs when mating fails as a result of one or a few genetic differences. Heterogenic incompatibility resulting in failure of vegetative mycelia to fuse has already been considered (pp. 102–103). In the ascomycete *Podospora anserina*, some of the interactions which prevent vegetative cell fusions also act to prevent mating. In some fungi, some mechanisms of heterogenic incompatibility may act solely in the sexual phase. The possible role of heterogenic incompatibility is returned to when considering the parasexual cycle (below).

Amixis

With amixis, all of the usual morphological features of the sexual process are present, but nuclear fusion and meiosis do not occur. This is equivalent to loss of sexuality and obligate inbreeding. It can be distinguished from homothallism only by detailed cytological study. Amixis has been little studied in fungi but some examples are known: *Podospora arizonensis* (Ascomycota) produces spores that look like normal meiospores but have been produced by mitotic divisions. In the Chinese paddy straw mushroom, *Volvariella volvacea*, homokaryotic mycelium produces basidiocarps; two identical, haploid nuclei enter each basidium, fuse and undergo meiosis, producing four haploid nuclei, one being present in each of the four basidiospores. The progeny are genetically identical to the parent but it remains to be seen whether the process of meiosis could cause epigenetic changes (p. 134), as occur during meiosis in animals and other eukaryotes.

The Parasexual Cycle

Somatic (vegetative) incompatibility (pp. 102–103) usually prevents widespread heterokaryosis between non-self mycelia. However, VC is not a complete barrier to formation of heterokaryons between different individuals. Occasionally two haploid nuclei in a vegetative mycelium may fuse to give a somatic diploid nucleus. If the mycelium is heterokaryotic then the fusion may be between genetically unlike nuclei to give a heterozygous diploid nucleus. This is the start of the **parasexual** cycle (Figure 4.9). Such diploidization is rare, occurring perhaps once in a population of a million nuclei. Mitotic crossing over in these diploid nuclei may occur and can generate diversity. This is a rare event compared with meiotic crossing over, occurring perhaps once per 500 mitoses. Errors in mitosis are quite common, and often the mitosis of a diploid nucleus results in aneuploidy. Such abnormalities in chromosome number (aneuploidy, see p. 124) commonly lead to poor growth, and further change in chromosome number results in restoration of the haploid state. Before the advent of molecular biology, parasexual recombination was used to produce new strains of asexual fungi of commercial importance (e.g. to produce strains of *Penicillium chrysogenum* with increased penicillin production). It has also been exploited in the lab for linkage mapping in asexual fungi. Parasexual recombination is probably largely confined to laboratory culture, and does not appear to be an important phenomenon in nature, though there is some evidence that *Cryphonectria parasitica* in North America has recently recombined by a parasexual process. Fungi that are diploid for most of their lifecycle can also undergo a parasexual cycle. In *Candida albicans*, for example,

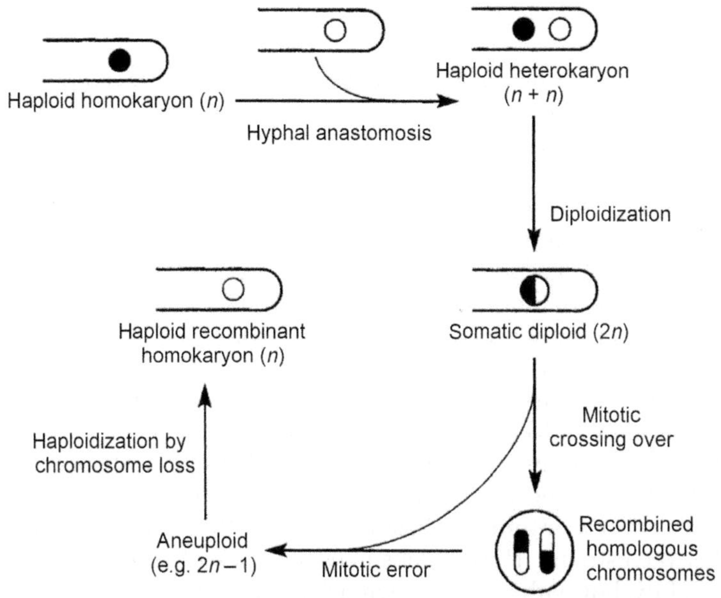

FIGURE 4.9 The parasexual cycle. Hyphal fusion between two genetically different homokaryons produces a heterokaryon. If the haploid nuclei fuse they produce a heterozygous diploid nucleus. Rather than undergoing meiosis, they divide mitotically. Recombination between homologous chromosomes (mitotic crossing over) can sometimes occur. Sometimes, mitotic errors can yield aneuploids, with abnormal chromosome numbers (e.g. $2n-1$) and commonly poor growth. The haploid state is restored by loss of chromosomes. Since the chromosomes that are lost can have come from either of the original homokaryons, a recombinant can result, even if mitotic crossing over has not occurred.

diploid cells fuse to form tetraploid ($4n$) cells, which undergo mitosis and random loss of chromosomes that returns them to the diploid state without meiosis occurring.

VARIATION, MICROEVOLUTION, AND SPECIATION

Variation Within Fungal Species in Nature and the Laboratory

As with plants and animals, including humans, there is a great deal of variation within fungal species. Such variation can affect all aspects of the biology of a fungus. Strains isolated from nature often differ in morphology and physiology. *Aspergillus nidulans* strains, for example, differ in colony growth form, in the amount of sexual and asexual sporulation, in extension rate and quantity of metabolic products. Many different VC groups occur in *Aspergillus nidulans* and in other ascomycetes (pp. 102–103). Natural populations of basidiomycetes (e.g. *Schizophyllum commune*) have many different mating type alleles (pp. 114–118). Strains of plant pathogens vary in their ability to attack different strains/varieties of their host species (Chapter 8), which is of considerable practical importance in plant breeding and agriculture. Strains of a pathogen species that can be distinguished from other members of the same species by their ability to be pathogenic on a specific host species, host variety, or group of hosts (i.e. have the same combination of virulence genes), are termed variously **pathotype**, **biotype**, or **race**.

Strains may also differ in chromosome number. In some species, there are strains that are polyploid, with haploid chromosome numbers that are a multiple (e.g. 2X, 4X) of the basic haploid number (X) (e.g., in the chytrid *Allomyces*, the yeast *Saccharomyces* and in most fungal groups). Alternatively, a strain may be aneuploid, with one or a few chromosomes duplicated or lost (e.g. X+1, X−1, X−2) (e.g. in *Neurospora crassa* and *Aspergillus nidulans*). Also, many pathogenic fungi possess complements of accessory (also called dispensible or 'B') chromosomes, enriched in gene duplications, and transposable elements (TEs) (pp. 132–134), that may play a role in pathogenicity or fungicide resistance. *Nectria haematococca* accessory chromosome 14, for example, has a cluster of genes that encode detoxification of a phytoalexin (p. 256).

Genetic variation may be continuous or discontinuous. **Discontinuous variation** refers to the situation where a feature is either present or absent. A dispensable chromosome, for example, is either present or lacking, and a strain either belongs to a particular mating type or it does not. Genes of plant pathogens encoding avirulence or host-specific toxins often exhibit presence/absence polymorphisms. Discontinuous variation occurs where a single character has a decisive effect. **Continuous variation** refers to the situation where a feature varies in magnitude between strains. Thus, strains may show a wide range of growth rates, some differing only slightly, but with a large difference between the fastest and slowest growing strains. Continuous variation is seen where the magnitude of a feature is influenced by many genes, each with a relatively small effect. However, it has recently been shown that sometimes different alleles can provide a continuous (quantitative) phenotype.

In the laboratory, even the same isolate of some species (e.g. *Fusarium*) are highly variable in culture, morphological changes occurring with subculture. Others (e.g. *Aspergillus*) are regarded as stable. So some species exhibit greater phenotypic plasticity than others. Phenotypic differences can reflect the drastically different conditions in culture compared with nature. Many *Fusarium* species, for instance, live in very dilute solutions in the water-conducting vessels of plants, but in laboratory culture, nutrient concentrations are typically much higher. Hence, with *Fusarium* there may be intense selection favouring any variant capable of coping with the unnatural conditions in culture more efficiently. Subculture by means of pieces of agar bearing mycelium can propagate mutated cells rather than the original. Also, as it can give a culture arising from the hyphae present rather than spores, the ability to sporulate is sometimes rapidly lost, as there is then no selection against mutants defective in sporulation. Plant pathogenic fungi often lose the ability to sporulate and lose virulence in culture, but the loss in phenotype can often be traced back to methylation (epigenetic modification – see pp. 134–135) rather than to mutations. Thus both prolonged growth in artificial conditions and repeated subculture can lead to genetic or epigenetic change. It is thus desirable to maintain stocks under conditions that permit prolonged preservation in a dormant state, such as liquid nitrogen refrigeration, or to use freshly isolated material.

Microevolution

Variation in fungal populations has been discussed above. Here we consider how this variation is generated, and how it changes over time. The main sources of genetic variability result from mutation, recombination, transposons, and horizontal gene transfer (HGT), all of which alter the composition of an organism's DNA, which encodes most of an organism's heritable characteristics. However, other mechanisms are also responsible for heritable phenotype; epigenetic changes alter phenotype by causing changes to DNA, but not to the nucleotide sequences.

Over time, allele frequencies at or below the species level change – a process termed **microevolution**. An allele may diminish in frequency, and even disappear from a population. On the other hand, it could increase in frequency and sometimes replace all alternative alleles in a process called a **selective sweep**. These changes in allele frequencies can occur due to mutation rate, selection, gene flow, and genetic drift, and are explained first below, followed by consideration of recombination, TEs, HGT, and epigenetics.

Mutation

Mutation is a permanent change to the nucleotide sequence of a genome. It can be a point mutation at a single locus, or a deletion, insertion, or rearrangement of a larger section of DNA. It only weakly affects allele frequencies, but is a major factor for introducing new alleles. Gene mutation is the ultimate source of genetic variability. Genes vary in their mutation frequency, but one mutation per million copies of a gene per generation can be taken as an average value. The genome size of different fungi varies, but taking a genome of 10,000 genes as typical, in each generation one cell in a hundred could in some respect carry a new mutation. Most of the compartments/cells in a fungal colony are capable of giving rise to further vegetative cells or to spores, so the number of cells in a fungal population that can undergo mutation and leave progeny can be very large. Furthermore, many fungi are haploid for a large part of their life cycle, so mutations will be expressed immediately, and if beneficial will spread through a population by natural selection (next section).

An example of the effectiveness of mutation in producing variation is provided by yellow rust (*Puccinia striiformis*). Most populations in Europe and North America are clonal, yet new races still emerge quickly, as a result of mutation. The same is true for several other rusts and powdery mildew fungi (Chapter 8).

Large populations often have more alleles than small populations, because there are more mutations upon which selection can operate. Also, large populations experience less genetic drift (p. 130), and are less likely to lose alleles. So with regard to plant pathogens in agroecosystems, it is important to keep population sizes low so that there are fewer mutations in avirulence genes or genes encoding fungicide resistance. With regard to breeding plants for resistance to pathogens, if two (or more) genes are introduced simultaneously into a host genotype, then the pathogen will need two (or more) simultaneous mutations from avirulence to virulence to overcome resistance to infection. If the mutation rate is 10^{-6}, then the probability of two genes mutating simultaneously would be 10^{-12}, and the probability of three genes mutating 10^{-18}. The mildew *Blumeria graminis* f. sp. *hordei*, produces around 10^{13} spores ha^{-1} d^{-1}, so about 10 double mutants would be produced each day in 1 ha of barley, and one triple mutant in 10^5 ha. Often 10^6 ha of crop are planted, but since the vast majority of spores fail to land in a suitable environment, the rare mutants have little chance to establish. Hence plant breeders are now keen to employ **resistance gene pyramids** composed of novel genes that have not yet been defeated.

In summary, by creating new alleles, mutation is an important first step of evolution. Recombination (pp. 130–132) can also create new alleles by intragenic recombination, seen for example, in genes encoding fungicide resistance and host specific toxins in some plant pathogenic fungi. Over an extremely long time, mutation can result in significant changes in allele frequencies. However, if mutation was the only factor acting on populations, then the rate of evolution would often not be observable.

Natural Selection

Darwin was impressed by the success of plant and animal breeders in producing, by means of **artificial selection**, new varieties with features that were considered desirable. This involves selecting individuals showing the required features to a greater extent than other individuals, and breeding from them, the process being repeated over many generations. Disease-resistant crop plants are obtained in this way, by breeding from the most resistant plants. Darwin also realised that in natural populations many individuals did not survive to reproduce. He proposed that any features that favoured survival would be **naturally selected**, and tend to spread through a population, ultimately leading to evolutionary change. Natural selection of favourable variations is now generally accepted as being the major, and probably the only, basis for the acquisition of features that result in a population becoming better adapted to its environment, the process of **adaptive evolution**.

Natural selection is largely responsible for adaptive genetic changes in a population, including stabilising, directional (in which variation decreases) and disruptive changes (in which variation increases) (Figure 4.10). Most commonly it is a conservative force, eliminating deviants and maintaining the status quo. This is because a population is normally well adapted to the environment in which it occurs, and genetic change, whether from mutation, recombination, or gene flow, is more likely to result in a decrease rather than in an increase in fitness. Natural selection acting to eliminate variants that differ markedly from the average is termed **stabilising selection**. For example, heterokaryotic isolates of the basidiomycete *Schizophyllum commune* obtained from fruit bodies, and thus representing mycelium that has survived natural selection, have a limited range of radial extension rates. Heterokaryons produced in the laboratory by mating homokaryons have a much wider range of extension rates but with approximately the same mean. Hence in the natural population, individuals with extension rates much lower or much higher than the mean have been eliminated by stabilising selection. Hybridisation and artificial selection can give variants differing widely from the forms occurring in nature, indicating the widespread operation of stabilising selection.

Natural selection can also act to produce change – termed **directional selection**, both to characters showing continuous variation and those determined by single genes. Directional selection operating on a continuously varying feature will favour individuals at one extreme (Figure 4.10). Acting on features determined by single genes, it will either diminish or increase the frequency of an allele, in extreme instances bringing about the extinction of the allele or causing it to be 'driven to fixation', completely replacing alternative alleles. The appearance in the course of a few decades of several pathotypes in an Australian population of *Puccinia graminis* f. sp. *tritici* of clonal origin, is an example of directional selection occurring. Mutation pressure could not have affected the spread of the pathotype mutants in so short a time. In the higher termite/*Termitomyces* mutualism, the termites feed on nutrient-rich fungus nodules – immature fruit bodies containing asexual spores, some of which survive passage through the gut and act as inocula when new food is brought into the nest (p. 332). Since only a few spores successfully form mycelia, this leads to reduced variation within a nest. Since genotypes that

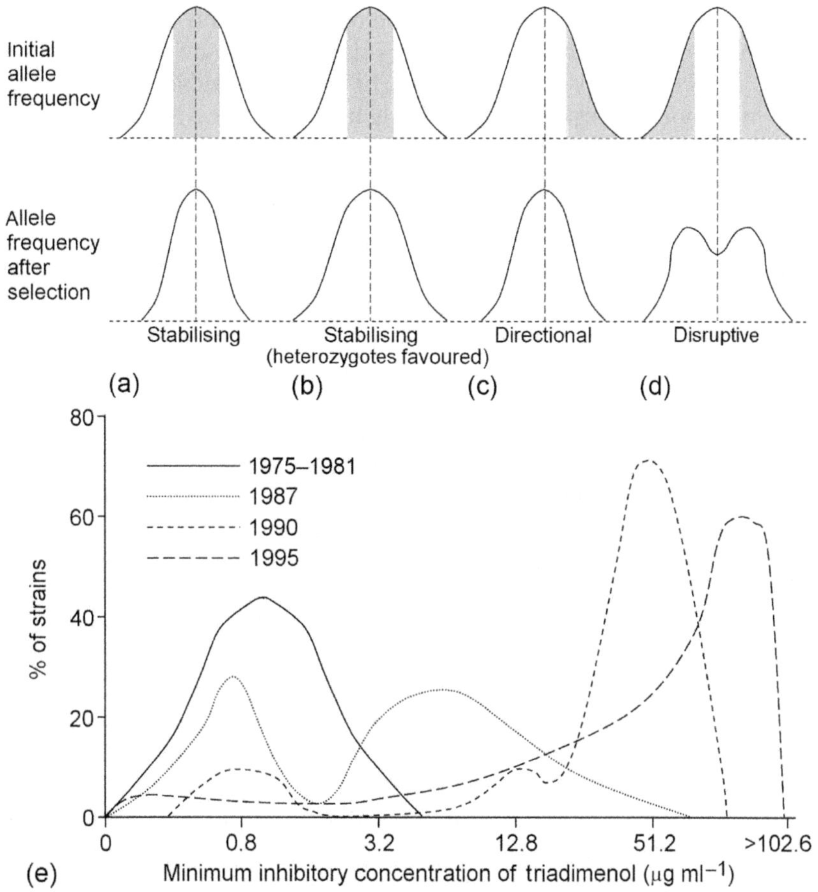

FIGURE 4.10 Natural selection. Selection can result in (a, b) stabilising allele frequencies in populations or in (c) directional or (d) disruptive changes. The upper row indicates the initial allele distribution, with the stippled areas representing favoured alleles. The bottom row shows the distribution of allele frequency following one or a few generations of selection. (e) Directional selection in loss of sensitivity of *Rhynchosporium secalis* populations, on barley in the UK, to the fungicide Propiconazole. The mean minimum inhibitory concentration increased by about 10-fold in 3 years. *Source: Brent, K.J., Hollomon, D.W., 1998. Fungicide resistance: the assessment of risk. FRAC Monogr. 2, 1–53.*

produce many nodules are most likely to be inoculated onto fresh organic matter, the termites effectively select for high nodule production. If a mutant arises that produces few nodules, it will be underrepresented in inoculation of fresh organic matter and hence 'automatically' selected against.

Disruptive (or **diversifying**) **selection** favours individuals at the two extremes of a spectrum of variability and acts against intermediate forms. It occurs where there are two possible environments for exploitation, and success results from being highly effective in one rather than moderately effective in both (Figure 4.10). It accounts for the origin within a pathogenic species of forms specialised for different hosts, such as *Puccinia graminis* f. sp. *tritici*, which attacks wheat, and *Puccinia graminis* f. sp. *avenae* which attacks oats. The two forms can still hybridise and retain a limited ability to infect the other form's host, implying a relatively recent origin by disruptive selection from a less specialised common ancestor.

Though natural selection may drive an allele to fixation or extinction, it may act to maintain a balance with two or more alleles occurring at fairly high frequency. Natural selection operating in this way is referred to as **balancing selection**. It can occur where, in the diploid or dikaryotic phase, the heterozygote is fitter than either homozygote (heterozygous advantage – p. 110). Balancing selection can also take the form of **frequency dependent selection**, in which selection increases the frequency of an allele if it is rare and decreases its frequency if it is abundant. A rare mating type allele, for example, will increase in frequency, since individuals that possess it will be able to mate with most other individuals. If, however, the allele becomes abundant, then individuals possessing it will be very likely to encounter individuals of identical mating type, with resulting failure of mating. Frequency dependent selection will also occur with the gene-for-gene relationship that determines pathogenicity (p. 256). A rare virulence allele will increase in frequency since there will have been little selection for host resistance. If, however, it becomes abundant, then the corresponding gene for host resistance will also become abundant, and selection will operate against individuals carrying the now nearly useless virulence allele.

Natural selection may act differently in haploid and diploid/heterokaryotic fungi. In haploid fungi, a gene will be expressed either throughout or at some stage in the life cycle, and natural selection can operate to influence the frequency of its alleles. In diploid fungi, a recessive allele will have no effect on phenotype, and thus will be partly sheltered from natural selection; if deleterious, it will decrease in frequency more slowly than it would if partly or completely dominant or present in a haploid fungus. Its extinction is likely to be long delayed, since when low frequencies are reached it will rarely be present in the homozygous state in which it would be exposed to selection. In heterokaryotic basidiomycetes, the arguments for haploid fungi are relevant to the homokaryon phase, and the arguments for diploid fungi are relevant to the heterokaryon phase. Hence, fungi that are diploid or heterokaryotic through most of their life cycles are likely to carry many currently deleterious recessive alleles which, in a different genome or environment, might prove advantageous.

Gene Flow

For the geneticist a population consists of individuals that readily exchange genes. The transfer of genes between populations that are spatially separated, and are hence less likely to exchange genes, is termed **gene flow**. Extent of gene flow can be ascertained by examining allele frequencies within and across populations. Potential gene flow can be assessed by quantifying spore recruitment by spore trapping. Species specific spore trapping methods, such as using a homokaryotic mycelium as bait for basidiospores of compatible mating type (pp. 114–118), usually show that spores reflect the local established population, and are rarely found > 20 km from known locations. Nonetheless, spores sometimes travel long distances. For example, the rare basidiomycete *Peniophora aurantiaca* was detected in Göteberg, Sweden, over 1000 km from its closest known occurrence. Also, there are examples of some plant pathogens that have moved from one continent to another as airborne spores (pp. 263–264).

The frequency of alleles in a population can be affected by gene flow from other populations (i.e. migration). The magnitude of the effect will depend on the numbers of immigrants compared with numbers in the native population, and the extent to which the immigrants differ from the natives in allele frequency. It is likely that allele frequencies in small populations adjacent to large ones will be influenced strongly by gene flow. Between distant populations, gene flow is likely to be sporadic, but may be facilitated by intervening populations acting as stepping stones. The effect of gene flow will be to reduce genetic differences between populations, and hence may delay or prevent the divergence of populations in different geographical areas into separate species. A level of gene flow as low as one migrant per generation is sufficient to prevent population divergence. This is true even in large populations as although one migrant has a diluted effect, genetic drift (p. 130) is also smaller.

The presence, extent, or absence of gene flow varies amongst fungi. With *Neurospora crassa*, individuals collected from what was thought to be one population spanning the Caribbean Islands to the southern United States turned out to be at least two distinct populations, which diverged approximately 0.4 mya but are not yet fully differentiated into species, still sharing 90% Single nucleotide polymorphisms (SNPs). Genomic 'islands' of differentiation include genes related to temperature and circadian rhythm. The subtropical Louisiana population has a higher fitness at low temperature (10 °C), with several of the genes within the distinct genomic islands having functions related to response to low temperature. Southern Louisiana has an average yearly minimum temperature 9 °C lower than the Caribbean, so divergence is probably due to local adaptation to temperature. Another difference was in the circadian oscillator frequency gene, which suggests that the difference in latitude may be another important environmental factor differentiating the populations. The populations are not completely isolated, and there is still gene flow, but the presence of these genomic islands of differentiation, implies the environment exerts a strong enough selection pressure to keep the two populations differentiated despite the ongoing gene flow. In contrast to *Neurospora crassa*, genetic differences in the cosmopolitan basidiomycete *Schizophyllum commune* are mainly in allele frequencies, with genetic distance correlating with geographic distance and a lack of sharp boundaries between populations.

Gene flow by long-distance spore dispersal is implied, with even a limited intercontinental gene flow. Genetic differences between populations are not always correlated with geographic distance, but with the distance that a fungus has 'travelled' to get to a new location. The distance travelled is not always equivalent to the geographical distance, for example, if there is a geographical feature that has been circumvented rather than crossed over, the travelled distance is greater than the geographical distance.

Gene flow can be important even over long distances. Almost every naturally occurring *Neurospora* colony that is sampled is genetically distinct, indicating that the colonies have arisen from single ascospores, the products of meiosis, and not from conidia. It is hence reasonable to infer that the main role of conidia, which compared with ascospores are produced in massive amounts, is to fertilise protoperithecia. Studies on the mitochondrial DNA of *Neurospora crassa* show regional differences. Studies on the nuclear genome, however, show that regional diversity is no greater than that within populations. It seems, therefore, that gene flow resulting from the dispersal of conidia has helped maintain homogeneity of the nuclear genome throughout the species. In fertilisation, however, the conidia do not contribute mitochondria, allowing geographically distinct populations of mitochondrial DNA to evolve.

The wheat stem rust provides an example of the potential importance of gene flow as a source of genetic novelty in a population. Major changes occurred in 1954 in the ability of *Puccinia graminis* sp. *tritici* in Australia to attack different host varieties. Study of rust collections, maintained from that period, indicates that a change in isozyme pattern also occurred.

The change was due to a pathogen strain hitherto lacking in Australia but found in African populations. The implication was that an introduction from Africa occurred, either by long-distance spore dispersal or inadvertently by humans – the latter possibility being an increasing hazard.

Genetic Drift and the Founder Effect

Genetic drift is the change in frequencies of alleles in a population due to chance. If a population is small then chance could determine whether a neutral allele becomes extinct or increases in frequency to fixation. If a population is very small then such random genetic drift could determine the fate of an allele even in the presence of moderately strong natural selection. In nature, however, it may be unusual for a population to stay small long enough for drift to occur – the population could become extinct, grow, or merge with another population. Tendencies to genetic drift will be opposed by gene flow. Hence if a fungus is abundant and widespread with copious spores capable of long distance dispersal, gene flow is likely to counteract any tendency to genetic drift. There is evidence for this in the cosmopolitan and abundant fungi *Neurospora crassa*, *Puccinia graminis* f. sp. *tritici*, and *Schizophyllum commune*.

There are, however, ways in which random events could determine the genetic structure of a population and the course of microevolution. One or a few individuals will not cover the genetic diversity in the population; many alleles present in the whole population will be absent from such a small set of individuals. A small set of individuals could occur as the result of a catastrophe almost destroying a population or by the dispersal of one or a few individuals to a new environment. The population resulting from such a **founder effect** will be genetically different from the one from which it originated. Many fungi live in environments that are highly favourable but transient, and will hence be liable to colonisation from one or a few spores when they arise, and population crashes when they disappear. Founder effects are likely to occur with such fungi and, if the fungi are not highly abundant, may not subsequently be overwhelmed by gene flow.

Australia provides lots of examples of single founder events: *Puccinia striiformis* – cause of yellow (stripe) rust of wheat – was introduced into Australia in 1979 (p. 263), as a single race from Europe, but mutations have now resulted in new pathotypes which differ from those in Europe. Similarly, *Cryphonectria parasitica* – cause of chestnut blight (pp. 287–289) – in North America has much less genetic diversity than in Asia, probably reflecting a founder effect. The dry-rot fungus, *Serpula lacrymans*, originated in northeast Asia, where it has most genetic variations. However, there is very little genetic variation in the founder populations across the globe (Figure 4.11). In some areas the indoor genetic populations are unique (e.g. Japan), representing a single founder event, whereas elsewhere (e.g. Australia), there is slightly more variation representing founder events from Japan and from Europe.

Recombination

Although mutation can result in new genotypes that become established in a population, such genotypes may, as indicated above (pp. 125–126), differ in only a few ways from the strains from which they were derived. On the other hand, the mating of two strains that differ at many loci will result, through recombination at meiosis, in a wide variety of new genotypes. Continuing with the example, from a previous section (p. 125), of *Blumeria graminis* f. sp. *hordei*, infecting three fields with different barley cultivars, one with resistance gene *R1*,

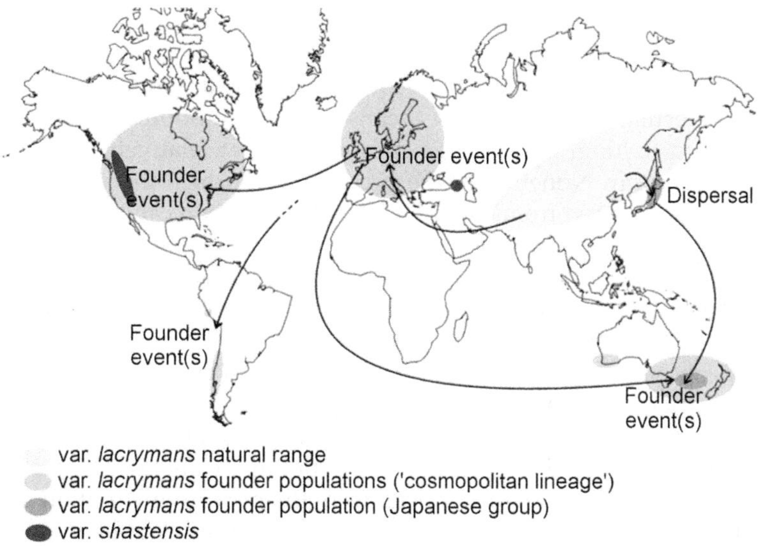

FIGURE 4.11 Worldwide spread of the dry rot fungus, *Serpula lacrymans*, from its origins in northeast Asia. The Japanese indoor population represents a single founder event. From there it was spread to southeast Australia and New Zealand. A genetically highly homogeneous population is present in Europe, the Americas and Australia, and New Zealand. It was probably first spread to Europe from Asia in infected wood, and from there to the other areas in a similar way, perhaps in wooden ships. *Source: Kauserud, H., Knudsen, H., Högberg, N., Skrede, I., 2012. Evolutionary origin, worldwide dispersal, and population genetics of the dry rot fungus Serpula lacrymans. Fungal Biol. Rev. 26, 84–93.*

a second with *R2*, and a third with no resistance, if two mutants for virulence *vr1* and *vr2* arise separately in the non-resistant cultivar, those spores with *vr1* could successfully establish in the *R1* cultivar, and those with *vr2* in the *R2* cultivar. After epidemics in *R1* and *R2*, the spores could move back into the non-resistant crop, where sexual reproduction between the two could occur. If the genes are unlinked, recombination would result in 25% of the progeny having both *vr1* and *vr2* alleles, which would allow infection of cultivars with both *R1* and *R2* genes.

A classic example of genetic effects of the sexual process in natural populations is provided by early studies on the amount of variability of *Puccinia graminis* f. sp. *tritici* populations east and west of the Rocky Mountains in the United States. The sexual stage in the rust's life cycle occurs on barberry (*Berberis*). This was eradicated east of the Rocky Mountains in the late 1930s, preventing sexual recombination in the eastern populations since that time. West of the Rocky Mountains barberry, and hence sexual recombination in rust populations, persist. The eastern rust populations have relatively few pathotypes and isozyme patterns, and each pathotype is associated with a single isozyme pattern, different from the isozyme patterns associated with other pathotypes, showing that genetic recombination has not occurred. West of the Rocky Mountains pathotypes and isozyme variants are numerous, and the association between pathotype and isozyme pattern is random, indicating that genetic recombination has occurred. Hence the sexual process has generated virulence and isozyme diversity, and has broken any constant associations between pathotype and isozyme pattern. Another example is provided by *Puccinia striiformis* f. sp. *tritici*. Populations in Europe and Australia are clonal with limited molecular diversity, but in China there is considerable phenotypic and genotypic diversity and evidence of genetic recombination.

Whether populations are clonal, freely recombining or with limited recombination can be established by molecular methods, or by means of classical population genetics. However, the latter, as applied to fungi in nature, is so laborious that relatively little information has been obtained in this way. Nonetheless the procedure by which the presence or absence of recombination can be inferred from data on molecular variation can be illustrated by considering genes and alleles. If two genes (*a* and *b*) each have two alleles (1 and 2), there are four possible genotypes, a_1b_1, a_1b_2, a_2b_1, and a_2b_2. If only two genotypes are found in a population, for example a_1b_1 and a_2b_2, then recombination between the genes is not occurring, due either to the genes being closely linked on the same chromosome or because mating and genetic recombination is absent in the population. If all four genotypes are equally frequent, then it is likely that unrestricted recombination is taking place. Since there is the possibility that two genes may be linked, it is in practice necessary to consider a larger number of genes, and in a natural population most genes will have not two but many alleles. The presence of recombination (i.e. signature of sex) can be deduced from the pattern of variation in a population. If all three of the following are found after clone-correcting the dataset, it is almost certainly a sexual population: (1) high genotype diversity, usually assessed with molecular genetic markers or DNA fingerprints; (2) random associations among the alleles found at individual, neutral loci; and (3) both mating types present at equal frequencies (if an ascomycete).

Such molecular studies are now providing valuable information on recombination within and between populations in nature. For example, Californian populations of the human pathogen *Coccidioides immitis* (p. 303), a fungus for which a sexual stage is unknown, has been shown to have essentially unrestricted genetic recombination. However, another human pathogen that lacks a sexual stage, *Candida albicans* (p. 302, pp. 305–307), is clonal with little recombination, which is generated via a parasexual cycle (pp. 122–123). A sexual stage, as indicated earlier in this chapter, results in recombining populations in both heterothallic and homothallic fungi, but if mitospores are also produced, clonal as well as recombining populations, are likely to arise, as in *Puccinia graminis* f. sp. *tritici* mentioned above.

Transposable Elements

TEs, also known as **transposons** or more colloquially as 'jumping genes', are DNA sequences that can move from one location in a genome to another. Indeed, TEs are sometimes spread to other species by HGT (p. 134) TEs can cause changes in genome architecture and gene function, via recombination and expansion, resulting in rearrangement of chromosomes and new gene neighbourhoods. Transposition often results in duplication of a DNA sequence and the insertion of one copy at a different site in the genome termed **Duplicative transposition**. Comparative phylogenomics has shown that gene duplication occurs frequently in fungi. It allows genes to acquire new functions by mutation (pp. 125–126), since one copy remains to maintain the original function, leaving the second copy free to mutate without compromising the organism's fitness.

As with other eukaryotes, there are two classes of TEs: class I (termed **retrotransposons**) transpose by synthesising cDNA copy based on an RNA intermediate using reverse transcriptase; class II (**DNA transposons**) transpose the DNA directly (Table 4.5). The copy numbers of fungi are much lower than in other eukaryotes, only a small proportion of the genome

TABLE 4.5 Examples of Transposons in Fungi

Type	Characteristics	Fungus	Transposon name
Class I: LTR retrotransposons	Have long terminal repeats (LTRs) flanking the polyprotein genes. They usually encode two open reading frames (ORFs). Have *Gag* (encode for viral coat proteins) and *pol* genes (encode for integrase, protease, reverse transcriptase and RNase)	*Aspergillus fumigatus*	*Afut1*
		Botrytis cinerea	*Boty*
		Fusarium oxysporum	*Foret-1, Skippy*
Class I: LINE (long interspersed nuclear elements) retrotransposons	Have poly-A-tails but no LTRs, *gag* or *pol* genes	*Fusarium oxysporum*	*Palm*
		Neurospora crassa	*Tad1-1*
Class I: SINE (short interspersed nuclear elements) retrotransposons	Derive from RNA polymerase transcripts; no special structural features; no *gag* or *pol* genes. They are transactivated (increased rate of gene expression) by reverse transcriptases provided by retrotransposons or LINE-like elements	*Erysiphe graminis* f. sp. *hordei*	*EGH24-1, Eg-R1*
		Nectria haematococca	*Nrs1*
Class II (also called DNA transposons)	*Fot1/Pogi*-like	*Aspergillus nidulans*	*F2P08*
		Botrytis cinerea	*Flipper*
		Fusarium oxysporum	*Fot1, Fot2*
	Tc1/mariner superfamily	*Aspergillus niger*	*Ant1*
		Fusarium oxysporum	*Impala*
		Phanerochaete chrysosporium	*Pce1*

Two other class I transposons have recently been introduced DIRS and Penelope-like (PLE).
Source: Kempken and Kück, 1998

being repetitive DNA (e.g. 1.1% and 7.3 %, respectively) in the plant pathogens *Ustilago maydis* and *Magnaporthe oryzae*, and 5% in the human pathogen *Cryptococcus neoformans*. Perhaps the most studied fungal transposon is *Ty* ('transposon yeast'), a retrotransposon, which occurs in *Saccharomyces cerevisiae*. A terminal portion of the *Ty* sequence is known as δ, and facilitates insertion. *Ty* acts as a retrovirus. It is transcribed into RNA and encapsidated. Part of the RNA sequence encodes a reverse transcriptase which copies the RNA into a DNA sequence which then inserts into the yeast genome. *Saccharomyces cerevisiae* commonly carries about 35 copies of *Ty-1*, one form of *Ty*, about 6 copies of *Ty-917*, a mutant form, and about 100 solo δ sequences.

The genomes of plant pathogenic fungi contain a larger fraction of TEs than saprotrophs. They often form clusters with rapidly evolving genes involved in host–pathogen interactions (e.g. host specificity genes in *Fusarium* spp., *Magnaporthe oryzae*, and *Verticillium* spp.), and oomycete *Phytophthora infestans* effector genes. In contrast, TEs are not clustered in *Blumeria graminis*, but rather are distributed throughout the genome. Evolution of new lineages able to colonise new hosts is associated with TEs in the above examples, and in the case of *Blumeria graminis* increase in TEs and genome size coincided with a change in trophic behaviour to obligate biotrophy.

The expression and mobility of TEs are mostly silenced epigenetically (below) but this may be disrupted under physiological stress, including that posed by climate change and invading new habitats/hosts, leading to rapid restructuring of the genome and altered gene expression, with associated changes in morphology and physiology, and speciation (pp. 135–137). For example, the fungus-like oomycete *Phytophthora ramorum* (the cause of sudden oak death), which comprises only a few clonal lineages in California, has different phenotypes in live oak (*Quercus agrifolia*) and bay laurel (*Umbellularia californica*) despite having identical genotypes; the physiological stress caused by colonising the oak is thought to result in disruption of epigenetic silencing of TEs. Comparison of symbiotic and non-symbiotic species of Agaricomycetes has shown that TEs are usually more common in symbiotic fungi than close relatives that are saprotrophs. Moreover, genes that have been identified as having crucial roles in the development of symbiotic tissues appear to be associated with TE's and also lack homologues in other organisms. Both of these observations suggest the possibility that TEs have played a part in the genetic changes that underlie a move from saprotrophic to biotrophic nutrition.

Horizontal Gene Transfer

Comparing whole genomes has led to greatly increased understanding of gene evolution, including the discovery of the importance of HGT in which genes have been directly transferred across fungal lineages, without sexual recombination. Not only has there been transfer between fungal species, but genes have also been acquired from bacteria and plants. Examples are given in Chapter 6 (p. 201).

Epigenetics

DNA sequences encode most of an organism's heritable characteristics, but other mechanisms are also responsible for heritable phenotype. Functionally relevant modifications to the genome can occur which do not involve changes to the nucleotide sequences – termed epigenetic. These changes are reversible and inheritable. Epigenetic phenomena include changes in DNA methylation, histone modification, centromere location, and RNA silencing systems.

DNA methylation occurs in many fungi (e.g. *Neurospora crassa*), but not all (e.g. *Aspergillus nidulans*). Most filamentous fungi have DNA methyl transferases (DMTs) which form 5-methylcytosine by methylating C5 of cytosine. DNA methylation is a genome defence that blocks transcription of TEs (pp. 132–134) and other 'selfish DNA' – gene silencing. Other genes can be silenced too. Repeat induced point mutation (RIP) is a process that has evolved uniquely in fungi as a defence against TEs. During the RIP process transposons, and other duplicated sequences, are identified and point mutations are introduced by conversion of C:G pairs to T:A pairs in both sequence copies. These mutations, and sometimes also additional DNA methylation, lead to gene silencing. RIP was seen first in *Neurospora crassa*, and was the first eukaryotic gene silencing mechanism discovered, but has now been demonstrated to occur in many other fungi (e.g. *Aspergillus nidulans*, *Aspergillus oryzae*, *Colletotrichum cereale*, *Ophiostoma ulmi*, and *Penicillium chrysogenum*).

While the effects mentioned so far relate to DNA, modification of core histones (including H2A, H2B, H3, and H4) and replacement of canonical histones with variants can occur after translation, resulting in epigenetic phenomena. Gene silencing can also result from RNA-based silencing mechanisms: small interfering RNA (siRNA) can alter expression from invading viruses or selfish elements; the plant pathogen *Botrytis cinerea* produces siRNA that silences genes in host plants allowing it to colonise (pp. 274–276); normal gene expression can be regulated by micro RNA (miRNA) in dedicated pathways.

Speciation

Speciation is the situation where new species arise from preexisting species. It is more or less impossible to know exactly when a new species has arisen, not least because there is a lack of agreement about how different two taxa must be to qualify as being different species. There is, however, abundant evidence of speciation in progress or recently accomplished. This evidence comes from the comparison of closely similar taxa, and from their attempted or successful hybridisation.

Speciation is, most commonly, a splitting of an existing species into two, although one of these species may subsequently replace the other. For example, *Rhynchosporium* on barley (*Hordeum vulgare*; causing 'scald'), rye (*Secale cereale*) and wild *Agropyron* species is actually three different species that evolved independently and diverged by host specialisation. Each has a unique haplotype (based on sequences of four independent loci and pathogenicity tests). There is no gene flow between populations on rye and barley (evidenced by population genetic data).

New species can arise in different ways: through adaptation, genomic change, or hybridisation with other species. Three main modes of speciation are recognised for eukaryotes: (1) **allopatric** or **geographical** speciation which occurs where populations have become geographically isolated. Genetic differences accumulate in the isolated populations, and these eventually become reproductively isolated. (2) **Sympatric** speciation where gene flow between populations in the same area ceases. (3) **Abrupt** speciation which arises mainly when polyploidy or chromosomal rearrangement causes reproductive isolation. These categories are not quite as useful for fungi as they might superficially appear, as although fungi can be distributed widely they occupy microhabitats; significant separation might be as little as different locations within the same fallen tree trunk or, for a rust fungus, wheat (*Triticum*) and oak (*Quercus*) plants in the same area. Nonetheless there are many examples, revealed by multiple gene genealogy, of cryptic species that have non-overlapping geographical ranges, suggesting that they have evolved by allopatric speciation. For example, the morphological species *Neurospora crassa* has one phylogenetic species located in the Congo (Africa), another in the rest of Africa (but not Congo) and the Caribbean, and a third in India. There are at least nine species within the morphological species *Fusarium graminearum*, a cause of scab on wheat and barley, four endemic to South America, one in Central America, one in Australia and one in India. Providing evidence of sympatric speciation is also difficult as it is almost always impossible to exclude a past period of allopatry.

Given the above difficulties, we will just discuss factors that predispose to speciation and factors that result in reproductive isolation. Though reproductive isolation is the criterion upon which the biological species concept is based (p. 107), it is just one of the many stages of speciation. It can occur at early or late stages in the speciation process, and can be a critical stage, as in sympatric speciation, or simply a by-product of genetic divergence, as in allopatric speciation. It is clearly not essential to all modes of speciation as asexual fungi can also speciate.

Speciation in Fungi with a Sexual Phase

Genetic divergence can occur by non-selective means such as drift (p. 130), or when the selective forces acting on the two populations differ. Gene flow (pp. 128–130) between populations will oppose and limit genetic divergence, but gene flow will be of little importance if the two populations are distant from each other and dispersal is inefficient. Then, increasing divergence is likely to result in **genetic disharmony** (p. 121) between the populations, which may affect the mechanisms involved in mating, so that if the two populations come into contact again, mating may fail, or if progeny do result, they may be few or of low viability.

Fungal populations may come to differ from each other in chromosome number, through **polyploidy or aneuploidy** (p. 124). Differences in chromosome number between strains usually reduces the chances of hybridisation and genetic exchange, because of chromosome pairing difficulties at meiosis, though some fungi (e.g. *Zymoseptoria tritici*) have a high level of sexual reproduction among strains that have highly imbalanced numbers of chromosomes. Thus, differences in chromosome number will often increase the likelihood of strains diverging into separate species. On the other hand, many species are not completely intersterile, allowing the possibility of hybridisation. If **hybridisation** is successful, the resulting hybrid may differ from both parental types in morphology and chromosome number and may also prove to be competitive. Chromosome counts and hybridisation experiments indicate that the chytridiomycete *Allomyces javanicus* originated as a hybrid between two species with considerable morphological differences, *Allomyces arbuscula* and *Allomyces macrogynus*. Many emerging plant pathogens are hybrids (pp. 285–290). The rust *Melampsora* × *columbiana* emerged in 1997 from hybridisation between *Melampsora medusa* and *Melampsora occidentalis*, which are respectively pathogens of *Populus deltoides* and *Populus trichocarpa*. It was a homoploid (i.e. had the same number of chromosomes as its parents). Allopolyploids (i.e. with different numbers of chromosomes to the parents) are instantly reproductively isolated, but homoploids are not. *Melampsora* × *columbiana* emerged as a new species pathogenic on a *Populus* hybrid resistant to the two parents.

Chromosomal rearrangements can theoretically result in speciation – **chromosomal speciation**. However, showing that rearrangements are a cause rather than a consequence of speciation is difficult. Until recently it has been almost impossible to make detailed analyses of fungal chromosomes, but population genomics projects are likely to expand understanding of this area dramatically in the coming decade.

Reproductive isolation which, as already mentioned, can be a cause or consequence of speciation, can result from two types of barriers – pre and postmating. Premating isolation can result from several different barriers: (1) specialisation for a certain host; (2) specialisation of biotic vectors can prevent contact between populations (e.g. *Botanophila* flies preferentially select the endophyte *Epichloë typhina* rather than *Epichloë clarkia*); (3) differences in time of reproduction; (4) a high rate of selfing (e.g. in the anther smut *Microbotryum violaceum*); (5) assortative mating where individuals are able to discriminate between the same and different species, especially in Agaricomycetes. Postmating isolation barriers result from sterility or non-viability of progeny derived from matings between species (e.g. crosses between *Neurospora* species result in abnormal or only a few progeny, or only a few viable ascospores). Epigenetic mechanisms (pp. 134–135) may also contribute to postmating isolation.

The inability of two populations to mate does not necessarily entirely prevent gene flow between them, as both may be able to mate with a third population. For example, Californian isolates of the basidiomycete *Serpula himantioides* are unable to mate with Swedish isolates, but both are able to mate with Indian isolates. Where populations adapted to different habitats are in close proximity, mating between them could result in progeny ill-adapted to either environment, or, in progeny of low viability due to genetic disharmony, if considerable divergence has already occurred. Under such circumstances there will be selection for the evolution of mating barriers to prevent the production of progeny of low fitness. One or a few mutations can suffice to bring about such **heterogenic incompatibility**. It is hence likely that barriers to mating can arise both fortuitously and by natural selection. Gene flow can also be limited, and divergence promoted, by self-fertility (pp. 119–120) and amixis (p. 122).

Speciation in Asexual Fungi

Speciation in asexual fungi, if there is such a thing as a truly asexual fungus (p. 114), will occur in ways similar to that in acknowledged sexual species. However, there is no sexual recombination to break down successful combinations of multiple alleles. The selective pressure on one gene will have an effect on the whole genome. The human pathogen *Penicillium marneffei* (p. 308) can spread long distances by aerially dispersed spores, however, it is highly endemic. DNA multilocus typing has shown that different clones are found in different environments, implying that adaptation to these environments constrains the fungus' ability to disperse successfully. Similarly, many species in the plant pathogenic *Magnaporthe grisea* complex are host specific, with different recently evolved lineages in cutgrass, millet, rice, and torpedo grass. The rice pathogen *Magnaporthe oryzae* (pp. 265–267) arose recently (about 7000 years ago) probably when rice was domesticated.

However, molecular methods have shown extensive genetic recombination in a few apparently asexual fungi, such as *Coccidioides immitis* (p. 303), though this could be due to hitherto undetected mating. Nonetheless, in fully asexual species genetic recombination can occur via the parasexual cycle (if it is not simply a laboratory artefact – pp. 122–123), but will be limited by vegetative incompatibility. Hence a mitosporic fungal species defined on traditional morphological criteria is likely to consist of a number of biological species, with some clonal and some showing genetic recombination, as has been found in the banana pathogen *Fusarium oxysporum* f. sp. *cubense*. It seems likely that in mitosporic fungi, when genetic exchange cannot occur due to vegetative incompatibility, the resulting clones constitute biological species, and through subsequent evolution may develop into morphologically distinguishable species. The different VC groups are genetically distinct lineages, for example in *Aspergillus flavus*.

Further Reading

General

Burnett, J., 2003. Fungal Populations and Species. Oxford University Press, Oxford.

Gladieux, P., Ropars, J., Badouin, H., Branca, A., Aguileta, G., de Vienne, D.M., Rodriguez de la Vega, R.C., Branco, S., Giraud, T., 2014. Fungal evolutionary genomics provides insight into the mechanisms of adaptive divergence in eukaryotes. Mol. Ecol. 23 (4), 753–773.

Moore, D., Ann, L., Frazer, N., 2002. Essential Fungal Genetics. Springer, New York.

Worrall, J.J. (Ed.), 1999a. Structure and Dynamics of Fungal Populations. Kluwer Academic Publisher, Dordrecht.

Defining Individuals, Populations, and Species

Glass, N.L., Dementhon, K., 2006. Non-self recognition and programmed cell death in filamentous fungi. Curr. Opin. Microbiol. 9, 553–558.

Glass, L.N., Kaneko, I., 2003. Fatal attraction: nonself recognition and heterokaryon incompatibility in filamentous fungi. Eukaryot. Cell 2, 1–8.

James, T.Y., Moncalvo, J.M., Li, S., Vilgalys, R., 2001. Polymorphism at the ribosomal DNA spacers and its relation to breeding structure of the widespread mushroom *Schizophyllum commune*. Genetics 157, 149–161.

Malik, M., Vilgalys, R., 1999. Somatic incompatibility in fungi. In: Worrall, J.J. (Ed.), Structure and Dynamics of Fungal Populations. Kluwer Academic Publishers, Dordrecht, pp. 123–138.

Read, N.D., Goryachev, A.B., Lichius, A., 2012. The mechanistic basis of self-fusion between conidial anastomosis tubes during fungal colony initiation. Fungal Biol. Rev. 26, 1–11.

Read, N.D., Roca, M.G., 2006. Vegetative hyphal fusion in filamentous fungi. In: Baluska, F., Volkmann, D., Barlow, P.W. (Eds.), Cell-Cell Channels. Landes Bioscience, Georgetown, Texas, pp. 87–98.

Stenlid, J., 2008. Population biology of forest decomposer basidiomycetes. In: Boddy, L., Frankland, J.C., van West, P. (Eds.), Ecology of Saprotrophic Basidiomycetes. Elsevier, Amsterdam, pp. 105–122.

Taylor, J.W., Turner, E., Townsend, J.P., Dettman, J.R., Jacobson, D., 2006. Eukaryotic microbes, species recognition and the geographic limits of species: examples from the kingdom Fungi. Philos. Trans. R Soc. Lond. B Biol. Sci. 361, 1947–1963.

Worrall, J.J., 1999b. Fungal demography: mushrooming populations. In: Worrall, J.J. (Ed.), Structure and Dynamics of Fungal Populations. Kluwer Academic Publishers, Dordrecht, pp. 175–194.

Life Cycles and the Sexual Process

Heitman, J., Sun, S., James, T.Y., 2013. Evolution of fungal sexual reproduction. Mycologia 105, 1–27.

Jones, S.K., Bennett, R.J., 2011. Fungal mating pheromones: choreographing the dating game. Fungal Genet. Biol. 48, 668–676.

Lee, S.C., Ni, M., Li, W., Shertz, C., Heitman, J., 2010. The evolution of sex: a perspective from the Fungal Kingdom. Microbiol. Mol. Biol. Rev. 74, 298–340.

Raudaskoski, M., Kothe, E., 2010. Basidiomycete mating type genes and pheromone signalling. Eukaryot. Cell 9, 847–859.

Riley, R., Corradi, N., 2013. Searching for clues of sexual reproduction in genomes of arbuscular mycorrhizal fungi. Fungal Ecol. 6, 44–49.

Taylor, J.W., Hann-Soden, C., Branco, S., Sylvain, I., Ellison, C.E., 2015. Clonal reproduction in fungi. PNAS 112, 8901–8908.

Wilson, A.M., Wilken, P.M., vander Nest, M.A., Steenkamp, E.T., Wingfield, M.J., Wingfield, B., 2015. Homothallism: an umbrella term for describing diverse sexual behaviors. IMA Fungus, 6, 207–214.

Sources of Variability and Evolution in a Population

Croll, D., McDonald, B.A., 2012. The accessory genome as a cradle for adaptive evolution in pathogens. PLoS Pathog. 8. e1001608.

Daboussi, M.J., Capy, P., 2003. Transposable elements in filamentous fungi. Annu. Rev. Microbiol. 57, 275–299.

Helgason, T., Fitter, A.H., 2009. Natural selection and evolutionary ecology of the arbuscular mycorrhizal fungi (Phylum Glomeromycota). J. Exp. Bot. 60, 2465–2480.

Kasuga, T., Kozanitas, M., Bui, M., Hüberli, D., Rizzo, D.M., Garbelotto, M., 2012. Phenotypic diversification is associated with host-induced transposon derepression in the sudden oak death pathogen *Phytophthora ramorum*. PLoS One 7. e34728.

Kempken, F., Kück, U., 1998. Transposons in filamentous fungi – facts and perspectives. Bioessays 20, 652–659.

Martienssen, R.A., Colot, V., 2001. DNA methylation and epigenetic inheritance in plants and filamentous fungi. Science 293, 1070–1073.

McDonald, B.A., 2004. Population genetics of plant pathogens. The Plant Health Instructor. http://dx.doi.org/10.1094/PHI-A-2004-0524-01

Muszewska, A., Hoffman-Sommer, M., Grynberg, M., 2011. LTR retrotransposons in fungi. PLoS One 6 (12). e29425.

Novikova, O., Fet, V., Blinov, A., 2009. Non-LTR retrotransposons in fungi. Funct. Integr. Genomics 9, 27–42.

Richards, T.A., 2011. Genome evolution: horizontal movements in the fungi. Curr. Biol. 21, R166–R168.

Richards, T.A., Talbot, N.J., 2013. Horizontal gene transfer in osmotrophs: playing with public goods. Nat. Rev. Microbiol. 11, 720–727.

Smith, K.M., Phatale, P.A., Bredeweg, E.L., Connolly, L.R., Pomraning, K.R., Freitag, M., 2012. Epigenetics of filamentous fungi. In: Meyers, R.A. (Ed.), Encyclopedia of Molecular Cell Biology and Molecular Medicine. Wiley-VCH Verlag GmbH, Weinheim, Germany, pp. 1063–1107. http://dx.doi.org/10.1002/3527600906.mcb.201100035

Stukenbrock, E.H., Croll, D., 2014. The evolving fungal genome. Fungal Biol. Rev. 28, 1–12.

Stukenbrock, E.H., McDonald, B.A., 2008. The origins of plant pathogens in agro-ecosystems. Annu. Rev. Phytopathol. 46, 75–100.

Weiberg, A., Wang, M., Lin, F.-M., Zhao, H., Zhang, Z., Kaloshian, I., Huang, H.-D., Jin, H., 2013. Fungal small RNAs suppress plant immunity by hijacking host RNA interference pathways. Science 342, 118–123.

Speciation

Albertin, G., Hood, M.E., Refrégier, G., Giraud, T., 2009. Genome evolution in plant pathogenic and symbiotic fungi. Adv. Bot. Res. 49, 151–193.

Giraud, T., Refrégier, G., Le Gac, M., de Vienne, D.M., Hood, M.E., 2008. Speciation in fungi. Fungal Genet. Biol. 45, 791–802.

Kohn, L.M., 2005. Mechanisms of fungal speciation. Annu. Rev. Phytopathol. 43, 279–308.

Schardl, C.L., Craven, K.D., 2003. Interspecific hybridization in plant-associated fungi and oomycetes: a review. Mol. Ecol. 12, 2861–2873.

Zaffarano, P.L., McDonald, B.A., Linde, C.C., 2008. Rapid speciation following recent host fungus shifts in the plant pathogenic fungus *Rhynchosporium*. Evolution 62, 1418–1436.

CHAPTER

5

Physiology and Adaptation

Sarah C. Watkinson

University of Oxford, Oxford, UK

INTRODUCTION

Fungi have evolved to use every kind of carbon source, from dead trees to living tissues. In this chapter we consider the physiological adaptations that fit the huge diversity of species for their different ecological niches. The first section deals with nutrient acquisition: how fungi reach and assimilate their sources of carbon, nitrogen, and other elements. Secondary metabolites, used by fungi in competition and development, include many bioactive compounds, including vital antibiotics, and genetic analysis is proving fruitful in their synthesis and discovery. Cellular sensing, signalling, and response – including development in response to environmental cues – are central to the opportunistic life of fungi. We describe sensing mechanisms for extracellular and intracellular nutrients, and for physical factors including water potential, temperature, and light. We conclude by describing the ways in which fungal activities affect minerals and the soil, including the biotechnical potential of fungi in xenobiotic detoxification.

NUTRIENT ACQUISITION, UPTAKE, AND ASSIMILATION

All fungi require an organic source of carbon and energy, as well as combined nitrogen, phosphorus and sulphur, cations potassium, calcium and magnesium, and many other elements in trace amounts. Most culturable saprotrophic fungi will grow on a synthetic defined medium such as that shown in Table 5.1. The core pathways of primary metabolism are largely the same in fungi as in animals. Energy and carbon skeletons for biosynthesis are supplied from sugars, via the glycolytic, tricarboxylic acid, and pentose phosphate pathways. Under carbon starvation, fungi can assimilate carbon dioxide to supplement the supply of carbon skeletons, but this is an energy-requiring process that cannot sustain growth indefinitely. Oxygen is the normal terminal electron acceptor in respiration, as in plants and animals, but nitrate can be used as an alternative in low oxygen conditions.

The Fungi
http://dx.doi.org/10.1016/B978-0-12-382034-1.00005-0

TABLE 5.1 Composition of a Synthetic Growth Medium

Mineral base	KH_2PO_4	1g
	$MgSO_4 \cdot 7H_2O$	0.5g
	KCl	0.5g
	$FeSO_4 \cdot 7H_2O$	0.01g
Carbon and energy source[a]	Sucrose	30g
Nitrogen source[b]	$NaNO_3$	2g
Water		1L
If a solid medium is required	Agar	20g

[a]*An organic form of carbon is always required, as all fungi are heterotrophic. Glucose, sucrose, or starch are commonly used.*
[b]*Most fungi can grow with inorganic sources of nitrogen, usually provided as ammonium or nitrate salts. Some need amino acids.*

A detailed account of the pathways of respiratory metabolism is available from specialised texts on fungal physiology (see Further Reading). Although glucose is the primary precursor for these pathways, fungi rarely encounter free glucose in nature. Instead, they must acquire it from widespread and reliable sources, typically insoluble glucan polymers such as cellulose from plant residues, or the tissues of living hosts. The fungal kingdom includes species with enzymes able to break down and feed on most naturally occurring carbon polymers (Table 5.2), and even some manmade ones including plastics.

TABLE 5.2 Some Fungal Enzyme Activities Which Hydrolyse Polymers

Substrate	Enzyme
Arabinans	Arabinofuranosidase
Callose	1→3 Glucanases
Cellulose	Endoglucanases, cellobiohydrolases
Chitin	Chitinases
Cutin	Cutinases
DNA	Deoxyribonuclease
Hemicellulose	Hemicellulases
Lignin	Ligninases
Mannans	Mannanase
Pectic substances	Pectin methylesterase, pectate lyase, polygalacturonase
Proteins	Proteinases
RNA	Ribonuclease
Starch	Amylases
Xylans	Xylanase

Carbon and Energy Sources

Sugars Fungi are extremely well equipped for sugar uptake, with both active and passive uptake systems. Transport proteins may be constitutive or induced. Generally, the sugars taken up preferentially from a mixture are those which require the least energy to assimilate and are most widespread in the environment. For example, the soil ascomycete *Chaetomium* takes up glucose as soon as it is supplied in a culture medium, but the relatively rare fructose, given as sole carbon source, is taken it up only after an induction period of several hours. Most culturable fungi will rapidly assimilate common monosaccharides and disaccharides from solution. Genomic analysis (see Chapter 6) reveals the presence in fungi of multiple genes encoding sugar transport proteins, although experiment is needed to establish their functional significance. Almost a hundred different genes of *Saccharomyces cerevisiae* encode transmembrane transporter proteins, mostly from the major facilitator superfamily MFS, with the sugar porter (SP) predominating. Many monosaccharide membrane transporters operate by facilitated diffusion, the passive energy-independent movement of a solute down its concentration gradient. Other monosaccharide transporters, active when sugar is scarce in the environment, are energy requiring and transport the sugar molecule against its concentration gradient, coupled to the simultaneous movement of one or more protons.

In competitive environments, rapid responses confer critical selective advantage. Global transcription factors act as master switches for metabolism, controlling the expression of many suites of pathways simultaneously in response to changing nutrient levels, both in the cell and in its environment. Uptake systems can thus be regulated so as to prefer better value carbon/energy sources from mixtures, and scavenge nutrients efficiently from dilute solutions. Carbon assimilation is under the control of the global transcription regulator CreA, which regulates the expression of whole suites of genes encoding catabolic and anabolic pathways of carbon metabolism, as well as cell membrane transporters.

Polymeric Forms of Carbon

Cellulose is ubiquitous in plants and plant remains and is the most common carbon compound on the planet. Most saprotrophic fungi can utilise it as a carbon source, and fungi are its principal decomposers.

Naturally occurring cellulose as it occurs in the plant cell wall (Figure 5.1) is insoluble, strong, fibrous, and resistant to hydrolysis because its β1-4 linked glucan chains, up to 7 μm in length, are held in parallel alignment by hydrogen bonding to form long regular, crystalline bundles (Figure 5.2). The degree of crystallinity varies along the length of a bundle, with amorphous, less regular regions at recurring intervals. The bundles are themselves aligned to form microfibrils, 4–10 nm in diameter, which are visible in electron micrographs. This closely packed structure helps to protect the glycosidic bonds from hydrolysis, and the fungi that are able to attack crystalline cellulose possess a specialised array of enzyme activities. Physiological studies of cellulolytic fungi in culture have characterised the activities of secreted hydrolytic enzymes as either **endoglucanases** that break the bonds in the middle of the chain, releasing oligosaccharides and opening up the microfibril to further attack, or **exoglucanases** that act on the resulting chain ends. Cellobiohydrolase cleaves off dimers, cellobiose units, which are further hydrolysed to glucose by cellobiase.

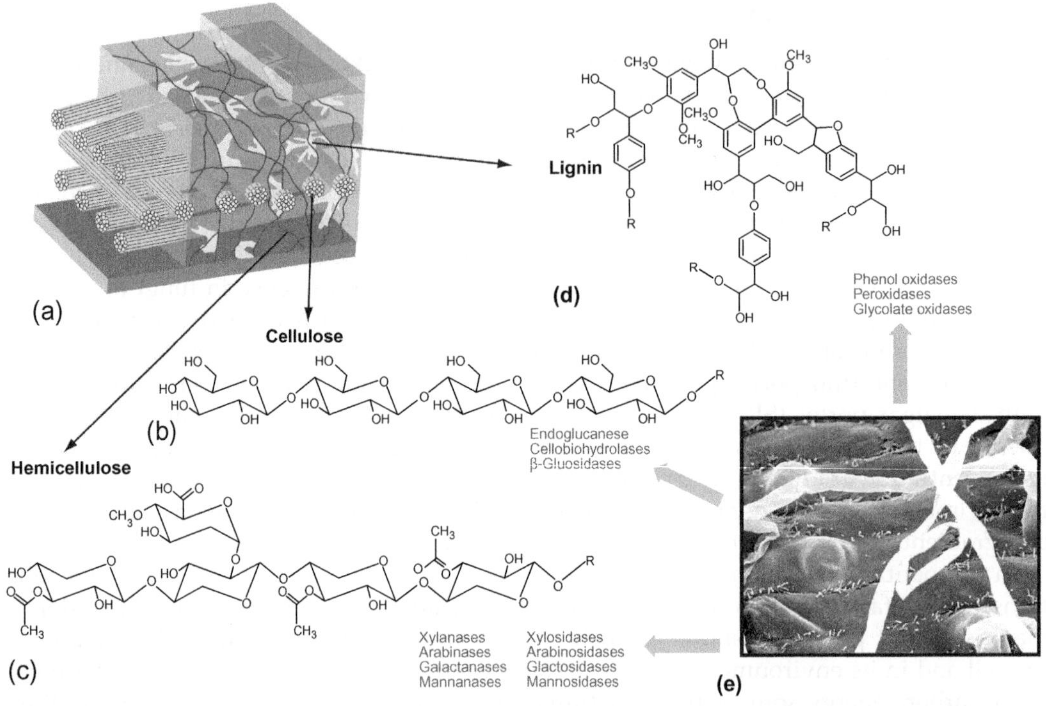

FIGURE 5.1 The plant cell wall **(a)** and the location and chemical composition of cellulose **(b)**, hemicellulose **(c)** and lignin **(d)**. Hemicellulose refers to a variety of carbohydrates and shown is *O*-acetyl-4-*O*-methyl-d-glucuronoxylan, which is common in angiosperms. Listed beneath each compound are the extracellular enzymes used by soil microorganisms to depolymerize the compounds during the process of plant-litter decay. The process of decomposition is mediated by a community of microorganisms (e), some of which are more or less active as plant detritus enters soil and particular compounds are successively depleted from that material. *Source: Frank Dazzo. From Zak et al., 2006.*

Genomic analysis reveals that fungi which utilise plant remains as carbon resources typically contain multiple genes encoding carbohydrate-active enzymes, in dozens of families. These are catalogued in the CAZy database of carbohydrate-active enzymes, described in Chapter 6. The categories listed, with their activities, are: 'Glycoside Hydrolases (GHs): hydrolysis and/or rearrangement of glycosidic bonds; GlycosylTransferases (GTs): formation of glycosidic bonds; Polysaccharide Lyases (PLs): nonhydrolytic cleavage of glycosidic bonds; Carbohydrate Esterases (CEs): hydrolysis of carbohydrate esters; and Auxiliary Activities (AAs): redox enzymes that act in conjunction with CAZymes'. Enzymes that act on plant polysaccharides often also contain a carbohydrate-binding module (CBM) that helps to attach the enzyme molecule to the glucan chain.

The highly cellulolytic soil ascomycete *Trichoderma reesei* (*Hypocrea jecorina*) is used for the industrial production of enzymes that hydrolyse plant polysaccharides. Interestingly, although its genome contains genes in the CAZy categories *GH*, *GT*, *PL*, *CE*, and *CBM*, as expected, these genes are no more numerous in *Trichoderma* than in the genomes of other, less aggressive cellulose decomposing species. *Trichoderma* appears to owe its exceptional

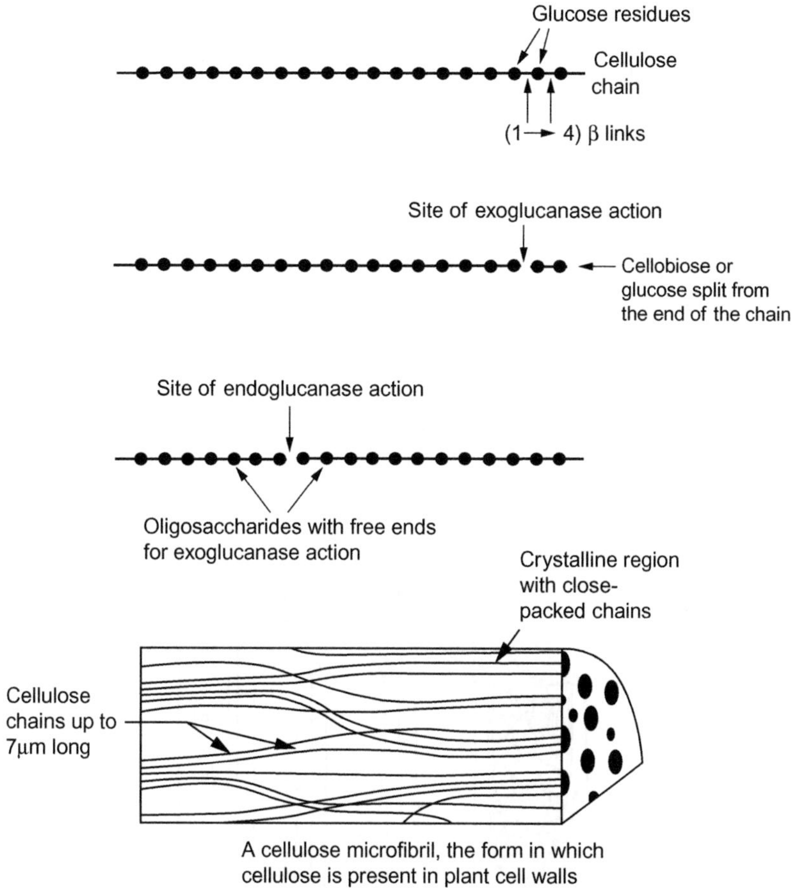

FIGURE 5.2 The structure of cellulose as it occurs in the plant cell wall, and the action upon it of different categories of hydrolytic enzyme. Endoglucanases break the bonds in the middle of the chain, opening up the microfibril to further attack and increasing the numbers of exposed ends of cellulose chains on which exoglucanases can act. Exoglucanases remove monomers or dimers from the ends of the cellulose molecule.

capacity for cellulolysis to the efficient secretion of its enzymes, rather than to an unusually extensive enzyme repertoire.

Genes for carbohydrate active enzymes are clustered in the *Trichoderma* genome, and found close to others for secondary metabolism. This probably facilitates coordinated expression of cellulase and antibiotic genes, helping *Trichoderma* to compete with the many bacteria, fungi, and animals that also utilise fragments of cellulose in the soil. Hundreds of saprotrophic and weakly parasitic microscopic ascomycete soil fungi occupy similar niches and produce secondary metabolites that act as plant toxins and antibiotics as part of their competitive armoury. They are thus a fruitful source of exploitable bioactive compounds and biosynthetic pathway genes (see Secondary Metabolism). Some, such as the genera *Chaetomium*, *Fusarium*, and *Paecilomyces*, cause 'soft rot' of wood by decomposing cellulose, hemicellulose, and pectins on its surface, although they do not penetrate sufficiently to cause structural damage.

Others, including species of *Cochliobolus*, *Fusarium*, and *Gaeumannomyces* are plant pathogens (see Chapter 8) which parasitize plant roots, killing plants with toxins. These are well known for causing losses in cereal crops.

Lignocellulose

The most durable source of cellulose in nature is wood. Comparative phylogenomics and molecular clock data suggest that fungi have been using wood as food since woody plants first evolved 400–500 million years ago (Figure 7.11). Fungal decomposition of wood plays an essential role in ecosystem and global carbon cycling. Wood contains 40–45% cellulose but decomposes only slowly because the cellulose fibres are embedded in lignin, a hydrophobic phenolic polymer which is resistant to microbial attack and cannot be utilised by fungi. Figure 5.3 shows the generalised cellular structure of wood, and the spatial distribution of cellulose, hemicellulose, and lignin. In the wall of woody cells (tracheids and vessels) the cellulose-rich S2 layers are sandwiched within lignified S1 and S3 layers. To feed on wood, hyphae must be able either to dissolve or penetrate lignin.

The so-called **white rot** wood decay fungi destroy lignin to gain access to cellulose and hemicelluloses in lignocellulose. White rot is the ancestral form of fungal wood decay and

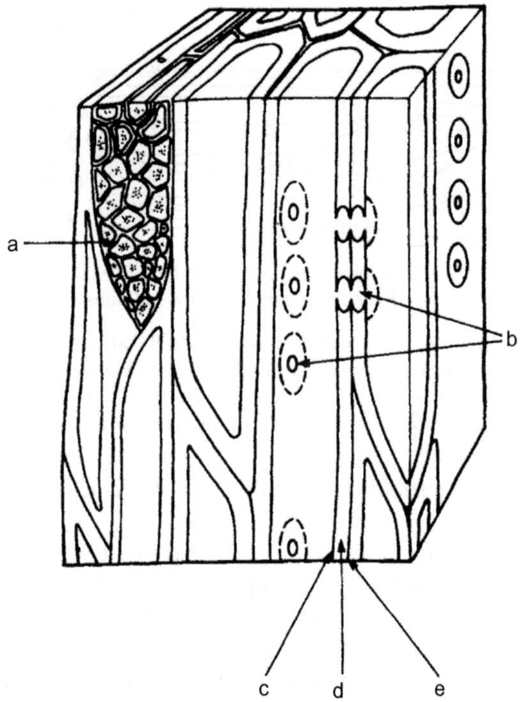

FIGURE 5.3 Diagram of the cellular structure of wood: (a) medullary ray consisting of cells which are alive in living trees, and provide a source of readily available nutrients for fungi invading timber; (b) pits, perforations in the tracheid wall, usually coinciding in position in adjacent tracheids, and providing a pathway for hyphae to grow from one tracheid into another; (c) S3 wall, with abundant lignin added to cellulose framework; (d) S2 wall, mainly cellulosic and relatively less lignified; (e) S1 wall, lignified.

is performed by both basidiomycetes and ascomycetes. Well known examples are the honey fungus, *Armillaria mellea*, and *Phanerochaete chrysosporium* used as an experimental model for white rot wood decomposition. Like many useful categories of biological activities, the distinction between white and brown rot fungi is not sharp in nature, and comparative ecological genomics shows species that do not fit either description precisely and share characteristics of both or neither. While some fungi decompose lignin and cellulose simultaneously, others remove the lignin first to reveal the whitened cellulose in a **selective delignification** (Figure 5.6a). This fungal action of freeing wood cellulose from its protective coat of lignin is important in ecosystem carbon cycling, because the delignified cellulose can be consumed by many different animal phyla (Chapter 9, p.325) including protozoa, nematodes, molluscs, crustacea, insects, and arachnids. This phylum-level diversity led the evolutionary biologist W.D. Hamilton to suggest that fungal decay of trees might have contributed significantly to the early diversification of invertebrates. Vertebrates may also benefit from fungal delignification. Delignified white rotted dead wood (*palo podrida*) is used as cattle forage in the Andes, and fungi are being investigated for use in rendering the woody agricultural wastes of leguminous crops suitable for feeding ruminant farm animals.

The recalcitrance of lignin arises from its polymeric structure as a three-dimensional heteropolymer, formed through random linkage of phenylpropanoid monomers through several types of covalent bond (Figure 5.4). Lignin has no repeating pattern of bonds, and is highly hydrophobic. It cannot, therefore, be broken down by enzymic hydrolysis where an active site of an enzyme binds to a single type of chemical bond. Instead, oxidases and peroxidases act in a non-sterically specific manner to break the molecule into a complex mixture of fragments. Oxidative lignin breakdown is a violent reaction which could not occur inside a living cell, and has been termed 'enzymic combustion'. The reactants are generated extracellularly in contact with the substrate as the hypha lies against the cell walls that form the wood. Peroxidases that use hydrogen peroxide as oxidant, and enzymes that generate hydrogen peroxidase, act together to oxidise and destabilise the lignin molecule, which then breaks down chemically into smaller fragments (Figure 5.5). Lignin peroxidases (LiP), manganese-dependent peroxidases (MnP), and versatile peroxidases (VP) are haem glycoproteins belonging to a monophyletic gene family, termed Class II peroxidases, which has diversified within basidiomycetes. The MnP enzyme forms stable diffusible complexes with organic acids such as oxalate, exuded by many fungi, and these complexes oxidise lignin phenols to phenoxy radicals. MnP is found in all white rot basidiomycetes (unlike LiP, which has arisen only once, in Polyporales) and thus has a key role in the breakdown of lignin in ecosystems as the main enzyme associated with basidiomycete white rot wood decay (Figure 5.5). The LiP enzyme molecule becomes oxidised in the presence of hydrogen peroxide to an active state in which it abstracts an electron from a non-phenolic aromatic compound to generate an unstable aryl cation radical which then undergoes a variety of non-enzymic intramolecular reactions including ring cleavage to produce a mixture of small molecules. Lignin peroxidases were added to the CAZy database in 2013 in recognition of the essential role that they play in the decomposition of lignin–carbohydrate complexes in lignocellulose. They also have their own database, the FOLy database at http://foly.esi.univ-mrs.fr.

Accessory enzymes involved in lignin degradation include hydrogen peroxide-generating enzymes such as aryl alcohol oxidase, glyoxal oxidase, and pyranose-2 oxidase. Laccases are blue copper oxidase enzymes which use oxygen to oxidise *p*-diphenols. They are found in

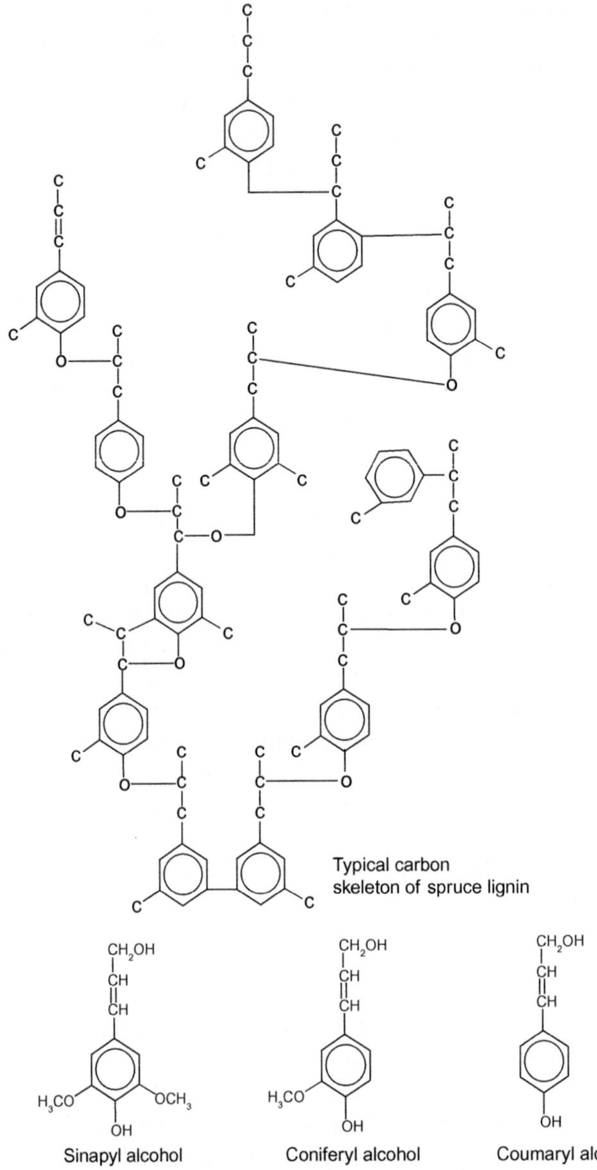

FIGURE 5.4 The molecular structure of lignin, a complex three-dimensional polymer formed from phenylpropanoid subunits. Coumaryl, sinapyl, and coniferyl alcohols are the commonest of the phenylpropanoid subunits, and they are joined randomly in a three-dimensional, asymmetric network by various different linkages, the commonest of which are illustrated. Coniferyl alcohol is the most frequent subunit in the wood of coniferous trees.

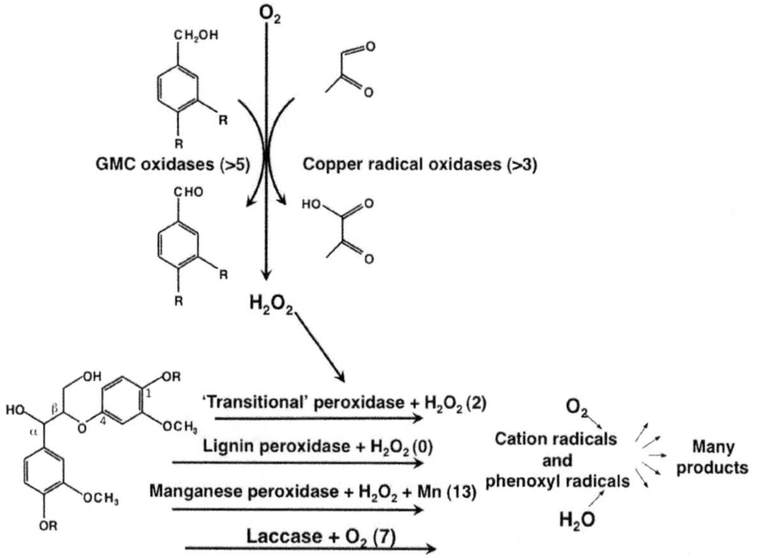

FIGURE 5.5 The role of peroxidases in lignin breakdown. Extracellular processes involved in lignin degradation by a white rot fungus. Hydrogen peroxide generated by reactions catalysed by copper radical oxidases and glucose–methanol–choline (GMC) oxidases acts as substrate for a wide array of oxidations that yield many different products. *Source: Cullen, Wood Decay. pp. 43–62, Section 2, Saprotrophic Fungi, in Martin (2014).*

most white rot fungi and have been used as indicators of white rot capacity, but cannot attack whole lignin. Lignin fragments cannot provide an energy source for fungi but soluble molecules resulting from oxidative solubilisation can be taken up and degraded intracellularly through the cytochrome P450 system, well represented in fungal genomes and associated with xenobiotic metabolism (see below). The genome of the model white rot basidiomycete *Phanerochaete chrysosporium* has 154 cytochrome P450 genes.

By contrast with white rot, in the **brown rot** mode of wood decomposition, fungi remove cellulose from within the lignocellulose complex, without completely removing the lignin, so that a brown residue of the polymer remains (Figure 5.6b). Most of the mass loss caused by brown rot decay is from the digestion of polysaccharide. However, the lignin is changed, with minor side chain oxidation, demethylation, and some depolymerisation. The capacity for brown rot, a more energy efficient utilisation of wood carbohydrate than white rot, has evolved convergently in at least six different clades of basidiomycetes, including examples in both Polyporales and Boletales (Figure 7.10). Brown rot does not appear to be a feature of ascomycete wood decay. Comparing gene evolution across a range of white and brown rot basidiomycete species indicates that the transition from white to brown rot wood decomposition involved not only the loss of genes for oxidases but also an overall reduction in numbers of some gene families and gene copies of some glycoside hydrolases. However, there has been duplication of genes in other glycoside hydrolase gene lineages, resulting in a reduced and refined collection of enzymes for cellulose depolymerisation.

Brown rot decay of wood is characterised by early reduction in strength, before there has been significant loss in volume or biomass. Timber decay caused by *Serpula lacrymans* in a building can go unnoticed up to the point of collapse. Cellulose fibres provide the tensile

FIGURE 5.6 The appearance of wood decayed by: (a) a white rot fungus. This branch has been selectively delignified, leaving a pale, fibrous, cellulose-rich residue. (b) The remains of a fallen tree decomposed by a brown rot fungus. The decomposed wood breaks apart in a cubical pattern and then crumbles to a brown, lignin-rich residue. Inset figure: close-up of brown rotted wood. *Source: (c), Stuart Skeates.* (See the colour plate.)

strength of wood and the early stages of depolymerisation involve chain breakages in the amorphous regions of crystalline cellulose, thereby weakening the cellulose fibrils. A problem in understanding the role of enzymes in this process is that the micropore diameter within the wood composite structure, at the stage when cellulose depolymerisation is first evident from physical analyses such as X-ray diffraction, appears too small to allow diffusion of hydrolytic enzymes. Microscopy reveals that hyphae are sparsely distributed in decayed wood, and where hyphae can be seen, they are not in close contact with the cell wall, implying that decay is initiated at a distance from the surface of the fungal cell. It appears that the primary attack on the cellulose molecule is by a free radical oxidation rather than a hydrolytic enzyme. Free radical generation outside the hyphae is facilitated by secretion of small molecules including oxalic acid, and phenolates that act as electron donors. Under the low pH conditions generated by these chemicals, the strongly oxidative hydroxyl radical ·OH is generated by Fenton's reaction, in which organic acids and hydrogen peroxide combine in a redox reaction with ferrous iron:

$$Fe^{2+} + H_2O_2 + H^+ \rightarrow Fe^{3+} + \cdot OH + H_2O$$

Phenolates, which may be secreted fungal secondary metabolites or derived from lignin decomposition, act both to chelate and also reduce iron, with accompanying oxidation of phenolic groups to quinones.

The hydroxyl radical has a nanosecond half-life; hence it must be produced close to its site of action on the cellulose crystal. The means by which it is targeted is an intriguing question. One answer is suggested by the discovery of compound genes, in the genomes of the white rot fungus *Phanerochaete chrysosporium* and the brown rot species *Serpula lacrymans*, which appear to encode both a CBM, and an iron reductase. This protein, by binding to the cellulose molecule and localising iron reduction for the radical-generating Fenton's reaction close to it, may help to target hydroxyl radical attack to the cellulose molecule. This theory is supported by transcriptional analysis which shows that the compound gene is expressed over a hundred times more when the fungus is grown on wood than when glucose is its sole carbon source.

While the sequenced brown rot species so far investigated have many glucanase enzymes that break bonds in the middle of the glucan chain of cellulose, they all lack the **exoglucanases** which further degrade the oligomeric carbohydrate produced by these reactions. These exoglucanases are 'processive' glucanases that snip glucose or cellobiose units from glucan chain ends. They are present in white rot fungi, but appear to be absent from the genomes of brown rot species that have been sequenced. How exoglucanase action is produced in brown rot fungi is unknown at the time of writing. Brown rot wood decomposer fungi mainly decay softwoods and are the most common cause of wood decay affecting the built environment and timber structures in temperate climates. The dry rot fungus *Serpula lacrymans* (Boletales) is the most common cause of wood decay in buildings in Northern Europe, followed by the cellar fungus *Coniophora puteana*. *Serpula incrassata* and *Postia* (*Poria*) *placenta* are responsible for wood decay in buildings in the United States. Molecular clock data (see Chapter 7; Figure 7.11) show that the evolution of white and brown rot fungi coincided with the evolution of angiosperm and gymnosperms, which presumably supplied new nutritional niches for them. In turn, fungi are believed to have played a part in the development of forest soils by generating the humus that conditions these carbon rich, nitrogen-poor soils.

Protein Secretion

Protein secretion is an important feature of fungi that is central to their ecology and is exploited in biotechnology. The release of enzymes is regulated in time and space according to the substrates available and the condition of the fungus. Protein secretion has been investigated intensively in yeasts, because they are experimentally tractable and are used for the heterologous expression of imported genes in biotechnology. Much less is known about secretory processes in filamentous fungi, even though their levels of protein secretion are typically 10-fold higher than those of yeasts. Filamentous fungi are generally better adapted for protein secretion than yeasts. Because their hyphae feed by extracellular digestion of chemically diverse solid substrates, rapid and efficient enzyme secretion confers strong selective advantage. There is, therefore, a practical impetus behind the investigation of protein secretion in filamentous fungi, with huge potential benefit from transforming yeasts to secrete high levels of heterologous proteins.

In the eukaryotic cell, proteins destined for secretion into the external environment are processed for export in the endoplasmic reticulum Chapter 2, p. 45, where they are folded and glycosylated. In a series of interactions, the nascent protein is combined with a binding protein (BiP), with protein disulphide isomerase (PDI), which catalyses disulphide bond formation and acts like a chaperone, and with peptidyl prolyl isomerase (PPIase), which interacts with other proteins and may accelerate protein folding. Correctly folded proteins are targeted to the Golgi apparatus where they are glycosylated and are then either exported to the exterior via the plasma membrane or targeted to the vacuole. Based on GFP imaging in *Aspergillus niger*, export is believed to be confined to the apical region of the hypha.

Aspergillus niger and other filamentous fungi are of interest for industrial production of a range of heterologous secreted proteins. The secretory pathway of *A. niger* has been successfully modified to secrete proteins that are normally intracellular, to facilitate the harvesting of new heterologous proteins produced by transformants. The intracellular protein for export is targeted into peroxisomes by adding a peroxisome import signal tag. The secretory signal proteins on peroxisomes are also modified so that they fuse with the plasma membrane. The proteins imported into the peroxisomes are found to be released to the external environment through this artificial secretion pathway, termed '**peroxicretion**'.

Identification and purification of secreted proteins makes it possible to carry out imaging to determine the time and place of enzyme release in relation to ecologically relevant substrates. Proteins can be labelled *in situ* serologically, using antibodies conjugated with a visualisable label. However it is now more common to use GFP fusions and image the protein's cellular location by confocal microscopy.

Nitrogen Sources

Fungi share most pathways of amino acid metabolism and protein biosynthesis with other eukaryotes, including animals. However, unlike animals, fungi do not require amino acids from their environment to synthesize protein, but can assimilate inorganic compounds widely available in soil. The ability to make their own amino acids using soil nitrate and ammonia, providing they have a carbon source, enables saprotrophic fungi to convert dead wood and other nitrogen-poor, carbohydrate-rich plant remains into protein. This gives fungi a key role in the trophic webs of terrestrial ecosystems and is also the basis of various processes for converting agricultural waste into animal feed and human foods.

Fungi, unlike many prokaryotes, cannot utilise molecular nitrogen and so must scavenge the element from an unpredictable variety of soluble and insoluble environmental sources, competing with plants and all the other microbes in the habitat. Enzymic, transport, metabolic, and developmental systems for taking up and assimilating nitrogen are expressed according to the ever-changing opportunities for nitrogen acquisition. Fungi respond to nitrogen abundance or starvation conditions with 'global' changes in metabolism, with transcription factors operating to activate whole pathways for assimilation or dissimilation. Conserved regulator genes encode transcription factors that in turn, control and coordinate up- and down-regulation of gene clusters, encoding whole pathways for nitrogen uptake and assimilation. The global transcription factor AreA, conserved across the fungal kingdom, is a positive regulator that switches on pathways of nitrogen catabolism under nitrogen depletion, while nt2 is a negative regulator and represses nitrogen assimilation when nitrogen is available. Both are homologous GATA factors with a shared DNA-binding sequence.

Soil, water, and living tissues provide a wide variety of inorganic and organic nitrogen sources for fungi. In agricultural soils, nitrate derived from bacterial nitrification, fertiliser, and atmospheric deposition, is the most abundant. By contrast, in natural and semi-natural soils with low disturbance and a large competing microbial population, nitrate and ammonium occur only sporadically and briefly, and fungi in these habitats can assimilate many different organic nitrogen compounds.

Nitrate

Nitrate uptake in fungi occurs via transport proteins encoded by the *NRT1* and *NRT2* gene families. These membrane proteins mediate transport of NO_3^- together with protons (H^+) in a symport driven by the pH gradients across membranes (see Chapter 2, p.42). They are encoded within a coordinately regulated cluster, *fHANT-AC*. This consists of three genes: *nrt2*, which codes for a high-affinity nitrate transporter; *euknr*, which codes for nitrate reductase that keeps cytosolic nitrate low, maintaining a gradient; and *NAD(P)H-nir*, which codes for nitrite reductase. While these individual genes are found in all eukaryotes, the cluster is seen only in fungi and in *Phytophthora*. Phylogenetic analyses suggest that the cluster was first assembled in a lineage leading to oomycetes and subsequently transferred horizontally to an early ancestor of the fungi. The acquisition of *fHANT-AC* may even have been the crucial evolutionary step that enabled higher fungi to colonise land by allowing exploitation of nitrate in aerobic soils. Functional diversification of the nitrate assimilating cluster was demonstrated in a phylogenetic analysis of basidiomycete species with ecological niches representing saprotrophy and biotrophy in three separate lineages. Fungi have two paralogs of *NRT2* encoding a high and a low affinity transporter. Transcription analyses suggest that the high-affinity transporter, which enables fungi to concentrate nitrate from environments in which it is present at less than micromolar concentrations, may be adaptive to environments such as pristine boreal forest soils and water with very low levels of nitrate. In such forest soils, ectomycorrhizal fungi play a crucial role in nitrogen cycling by assimilating nitrogenous compounds from soil and transferring them to tree hosts. Accordingly, genomes of forest mycorrhizal fungi (Chapter 7) are being examined for evidence of nitrogen acquisition mechanisms adapted to the limiting levels of biologically available forms of nitrogen in these soils. Truffles grow as mycorrhizas extending into forest soils from tree roots. Their nrt2 protein is a very high-affinity transporter ($K_m = 4.7\,\mu M$ nitrate) that is bispecific for nitrate and nitrite. It is expressed in free-living mycelia and in mycorrhiza of the truffle, *Tuber borchii*, where it preferentially accumulates in the plasma membrane of hyphae that contact roots. *Laccaria bicolor*, which is also ectomycorrhizal, but common and widespread compared with truffles, takes up nitrate even when glutamine is present, suggesting that regulation varies according to niche adaptation. The highly efficient nitrate scavenging of fungi is exploited by plants whose own capacity for uptake is inferior. For example, the arbuscular mycorrhizal *Rhizophagus irregularis* (Chapter 7) helps cucumber roots to scavenge nitrate from soil.

In the cell, nitrate is reduced to ammonium in a two-step process via nitrite. Nitrate reduction places an energy demand on the organism, supplied by the coenzyme NADPH produced in the pentose phosphate pathway of glucose metabolism. Many fungi are known to utilise nitrate as an alternative terminal electron acceptor in the respiratory chain when oxygen is limiting, as in temporarily anaerobic soils, where they may therefore contribute to soil denitrification. Nitrate at high concentrations is damaging because it may impose oxidative stress on the cell, triggering responses that include alterations in secondary metabolism and autolysis.

Ammonium

Ammonium ion is taken up by a family of transporter proteins, the ammonium transporter/methylammonium permease family (MEP family), unique to fungi but homologous with the conserved AMTP family found in all eukaryotes. The fungal MEP proteins probably originated as a horizontal gene transfer from a prokaryote early in fungal evolution. They are membrane-spanning proteins composed of 11 highly conserved transmembrane domains that fold into a pore through which ammonia or ammonium pass passively. They are believed to function as modified gas channels, allowing influx of a gas rather than the dissolved ion, conducting ammonia (NH_3). The ammonium proton is lost at the entrance to the pore, ammonia moves through the channel, and is re-protonated at the cytoplasmic side. Ammonia is preferred to nitrate in uptake from mixtures, through repression of nitrate uptake pathways.

Metabolic incorporation of nitrogen occurs by combination of ammonium with precursors from the pathways of respiratory metabolism, forming amino acids. Pyruvate from glycolysis is aminated to alanine, and oxoglutarate from the TCA cycle to glutamate. These serve as intermediates in complex networks of amino acid catabolism and anabolism. Several alternative pathways of ammonium assimilation operate in fungi according to the extracellular concentration of ammonium. Under nitrogen starvation, the so-called GS-GOGAT pathway is activated by expression of genes encoding the component enzymes. The first, glutamine synthase, with a high affinity for ammonium, catalyses its combination with glutamate and the resulting glutamine is used by an amino transferase to aminate 2-oxoglutarate. This reaction is powered by ATP with NADPH as electron donor. When ammonium is more readily available it is assimilated through the energetically cheaper reaction of ammonium with 2-oxoglutarate, which also uses NADPH but does not need ATP.

Amino acids

Amino acids are taken up from the extracellular solution by a variety of constitutive and inducible membrane transport proteins as described in Chapter 2. Core metabolic amino acids such as alanine, glycine, glutamine, aspartate, and glutamate, derived from primary carbon metabolic intermediates by a single amination step, are readily assimilated. Other amino acids, including methionine, are used more slowly and are inferior as sole nitrogen sources. The preferences of fungi for various nutrient sources in culture have been intensively studied by brewing industries, and *Saccharomyces cerevisiae* presented with a cocktail of amino acids will exhaust them one at a time in a clear order of preference.

Not all amino acids taken up by the cell can be metabolised. The non-protein amino acid α-aminoisobutyric acid (AIB) is a product of secondary metabolism in some fungi and bacteria and is combined into peptides with antibiotic function termed **peptaibols**. AIB is preferred even to glutamate by fungal amino acid transporters, but, since it is not metabolised, accumulates unchanged in the cell and is used in biochemical amino acid transport studies. Its inhibitory effect on cell growth has been investigated for tumour suppression. At high concentrations it inhibits mycelial spread in basidiomycetes through competitive inhibition of glutamate uptake. At sub-toxic intracellular concentration it is used a ^{14}C-labelled marker for amino acid translocation in mycelial networks (Figure 5.7), because it is not metabolised and remains unchanged in the free amino acid pool.

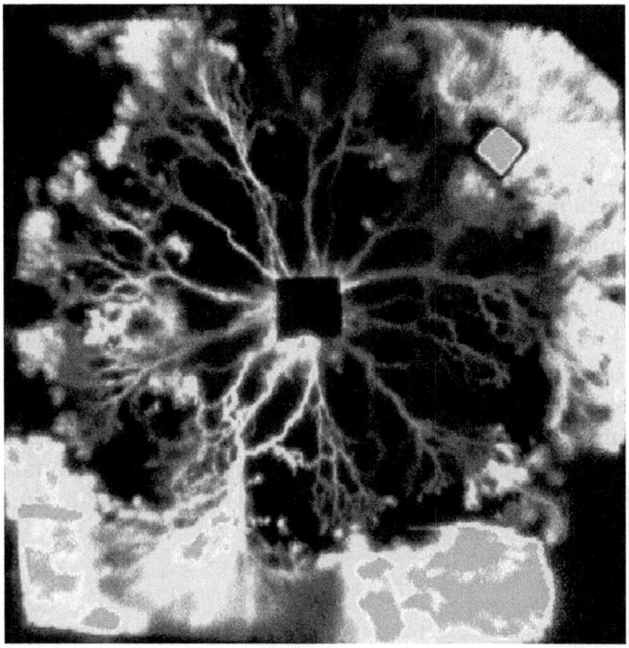

FIGURE 5.7 Photon counting scintillation imaging of ^{14}C-AIB in the mycelium shown in Figure 5.18, during new wood resource capture (Further Reading in Tlalka et al., 2008, Video S2) Video link, see Further Reading. The mycelium shown in Figure 5.18 was supplied with the non-metabolised tracer amino acid AIB (α-aminoisobutyric acid) labelled with ^{14}C. After incubation to allow the AIB to become evenly distributed within the mycelium, a fresh wood block was placed near the colony margin, top right, and the system was imaged by PCSI for the ensuing 10 days. The video shows AIB redistribution from the whole mycelium into the site of colonisation of fresh wood. This is consistent with an ability of the mycelium to sense and respond to a localised fresh carbon supply by importing nitrogen, to achieve a balanced internal C/N ratio for biosynthesis. *Source: Tlalka et al. (2008). http://dx.doi.org/10.5072/bodleian:d217qq90r* (See the colour plate.)

The mycelium stores free amino acid as a labile expandable pool that responds to internal nitrogen supply and demand, as well as availability in the external environment. Free amino acid levels in the cell can vary 10-fold or more, according to nitrogen availability in the environment. The intracellular pool is composed mainly of core amino acids including glutamate, glutamine, serine, proline, aspartic acid, glycine, and alanine, with smaller amounts of arginine, ornithine, lysine, and histidine. Low levels of non-protein amino acids, including GABA may also be detected. Kinetic studies indicate that separate pools with relatively fast and slow turnover may be present in cytoplasmic and vacuolar locations. Transporters with different affinities may accumulate dibasic arginine, ornithine, and lysine in the vacuole. In addition to its many other functions (Chapter 2), the vacuole can thus act as a short-term mobile nitrogen storage and distribution compartment. Vacuolar movement may be involved in the transfer of nitrogen from mycorrhizal mycelium into the root. Here amino nitrogen, assimilated into arginine in the extramatrical mycelium in soil, is translocated into the root where the arginine provides a nitrogen source. This process has been described as a spatially separated urea cycle. The plastic and dynamic vacuolar system contributes to nutrient translocation in the hypha. Predictive models show that amino acids for hyphal tip growth may be imported to the tip in this way although such transport is only efficient over distances of a few micrometers.

Other Organic N Compounds

A variety of organic nitrogen compounds are utilised by fungi adapted to habitats where readily assimilable forms of nitrogen are in short supply. Fungi in low-disturbance habitats such as boreal forest soils exist in a state of nitrogen limitation and have evolved strategies to scavenge, store, and recycle nitrogen within their mycelia. Over a thousand organic nitrogen compounds, some of which are highly resistant to biological breakdown, are found in such soils. Up to 50% of soil nitrogen may be present in the form of phenolic nitrogen compounds. Forest fungi mine soil for nitrogen, using oxidising enzyme systems such as the class II peroxidases used in lignin depolymerisation, to break down recalcitrant complexes. The extensive mycelium of woodland basidiomycetes is a repository of nitrogen in the form of chitin, proteins, polyamines, and amino acids.

Protein

Protein is depolymerized by fungi using multiple secreted proteases and peptidases that enable them to hydrolyse most proteins including those in human tissues. Fungal proteinases are of practical interest in the etiology of fungal infections and for industrial applications. The virulence of ringworm fungi that grow on nails, hair, and skin is due to proteinases adapted to keratin hydrolysis.

Seven categories of common protease, and aspartic, serine, metalloproteases, and cysteine protease (characterised by active site and identified biochemically by the effect of standard inhibitors on their activity) are present in fungi. Many others are inferred from biochemical activity or gene sequence but are as yet uncharacterised. Genomes reveal a multiplicity of putatively proteolytic enzymes, far greater than expected from biochemical studies. Proteases destined for the extracellular environment are characterised by the presence of a signal peptide that targets them to the secretory pathway, and an N-terminal propeptide, comprising as much as 50% of the peptide chain, that orchestrates correct protein folding and stability. These are removed by specific proteolysis to activate the secreted enzyme. Enzymes of saprotrophs are typically glycosylated, which contributes to their robustness and durability in soil where proteins are liable to be consumed by other microorganisms.

Fungi parasitic on animals have a wider range of secreted proteinases than saprotrophs, and the particular suite of these found in a species may confer specificity for types of tissue. Proteases conferring virulence on parasitic fungi, but also present in saprotrophs, are categorized as subtilisins and fungalysins. *Aspergillus fumigatus*, cause of aspergillosis in the human lung, secretes fungalysins that attack the interstitial mucopolysaccharides collagen, elastin, and laminin. Dermatophytes, which attack keratin (hair, nails, and skin), secrete not only multiple fungalysins and subtilisins, but also sulphite, which is required to break the disulfide bridges of the insoluble protein to allow access for enzymes. Comparison of the genomes of important fungal pathogens of humans, including *A. fumigatus*, and dermatophytes, has shown the path of gene evolution of proteases. Dermatophytes have many more fungalysins and alkaline proteases than *A. fumigatus*. The two genes are shared and ancestral to both *Aspergillus* species and dermatophytes, but in dermatophytes there have been multiple expansions in these gene families. Different pathogens express particular suites of protease activities, of which, a subset are virulence factors. From phylogenetic tree topology, gene evolution preceded species divergence, suggesting that the acquisition of new proteolytic capacities opened the way for diversification into new niches. Disease symptoms caused by fungal protease action on animal tissues may be due not solely to proteolytic attack, but also to the allergenic properties of the proteins that induce a localised immune response and cause inflammation.

Other Major Nutrients

Phosphorus

As a component of nucleic acids and cell membranes, and central to energy metabolism, phosphorus is continually required for fungal growth. However, its availability and chemical form in the environment vary in space and time. Fungal systems for phosphorus acquisition are highly adapted for efficient scavenging.

Phosphate ions are assimilated from solution in soil and water by proton-coupled symporters in the cell membrane. The characteristics of phosphate transporters are intensively studied in mycorrhizal fungi because of their key role in plant health. Typically, several transport proteins differing in substrate affinity are encoded in the genome, allowing the fungus to adjust to varying levels of environmental phosphate by differential transcription. The ectomycorrhizal basidiomycete *Hebeloma cylindrosporum*, for example, has two genes, *HcPT1* and *HcPT2*. Under phosphorus limiting conditions, *HcPT1* is up-regulated, suggesting a special role in scavenging scarce phosphate. Internal phosphate levels also regulate the transcription of phosphate transporters, depressing expression when intracellular phosphate increases. The cellular level is also affected by conversion of phosphate to a polymeric form, polyphosphate, consisting of chains of three to 1000 phosphate residues. Polyphosphate is sequestered in the vacuole. The physiology of the fungus thus accommodates to both external supply and internal demand. Phosphate homeostasis within the mycelial network is also regulated by translocation. Intracellular phosphate is shown by ^{32}P-labelling to be transported in mycelial networks to sites of active growth. The rate can be faster than diffusion, but the mechanism and anatomical pathways remain obscure. Phosphorus accumulation and redistribution through fungal networks is central to forest ecosystem nutrient dynamics (Chapter 7).

Uncultivated soil typically contains less than 10 μM levels of free phosphate, and fungi are adapted to utilise other sources of phosphorus in soil, including relatively insoluble minerals like apatite, and organic compounds and complexes. This phosphorus scavenging capacity, together with phosphate translocation, gives mycorrhizal fungi their central role in importing phosphorus from rocks to the biosphere (Table 5.4). Hyphae solubilise phosphate by lowering the pH in their vicinity through proton release and by secreting oxalic, citric, and malonic acids from hyphal tips and along the length of the hypha. Oxalate is produced in large amounts and crystals are common on the surface of aerial hyphae. The dense agglomerations of hyphae formed by some mat-forming ectomycorrhizal species of forest fungi can raise oxalate levels in the surrounding soil to forty times that of uncolonised soil.

In mature natural ecosystems most soil phosphorus is in the form of organic compounds. Simpler organic forms include monoesters (mononucleotides and sugar phosphates) and diesters (nucleic acids and phospholipids). Monoesterases and diesterases act on these compounds to release assimilable phosphate. Monoesters are more abundant and can comprise up to 50% of the total soil organic phosphorus pool. Inositol phosphates exist in soil in the form of complexes with organic and inorganic material. Phosphate is released from these sources by **phytase** enzymes. Such lysis is an important ecological function of mycorrhizal fungi (Chapter 7) and there is evidence of synergy between the varying phospholytic activities of different ectomycorrhizal taxa, with host phosphorus uptake enhanced when a diversity of fungal symbionts is present. Phytase synthesis is induced under phosphorus-limiting conditions and repressed by intracellular phosphate.

Sulphur

Most fungi are able to utilise sulphate. It is taken up via membrane sulphate transporters, and phosphorylated in the cell via adenosine triphosphate in two steps to give 3′-phosphoednosine-5′-phosphosulphate. This is reduced to sulphite then to sulphide, which is condensed with *O*-acetyl serine to give cysteine for incorporation into proteins and as an intermediate for the synthesis of methionine. Organic sources of sulphate can also be used, such as choline-*O*-sulphate, common in plants and fungi, as an internal sulphur store, as well as aromatic sulphate esters. These are transported into the cell via specific transporters, which, along with sulphate transport proteins, are strongly induced in sulphur-limited conditions, and repressed by sulphur-containing amino acids.

Sulphur metabolism is under the control of global transcription factors. Sulphur limitation induces transport systems for methionine or sulphate, and enzymes including aryl sulphatase and choline sulphatase to release sulphur from internal reserves and extracellular organic compounds including protein. All the sulphur acquiring systems are repressed under sulphur sufficiency, which is reached in experimental cultures of yeast and *Aspergillus* at a 5 mM concentration of methionine.

Sulphur is essential not only for protein synthesis but also in the glutathione-based system for combatting stress. Glutathione, a non-protein thiol with a very low redox potential, is present at up to 10 mM in yeasts and filamentous fungi. It is an important antioxidant, reacting non-enzymatically with reactive oxygen species as well as participating in detoxification of xenobiotics. Elemental sulphur, which is cytotoxic, can be reduced extracellularly to hydrogen sulphide by the soil fungus *Fusarium oxysporum*, via the glutathione reductase/glutathione couple. Under experimentally imposed stress, *Saccharomyces cerevisiae* activates a sulphur-sparing system which conserves intracellular sulphur for glutathione synthesis by replacing sulphur-rich proteins with isozymes containing lower levels of sulphur.

Essential Metals

All fungi require substantial amounts of potassium and magnesium, and around 50 mM concentrations of Mg and K salts are used in defined culture media. These cations are present at high concentrations in fungal cells and are structural components of membranes. Iron, copper, calcium, manganese, zinc, and molybdenum are required by all organisms as cofactors for enzymes and other proteins. The amounts needed are relatively low, less than 6–10 μM for iron and molybdenum. Some of the trace elements, including copper and zinc, become toxic for fungi at levels only a few times greater than those required for optimal growth.

In spite of the small quantities of trace metals required, their presence does not always ensure their availability for the fungus. Iron presents a particular problem. Except under strongly acidic conditions, ferrous iron undergoes rapid spontaneous oxidation to the ferric form, followed by precipitation as the highly insoluble ferric hydroxide. Fungi address this problem by synthesising iron chelating agents, **siderophores**, with which iron forms a soluble complex (Figure 5.8). Although the chemical equilibrium strongly favours complex formation, some free metal ions and chelating agent are present, and utilisation of the free ion by the organism results in further liberation of metal ions from the complex. The chelating agents act as metal buffers in a way analogous to pH buffers, maintaining a constant free ion concentration.

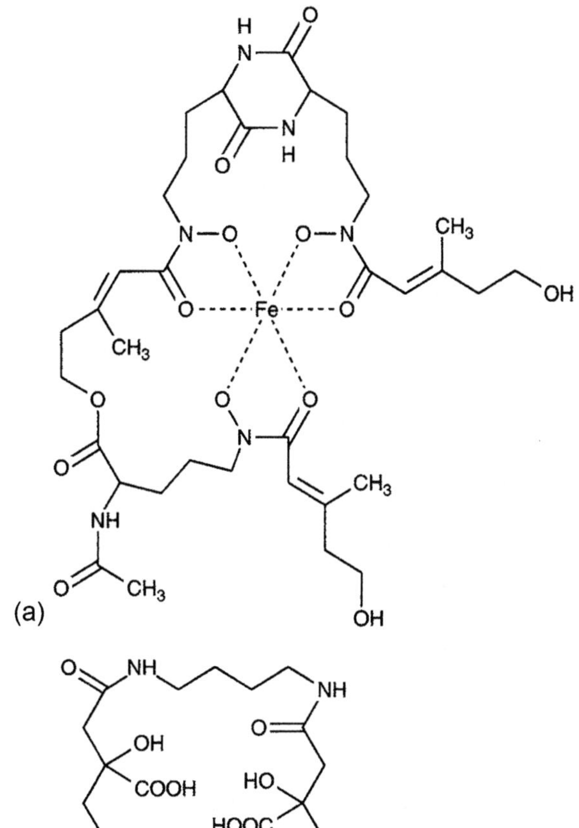

FIGURE 5.8 Fungal siderophores: (a) coprogen, produced by *Neurospora crassa* and some species of *Penicillium*. It is formed by the hydroxylation and acetylation of three L-ornithine molecules, and is an example of a hydroxamate siderophore. These are varied in structure and may originate from 1, 2, or 3 L-ornithine molecules. They are widespread in ascomycetes and basidiomycetes. The figure shows the coprogen-iron complex; (b) Rhizoferrin, produced by Mucorales. It consists of two citric acid molecules linked via amide bonds to putrescine.

Siderophores operate to maintain iron homeostasis. They are released from hyphae when iron availability is limiting growth, and chelate ferric ions in the surrounding solution. Metabolic energy is utilised in transporting the iron–siderophore complex across the plasma membrane into the cell. Siderophores are also involved in the storage of iron in the cell. In some fungi, these siderophores have the same structure as those involved in transport, but in others they differ. Members of the Mucorales (zygomycetes) store iron in their cells by using the iron-binding protein, ferritin, as do plants and animals. Some fungi do not produce siderophores but transport iron into the cell utilizing a ferric reductase at the cell surface. Other fungi release large amounts of the relatively weak chelating agent citric acid into the environment when iron is scarce. Siderophores are essential for virulence of fungal pathogens against animals and plants. Fungi that do not themselves produce siderophores may utilise those produced by other fungi when they are available.

Metals and metalloids can be acquired by fungi from soil minerals and from rock. Secreted organic acids and mechanical action by hyphae combine to solubilise and release cations to make them available for absorption. Metallic cations are strongly adsorbed by fungal cell walls by the physicochemical process of ion exchange. Binding is affected primarily by the valency of the cation, so that trivalent ions displace divalent ions, which in turn displace monovalent cations. Heavy metal pollution can be monitored by analysing fungi or lichens growing in particular environments. Contaminated fungi or lichens can threaten health when used as food by humans.

Some radioactive cations including potassium^{40} and the radioactive element caesium are readily taken up by fungi. For example, the major radionuclide of long-term concern that was released to the atmosphere by the Chernobyl nuclear accident in 1986 is caesium^{137}, with a half-life of 30 years. Caesium is accumulated by all living organisms, because of its similarity to potassium. In a survey in 1994–1995 of sites in Russia about 200 km northwest of Chernobyl, levels of ^{137}Cs in wild mushrooms were up to a 1000-fold higher than the peak levels in agricultural products. This was reflected in tissue levels of the isotope in the local population, which were 10-fold higher in individuals who ate wild mushrooms regularly compared to those who never did. Even higher levels have been found among the reindeer-herding Samis of northern Scandinavia, where there is a very short food chain from contaminated lichen to reindeer to human. Unexpectedly, analyses in successive years have shown little diminution in isotope levels in lichens, so the problem will be with us for many years to come. Some fungi selectively concentrate particular heavy metals. An example is the common fly agaric, *Amanita muscaria,* which accumulates extraordinarily high levels of vanadium of up to 200 mg per kg.

Growth Factors and Vitamins

Many fungi have rather simple nutrient requirements. Some, as indicated above, need no organic compounds other than a carbon source, and many others have only a few additional needs. It is, however, those organisms which have the simplest needs which are most readily grown in pure culture on defined media. Others have more elaborate requirements, including rarer amino acids and vitamins (substances needed in very small quantities). Many fungi require the vitamins thiamine (e.g. *Phycomyces*) or biotin (e.g. *Neurospora*), water-soluble B vitamins which are important in animal nutrition and were originally isolated from yeast extract. Instances of requirements for most of the other B vitamins are known in the fungi but are less common. Sterols may be necessary, especially under anaerobic conditions. Other requirements may include fatty acids, purines and pyrimidines and inositol.

Autolysis, Autophagy and Apoptosis

These are processes in which cellular protein is broken down to release amino acids for biosynthesis of new proteins. Starved fungal cultures decrease in biomass owing to degradation of cell material by autolysis, and will eventually die if nutrients become exhausted. However, before irreversible autolysis occurs, the cell may show starvation induced changes including the activation of secondary metabolic pathways and the onset of sporulation (see below). In natural environments, starvation may occur in parts of a mycelial system while other parts are well resourced. Foraging systems of basidiomycete mycelial cords can be manipulated in culture to access localised fresh resources, inducing the shut-down of under-resourced parts of the mycelial network. Such nutrient recycling in filamentous fungi can be aided by the process of **autophagy**, a controlled breakdown of protein and organelles that occurs within specialised double membrane-bound vesicles, the **autophagosomes**. These deliver their contents to the vacuole which acts as a lysosomal compartment where proteins are degraded.

Autophagy is conserved across eukaryotes and involves over 30 genes, but only a few of these have been investigated for their function during the growth and development of filamentous fungi. It appears that autophagy serves not only as a pathway for nutrient recycling, but can also be essential for hyphal morphogenesis. The autophagy gene *Atg8* plays a key role when germinating spores of the rice blast pathogen, *Magnaporthe oryzae*, infect leaves (Chapter 8, p.266). Autophagy of the appressorium contents enables cytosolic materials to be relocated from the conidium on the leaf surface into the growing germ tube. The Atg8 protein has been visualised in fungi using GFP fusions, appearing as dots in the cytoplasm that subsequently move to the vacuole. Caspases, a specialised group of peptidases that operate the process of **apoptosis** or programmed cell death, have recently been found to be involved in antagonistic interactions between different fungal species, and in the development of spores and sporophores.

SECONDARY METABOLISM

Both fungi and plants synthesise innumerable chemically complex substances by means of secondary metabolic pathways that differ from those of primary metabolism in the low substrate specificity of their enzymes. Secondary pathways tend to be branched and interact with one another. One enzyme may catalyse reactions with multiple products, and the products from one reaction may act as substrates for another. This results in the formation of an enormous variety of chemically complex products. Pathways that produce substances with biological activities, such as antibiotics and toxins (Chapter 9, p. 297), appear to have been have been selected for in habitats where they have conferred some advantage in the interaction of an organism with its neighbours. The genes specifying these pathways have become duplicated with minor changes during species diversification, producing reactions that generate yet more chemical products. Field mycologists identify species according to tiny colour differences – whether, for example, a mushroom is the colour of cinnamon or tobacco – illustrating the taxon-specificity that characterizes many secondary metabolites. Suppressing competitors and attracting dispersing agents are biologically important roles of fungal secondary metabolites. Others that are produced by pathogenic fungi act as virulence factors in disease (Chapter 8). Penicillin and other antibacterial compounds are probably advantageous to soil ascomycetes that produce them, as competitive inhibitors of peptidoglycan synthesis in cell wall synthesis by soil bacteria that compete for carbon substrates. The unique scent and taste of truffles give these ascomycete fruiting bodies a market value of £3000 per kilogram. These fungi have evidently succeeded in attracting mammalian fungivores.

Psilocybin from 'magic mushrooms' is a hallucinogen that has long intrigued humans. Recreational use of magic mushrooms is a widespread cultural phenomenon that affects the public perception of mycology. One hundred or so species of *Psilocybe*, including the liberty cap, *Psilocybe semilanceata*, contain an alkaloid called psilocybin whose structure resembles the neurotransmitter serotonin. Psilocybin is converted into psilocin after ingestion, which is thought to be the psychoactive compound. When psilocin binds to a pair of serotonin receptors in the central nervous system, it causes a temporary reduction in blood flow to certain parts of the brain that diminishes neurological activity. This stimulates cross-talk between different regions of the brain resulting in the perception of colours when listening to music. Other common experiences include visions of geometric patterns, altered perception of the passage of time, and stimulation of memory. In a clinical setting, the majority of test subjects who ingested purified psilocybin reported very positive experiences and some ranked their hallucinations among the most elevating experiences of their lives. This finding has encouraged researchers to investigate the potential uses of psilocybin in the alleviation of anxiety and the treatment of clinical depression.

Magic mushrooms have been used in Mesoamerican religious practices. In the Aztec religion, hallucinogenic mushrooms were regarded as the flesh of the gods, or teonanácatl, which fostered communication between the gods and their priests. Descriptions by eighteenth century explorers suggest that the fly agaric mushroom, *Amanita muscaria*, was used as an intoxicant in the Russian Far East. The primary psychoactive compound in this species is muscimol, which has different effects on the central nervous system from psilocybin. Muscimol crosses the blood-brain barrier, binds to the $GABA_A$ receptor and interferes with the transmission of nerve impulses. The resulting elevation of serotonin and dopamine levels induces feelings of weightlessness and changes in size perception. Claims about more widespread ritual uses of hallucinogenic mushrooms, including their worship by Neolithic cultures in Europe, are not supported by critical archaeological and anthropological research.

Psilocybin and muscimol synthesis have both evolved independently in a number of mushroom lineages. Outside the genus *Psilocybe*, which is classified in the family Hymenogastraceae, psilocybin is produced by some species of *Panaeolus* (family Psathyrellaceae) and *Inocybe* (Inocybaceae), and occurs in at least five other families in the Agaricales. Muscimol is similarly widespread among unrelated basidiomycetes. The repeated emergence of psilocybin and muscimol synthesis suggests that these alkaloids have some adaptive significance, but their role in fungal physiology is unknown. The most compelling idea is that they act as antifeedants that protect fruit bodies from insect damage, but this is not bolstered by any experimental data.

Despite their huge variety, most secondary metabolites fall into three broad chemical types: polyketides, non-ribosomal peptides, and terpenes. Some examples of bioactive secondary metabolites are shown in Figure 5.9. The low diversity of synthetic pathways contrasts with the enormous diversity of products. A possible evolutionary explanation for this paradox is that any bioactive product that is only a step away from primary metabolism would be more likely to interfere with metabolism and be eliminated by natural selection. However, once a

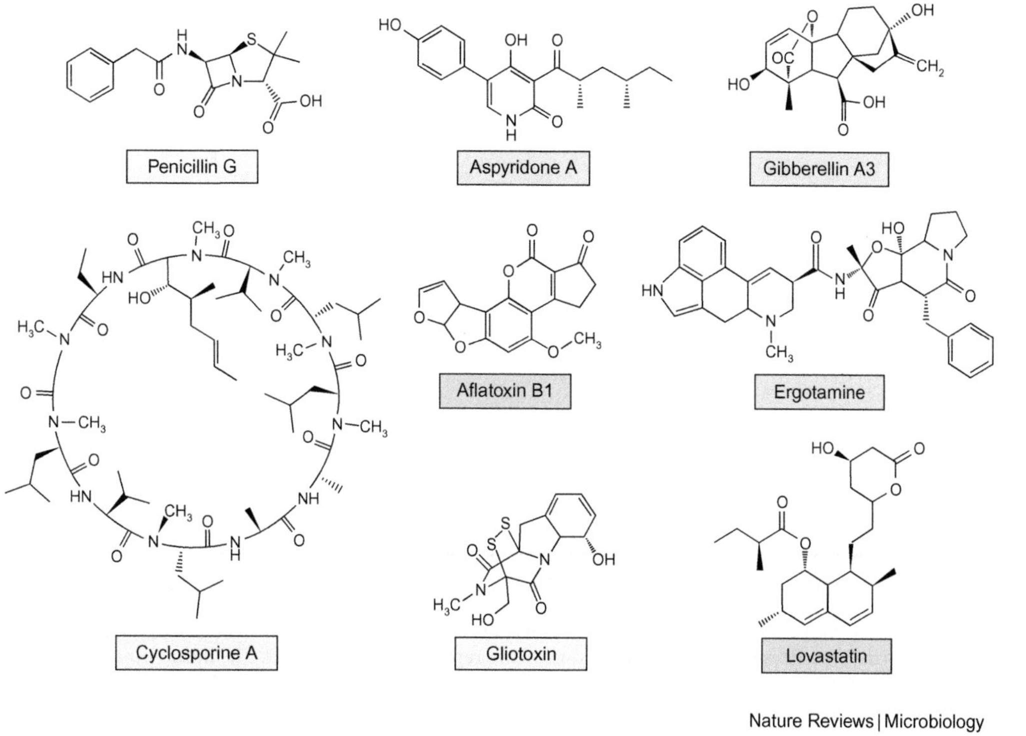

FIGURE 5.9 Examples of secondary metabolites produced by fungi. The clinically used antibiotic penicillin is produced by *Penicillium chrysogenum*. Other clinically used secondary metabolites include immunosuppressants such as the cyclosporines and the cholesterol-reducing compound lovastatin, produced by *Tolypocladium inflatum* and *Aspergillus terreus*, respectively. Many secondary metabolites have adverse toxic activities, such as the aflatoxins produced by *Aspergillus flavus* and gliotoxin, produced by *Aspergillus fumigatus*. The toxicity of gliotoxin has been attributed to a disulphide bridge that is the functional motif of this metabolite. Aspyridones, from *Aspergillus nidulans*, have moderate cytotoxic activity. Gibberellins are plant hormones that are also produced by fungi such as *Fusarium fujikuroi*. Ergot alkaloids are produced by several fungi, including (most prominently) *Claviceps purpurea*, which produces ergotamines. Light grey indicates non-ribosomal peptide derivatives; dark grey represents a non-ribosomal peptide derivative that requires a tryptophan dimethylallyltransferase for synthesis; red (dark grey in the print version) represents polyketide derivatives; blue (dark grey in the print version) represents a mixed polyketide–non-ribosomal peptide compound; green (dark grey in the print version) represents a gibberellin, for which synthesis involves terpene cyclase but no non-ribosomal peptide synthetase or polyketide synthase. *Source: Brakhage (2013).*

secondary metabolic pathway is established in a lineage of organisms, selection is likely to favour its elaboration, because any bioactive intermediates will then be less likely to interfere with primary metabolism.

Genes encoding secondary metabolic pathways are clustered in the genome, enabling their expression and cellular function to be coordinately regulated. Figure 5.10 shows the steps in synthesis of atromentin, a secondary metabolite characteristic of Boletales, and Figure 5.11 shows the arrangement of the gene cluster governing this biosynthesis. A variety of atromentin-derived molecules are produced by Boletales (Figure 5.12), and endow the sporophores of this group, including those of porcini, *Boletus edulis*, and also the dry rot fungus, *Serpula lacrymans*, with

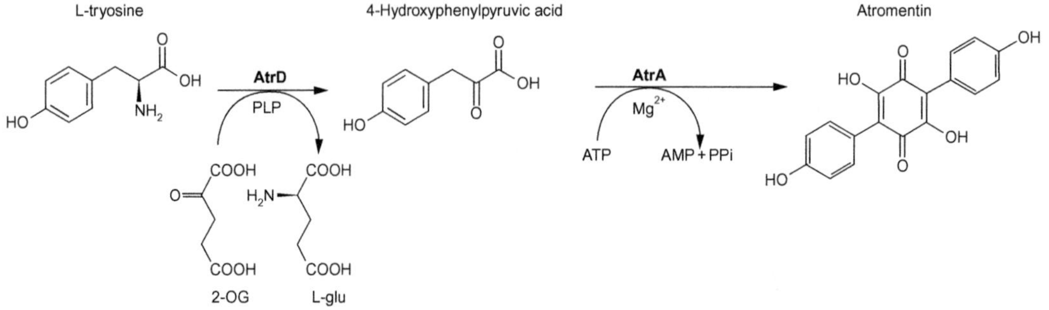

FIGURE 5.10 The pathway of atromentin synthesis in *Tapinella* (Boletales). *Tapinella panuoides* (Boletales) synthesises the secondary metabolite atramentin from L-tyrosine derived from the shikimic acid pathway. *Source: Schneider et al. (2008).*

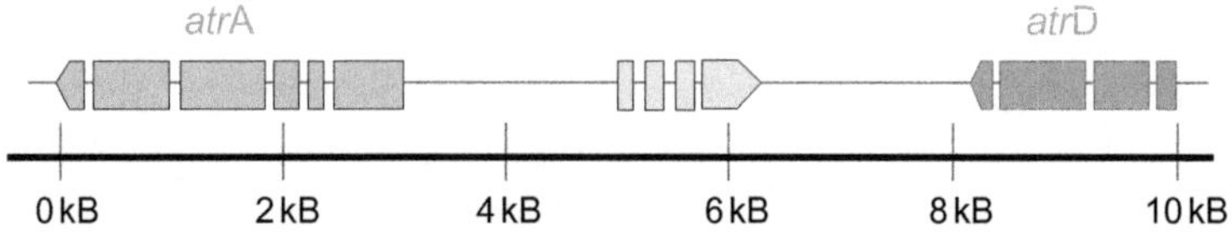

FIGURE 5.11 Genes encoding biosynthetic enzymes for atromentin in *Tapinella*. Genetic map of atromentin biosynthetic genes in *Tapinella panuoides*. The dark grey arrows represent atrA and atrD. The intron positions within the genes are indicated as spaces between arrow segments. The reading frame between atrA and atrD, shown in light grey, codes for a putative alcohol dehydrogenase. *Source: Schneider et al. (2008).*

their characteristic yellow and brown coloration. The expression of the antibiotic secondary metabolite bikaverin has been investigated in the ascomycete genus *Fusarium*, which includes soil saprotrophs and plant pathogens. Many *Fusarium* species synthesise this red polyketide pigment and have multiple genes encoding its synthesis. In an analysis of the *Fusarium fujikuroi* genome, the bikaverin gene cluster was found to include genes not only for bikaverin synthesis, but also for regulating its biosynthesis in response to nitrogen starvation and acid pH, and for secreting bikaverin via an efflux pump. Fungal endophytes of plants (described in detail in Chapter 7) produce a plethora of bioactive secondary metabolites. In some cases these are also produced by the host plant, the gene cluster encoding their biosynthesis having apparently undergone horizontal gene transfer (see Chapter 4). The gene cluster for biosynthesis of the diterpenoid tumour suppressant taxol is present and expressed both in yew trees, (*Taxus* sp.), and also in their fungal endophyte. The presence of a plastid-targeted signal sequence in one of the enzymes indicates that the pathway for taxol synthesis originated in the plant.

Secondary metabolic pathways are typically expressed in response to environmental cues (Figure 5.13). Developmental processes may be simultaneously regulated, for example when sporophore development is accompanied by the synthesis of new spore wall components, and when fruit bodies acquire pigments, flavours, and toxins. The genetics underlying regulation has been intensively studied for decades in the model fungus *Aspergillus nidulans*, which shows light-induced asexual sporulation, while in the dark it produces both the sexual fruiting bodies and the carcinogenic **aflatoxin** precursor sterigmatocystin. Development is under the control of the velvet gene family, conserved across all fungi, which encode transcription factors. From genetic and experimental investigation, it emerges that development and secondary

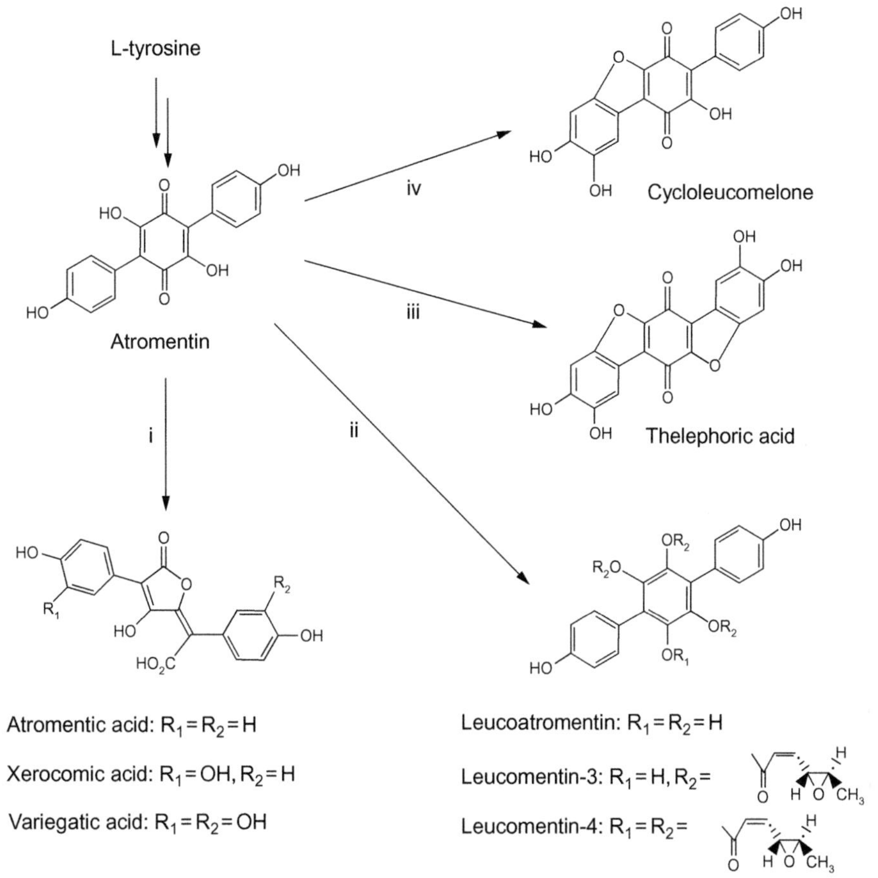

FIGURE 5.12 Synthesis of atramentin derivatives in Boletales. Atromentin and examples of derivatives. Atromentin can undergo modifications such as (i) oxidative ring splitting into atromentic acid, which upon sequential hydroxylation yields xerocomic, and variegatic acid, (ii) reduction and esterification to produce leucoatromentin and leucomentins, (iii) dihydroxylation followed by symmetric heterocyclization for production of thelephoric acid, or (iv) hydroxylation and single heterocyclization to yield cycloleucomelone. *Source: Schneider et al. (2008).*

metabolism interact through velvet and LaeA proteins, which bind to form a trimeric heteropolymer. In this form, LaeA can cross the nuclear membrane to act directly on DNA.

Mining the full genome sequences of fungi indicates that their potential to produce secondary metabolites is greatly underestimated. Gene clusters related to secondary metabolism are being explored for genes potentially exploitable in biotechnology for the production of bioactive products. Most of the biosynthesis gene clusters are silent under laboratory conditions, leading to a search for the physiological conditions that activate these genes. A direct experimental demonstration of the function of secondary metabolism in an ecological interaction was the finding that silent clusters of biotechnologically interesting *PKS* genes in the genome of *A. nidulans*, which are not transcribed in sterile conditions, could be induced by contact with soil streptomycetes. Substances induced by this microbial interaction include the polyketide orsellinic acid, the lichen metabolite lecanoric acid, and cathepsin K inhibitors.

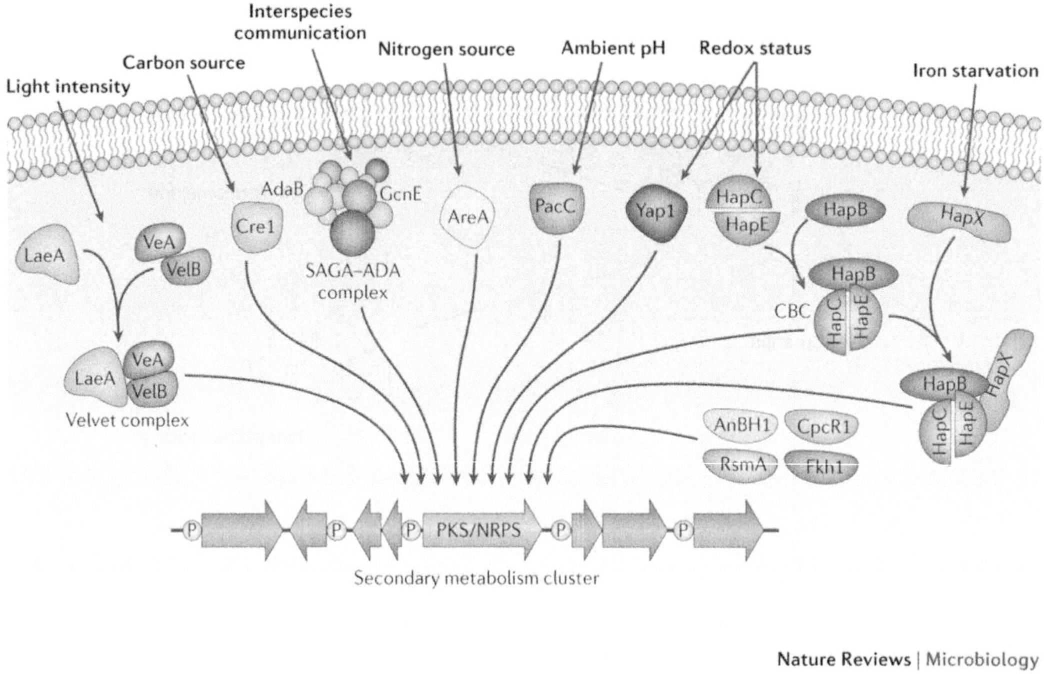

FIGURE 5.13 Regulation of secondary metabolism by environmental factors. Environmental signals can influence the regulation of various secondary metabolism gene clusters through regulatory proteins that respond to these environmental stimuli and, in turn, modulate the expression of the clusters. Shown is a model secondary metabolism gene cluster containing a gene encoding a central non-ribosomal peptide synthetase (NRPS), a polyketide synthase (PKS), or a hybrid PKS–NRPS enzyme. CBC, CCAAT-binding complex; CpcR1, cephalosporin C regulator 1; LaeA, loss of aflR expression A; RsmA, restorer of secondary metabolism A; SAGA–ADA, Spt–Ada–Gcn5–acetyltransferase–ADA. *Source: Brakhage (2013).* (See the colour plate.)

SENSING AND RESPONDING TO THE ENVIRONMENT

Fungi sense their environment with cellular systems homologous with animals, and can perceive most of the same stimuli as human cells, including levels of extracellular and intracellular nutrients, pheromones, pH, water and oxidative stress, temperature, light intensity and spectral composition, gravity, and touch.

Nutrient sensing

Nutrient sensing is essential for the opportunistic lives of fungi. In its ever-changing environment the cell and mycelium must be able to respond to the external availability of the different nutrients it requires. Internal homeostasis requires the cell to sense and respond to fluctuating intracellular levels of metabolites, and to physical variables including water potential and pH.

Sensors for extracellular glucose enable hyphae to reduce wasteful enzyme release by repressing the synthesis and secretion of cellulases and other enzymes. Several mechanisms sense ambient glucose. The membrane protein Gpr1, a G-protein coupled receptor (GPCR), is probably conserved across fungi and acts as a receptor to sense glucose and sucrose. It initiates an intracellular signalling pathway that activates adenylyl cyclase to raise cellular cAMP and activate protein kinase. Fungi also have the hexose transporter gene family *HXT*, which encodes sugar transporters with varied affinities and expression patterns that respond to the amount and type of available sugar. Homologues without transport function, conserved across ascomycetes and basidiomycetes, also act as sugar sensors, or **tranceptors**, that regulate the expression of ***HXT*** genes through the transcription suppressor protein Rgt1. Genes involved in trehalose metabolism also mediate fungal responses to sugars. The Tps1 enzyme protein, trehalose-6-phosphate synthase, is found to act as a central regulator of plant infection in the rice blast fungus, *Magnaporthe oryzae*, (Chapter 8) where it functions as a sugar sensor. Moreover, in this plant pathogen, Tps1 was found to have an additional role in integrating carbon and nitrogen metabolism, via control of the pentose phosphate pathway. This equips the fungus to adapt to nutritional and redox conditions inside the host cell during infection. GPCRs also sense pheromones. In ascomycetes and basidiomycetes, GPCR binding of peptide pheromones initiates signal cascades leading to sexual morphogenesis, taxes, and tropisms to bring mating partners together and orchestrate nuclear fusion (Chapter 4).

Amino acids are sensed with a three gene encoded system, *SPS*, consisting of a tranceptor *Ssy1*, which on binding an amino acid activates a protease, *Ssy5*. *Ssy5* activates transcription factors, leading to the expression of amino acid transporters and pathways for amino acid metabolism. In this way, the cellular machinery for uptake is coordinately activated but only when it is required. In the human pathogen *Candida albicans*, this amino acid sensing pathway is required for virulence, presumably because it alerts the fungus to the presence of host tissue. Specific amino acids may act as cues, sensed via a G-protein coupled receptor Gpr1, controlling the switch from yeast to hyphal morphology that allows the fungus to invade tissue. Nitrogen availability is sensed by Mep cell surface ammonium transporters described above. Mep2 is expressed under conditions of low ammonium supply and is believed to act as a transporter-driven sensor which triggers filamentation (Figures 5.14 and 5.15).

Intracellular nutrient levels

Intracellular nutrient levels change continuously as fungi forage in a heterogeneous environment. Changes in cellular nutrient status not only induce coordinated regulation of cellular systems involved in nutrient acquisition and metabolism, but also have developmental effects that include changes in cell shape, cell cycle progression, the initiation of spore development, the formation of sporophores and multicellular tissues, and starvation-induced autolysis and autophagy. Such responses are central to fungal niche adaptation, allowing the mycelium to optimize the spatial and temporal allocation of its available resources to accommodate to its changing environment.

Intracellular carbon nutrients are sensed via the cAMP-PKA pathway, and nitrogen nutrients by the *Tor* (Target of rapamycin) signalling pathway. The central role of the Tor protein in cellular responses to nitrogen supply was discovered through studies of the antibiotic rapamycin, which was originally investigated as a tumour suppressant. The *Tor* gene encodes a kinase that controls protein phosphorylation cascades that regulate numerous growth processes. Fungi have two homologues of *Tor*, *Tor1* and *Tor2*. Each is a multiprotein complex. Tor-mediated signalling is activated in the nitrogen replete cell and upregulates multiple cell functions including protein synthesis, ribosome biogenesis, yeast dimorphic transitions, and cell cycle progression. Inactivation of Tor signalling by rapamycin or starvation arrests growth and activates the autophagy pathway. Regulation is by control of the entry of transcription factors to the nucleus, mediated by phosphorylation and dephosphorylation reactions. Tor signalling is likely to be involved in mediating morphogenetic responses to changes in external nitrogen availability.

Physical Factors

Water

All fungal activities are dependent on the regulation of cellular water content. The hydration and dehydration of fungal colonies is determined by differentials in water availability between the organism and its surroundings. Water potential, ψ, is the recommended term for quantifying water availability, defined as the potential energy of water per unit volume. The benchmark of zero water potential refers to pure water at atmospheric pressure. Water potential is reduced to negative values by dissolved solutes; water potential is increased by hydrostatic pressure. The difference in hydrostatic pressure between a cell and its surroundings is called turgor pressure (also known as pressure potential, ψ_p). The water potential of a cell is determined by its osmotic potential (ψ_π), proportional to the concentration of dissolved solutes, and its turgor pressure according to the following equation:

$$\psi = \psi_\pi + \psi_p$$

All three terms are expressed as pressures with the Pascal ($Pa = Nm^{-2}$) as the SI unit. These simple considerations are often complicated by the use of the term osmotic pressure, which is equal in magnitude to osmotic potential but opposite in sign, and its confusion with turgor: osmotic pressure is not synonymous with turgor. Water potential is also affected by interactions at solid–liquid interfaces such as the surface of colloids. The term matric potential has been used to represent these effects, but, in most cases, the influence on the solid phase contributes to the osmotic potential and turgor of the cell, and is not measured separately. The matric potential of liquid or solid culture media is insignificant, but it becomes an important parameter in soils and can be measured with an instrument called a tensiometer.

Water moves into a hypha by osmosis in response to a differential in water potential. When the osmotic potential of the cytoplasm is lower than the fluid surrounding the hypha, there is a net influx of water and turgor pressure rises until the water potential of the hypha matches its surroundings. As the hypha grows, its turgor pressure tends to fall, but physiological adjustment of the osmotic potential maintains relatively constant turgor. Osmotic adjustment, or regulation, involves solute uptake and excretion as well as the synthesis of compatible solutes including sugar alcohols.

Under most circumstances, the water potential of an active mycelium is closely matched to the water potential of its surroundings. The turgor pressure of the constituent hyphae can be measured directly using a pressure probe (see Chapter 2), or estimated from the difference between the osmotic pressure of the cytoplasm and the surrounding fluid phase. For fungi growing at atmospheric pressure, the external water potential is equal to osmotic potential because the pressure potential term is zero:

$$\psi_{(\text{outside})} = \psi_{\pi(\text{outside})} = \psi_{(\text{inside})} = \psi_{\pi} + \psi_{\text{p}}$$

$$\text{and, mycelial turgor pressure} = \psi_{\pi(\text{outside})} - \psi_{\pi(\text{inside})}$$

Osmotic potential of fungal samples can be measured very accurately using osmometers. Some of these determine osmotic potential from the depression of the freezing point of a sample of cytoplasm, or other fluid, relative to pure water. However, vapour pressure deficit osmometers are superior for most applications. These measure the vapour pressure of water in equilibrium with the fungal sample, from the lowering of the temperature necessary to reach the dew point.

Research on fungal water relations has been complicated by the use of alternative terminology. Food microbiologists, for example, prefer to use the term water activity (a_{w}) to quantify water availability. This is the ratio of water vapour pressure between the sample (p_{s}) and pure water (p_{w}):

$$a_{\text{w}} = \frac{p_{\text{s}}}{p_{\text{w}}}$$

Water activity varies from zero (no water) to 1.0 (pure water). Water activity is related to water potential using the following logarithmic equation:

$$\text{water potential}(\psi) = k \,\text{In}\, a_{\text{w}}$$

k is a temperature-dependent constant; $k = 1.35$ at 20 °C and 1.37 at 25 °C.

The availability of water in the gas phase is expressed in terms of relative humidity. Relative humidity is equivalent to the water activity represented as a percentage (e.g. $a_{\text{w}} = 0.75 = 75\%$ relative humidity).

WATER POTENTIAL AND THE FUNGAL ENVIRONMENT Fungal activity, whether it is manifested as plant disease, mouldiness of materials, rotting of wood, or the appearance of mushrooms in fields and woodlands, is most obvious in damp conditions. This is consistent with experiments demonstrating that most fungi grow best at high water potentials, above −1 MPa. Commonly used media, including those containing 2% sucrose, have a water potential in this range; 0.4 M sucrose, equivalent to a 14% solution, has a water potential of −1 MPa. At lower water potentials, rates of hyphal growth diminish, until values are reached at which growth does not occur (Table 5.3). Most wood-destroying fungi, for example, cannot grow at water potentials below −4 MPa. The inhibition of fungal growth at low water potentials is the basis for such traditional methods of food preservation as drying and the addition of salt or sugar. A few fungi, however, are adapted for growth at very low water potentials. These are usually termed **osmophiles** or **xerophiles**, although most are really osmotolerant, growing best at relatively high water potentials. Many yeasts and species of *Aspergillus* (teleomorph *Eurotium*) and *Penicillium* (*Talaromyces*) are osmotolerant and are important agents of biodeterioration. Metagenomic analysis of very dry habitats in the high Andes revealed a group of fungi nested within the Spizellomycetales, an order of chytrids. This was particularly

TABLE 5.3 Water Availability in Different Environments and Approximate Lower Limits for Growth of Some Fungi

Water activity	Water potential (MPa)	Examples
1.0	0	Pure water
0.996	−0.5	*Phytophthora cactorum*, lower limit
0.995	−0.7	Typical mycological media
0.98	−2.8	Sea water
0.97	−4	Most wood-destroying fungi, lower limit
0.95	−7	Bread. Leaf-litter Basidiomycetes, lower limit
0.90	−14	Ham. *Neurospora crassa*, lower limit
0.85	−22	Salami. *Saccharomyces rouxii* in NaCl solution, lower limit
0.80	−30	*Aspergillus nidulans* and *Penicillium martensii*, lower limits
0.75	−40	Saturated NaCl solution, *Aspergillus candidus*, lower limit
0.65	−60	22 molal glycerol
0.60	−69	Limit for cell growth – *Zygosaccharomyces rouxii* in sugar solutions and the mould *Monascus* (*Xeromyces*) *bisporus*
0.58	−75	Spores of some *Eurotium*, *Aspergillus*, and *Penicillium* species are able to survive for several years
0.55	−80	Saturated glucose solution. DNA denatured
0.48	−90	Antarctic dry valleys

Data from various sources. Molality (molecular weight in grams per 1000 g solute) and not molarity (MW in g per final volume of 1000 ml) is used in dealing with osmotic potentials. The lower limits of growth are those obtained at optimal temperature and nutrition; when these factors are sub-optimal, the limits are not so low. As indicated with *Saccharomyces rouxii*, organisms are usually more tolerant of high sugar than high salt concentrations.

unexpected because most Chytridiomycota are freshwater organisms. It is tempting to speculate that this discovery of apparently xerotolerant chytrids came from resistant spores deposited in herbivore dung. The most osmotolerant fungi known can grow at a water potential of −69 MPa. The dry rocky valleys of Antarctica, with a soil water potential of about −90 MPa, are sterile, except for transient microbial growth following snowfall and in relatively humid microenvironments such as cracks in rocks. The spores of some fungi are able to escape desiccation through development of impermeable walls, and can survive bone dry conditions until moisture allows growth to resume.

ADAPTATION TO CHANGES IN EXTERNAL WATER POTENTIALS The water potential of a growing fungus is a little lower than the water potential of its surroundings, driving water influx and permitting cell expansion. The external water potential may fall, however, through evaporation, increasing the concentration of solutes dissolved in the ambient medium. If this reverses the gradient in water potential, water will leave the cell and growth will cease. The cell membrane may detach from the inner surface of the cell wall, an effect known as plasmolysis, and desiccation and cell death may occur. If dew or rain increases the external water potential, the cell will absorb water and its hydrostatic pressure, or turgor, will rise. If this occurs swiftly and the cell is poorly adapted to low external water potential, the increase in turgor pressure may rupture the cell wall. Fungi that live in some habitats are exposed to wide fluctuations in water potential and must adjust their cytoplasmic osmotic potential to cope with these environmental circumstances. Lichens are especially well adapted for life in habitats that experience extreme changes in environmental water potential. Desert lichens, for example, are active photosynthetically for a brief time at dawn, when their thalli containing the photobiont cells are damp with dew, but become desiccated in the morning and remain inactive until dampened again the next day.

One way in which the osmotic potential of a fungal cell can be lowered is by the uptake of dissolved solutes from the environment. *Thraustochytrium aureum*, a chytrid that lives in brackish water and in the sea, can adjust its internal osmotic potential by the uptake of inorganic ions. At high concentrations, however, inorganic ions and many other solutes can change the configuration and catalytic activity of enzyme molecules. Hence, where very low cytoplasmic osmotic potentials are required, many fungi synthesise polyhydric alcohols (polyols), which are 'compatible solutes' that have little effect on enzyme activity even at high concentrations. These polyols may be produced from sugars taken up from the growth medium or from the breakdown of polymeric reserves. The polyol concerned with osmotic adjustment in the moderately osmotolerant yeast *Saccharomyces cerevisiae* as well as the highly osmotolerant species *Zygosaccharomyces rouxii* is glycerol. Other polyols important in osmotic adaptation in fungi are mannitol and arabitol. When an increase in the internal osmotic potential of a fungus is needed, this can be brought about by the loss or export of solute to the environment, or by conversion of the solute to an insoluble reserve material.

Molecular genetic responses of fungi to hyper- and hypoosmotic stress (drying and wetting, respectively) were first investigated in *Saccharomyces cerevisiae*. Yeast cells respond to hyperosmotic stress by temporary cessation of growth, with disassembly of the actin cytoskeleton and loss of cell polarity, decrease in cell wall porosity and membrane permeability to glycerol, and accumulation of glycerol. The chitin content of the cell wall increases too. Yeast cells detect and respond to high and low extracellular osmolarity by activating two different MAPK signalling pathways. High osmolarity activates the HOG (high osmolarity glycerol) pathway. This pathway exhibits multiple redundancies, starting at the sensor level, where two independent branches activate the MAPK signalling cascade: a putative transmembrane osmosensor, and a protein phosphorylation relay system which is inhibited by hyperosmotic stress. In both cases, the result is activation of the MAPK cascade, which in turn activates the *HOG1* gene. The HOG1 protein in turn induces transcription of several genes, including *GDAP1*, the structural gene for glycerol 3-phosphate dehydrogenase involved in glycerol synthesis. The cell then accumulates glycerol which lowers its osmotic potential to restore water influx. Accompanying responses include a decrease in membrane permeability to glycerol. Hypoosmotic stress activates the protein kinase C (PKC) signal transduction pathway. The PKC pathway is activated by a range of other stimuli such as nutrient stress, and it is thought that a major role for this molecular response is the maintenance of cell integrity by controlling the assembly of the cell wall and plasma membrane synthesis.

It has already been emphasised that the cell wall is a dynamic structure that must be maintained to resist mechanical and chemical stress. The modifications required for this entail activation of enzymes that synthesise cell wall polymers, and the vectorial transport of vesicles that carry wall and membrane components. The stretching of the cell membrane caused by hypoosmotic stress is thought to be detected by a mechanosensor, which activates a small GTP-binding protein, Rho1, which controls the activity of the Pkc1 protein and a $\beta(1 \rightarrow 3)$ glucan synthase. Pkc1 activates the MAPK cascade, and a result is the activation of chitin synthase genes for cell wall biosynthesis, proteins involved in mannosylation and a GPI-anchored membrane protein. Activities of these proteins produce a stronger wall with higher chitin content. Chitin synthases from a range of fungi are also activated quickly following hypoosmotic stress, probably from preexisting inactive forms, giving a rapid increase in wall strength. Genomic analysis has allowed comparison of HOG pathway component genes across a number of fungal species. This reveals subtle differences between the proteins comprising the pathway, many of which are likely to have adaptive significance. Apart from raising glycerol content, responses mediated via the HOG pathway in filamentous fungi include the onset of conidiation, the ability to infect and invade plant cells, and secondary metabolite production, including the induction of aflatoxin synthesis in *Aspergillus*.

Fungal cells control their water potential not only as a stress response but also as an essential part of 'normal' growth and development. In addition to regulating their levels of compatible solutes, fungi can also control their water content via changes in the permeability of the cell membrane and cell wall. Gene expression associated with developmental water partitioning between cells has been analysed in the plant pathogen *Magnaporthe oryzae*. Development accompanying leaf infection involves a series of controlled changes in the water potential of adjacent cells. First the appressorium inflates and drives the infection peg through the plant cell wall, and then it deflates as the cytoplasm moves into the mycelium that develops in the plant (Chapter 7). **Aquaporins**, membrane proteins that channel water and small solutes, have been investigated in *Saccharomyces* and the mycorrhizal fungus *Laccaria bicolor* (see Chapter 7). *L. bicolor* has seven aquaporins that differ in substrate specificity and regulation, and are likely to be important in integrating host and symbiont physiology.

Light

Light is a vital environmental cue for development in all organisms. In the fungal kingdom, light responses are particularly important to ensure that spores for dispersal are produced in the open air and not buried in the substratum. Fungal physiology is highly responsive to light, with light-induced effects including change in asexual conidiation, the circadian clock, sexual development, and secondary metabolism. Light is sensed over the whole spectrum, from ultraviolet to far-red, and over a range of intensity from starlight to bright sunshine. Detailed investigation at molecular level has revealed three light sensing systems in fungi, based respectively on flavin-based photoreceptors for blue light, phytochrome for red light, and opsins related to the rhodopsin of animals and some archaea.

Blue-light responses are found in all organisms including prokaryotes and eukaryotes. The fungal blue-light response has been investigated in detail in *Neurospora crassa*, in which effects include carotenoid synthesis, induction and phototropism of protoperithecia, induction of hyphal growth, asexual sporulation, and the entrainment of the circadian clock. The best-described blue-light receptor in the fungal kingdom is the protein White Collar 1 (WC-1). The *WC* (white collar) genes *WC-1* and *WC-2* were discovered in from *Neurospora* mutants unable to produce mycelial carotenoids in the light, resulting in a 'white collar' round the colony margin. *WC*-like proteins are conserved at sequence and functional level across basidiomycetes, ascomycetes, and zygomycetes. They incorporate a flavin-based photoreceptor which absorbs blue light causing a reaction from the protein. From sequence analysis, WC-1 and WC-2 proteins are GATA-type transcription factors, and share further domains that interact during the response to light. *WC* genes are essential components of the *Neurospora* circadian system which regulates cellular processes so that they follow a 24-h cycle. The **circadian** system is a biological clock that enables organisms to measure time, anticipate diurnal changes in environmental factors, and regulate growth, sporulation, and associated patterns of gene expression. Biological rhythms are termed circadian when the periodicity is set endogenously and is not affected by changes in nutrient availability or temperature variations. The rhythm can be entrained by manipulating light or temperature conditions to set the clock. WC proteins interact with a very large number of clock control genes including a central timekeeping oscillator in the form of rhythmically expressed *frq* genes. This system has been intensively investigated in the *Neurospora* clock, producing a detailed model of the circadian system of intracellular feedback loops between environmental cues and cellular processes.

Red-light responses mediated by phytochrome are widespread in plants, which use phytochromes to respond to the balance between red and far-red light. Phytochromes are photoreceptor proteins with a linear tetrapyrrole as the chromophore. Similar red light sensing has been found in fungi. *Aspergillus nidulans*, as described above, forms abundant asexual conidia in light at wavelength 680 nm, but at 740 nm, or in the dark, produces sexual spores and toxic secondary metabolites. Phytochromes have been identified in the genomes of several fungal species, and have been analysed in detail in *Neurospora* and *Aspergillus*. *Aspergillus nidulans* has a single phytochrome gene, *fphA*, which appears to be involved in suppressing sexual development, because deletion results in a mutant phenotype which produces sexual spores in the light.

Fungi also have light-absorbing pigments, **opsins**, related to rhodopsin, the transmembrane-light-absorbing proteins used for light sensing in the animal retina and for energy transduction in archaeal prokaryotes. Opsins are conserved across ascomycetes and basidiomycetes and are believed to have originated from prokaryotes by horizontal gene transfer. Their role in fungi is not well understood. The opsin NOP-1 from *Neurospora crassa* absorbs green light, but deletion of *NOP-1* has no known phenotypic effect.

Apart from responding to light as an environmental cue, fungi have mechanisms for protecting the cell from its damaging effects. Fungi are exposed to potentially damaging ultraviolet light when sporulating on exposed surfaces and when spores travel long distances through air. Spore walls are protected by light-absorbing secondary metabolites including melanin, carotene and sporopollenin. Melanin is also produced as a response to ionising radiation.

Temperature

Fungi are exposed to a wide range of temperatures in natural environments that affect water availability. Many species cannot grow at temperatures above 30–40 °C. Fungi that can grow near freezing point or even a little below are termed psychrotolerant (from the Greek, *psychros*, cold), and if incapable of growth above 20 °C, psychrophilic. A group of fungi called snow moulds grow on vegetation such as grass or unharvested crops when these are buried by snow. These can be problematic when they kill grass or produce mycotoxins. Many cold-tolerant yeasts are known and some of them, including some basidiomycete yeasts, are psychrophilic. Yeasts are among the few microorganisms found in the cold, dry valleys of Antarctica. Temperatures well above 40 °C are often encountered in accumulations of decomposing vegetation and compost. *Mucor pusillus*, *Chaetomium thermophile*, and *Thermoascus aurantiacus* are examples of species that play an important role in the succession of organisms involved in successful composting. Fungi are described as thermotolerant if they are capable of growth at 50 °C or higher temperatures, and thermophilic if incapable of growth below 20 °C. The highest temperature at which fungal growth has been recorded is 60 °C. The vast majority of fungi that are neither psychrophilic nor thermophilic, favour intermediate temperatures and are termed mesophiles. Mesophiles can be psychrotolerant or thermotolerant.

Phylogenetic analysis shows that thermophily is restricted to a few groups of fungi including ascomycetes in the Sordariales, Onygenales, and Eurotiales, and zygomycetes classified in the Mucorales. Culture-independent analyses of fungal communities in hot natural habitats may well discover others.

Carbon Dioxide

Carbon dioxide is involved in signalling as well as in respiration. Its concentration increases around mycelia respiring in confined spaces, and it can act as a cue for morphogenesis. The morphogenetic effect of CO_2, particularly on sporulation and sporophore development, is well known, but the cellular and molecular mechanisms have not been investigated until recently. In the human pathogens *Candida albicans* and *Cryptococcus neoformans*, carbon dioxide sensing is found to be essential for virulence. In host tissues the carbon dioxide level can vary a hundredfold. Carbon dioxide enters the cell passively, by diffusion. CO_2 / HCO_3^- homeostasis involves carbonic anhydrase, a zinc metalloenzyme which catalyses conversion of CO_2 to the bicarbonate used as a substrate in biosynthetic carboxylation reactions. High levels also mediate signalling pathways that induce *C. albicans* to form hyphae and *C. neoformans* to produce capsules resistant to phagocytosis. Bicarbonate stimulates adenylyl cyclase to produce cAMP, which then activates a protein kinase to induce further steps mediating development.

pH

While the pH of the surrounding environment may vary widely, cellular function demands a constant internal pH. Fungal cells operate a pH-homeostatic system. This system also senses ambient pH and adjusts gene expression so that secreted enzymes and bioactive metabolites are released only at pH levels that permit function. The genes and cellular processes involved have been characterised in *Aspergillus nidulans*, *Saccharomyces cerevisiae*, and *Candida albicans*. Ambient pH is sensed by a complex of membrane proteins and a further endosomal complex mediates signal transduction. A transcriptional regulator (*PacC* in *A. nidulans*, *Rim101p* in *S. cerevisiae*) activates genes expressed under acidic conditions and represses them under alkaline conditions. The pH regulatory system is crucial to fungal pathogenicity in animals and plants. In *C. albicans*, genome-wide transcriptional profiling identified 514 pH-responsive genes, including those expressed in iron acquisition and invasive hyphal growth.

Fungi can affect environmental pH by secreting protons and organic acids. Brown rot basidiomycetes that secrete oxalic acid can acidify a culture medium to pH as low as 2.5. It has been suggested that this ability aids wood decay by tipping the ionic equilibrium towards a preponderance of ferrous iron that participates in cellulose breakdown through a Fenton's reaction, as described above. Fungal acidification of the environment can cause mobilisation and leaching of soil cations, and mineral transformations in rock are mediated through pH changes as described below.

DEVELOPMENTAL ADAPTATION FOR NUTRIENT ACQUISITION

Fungi feed by growing into fresh food. Alterations in form are necessary to adapt to the resources available. Nutrient sensing and developmental responses are critical in niche adaptation. We illustrate this concept with two well-studied examples, at opposite ends of the fungal size range: dimorphism in a pathogenic yeast, and the development of vast mycelial networks by basidiomycetes inhabiting the forest floor.

Dimorphic yeasts

Dimorphic yeasts are able to switch from unicells to filamentous hyphae in order to penetrate solid materials by means of tip growth. The ascomycete yeast *Candida albicans* causes thrush, affecting the mucous membranes (Chapter 9, p.302). It can grow either as unicellular budding yeast, or in the filamentous form (Figure 5.14) in which it invades epithelial

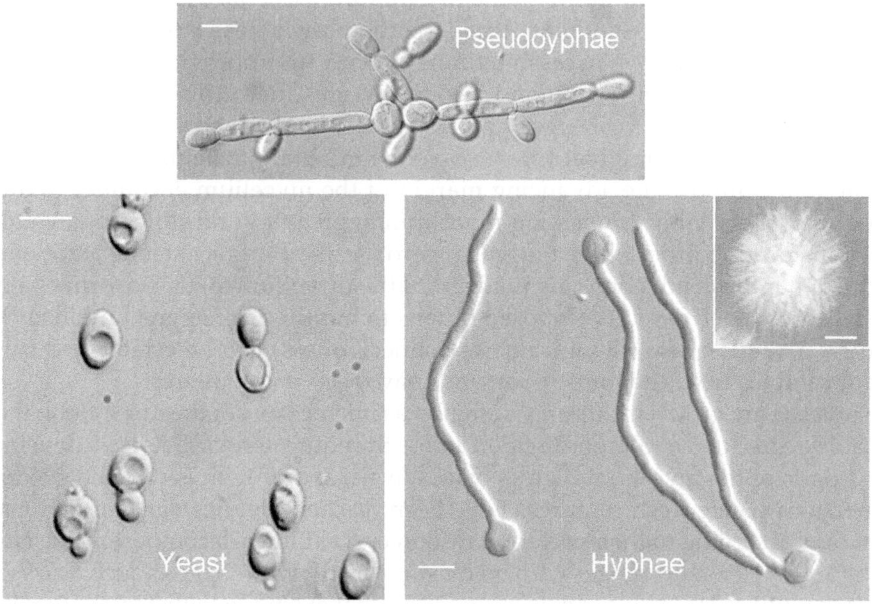

FIGURE 5.14 Yeast, pseudohyphal, and hyphal form of *Candida albicans*. *Source: Sudbery (2011).*

cells of the mucosa. It also uses the hyphal growth form to escape from macrophages that engulf the yeast phase. The switch between growth forms can be readily manipulated in liquid and solid culture, which makes it possible to analyse the process at the cellular and molecular level. A range of chemical and physical cues induce the switch from the yeast to hyphal form. These include starvation, growth in solid medium, and presence of *N*-acetylglucosamine. Cues encountered in mammalian tissue include the presence of serum and amino acids, microaerophilic conditions, neutral pH, a temperature around 37 °C, and carbon dioxide (at levels found in the bloodstream). Hyphal development is inhibited in cells growing close together, when farnesol acts as a cue for quorum sensing. Farnesol is a sesquiterpenoid secondary metabolite secreted by the cells. Molecular genetic methods have uncovered an integrated set of sensors and signal transducing pathways that act via positive or negative transcription factors on a gene, *Hgc1*, to regulate the expression of the cellular machinery for polarised hyphal growth (Figure 5.15). In addition, *N*-acetylglucosamine can induce hyphal growth directly, independently of *Hgc1*. Polarised hyphal growth, described in detail in Chapter 2, requires the assembly of a spitzenkörper and its accompanying apparatus of oriented cytoskeleton and secretory vesicle exocytosis, as well as control of nuclear division and septation as the cell elongates.

Cord-forming basidiomycetes

Cord-forming basidiomycetes, in contrast to yeasts, include the biggest and longest lived fungal individuals known. Honey fungus belonging to the genus *Armillaria* is well known to gardeners and foresters as it kills plants by colonising and killing woody roots. Its mycelium can spread over many metres by growing through soil to invade and destroy adjacent plants. Householders are familiar with the dry rot fungus, *Serpula lacrymans*, which infects damp softwood in buildings, and like *Armillaria* can grow across intervening non-nutrient spaces to colonise and decay successive timbers. Both fungi provide examples of a common mode of growth among woodland basidiomycetes that utilise living roots or fallen trees as massive carbon sources. Sugar translocation from the food base fuels the extension of the mycelium of these fungi and they scavenge mineral nutrients as they grow. When the hyphae that compose the advancing margin of the mycelium encounter and colonise fresh food, the intervening connecting mycelium aggregates and differentiates to form a nutrient-translocating pipe called a mycelial cord. As the fungus extends across the forest floor, more and more connections are made until the mycelium takes the form of a network of mycelial cords. Each cord is composed of tens to hundreds of aligned hyphae. The area occupied by a single clone (though not its connectedness) can be established by genetic analysis of samples from the forest floor, and may reach many metres.

When mycelia are grown in cultures designed as microcosms of the forest floor, they show a remarkable degree of overall coordination across the entire network. As they develop, older, exhausted parts of the mycelium regress by autolysis, and growth is directed preferentially in the directions where fresh resources have been reached. Species vary in the topology of their networks according to their preferred resources. Leaf litter decomposers like *Hypholoma facsiculare* forage over short ranges with many short cords, while species such as *Megacollybia platyphylla* that utilise large logs may produce cords that extend long distances from log to log with far fewer branches and connections (Figure 5.16).

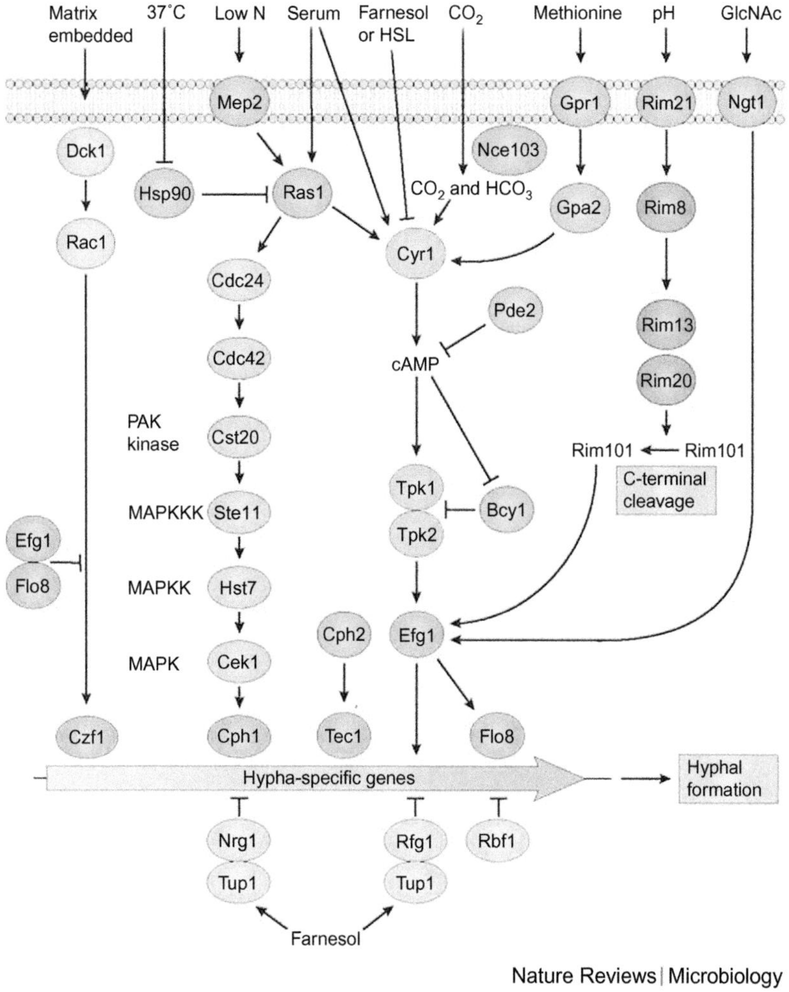

FIGURE 5.15 Signal transduction pathways leading to expression of hypha-specific genes in *Candida albicans*. Multiple sensing and signalling mechanisms that regulate the developmental response of the pathogenic yeast *Candia albicans* to its environment. Environmental cues feed through multiple upstream pathways to activate a panel of transcription factors. The cyclic AMP-dependent pathway that targets the transcription factor enhanced filamentous growth protein 1 (Efg1) is thought to have a major role. In this pathway, adenylyl cyclase integrates multiple signals in both Ras-dependent and Ras-independent ways. Negative regulation is exerted through the general transcriptional corepressor Tup1, which is targeted to the promoters of hypha-specific genes by DNA-binding proteins such as Nrg1 and Rox1p-like regulator of filamentous growth (Rfg1). Protein factors are colour-coded as follows: mitogen-activated protein kinase (MAPK) pathway (green, dark grey in the print version), cAMP pathway (turquoise, dark grey in the print version), transcription factors (orange, grey in the print version), negative regulators (yellow, light grey in the print version), matrix-embedded sensing pathway (light blue, grey in the print version), pH sensing pathway (brown, dark grey in the print version), other factors involved in signal transduction (mauve, dark grey in the print version), C-terminal, carboxy-terminal; Cdc, cell division control; GlcNAc, *N*-acetyl-D-glucosamine; Gpa2, guanine nucleotide-binding protein α-2 subunit; Gpr1, G-protein-coupled receptor 1; HSL, 3-oxo-homoserine lactone; Hsp90, heat shock protein 90; MAPKK, MAPK kinase; MAPKKK; MAPKK kinase; PAK, p21-activated kinase; Rbf1, repressor–activator protein 1. *Source: Sudbery (2011).* (See the colour plate.)

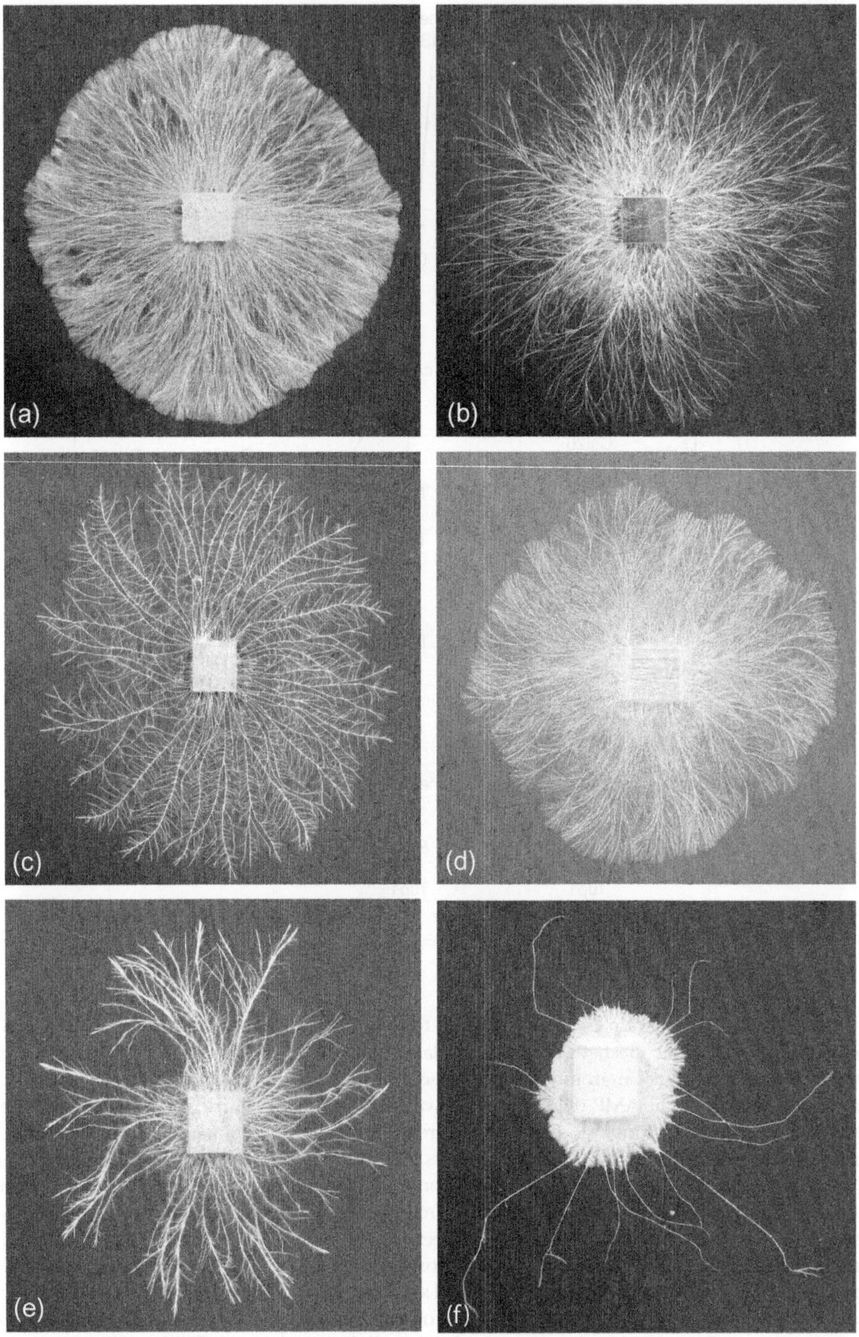

FIGURE 5.16 Mycelial cord-forming networks of six woodland basidiomycetes with varying network form related to foraging strategy. (a) *Hypholoma fasciculare*, (b) *Coprinopsis picacea*, (c) *Phallus impudicus*, (d) *Phanerochaete velutina*, (e) *Resinicium bicolor*, (f) *Megacollybia platyphylla*. Mycelium is shown growing from colonised wood blocks across soil, in microcosms designed to mimic natural conditions. *Source: Photo: A.'Bear.*

The corded systems of saprotrophic and mycorrhizal woodland fungi scavenge and accumulate mineral nutrients, as described in Chapter 7. They may retain most of the available nitrogen and phosphate in the ecosystem. Nitrogen equivalent to an annual agricultural fertiliser input may be held in fungal biomass. Network topology responds to nitrogen and phosphorus availability in the soil, with the mycelium showing diffuse growth of hyphae in patches where nutrients are abundant, and extending faster and more thinly in locations where nutrients are limited. Cords are differentiated and consist of three distinguishable hyphal types (Figure 5.17). Wide 'vessel' hyphae, apparently empty of contents, run in the centre, surrounded and interspersed with hyphae showing living cell contents. Fibre hyphae, with wall thickening that almost occludes the lumen, run longitudinally in a band around the periphery of the cord. By using radioactive tracers C^{14}, P^{32}, and K^{40} nutrient translocation has been shown to occur in both directions along a cord and to respond to supply and demand in different sites within the network. Figure 5.18 shows a mycelium of the cord forming fungus *Serpula lacrymans* extending over sand from a colonised block of wood. The whole mycelium is allowed to take up sub-toxic amounts of a radiolabelled, non-metabolisable amino acid, 2-aminoisobutyric acid, detected by photon counting scintillation imaging and recorded in a video trace. The labelled amino acid tracks the flow of

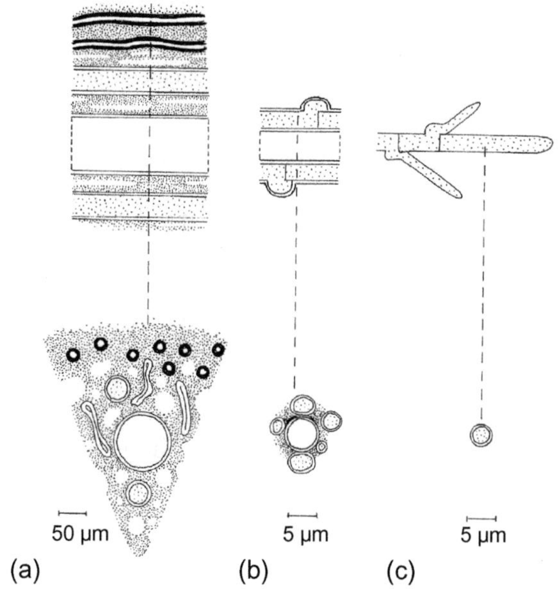

FIGURE 5.17 Cartoon showing the pattern of hyphal differentiation during cord formation and the relationship between different hyphal types in mature cords. (a) Structure of mature cord. Vessel hyphae apparently empty of contents and tendril hyphae with visible cytoplasm are run longitudinally in an extracellular matrix which is permeated by longitudinal spaces. Some hyphae appear to have collapsed. Thick-walled "fiber" hyphae are run longitudinally in the outer layers of the cord. (b) A cord starting to develop approximately 50 mm behind the mycelial margin. Wider and relatively empty "vessel" hyphae become surrounded by thigmotropic "tendril" hyphae with denser cytoplasm, which grow both acropetally and basipetally over the vessel hypha surface. Extracellular matrix material binds aggregations of vessel and tendril hyphae into cords. (c) Assimilating, extending hyphal tips forming a diffuse mycelium of separate hyphae at the advancing mycelial margin. *(Source: Drawing courtesy of Rosemary Wise.)*

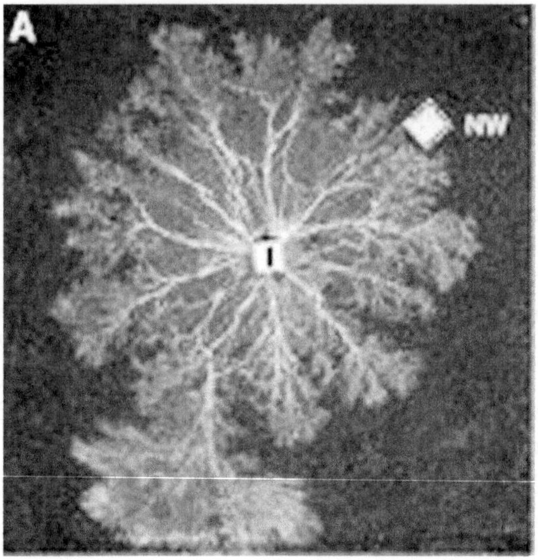

FIGURE 5.18 Mycelium of the cord forming fungus *Serpula lacrymans* growing out over sand from a colonised block of wood. See also Figure 5.7 which shows these mycelial cords functioning as channels for the rapid import of mycelial amino acid into the freshly-colonised wood, visualised by photon-counting scintillation imaging. I, the original inoculum block and food source for the mycelium, NW, a new wood block that the mycelium will meet and colonise. The video in Figure 5.7 shows the reallocation of amino acid within the mycelium following its capture of this new resource, and the rapid flow along mycelial cords.

the amino acid pool within the mycelium in real time. When a fresh wood block is presented to the edge of the colony it becomes colonised, and the majority of the organism's amino acids are rapidly translocated to the new source of food (Figure 5.18). The speed of translocation suggests that mass flow is significant, although the location of the mass flow pathway and the mechanism of loading and unloading remain unclear. Fungal accumulation and redistribution of plant nutrients through mycelial networks can affect plant growth, particularly of plants connected via a common mycorrhizal network (Chapter 7). Fungal translocation can facilitate the decomposition of nitrogen-poor plant remains, by importing nitrogen scavenged by hyphae in contact with soil (Figure 5.7). Sugar import can also provide the energy to enable hyphae to assimilate nitrogen. These observations show that the mycelium can act as a responsive resource-supply network with a key role in carbon and nitrogen dynamics in forests.

GEOMYCOLOGY

Geomycology is the study of fungal processes important in geology. Fungi are essential in global element cycling and soil fertility (Table 5.4).The effect of fungi on minerals has been studied mainly in aerobic terrestrial environments. Growing hyphae acidify their surroundings, as we have seen earlier. They release protons and exude carboxylic acids, and respiratory carbon dioxide produces carbonic acid in water. Together with the invasive abilities of hyphae, this makes some saprotrophic fungi capable of pioneering growth on rock. Many are

TABLE 5.4 Summary of Important Roles and Activities of Fungi in Geomycological Processes

Fungal attribute or activity	Geomycological consequences
GROWTH	
Growth and mycelium development; fruiting body development; hyphal differentiation; melanization	Stabilisation of soil structure Penetration of rocks and minerals Biomechanical disruption of solid substrates, building stone, cement, plaster, concrete, etc. Plant, animal and microbial colonisation, symbiosis and/or infection; mycorrhizas, lichens, pathogens Nutrient and water translocation Surfaces for bacterial growth, transport, and migration Mycelium acting as a reservoir of N and/or other elements
METABOLISM	
Carbon and energy metabolism	Organic matter decomposition and cycling of component elements, e.g. C, H, O, N, P, S, metals, metalloids, radionuclides Altered geochemistry of local environment, e.g. changes in redox, O_2, pH Production of inorganic and organic metabolites, e.g. H^+, respiratory CO_2, organic acids, siderophores Exopolymer production Organometal formation and/or degradation Degradation of xenobiotics and other complex compounds
Inorganic nutrition	Altered distribution and cycling of inorganic nutrient elements (e.g. N, S, P, essential and inessential metals, metalloids, organometals and radionuclides) Transport, accumulation, incorporation of elements into macromolecules Redox transformations of metal(loid)s and radionuclides Translocation of water, N, P, Ca, Mg, K, etc., through mycelium and/or to plant hosts Fe(III) capture by siderophores MnO_2 reduction Element mobilisation or immobilisation including metals, metalloids, radionuclides, C, P, S, etc.
Mineral dissolution	Mineral and rock bioweathering Leaching/solubilisation of metals and other components (e.g. phosphate) Element redistributions including transfer from terrestrial to aquatic systems Altered bioavailability of (e.g. metals, P, S, Si, and Al) Altered plant and microbial nutrition or toxicity Mineral formation (e.g. carbonates, oxalates, clays) Altered metal and nutrient distribution, toxicity, and bioavailability Mineral soil formation Biodeterioration of building stone, cement, plaster, concrete, etc.
Mineral formation	Element immobilisation including metals and radionuclides, C, P, and S Mycogenic carbonate formation Limestone calcrete cementation Mycogenic metal oxalate formation Metal detoxification Contribution to patinas on rocks (e.g. 'desert varnish')
PHYSICO-CHEMICAL PROPERTIES	
Sorption of soluble and particulate metal species, soil colloids, clay minerals, etc.	Altered metal distribution and bioavailability Metal detoxification Metal-loaded food source for invertebrates Prelude to secondary mineral nucleation and formation

TABLE 5.4 Summary of Important Roles and Activities of Fungi in Geomycological Processes—cont'd

Fungal attribute or activity	Geomycological consequences
Exopolymer production	Complexation of cations Provision of hydrated matrix for mineral formation Enhanced adherence to substrate Clay mineral binding Stabilisation of soil aggregates Matrix for bacterial growth Chemical interactions of exopolymers with mineral substrates
SYMBIOTIC ASSOCIATIONS	
Mycorrhizas	Altered mobility and bioavailability of nutrient and inessential metals, N, P, S, etc. Altered C flow and transfer between plant, fungus, and rhizosphere organisms Altered plant productivity Mineral dissolution and metal and nutrient release from bound and mineral sources Altered biogeochemistry in soil-plant root region Altered microbial activity in plant root region Altered metal distributions between plant and fungus Water transport to and from the plant
Lichens	Pioneer colonisers of rocks and minerals, and other surfaces Bioweathering Mineral dissolution and/or formation Metal accumulation by dry or wet deposition, particulate entrapment, metal sorption, transport, etc. Enrichment of C, N, P, etc. in thallus and alteration of elemental concentrations and distribution in local microenvironment Early stages of mineral soil formation Development and stimulation of geochemically-active microbial populations Mineral dissolution by metabolites including 'lichen acids' Biomechanical disruption of substrate
Insects and invertebrates	Fungal populations in gut aid degradation of plant material Invertebrates mechanically render plant residues more amenable for decomposition Cultivation of fungal gardens by certain insects (organic matter decomposition and recycling) Transfer of fungi between plant hosts by insect vectors (aiding infection and disease)
PATHOGENIC EFFECTS	
Plant and animal pathogenicity	Plant infection and colonisation Animal predation (e.g. nematodes) and infection (e.g. insects, etc.) Redistribution of elements and nutrients Increased supply of organic material for decomposition Stimulation of other geochemically-active microbial populations

These processes may take place in aquatic and terrestrial ecosystems, as well as in artificial and man-made systems, their relative importance depending on the species and active biomass present and physico-chemical factors. The terrestrial environment is the main site of fungal-mediated biogeochemical changes, especially in mineral soils and the plant root zone, decomposing vegetation, and on exposed rocks and mineral surfaces. There is rather a limited amount of knowledge on fungal geobiology in freshwater and marine systems, sediments, and the deep subsurface. In this Table, fungal roles have been arbitrarily split into categories based on growth, organic and inorganic metabolism, physico-chemical attributes, and symbiotic relationships. It should be noted that many if not all of these are linked, and almost all directly or indirectly depend on the mode of fungal growth (including symbiotic relationships) and accompanying chemoorganotrophic metabolism, in turn dependent on a utilisable C source for biosynthesis and energy, and other essential elements, such as N, O, P, S and many metals, for structural and cellular components. Mineral dissolution and formation are detailed separately although these processes clearly depend on metabolic activity and growth form. *(Gadd, 2011)* microcolonial, frequently melanized forms, capable of growing both as filaments and yeasts. They can grow only slowly, as they depend on scarce organic nutrient sources from the atmosphere and from other microbes. Their hyphae respond thigmotropically to microscopic pores and fissures, penetrating into the mineral structure and progressively eroding it with mechanical pressure and acid exudates. Fungal deterioration and bioweathering can affect rocks and minerals including carbonates, silicates, phosphates, and sulphides. These processes are part of the early stages of mineral soil formation, but also cause deterioration of building stone, cement, plaster, and concrete.

In mycorrhizas and lichens (Chapter 7), where symbiosis with a photobiont provides a continuous and reliable source of energy, hyphae can act continuously over long periods to dissolve nutrients from soil mineral particles and from rock. Nutrients obtained from minerals include anions, including phosphate and sulphate, and cations of essential metals, including potassium, calcium, and magnesium. In this way, ectomycorrhizal forest fungi can mine the underlying rock and minerals to supply phosphate to their host plants, in exchange for sugars produced by photosynthesis. This exchange is facilitated by the extensive translocating systems of many basidiomycetes, as described above, which enable them to access deep soil horizons and bedrock. The fungal partner in lichens growing on bare rock can similarly acquire mineral nutrients through solubilisation of nutrients, fuelled by the photobiont's photosynthesis.

Fungi can alter the spatial distribution of inorganic substances in soil. Acidification releases metals, including aluminium and iron, from complexes with clay particles in soil, allowing cations to be leached through soil horizons by water. Subsequent immobilisation at lower levels produces the layered horizons of acid **podsols** characteristic of forest soils over acid rock in regions of high rainfall. Fungal activity can also immobilise metal cations in exopolysaccharides deposited on the cell surface and lead to the formation of new minerals. Whewellite and weddellite have both been produced by fungi in culture and are considered to originate from fungal activity in nature. Fungal mobilisation and immobilisation of minerals is exploited in some approaches to removing toxic **xenobiotic** metals, metalloids and organometallic compounds from soil and water.

Xenobiotics

Water and soil are contaminated with increasing amounts of **xenobiotic** substances resulting from human activities. Organic xenobiotics include polyaromatic hydrocarbons, halogenated solvents, endocrine-disrupting agents and drugs, explosives, and agricultural chemicals. Inorganic xenobiotics are metals, metalloids including arsenic and selenium, organometallic compounds, and radionuclides. Metals are released by mining, smelting and the disposal of metal waste. Many xenobiotic substances are toxic. Recently there has been concern about endocrine-disrupting agents in wastewater, which are bioactive at low concentrations and are not retained by water treatment plants. Microbes are promising agents for bioremediation of xenobiotic pollution, but bacteria have been the main focus of research. Strains of bacteria have been engineered with suites of enzymes to degrade toluene and other organic pollutants in soil. The potential uses of fungi deserve more research. Exploratory hyphae that can penetrate soil micropores and bridge air gaps, seem ideally suited for scavenging pollutants from contaminated soil. Unlike bacteria, fungi can translocate carbon energy sources to sites of activity from distant points of uptake, enabling them to colonise the spatially heterogeneous environment of soil. Moreover, natural selection has favoured the evolution of fungi that secrete powerful oxidising enzymes of low substrate specificity to scavenge nutrients from recalcitrant compounds. This reduces the need to develop substrate-specific strains. Applications that have been explored include the use of white rot fungi to degrade polyaromatic hydrocarbons in soil and to decolorize wastewater from the Kraft process for paper manufacture. However, many other fungi can degrade toxic organic compounds in the environment and have not been tested as agents for xenobiotic bioremediation. Bioprospecting in the environment is likely to uncover many fungi with a capacity for dismantling resistant manmade molecules. Metals cannot be broken down but may be separated from susceptible biota or complexed into forms that are unlikely to be processed in natural food webs (see *Essential Metals*). Plants inoculated with selected strains of mycorrhiza can be used to reduce concentrations of heavy metal pollutants such as cadmium from soil.

Fungal enzymes that might be suitable for future applications in xenobiotic detoxification include those with intracellular activities: multiple mixed-function cytochrome P_{450} mono-oxygenases, phenol 2-mono-oxygenases, nitro reductases, quinone reductase, reductive dehalogenases, and miscellaneous transferases. Even though intracellular, these enzymes show wide substrate specificity. For example, fungal mixed-function cytochrome P_{450} oxidases can catalyse epoxidation and hydroxylation of numerous pollutants, including dioxins and polyaromatic hydrocarbons. They can also catalyse the degradation of anti-inflammatory drugs, lipid regulators, anti-epileptic and analgesic pharmaceuticals, suggesting the possibility of new environmental applications for the superlative scavenging activity of hyphae.

Further Reading and References

General Works

Anke, T., Weber, D. (Eds.), 2009. Physiology and Genetics, second ed. The Mycota, Springer, Berlin.
Deacon, J., 2006. Fungal Biology, fourth ed. Blackwell, Oxford.
Eriksson, K.-E., Blanchette, R.A., Ander, P., 1995. Microbial and Enzymatic Degradation of Wood and Wood Components (Springer Series in Wood Science). Springer, Berlin.
Boddy, L., Frankland, J., van West, P., 2007. Ecology of Saprotrophic Basidiomycetes. Academic Press/Elsevier, Amsterdam.
Gadd, G.M., Watkinson, S.C., Dyer, P.S., 2007. Fungi in the Environment. Cambridge. University Press, New York.
Hofrichter, M. (Ed.), 2010. Industrial Applications, second ed. The Mycota, Springer, Berlin.
Jennings, D.H., 1995. The Physiology of Fungal Nutrition. Cambridge University Press, Cambridge.
Jennings, D.H., Lysek, G., 1999. Fungal Biology, second ed. Bios Scientific Publishers Ltd., New York.
Martin, F. (Ed.), 2014. The Ecological Genomics of Fungi. John Wiley & Sons, Chichester.
Nowrousian, M. (Ed.), 2014. Fungal Genomics, second ed. The Mycota, Springer, Berlin.
Moore, D., Robson, G.D., Trinci, P.J., 2011. 21st Century Guidebook to Fungi. Cambridge University Press, Cambridge.

Nutrient Acquisition, Uptake, and Assimilation

Arantes, V., Milagres, A., Filley, T., Goodell, B., 2011. Lignocellulosic polysaccharides and lignin degradation by wood decay fungi: the relevance of nonenzymatic Fenton-based reactions. J. Ind. Microbiol. Biotechnol. 38, 541–555.

Ashford, A.E., Cole, L., Hyde, G.J., 2001. Motile tubular vacuole systems. In: Howard, R.J., Gow, N.A.R. (Eds.), The Mycota, VIII. Biology of the Fungal Cell. Springer Verlag, Heidelberg, pp. 243–265 (Chapter 12).

Bagley, S.T., Richter, D.L., 2002. Biodegradation by brown rot fungi. In: Osiewicz, H.D. (Ed.), The Mycota. Springer, Heidelberg, pp. 327–340.

Caddick, M.X., 2004. Nitrogen regulation in mycelial fungi. In: Brambl, R., Marzluf, G. (Eds.), The Mycota. Biochemistry and Molecular Biology. Springer, Berlin Heidelberg, pp. 349–368.

Cantarel, B.L., Coutinho, P.M., Rancurel, C., Bernard, T., Lombard, V., Henrissat, B., 2009. The carbohydrate-active enzymes database (CAZy): an expert resource for glycogenomics. Nucleic Acids Res. 37, D233–D238.

Cole, L., Hyde, G., Ashford, A., 1997. Uptake and compartmentalisation of fluorescent probes by Pisolithustinctorius; hyphae: evidence for an anion transport mechanism at the tonoplast but not for fluid-phase endocytosis. Protoplasma 199, 18–29.

Conesa, A., Punt, P.J., van Luijk, N., van den Hondel, C.A.M.J.J., 2001. The secretion pathway in filamentous fungi: a biotechnological view. Fungal Genet. Biol. 33, 155–171.

Cullen, D., 2014. Wood decay. In: Francis, M. (Ed.), The Ecological Genomics of Fungi. first ed. John Wiley & Sons, Inc, Hoboken, pp. 43–62 (Chapter 3).

Doidy, J., Grace, E., Kühn, C., Simon-Plas, F., Casieri, L., Wipf, D., 2012. Sugar transporters in plants and their interactions with fungi. Trends Plant Sci. 17 (7), 413–422.

Eastwood, D.C., et al., 2011. The plant cell wall–decomposing machinery underlies the functional diversity of forest fungi. Science 333, 762–765.

Floudas, D., et al., 2012. The paleozoic origin of enzymatic lignin decomposition reconstructed from 31 fungal genomes. Science 336, 1715–1719.

Goodell, B., 2003. Brown-rot fungal degradation of wood: our evolving view. In: Wood Deterioration and Preservation. American Chemical Society, Washington D.C. pp. 97–118.

Haas, H., Eisendle, M., Turgeon, B.G., 2008. Siderophores in fungal physiology and virulence. Annu. Rev. Phytopathol. 46, 149–187.

Hamann, A., Brust, D., Osiewacz, H.D., 2008. Apoptosis pathways in fungal growth, development and ageing. Trends Microbiol. 16, 276–283.

Hatakka, A., Hammel, K.E., 2010. Fungal biodegradation of lignocelluloses. In: Hofrichter, M. (Ed.), Industrial Applications. Springer, Berlin Heidelberg, pp. 319–340.

Klionsky, D.J., Herman, P.K., Emr, S.D., 1990. The fungal vacuole: composition, function, and biogenesis. Microbiol. Mol. Biol. Rev. 54, 266–292.

Levasseur, A., Piumi, F., Coutinho, P.M., Rancurel, C., Asther, M., Delattre, M., Henrissat, B., Pontarotti, P., Asther, M., Record, E., 2008. FOLy: An integrated database for the classification and functional annotation of fungal oxidoreductases potentially involved in the degradation of lignin and related aromatic compounds. Fungal Genet. Biol. 45, 638–645.

Martinez, D., et al., 2008. Genome sequencing and analysis of the biomass-degrading fungus Trichoderma reesei (syn. Hypocrea jecorina). Nat. Biotechnol. 26, 553–560.

Martinez, D., et al., 2009. Genome, transcriptome, and secretome analysis of wood decay fungus Postia placenta supports unique mechanisms of lignocellulose conversion. Proc. Natl. Acad. Sci. U.S.A. 106, 1954–1959.

McDonald, T.R., Dietrich, F.S., Lutzoni, F., 2012. Multiple Horizontal gene transfers of ammonium transporters/ammonia permeases from prokaryotes to eukaryotes: toward a new functional and evolutionary classification. Mol. Biol. Evol. 29, 51–60.

Monod, M., 2008. Secreted proteases from dermatophytes. Mycopathologia 166, 285–294.

Pollack, J.K., Harris, S.D., Marten, M.R., 2009. Autophagy in filamentous fungi. Fungal Genet. Biol. 46, 1–8.

Portnoy, T., Margeot, A., Linke, R., Atanasova, L., Fekete, E., Sandor, E., Hartl, L., Karaffa, L., Druzhinina, I., Seiboth, B., Le Crom, S., Kubicek, C., 2011. The CRE1 carbon catabolite repressor of the fungus Trichoderma reesei: a master regulator of carbon assimilation. BMC Genomics 12, 269.

Sagt, C., ten Haaft, P., Minneboo, I., Hartog, M., Damveld, R., Metske van der Laan, J., Akeroyd, M., Wenzel, T., Luesken, F., Veenhuis, M., van der Klei, I., de Winde, J., 2009. Peroxicretion: a novel secretion pathway in the eukaryotic cell. BMC Biotechnol. 9, 48.

Slot, J.C., Hibbett, D.S., 2007. Horizontal transfer of a nitrate assimilation gene cluster and ecological transitions in fungi: a phylogenetic study. PLoS One 2. e1097.

Tlalka, M., Fricker, M., Watkinson, S., 2008. Imaging of long-distance α-aminoisobutyric acid translocation dynamics during resource capture by Serpula lacrymans. Appl. Environ. Microbiol. 74, 2700–2708.

Xu, G., Goodell, B., 2001. Mechanisms of wood degradation by brown-rot fungi: chelator-mediated cellulose degradation and binding of iron by cellulose. J. Biotechnol. 87, 43–57.

Zak, D.R., Blackwood, C.B., Waldrop, M.P., 2006. A molecular dawn for biogeochemistry. Trends Ecol. Evol. 21 (6), 288–295.

Secondary Metabolism

Andersen, M.R., Nielsen, J.B., Klitgaard, A., Petersen, L.M., Zachariasen, M., Hansen, T.J., Blicher, L.H., Gotfredsen, C.H., Larsen, T.O., Nielsen, K.F., Mortensen, U.H., 2013. Accurate prediction of secondary metabolite gene clusters in filamentous fungi. Proc. Natl. Acad. Sci. U.S.A. 110, E99–E107.

Bayram, Ö., Braus, G.H., 2012. Coordination of secondary metabolism and development in fungi: the velvet family of regulatory proteins. FEMS Microbiol. Rev. 36, 1–24.

Brakhage, A.A., 2013. Regulation of fungal secondary metabolism. Nat. Rev. Micro 11, 21–32.

Firn, R.D., Jones, C.G., 2000. The evolution of secondary metabolism – a unifying model. Mol. Microbiol. 37, 989–994.

Fischer, R., 2008. Sex and poison in the dark. Science 320, 1430–1431.

Fox, E.M., Howlett, B.J., 2008. Secondary metabolism: regulation and role in fungal biology. Curr. Opin. Microbiol. 11, 481–487.

Hoffmeister, D., Keller, N.P., 2007. Natural products of filamentous fungi: enzymes, genes, and their regulation. Nat. Prod. Rep. 24, 393–416.

Lim, F.Y., Sanchez, J.F., Wang, C.C., Keller, N.P., 2012. Toward awakening cryptic secondary metabolite gene clusters in filamentous fungi. Methods Enzymol. 517, 303.

Schneider, P., Bouhired, S., Hoffmeister, D., 2008. Characterization of the atromentin biosynthesis genes and enzymes in the homobasidiomycete Tapinella panuoides. Fungal Genet. Biol. 45, 1487–1496.

Tlalka, M., Fricker, M., and Watkinson, S., 2008. Imaging of Long-Distance α-Aminoisobutyric Acid Translocation Dynamics during Resource Capture by Serpula lacrymans, Appl. Environ. Microbiol. 74, 2700–2708.

Responding to the Environment

Bahn, Y.-S., Mühlschlegel, F.A., 2006. CO_2 sensing in fungi and beyond. Curr. Opin. Microbiol. 9, 572–578.

Bahn, Y.-S., Xue, C., Idnurm, A., Rutherford, J.C., Heitman, J., Cardenas, M.E., 2007. Sensing the environment: lessons from fungi. Nat. Rev. Micro 5, 57–69.

Baker, C.L., Loros, J.J., Dunlap, J.C., 2012. The circadian clock of Neurospora crassa. FEMS Microbiol. Rev. 36, 95–110.

Connell, L., Rodriguez, R., Redman, R., 2014. Cold-adapted yeasts in antarctic deserts. In: Buzzini, P., Margesin, R. (Eds.), Cold-adapted Yeasts. Springer, Berlin Heidelberg.

Dadachova, E., Casadevall, A., 2008. Ionizing radiation: how fungi cope, adapt, and exploit with the help of melanin. Curr. Opin. Microbiol. 11, 525–531.

Dann, S., Thomas, G., 2006. The amino acid sensitive TOR pathway from yeast to mammals. FEBS Lett. 580, 2821–2829.

Darrah, P.R., Fricker, M.D., 2014. Foraging by a wood-decomposing fungus is ecologically adaptive. Environ. Microbiol. 16, 118–129.

Darrah, P.R., Tlalka, M., Ashford, A., Watkinson, S.C., Fricker, M.D., 2006. The vacuole system is a significant intracellular pathway for longitudinal solute transport in basidiomycete fungi. Eukaryot. Cell 5, 1111–1125.

Dunlap, J.C., Loros, J.J., 2006. How fungi keep time: circadian system in Neurospora and other fungi. Curr. Opin. Microbiol. 9, 579–587.

Fricker, M.D., Lee, J.A., Bebber, D.P., Tlalka, M., Hynes, J., Darrah, P.R., Watkinson, S.C., Boddy, L., 2008. Imaging complex nutrient dynamics in mycelial networks. J. Microsc. 231, 317–331.

Glass, N.L., Dementhon, K., 2006. Non-self recognition and programmed cell death in filamentous fungi. Curr. Opin. Microbiol. 9, 553–558.

Hohmann, S., 2002. Osmotic stress signaling and osmoadaptation in yeasts. Microbiol. Mol. Biol. Rev. 66, 300–372.

Lew, R.R., 2011. How does a hypha grow? The biophysics of pressurized growth in fungi. Nat. Rev. Micro 9, 509–518.

Maheshwari, R., Bharadwaj, G., Bhat, M.K., 2000. Thermophilic fungi: their physiology and enzymes. Microbiol. Mol. Biol. Rev. 64, 461–488.

Money, N.P., 2001. Biomechanics of invasive hyphal growth. Biology of the Fungal Cell. Springer, New York. pp. 3-17.

Morgenstern, I., Powlowski, J., Ishmael, N., Darmond, C., Marqueteau, S., Moisan, M.-C., Quenneville, G., Tsang, A., 2012. A molecular phylogeny of thermophilic fungi. Fungal Biol. 116, 489–502.

Nehls, U., Dietz, S., 2014. Fungal aquaporins: cellular functions and ecophysiological perspectives. Appl. Microbiol. Biotechnol. 98, 8835–8851.

Otsubo, Y., Yamamato, M., 2008. TOR signaling in fission yeast. Crit. Rev. Biochem. Mol. Biol. 43, 277–283.

Paul, M.J., Primavesi, L.F., Jhurreea, D., Zhang, Y., 2008. Trehalose metabolism and signaling. Annu. Rev. Plant Biol. 59, 417–441.

Peñalva, M.A., Tilburn, J., Bignell, E., Arst Jr., H.N., 2008. Ambient pH gene regulation in fungi: making connections. Trends Microbiol. 16, 291–300.

Purschwitz, J., Müller, S., Kastner, C., Fischer, R., 2006. Seeing the rainbow: light sensing in fungi. Curr. Opin. Microbiol. 9, 566–571.

Rohde, J., Bastidas, R., Puria, R., Cardenas, M., 2008. Nutritional control via Tor signaling in Saccharomyces cerevisiae. Curr. Opin. Microbiol. 11, 153–160.

Sudbery, P.E., 2011. Growth of Candida albicans hyphae. Nat. Rev. Micro 9, 737–748.

Wilson, R.A., 2007. Tps1 regulates the pentose phosphate pathway, nitrogen metabolism and fungal virulence. EMBO J. 26, 3673–3685.

Geomycology and Xenobiotic Metabolism

Gadd, G.M., 2013. Fungi and their role in the biosphere. Reference Module in Earth Systems and Environmental. Elsevier Sciences, Amsterdam.

Gadd, G.M., 2011. Geomycology. In: Reitner, J., Thiel, V. (Eds.), Encyclopedia of Geobiology. Springer, Heidelberg, pp. 416–432. Part 7.

Gadd, G.M., 2007. Geomycology: biogeochemical transformations of rocks, minerals, metals and radionuclides by fungi, bioweathering and bioremediation. Mycol. Res. 111, 3–49.

Harms, H., Schlosser, D., Wick, L.Y., 2011. Untapped potential: exploiting fungi in bioremediation of hazardous chemicals. Nat. Rev. Micro 9, 177–192.

Lah, L., Podobnik, B., Novak, M., Korošec, B., Berne, S., Vogelsang, M., Kraševec, N., Zupanec, N., Stojan, J., Bohlmann, J., Komel, R., 2011. The versatility of the fungal cytochrome P450 monooxygenase system is instrumental in xenobiotic detoxification. Mol. Microbiol. 81, 1374–1389.

Links

Assembling the Fungal Tree of Life: http://aftol.org/

MycoCosm: a fungal genomics portal (http://jgi.doe.gov/fungi), developed by the US Department of Energy Joint Genome Institute to support integration, analysis and dissemination of fungal genome sequences and other 'omics' data by providing interactive web-based tools. Grigoriev, I.V., Nikitin, R., Haridas, S., Kuo, A., Ohm, R., Otillar, R., Riley, R., Salamov, A., Zhao, X., Korzeniewski, F. 2014. MycoCosm portal: gearing up for 1000 fungal genomes, Nucl. Acids Res. 42, D699–D704.

The Carbohydrate-Active Enzyme (CAZy) database: http://www.cazy.org/ describes the families of structurally-related catalytic and carbohydrate-binding modules (or functional domains) of enzymes that degrade, modify, or create glycosidic bonds.

FOLy: An integrated database for the classification and functional annotation of fungal oxidoreductases potentially involved in the degradation of lignin and related aromatic compounds http://www.sciencedirect.com/science/article/pii/S1087184508000066

Figure 5.7. Photon Counting Scintillation Imaging of ^{14}C-AIB in a fungal colony during capture of a new wood source. Tlalka, M., Fricker, M.D., Watkinson, S.C., 2008. Imaging of long-distance {alpha}-aminoisobutyric acid translocation dynamics during resource capture by Serpula lacrymans, Appl. Environ. Microbiol. 74, 2700–2708. http://dx.doi.org/10.5072/bodleian:d217qq90r

CHAPTER

6

Molecular Ecology

Sarah C. Watkinson

University of Oxford, Oxford, UK

THE APPLICATION OF DNA TECHNOLOGY IN THE ECOLOGY OF FUNGI

Introduction

In this chapter we give an introduction to the molecular methods based on DNA technology which are transforming our understanding of the diversity and activities of fungi in the environment. Being mostly microscopic and often growing within opaque substrates or the living tissues of other organisms, fungi are hard to observe in their natural habitats, except when conspicuously sporulating. In laboratory culture, some may reveal more informative characteristics, but readily-culturable fungi are now known to constitute only a minority of those present in most habitats. A high degree of taxonomic expertise is needed to identify even sporulating fungi by morphology alone, because sporophore form, on which identification is usually based, is typically convergent so that distantly-related species can look the same. Using DNA sequences to identify fungal species has several advantages over identification based on morphological features. The bases in DNA provide a large number of non-adaptive characters. Environmental sequences can be placed within a clade of near neighbours, giving the predictive value of natural classification. By analysing the fungal nucleic acids in environmental samples, we can discover what kinds of fungi are present, where they are, and what substances they consume and change. Rapid technical developments in recent years mean that many processes in sequencing and bioinformatics can be automated.

Generic primers to amplify sequences in the internal transcribed spacer (ITS) region of ribosomal DNA (Figure 6.1) were first developed in the mid-1990s. Highly conserved ribosome-encoding flanking sequences facilitated primer design. For example, basidiomycete DNA could be amplified selectively from the mixture of fungal DNA extracted from soil samples. Species-specific ITS sequences or their restriction fragments were separated using gel-based methods and bacterial cloning. The value of DNA-based identification in fungal ecology was soon evident. For example, the fungi in ectomycorrhizal roots (Chapter 7) could be identified by matching their DNA to readily-identifiable aboveground mushrooms. This showed that the ectomycorrhizal species most conspicuous by their fruiting bodies were not necessarily

The Fungi
http://dx.doi.org/10.1016/B978-0-12-382034-1.00006-2

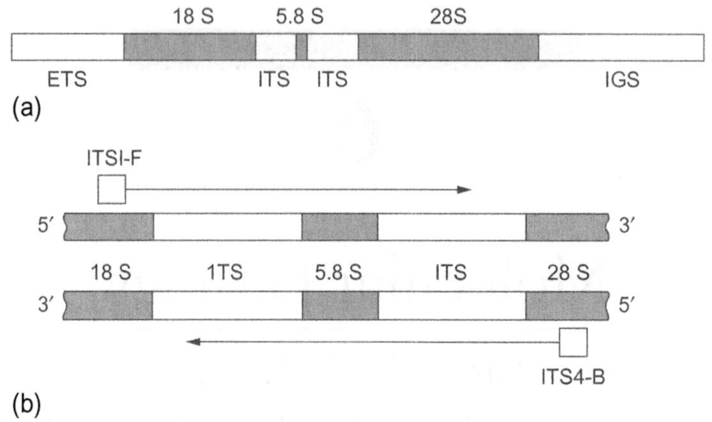

FIGURE 6.1 The utilisation of ribosomal DNA in fungal identification. (a) single copy of ribosomal DNA (rDNA), with coding and non-coding regions. An internally transcribed spacer (ITS) lies between the genes coding for 18S and 28S RNA, with that coding for 5.8S RNA embedded within it. Beyond the RNA genes are the externally transcribed spacer (ETS) and the intergenic spacer (IGS). (b) The complementary strands in the ITS region of DNA, with adjacent coding regions, showing the points at which the two primers, ITS1-F and ITS4-B, bind, and the direction of DNA replication towards the 3′ end of each strand.

the most abundant on roots, while some ectomycorrhizal fungi, although widespread on roots, were never recorded aboveground at all. The new molecular tools for investigating the diversity of underground fungi sparked an explosion of investigations in environmental fungal diversity. Not only soil fungi, but any fungal samples could be characterised in this way without the need for isolation into culture, including mycelial fragments, spores from air and water, and pathogenic fungi from tissues of diseased plant or animals. As fungal sequences accumulated in databases such as GenBank (www.ncbi.nlm.nih.gov/genbank/) sequence matching using the basic linear alignment search tool (BLAST) became the norm. Even when there is no exact match, BLAST searching, which provides an indication of the degree of similarity between sequences, may indicate the position of a fungal sequence in a phylogeny of relatives.

A new generation of automated high-throughput sequencing and bioinformatics technology has created a second revolution in molecular fungal ecology. Thousands, if not millions, of environmental sequences can now be separated and identified in parallel. High-throughput methods have largely replaced the labour-intensive method of bacterial cloning as a way of separating fungal ITS sequences in environmental samples, although the quick and inexpensive cloning and gel methods can still be useful in smaller studies (Figure 6.2).

Molecular fungal ecology has, however, not superseded long-established methods of microbiology. Direct observation and quantification at all scales, and culture on defined media, remain essential techniques in fungal ecology. However, these approaches are now generally combined with molecular methods in any investigation of the role of fungi in ecosystems.

FUNGAL DIVERSITY IN THE ENVIRONMENT

Community Analysis

The overall diversity of fungi present in an environmental sample can now be captured by sequencing the entire range of genetic material that it contains, an approach termed

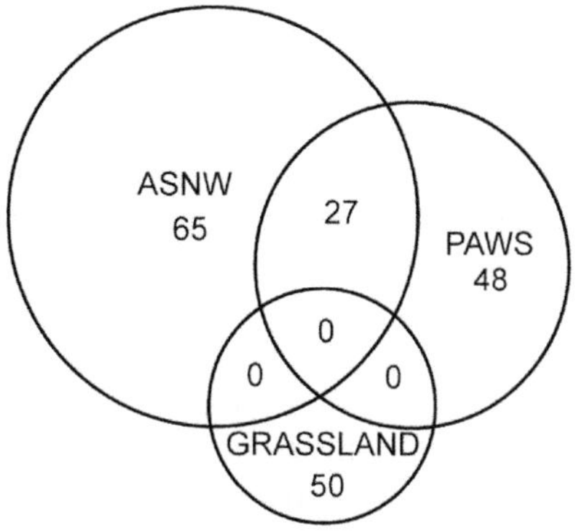

FIGURE 6.2 The use of small-scale sequencing and cloning methods to compare fungal community diversity under different vegetation types. The Venn diagram shows the numbers of unique fungal ITS types recovered from three main habitat types (ancient woodland, plantation, and adjacent grassland). Soil fungi from ancient semi-natural woodland exist in sites converted to non-native conifer plantations but not in grassland soil. *Source: Johnson et al. (2014).*

metagenomics. To obtain the **metagenome** of the sample, either DNA can be sampled, or alternatively RNA (which is then copied into the more stable cDNA for storage and sequencing). RNA sequences have the advantage of reflecting only the diversity and activities of fungi that are actively growing and synthesising new ribosomes and proteins, because RNA is generated only from genes being actively transcribed. In natural environments, the **metatranscriptome**, consisting of all the RNA present, reveals fungal genes expressed at the time of sampling. Sequencing systems such as 454-pyrosequencing, Illumina, and Ion Torrent generate sequences of all the nucleic acids present, in the form of short sequences of up to 400bp. The mass of sequence data is analysed and interpreted using dedicated computational tools which are continually being developed and refined. The extensive replication achievable with such high-throughput 'massively parallel' sequencing approaches makes it possible to extract robust data, even in the face of the multiple sources of biological and technical variation encountered in any investigation of microbes in real environments. The metagenome and metatranscriptome can reveal both microbial community diversity and patterns of gene expression.

Even without assignment to named taxonomic groups, the metagenome provides a snapshot of diversity that can point to patterns in fungal communities. It may also discover new fungal taxa (Chapter 1). Examples of this approach include analyses of fungal diversity in soils in the United States under prairie grassland and in French mountain forest. The prairie soils, from different sites across Kansas, were reported to contain an astonishing fungal diversity at all sampled sites, with over a million types of fungi and little phylogenetic overlap between sites, a finding which must merit further examination in view of the extent and importance of this North American grassland biome. In a study of forest soils in France, analysis was targeted at Ascomycete and Basidiomycete fungi growing symbiotically on tree roots as ectomycorrhiza, by using primers selective for Dikarya and using the curated mycorrhiza-specific database UNITE as well as GenBank for sequence matching. Around 1000 separate operational taxonomic units, OTUs, (described in Chapter 1) were found in each 4 g soil sample, a much higher diversity than had been predicted from previous observations. Of these, 81% were Dikarya with agaricomycetes predominating, with relatively few ectomycorrhizal taxa accounting for most of the species diversity.

Methodology

The technical complexity of highly-automated high-throughput sequencing inevitably introduces multiple potential sources of error. Understanding the likely pitfalls at every stage of the procedure (Figure 6.3) requires a range of expertise, from field sampling strategy to computational analysis, and collaboration between experts in a range of fields is essential to ensure that sources of error are understood and controlled for as far as possible.

DNA is often extracted from samples of a few grammes at most, so it is a challenge to design a scale and pattern of sampling that reflects real fungal diversity in a large field habitat. There is likely to be variation in size and abundance of fungal individuals, in their patterns of distribution, and in the kinds of materials to be sampled. Some individual fungi, such as those forming extensive mycelial networks in forest soils, extend over many metres, while other species will be present as scattered microscopic spores. Moreover, samples from most natural habitats will vary in texture, composition, and density. The process of collecting samples may itself cause rapid and confounding changes in nucleic acid composition, for example, fast-growing saprotrophic species may multiply preferentially following accidental damage, causing nuclear multiplication and consequent over-representation of their DNA, so

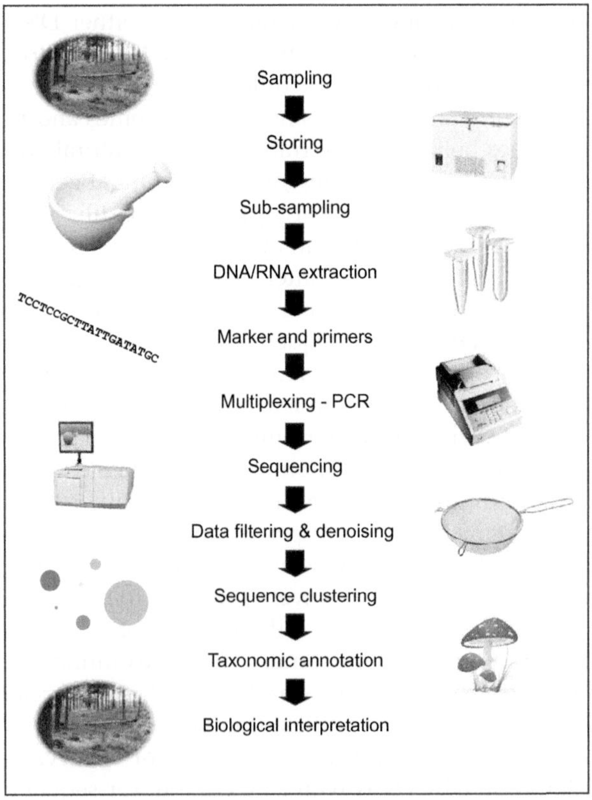

FIGURE 6.3 Overview of the steps involved in high-throughput sequencing of fungal communities. *Source: Lindahl et al. (2013).*

samples must be frozen or chemically preserved immediately on collection. If RNA is being sampled it must be frozen instantly to −80 °C and copied to the more stable *c*DNA for long-term storage. The efficiency of DNA extraction may vary between samples, and soils and plant material may contain inhibitors of downstream processes including the PCR.

Identification

The choice of sequence to be used for identification depends on the level of taxonomic discrimination required. The ITS region (Chapter 1, p. 5) is widely used to identify species and has been proposed as a universal 'barcode' locus. For higher-order taxa, the 28S subunit sequence is useful and the D1/D2 domains of the 28S characterise a taxonomic level approximating to the genus. Sequences in the long and short subunits – LSU and SSU—are used for the Glomeromycota, in which there are insufficient data available for other regions. These ITS sequences are combined with other conserved genes for phylogenetic species determination, with six gene regions generally used: 18S rRNA, 28S rRNA, 5.8S rRNA, elongation factor 1-(*EF1*), and two RNA polymerase II subunits (*RPB1* and *RPB2*). The choice of primers inevitably introduces bias, and to overcome this it is desirable to use many different primers, and primers with degenerate positions, so that the resulting amplicons represent as many as possible of the fungi present in the sample. Artefacts can arise during nucleic acid amplification, including the formation of chimaeras, mispriming, and formation of heteroduplexes, which can give overestimates of the real sequence diversity. Sequences are aligned using automated bioinformatics programmes to generate clusters of taxonomically meaningful groups based on similarity. Where sequences differ by less than a predetermined maximum, usually 3%, they are considered to represent a single OTU. The details of alignment programmes are ever-changing, as multiple programmes are developed to cluster similar sequences into taxonomically meaningful groups. The computational analysis of large sets of environmental sequences is non-trivial, and the reader is referred to specialist sources for detailed up-to-date information.

Identifying Cryptic Species, Clones, and Genets

Phylogenetic determination of individuals from environmental sequences has led to the recognition that many fungi that were assigned to a single named species on morphological features, exist in nature as multiple types with important differences in ecological function. For example, mycorrhizal fungi that are indistinguishable on the basis of morphology may consist of phylogenetically distinct clades differing in host plant specificity, preferred soil type, or in other functional aspects of the symbiosis (Chapter 7). These variants are termed **cryptic species**.

In studying the biology of single species of fungi in natural environments, it is frequently important to be able to distinguish individual organisms from each other, and to characterise mycelium or spores as members of the same or separate clones (see Chapter 4 for a discussion of fungal clones and genets). Environmental samples can be assigned to an individual clone using hyper-variable sequences such as microsatellite DNA or single nucleotide polymorphisms (SNPs) as a fingerprint. This has many practical applications in mycology. By tracing the spatial distribution of clones of ectomycorrhizal fungi it is possible to determine the underground extent and potential connectivity of the common networks (p. 226) of ectomycorrhizal basidiomycetes, some of which are capable of extending for several metres through the soil. Because they can act as channels carrying plant nutrients, water, and photosynthate between plants, their extent and host connections of these fungal channels may influence the structure of the aboveground ecosystem (see Chapter 7). Determining whether infective propagules are clonal or recombinant can provide valuable clues in the population genetics of pathogens (Chapter 8), pointing to the mode of spread of disease and hence the most effective targets for its control.

Diagnostic Methods for Known Species

Known species or strains of fungi can be quantified from environmental samples by **quantitative PCR**, in which the transcript level of species-specific sequences is assayed fluorimetrically. Both DNA and RNA (reverse-transcribed to cDNA) can be determined. This is a valuable diagnostic method in medicine and agriculture, being very sensitive and specific, and capable of being automated for the simultaneous analysis of hundreds of samples. Portable PCR machines enable on-site detection. Quantitative PCR has numerous applications in agriculture, including detecting and quantifying fungi present at very low levels. It can be used to detect crop and human pathogens, fungicide-resistant strains, and strains carrying biotechnologically-introduced traits. Imaging methods are used for diagnosing and visualising known species active *in situ*. Figure 6.4 shows the use of transformation with green fluorescence protein (GFP) to image active cells of the insect pathogen *Metarhizium* within the body of its mosquito larva host.

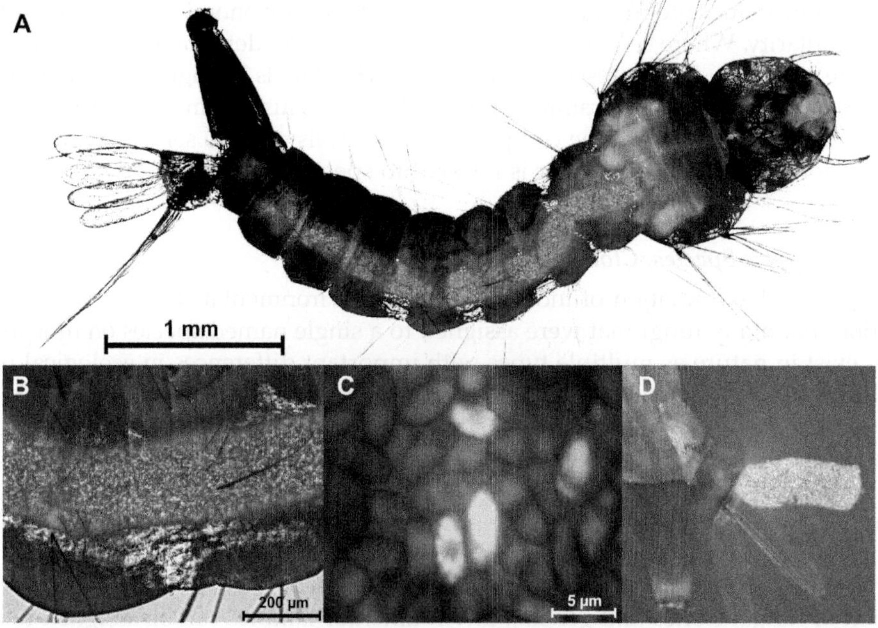

FIGURE 6.4 The insect pathogen *Metarhizium brunnei*, transformed with green fluorescence protein (GFP) and visualised by fluorescence microscopy within the body of its host, a larva of the mosquito *Aedes aegypti*. Species of *Metarhizium* are being investigated for biological control of mosquitoes that are vectors of human disease, including malaria, dengue, and yellow fever. (a) conidia in the gut lumen of the larva, (b, c) conidia expressing GFP in the gut lumen, and (d) active conidia in faecal pellets. *Source: Butt et al. (2013).* (See the colour plate.)

Classification and Nomenclature of Environmental Fungal Sequences

Molecular environmental sampling is now widely and successfully used for fungal species discovery and identification. This has created a problem: the rate of discovery of new species from their sequences is now too fast for them all to be assimilated into the established system of classification and nomenclature. To quote David Hibbett (see Further Reading): 'Fungal taxonomy seeks to discover and describe all species of fungi, to classify them according to their phylogenetic relationships, and to provide tools for their identification... to enable communication about fungal diversity, there is a pressing need to develop classification systems based on environmental sequences'.

Currently, when a potentially new fungal species is discovered, it is described and named according to the International Code of Nomenclature for algae, fungi, and plants. The name is then added to Index Fungorum (http://www.indexfungorum.org/), an ongoing list derived from world mycological literature. This system provides a registry of new taxonomic discoveries, and a standardised nomenclature for fungal species, essential for scientific communication in biology. The current system of identification and nomenclature based on specimens – herbarium and culture collections curated by taxonomists – makes it possible to identify species and test prior taxonomic hypotheses. However, the majority of fungal species or OTUs found as environmental sequences have no known fruit bodies, and have not been cultured so cannot be deposited as voucher specimens in herbaria, or cultures in a culture collection. Fungal taxonomists are exploring ways of integrating sequence-based and specimen-based fungal taxonomy into a single internationally recognised system. It has been estimated that naming and classifying the estimated five million or more undescribed fungal species using traditional specimen-based methods would take over a thousand years. Some mycologists therefore proposed that environmental sequences shown to be valid indications of the existence of new fungal taxa should be named and accepted into modern systems of taxonomy. For this, it is necessary to develop a system of nomenclature that will link fungal sequences in GenBank with specimen-based fungal herbarium taxonomy. The 2011 review by Hibbett and colleagues (listed below under Further Reading) reported 91,225 fungal ITS sequences in GenBank. These could be sorted into 16,969 clusters of 93% similarity. Of these, 37% were known only from environmental DNA, and only 13% of fungal sequences in GenBank matched sequences from described specimens. However, the proportion of remaining sequences with no matches to herbarium specimens does not necessarily represent fungi new to science, because many well described species of fungi listed in *Index Fungorum* are still not represented in GenBank. Most of the herbarium voucher specimens at the Royal Botanic Gardens at Kew, UK, when submitted to GenBank BLAST searches, produced no matches, or incorrectly named ones. While automated sequence identification can replace taxonomic expertise, it introduces new potential sources of error. For example, public databases such as GenBank may contain naming errors and wrongly-attributed sequences, which may not be corrected because only the original depositor – who need not be a taxonomist – can alter the GenBank record. For this reason, curated quality-assessed sequence databases are maintained for some groups of fungi where reliable identification is critical for a particular scientific community, for example, in mycorrhizal research UNITE, (http://unite.zbi.ee), SILVA (http://www.arb-silva.de), MaarjAM (http://maarjam.botany.ut.ee), and for plant and human pathology, Fusarium-ID (http://isolate.fusariumdb.org/index.php), and the Aspergillus Genome Database (http://isolate.fusariumdb.org/index.php) in plant and human pathology.

Efforts are currently under way to incorporate ITS sequences from professionally identified and curated herbarium voucher specimens and isolates into a single publicly-accessible database. These 'species barcodes' will in future cross-reference environmental sequences into a quality-controlled system of classification and identification, and provide taxonomic authentication for fungal sequence accessions in GenBank. In this way, the majority of fungi that are known only as environmental sequences might eventually be linked into a specimen-based list, such as Index Fungorum, to provide a single integrated catalogue.

LINKING FUNGAL DIVERSITY AND ECOSYSTEM PROCESSES

Characterising fungal diversity is only the first step in investigating the importance of fungi in an ecosystem. The next is to link fungal diversity to ecosystem function. Molecular methods are providing keys to what has been termed the 'black box' of microbial ecosystem processes. To illustrate this, we describe how a combination of molecular and chemical approaches is revealing the key role of fungi in the maintenance of boreal forest, the northern hemisphere conifer-dominated biome which is a major carbon sink in the global carbon cycle. These sub-arctic forests grow in areas of low sunlight and high rainfall, and their highly-leached soils are poor in available nitrogen and minerals. Soils are typically highly stratified, with a carbon-rich layer of recently-fallen litter that becomes gradually comminuted by animals and saprotrophic microbes to fragments and soluble organic material in the underlying humic layers. The mycelium of ectomycorrhizal and saprotrophic fungi accumulates and redistributes scarce nitrogen and minerals, and thus plays a central role in plant nutrition, and in the turnover of carbon and nitrogen. An analysis of pine forests in Sweden has shown that both the diversity of soil fungi and their nutritional mode – saprotrophic or ectomycorrhizal – differ between soil horizons (Figure 6.5). From sequence analysis of soil samples, the rapidly-decomposing carbon-rich litter layer is occupied mainly by saprotrophic species. Ectomycorrhizal species (Chapter 7) dominate deeper soil layers and nitrogen isotope ratios indicate that these are responsible for most nitrogen mobilisation in the soil. Carbon isotope ratios show that plant-litter-derived carbon can persist for as long as 45 years in lower soil horizons, in the form of humic and phenolic materials and organic nitrogen compounds which break down extremely slowly. Many of these long-lived substances are now known to be formed by fungi, and to represent an important contribution to soil carbon sequestration.

The process of fungal plant litter decomposition is a pivotal process in carbon and nitrogen cycles of terrestrial ecosystems (Chapter 9). Patterns of gene expression in environmental samples can reveal the production of specific enzymes in natural habitats and identify the fungi producing them. For example, an investigation of spruce (*Picea abies*) forest in Bohemia analysed the metagenome and metatranscriptome in the surface litter layer of the soil. Sequences of the fungal wood-decomposing enzyme, cellobiohydrolase (Chapter 5) were identified and attributed to the producing fungal species. RNA analysis, combined with isolation, culture and biochemical experiments, demonstrated that relatively few key species, for example, species of the agaricomycete *Mycena* and ascomycetes of the Xylariales, are responsible for most of the litter decomposition, with transcripts of ascomycete cellobiase predominating in the litter layer **transcriptome** (see below).

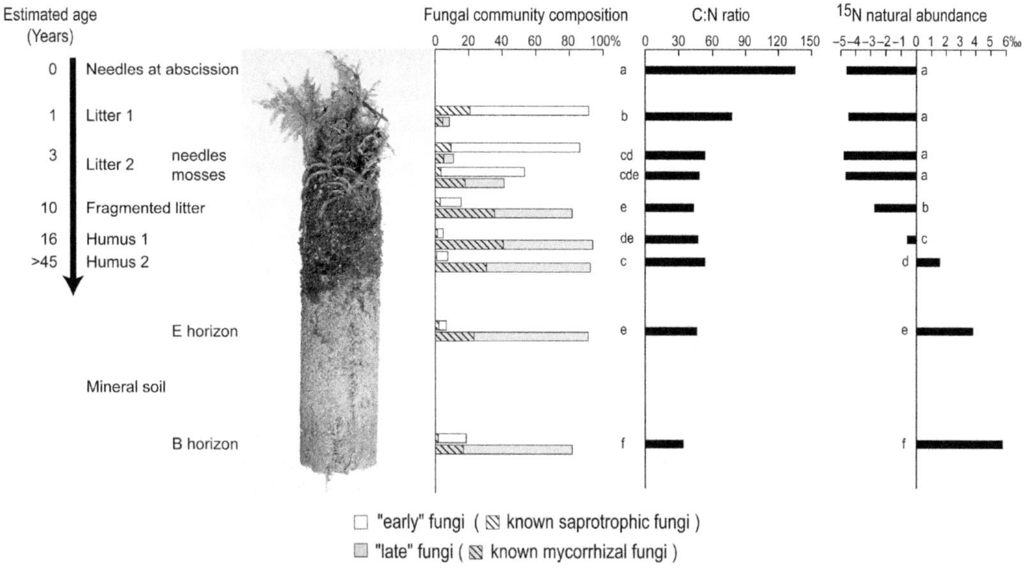

FIGURE 6.5 Fungal community composition, carbon:nitrogen (C:N) ratio and ^{15}N natural abundance throughout the upper soil profile in a Scandinavian *Pinus sylvestris* forest. Molecular methods for identification of fungi, combined with chemical analyses of carbon and nitrogen, indicated different roles of saprotrophic and ectomycorrhizal fungi in the carbon and nitrogen dynamics of separate soil horizons. *Source: Lindahl et al. (2007).* (See the colour plate.)

A specific gene or genes may have a key role in the ecology of a species, for example, virulent strains of a pathogenic fungus may differ from avirulent ones by the possession of a single gene (Chapter 8). Such genetic determinants of pathogenicity are identified by gene replacement (Figure 6.6). In the rice blast disease caused by *Magnaporthe oryzae*, a functional signal transduction pathway is necessary for germinating conidia of the fungus to penetrate into the tissue of a rice leaf and establish infection (see Chapter 8 for details of this plant pathogen). Experimental mutant strains in which a key gene is disrupted cannot infect the leaf, but virulence can be restored by transformation to reintroduce an undamaged copy of the gene. Gene replacement proved that components of the fungal signal transduction pathway – receptors and protein kinases – were required for infection in rice blast, indicating that sensing and response was involved in the fungus's entry into the leaf tissue.

Stable isotope probing (SIP) is used to identify naturally-occurring fungi that break down specific substrates. An example of its use might be the search for a decomposer to break down a xenobiotic pollutant compound (Chapter 5, p. 184). A heavy isotope such as ^{13}C, ^{15}N, or ^{18}O is incorporated into the target material by chemical synthesis or metabolism. The labelled target substrate is placed in the environment to be sampled until it becomes naturally colonised by decomposers present among the pool of soil microbes, including fungi. The DNA of organisms assimilating the target substrate becomes marginally heavier, and can be separated on a density gradient. The DNA bands can then be probed both for taxonomically diagnostic sequences, and for genes encoding enzymes likely to be expressed in assimilating the target compound. Organisms identified by SIP as using a specific substrate can be further characterised by imaging techniques that are being progressively developed and refined.

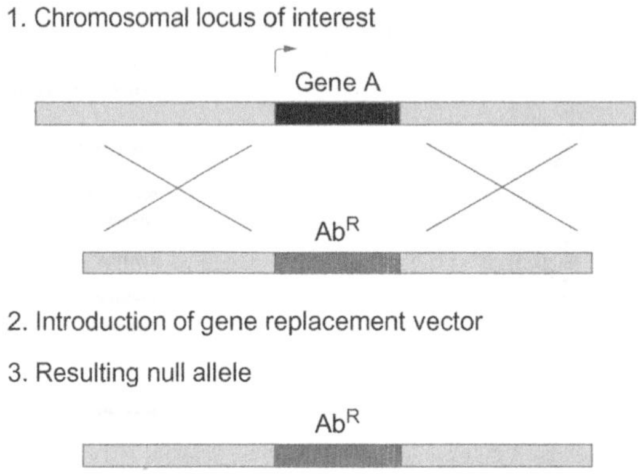

FIGURE 6.6 One-step gene replacement in fungi. In this process, a genomic clone spanning a gene of interest is first identified from a genomic library. A restriction fragment containing regions of at least one kilobase on either side of the gene of interest is cloned, and an antibiotic resistance gene cassette (Ab^r) is inserted into the space normally occupied by the protein-encoding part of the gene of interest. The antibiotic resistance gene cassette will contain a promoter and terminator sequence to allow high level expression in the fungus. The gene replacement vector is then introduced into the fungus by transformation. A double cross-over event can occur, allowing the selectable marker gene to replace the chromosomal copy of the gene. The resulting locus is known as the **null allele**, as it expresses none of the biological activity associated with the native gene. Targeted gene replacements provide a direct test of the role of specific genes in fungal pathogenicity. *Source: By Talbot N.J., reproduced from The Fungi, second edition, 2001.*

In the case of organisms for which taxonomically diagnostic sequences are known, fluorescent probes can be combined with micro autoradiography (fluorescent *in situ* hybridisation and micro autoradiography, FISH-MAR) to visualise individual cells, with a resolution of 0.5–2 μm – the range of diameters of hyphae and many spores. Raman spectroscopy can be used to enhance the resolution and power of cell imaging, by picking up the red shift in mass spectra of compounds that have incorporated a heavy atom. Instruments have been devised that image red shift effects in single cells. This isotope technique has the advantage that it labels cells as a whole, without being localised to a single marker compound such as DNA. In the future, ever higher resolution and more rapid procedures are expected to be developed to observe and identify microbial cells *in situ* and *in vivo*.

WHOLE-GENOME SEQUENCING AND COMPARATIVE GENOMICS

The whole genome sequence of a species contains the organism's entire nuclear genetic repertoire and is a valuable aid to knowledge of its ecology, evolution, and physiology.

The first whole fungal genome to be published was that of baker's yeast, *Saccharomyces cerevisiae* in 1996. Since then, improving sequencing technology has added scores of other species. A project under way at the U.S. Department of Energy Joint Genome Institute (JGI) aims to sequence the genomes of a thousand fungi: (http://genome.jgi.doe.gov/programs/fungi/1000fungalgenomes.jsf).

Such massive allocation of resources reflects recognition of the power of genomics for understanding and exploiting fungal biology. In the JGI Community Sequencing Program, mycologists from around the world are invited to make a case for the potential advantages of sequencing a given fungus, and species are selected on a peer-reviewed basis. The resulting genomic data, together with pipelines of computational procedures for their interpretation, are made available to the scientific community, providing a huge resource for investigating fundamental aspects of fungal biology. Once the whole genome sequence of a species has been assembled, it is scanned for open reading frames, sequences predicted to function as protein-encoding genes. These are annotated with their likely functions by matching to databases of known genes and proteins. For example, sequences may be identified as likely to encode enzymes associated with particular cell functions in nutrition and metabolism, or clusters of sequences may indicate that a species possesses the metabolic equipment to synthesise various classes of secondary metabolite.

Genome and Transcriptome

The genome sequence data alone show what genes a species possesses, but does not prove that they are functional. For this, the next step is to examine the **transcriptome**, the entire RNA produced by gene expression of a fungus. High-throughput methods, for example, Illumina and 454-pyrosequencing, show all the RNA (*c*DNA) sequences present and are used for large transcriptomic samples. Alternatively, where the whole genome sequence is known, microarrays can be used. This is a quicker and less expensive method for smaller samples, and has the advantage of giving immediate results. The DNA of all the genes annotated in the genome, or a selected subset, are arrayed on microscope slides, and transcripts of each gene are identified and quantified by hybridisation to its DNA template.

Changes in the transcriptome can show spatial and temporal patterns of gene expression associated with the functional adaptation of the organism, such as interactions with the environment and other organisms, developmental changes or nutrient acquisition. In this way, the fully-sequenced genome becomes a vast source of objectively-acquired information on cellular and metabolic aspects of niche adaptation. Figure 6.7 shows how the transcriptome of the timber dry rot fungus *Serpula lacrymans* changes according to whether it is growing on wood or on glucose as sole carbon/energy source. A wood substrate induces the expression of multiple carbohydrate-active enzymes as well as membrane transport proteins presumably active in enzyme secretion and nutrient uptake.

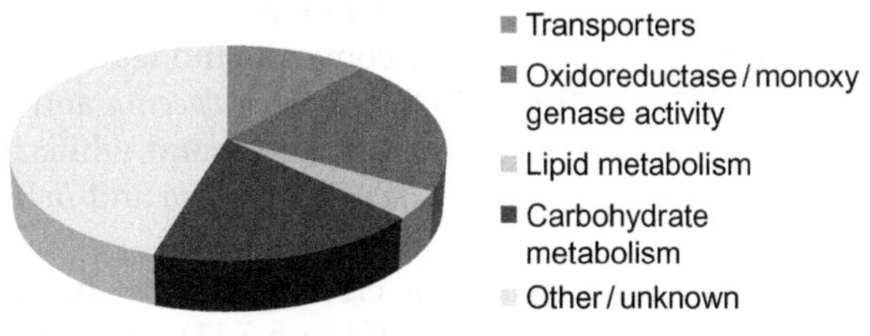

FIGURE 6.7 Functional characterisation of *Serpula lacrymans* transcripts with significantly increased gene expression when grown on wood compared with glucose-based medium, identified by microarray analysis. Expression of glycoside hydrolases involved in cellulose utilisation increased over a hundredfold in mycelium feeding on wood. *Source: Eastwood et al. (2011).* (See the colour plate.)

As progressively larger sets of fungal genomes become available for comparison, it is becoming possible to distinguish a core fungal genome from genes unique to individual species. At the time of writing, a study of 33 basidiomycetes and 30 other fungi were found to possess a core fungal genome of around 5000 genes. However, as many as half of the proteins in Basidiomycota lacked homologues in other groups of fungi, and 23% were unique to a single organism, with slightly lower figures for unique proteins in Ascomycota.

Gene Families and Niche Adaptation

Comparative genomics of multiple species can reveal not only their phylogenetic relationships, but also patterns of past lineage divergences that underlie the unique niche adaptation of individual species. Comparing the genomes of a number of different species from a range of ecological niches reveals expansions and contractions of gene families associated with different life styles, providing a clue to proteins which might give a species a selective advantage in a particular ecological niche. The agaricomycetes found in forest habitats include species with a range of fairly distinct nutritional modes. Physiology and biochemistry distinguish brown rot species that feed on cellulose and hemicellulose from wood and other forms of plant remains, with little or no lignin decomposition, white rot species that also remove most of the lignin from wood using ligninolytic peroxidases, and others that acquire sugars from host photosynthesis by mycorrhizal symbiosis with tree roots (Ectomycorrhizal fungi, see Chapter 7). Brown rot decomposition dispenses with some or all of the ligninolytic peroxidases that enable white rot fungi to break down lignin, and depolymerize crystalline cellulose partly by means of the oxidative Fenton's process (described in Chapter 5). From species phylogeny, the brown rot mode of nutrition has evolved convergently from lignin-decomposing white rot in several separate lineages. Ectomycorrhizal nutrition also shows convergence, with separate phylogenetic origins (Figure 7.10).

Evolutionary adaptation to these different nutritional niches has been accompanied by expansions and contractions in gene families. In agaricomycete phylogeny, lineages whose modern representatives utilise wood by brown rot decay are nested within wider clades of fungi mainly represented by white rotting species. It is inferred that brown rot wood decomposition evolved repeatedly as a feeding strategy that conferred advantages in some niches. The phylogeny of gene families associated with breaking down wood polymers (oxidases, including peroxidases and mono-oxygenases, hydrolases including a range of endo- and exoglucanases) shows that evolution of brown rot species from white rot ancestry was accompanied by the loss of many oxidative enzymes as well as a reduction and refinement in the suite of hydrolases. A similar loss of degradative enzymes occurred alongside the switch to symbiosis in ectomycorrhizal species. As the genomes of more and more species become known, it is, however, becoming clear that agaricomycete modes of wood decay encompass a continuous spectrum. While some fungi such as *Postia placenta* and *Phanerochaete chrysosporium* can be assigned to the traditional groups of brown and white rot fungi respectively, other more recently sequenced species such as *Jaapia argillacea* and *Botryobasidium botryosum* share features of both.

Reconciling gene phylogeny with molecular clock data in a **chronogram** can show the approximate dates of phylogenetic divergences (Figure 7.11). The results seem to present a picture of changes in early forest ecosystems. It has been suggested that the accumulation of wood, which formed the coal in carboniferous geological strata, might have been brought to an end when fungi evolved the capacity to feed on dead wood, returning its carbon to the atmosphere as carbon dioxide. Brown rot, which is the most common form of decay of conifer wood and leaves a lignin-rich residue, may have contributed to the development of the humus-rich, nitrogen poor acid soils of boreal forest.

Adaptation by Changes in Genome Architecture

Comparing whole genomes has led to greatly increased understanding of gene evolution, including the discovery of the importance of **horizontal gene transfer** in which genes have been directly transferred across fungal lineages, without sexual recombination. It is inferred when some genes, or a gene cluster, are found only in a single lineage nested within a phylogenetic tree. In fungi this appears to have been a key evolutionary process in opening up new niches and enabling adaptive radiation into new habitats. A whole set of genes encoding the enzymes and transport proteins required for nitrate utilisation, derived from several different ancestral lineages, may have enabled ascomycetes and basidiomycetes to colonise terrestrial soils in which nitrate is the main available form of nitrogen (Chapter 5). Genes that confer virulence may spread within a clade of fungi and generate new plant diseases. A recently emergent wheat disease is caused by *Septoria nodorum*. Its fully-sequenced genome revealed the presence of a gene for a protein toxin that acts as a virulence factor (i.e. necessary for virulence, see Chapter 8). This gene was found to be homologous with one encoding a protein toxin in an established wheat pathogen, *Pyrenophora tritici-repentis*, which causes yellow spot of wheat. The toxin gene appears to have been conveyed by horizontal transfer to the previously nonpathogenic *Septoria nodorum*, endowing this fungus with virulence against susceptible races of wheat. Horizontal transfer may involve not only small sequences but whole gene clusters. The recently sequenced genome of *Podospora anserina* contains a 23-gene cluster encoding the complete pathway for production of sterigmatocystin, a highly toxic aflatoxin precursor, which is presumed to have been horizontally transferred in its entirety from *Aspergillus nidulans*.

Transposable elements are autonomously-replicating, non-protein-encoding sections of the genome that can relocate around the genome (Chapter 4, p. 132). They have been recently recognised in fungi, where it is now believed that they may play an important part in the evolution of symbiosis. Transposable elements make up an exceptionally high proportion of the genome in some mycorrhizal species: around 60% in the truffle *Tuber melanosporum*, and 21–24% in *Laccaria bicolor*.

Transposable elements can cause changes in genome architecture and gene function. Genes can become silenced, or duplicated. Gene duplication allows a gene to acquire new functions by mutation, since one copy remains to maintain the original function, leaving the second copy free to mutate without compromising the organism's fitness. Inserted transposable elements may interfere with meiosis and thus remove sections of the genome from the possibility of recombination. Comparisons of symbiotic and non-symbiotic species of agaricomycetes has shown that transposable elements are usually more common in symbiotic fungi than their saprotrophic close relatives. Moreover, genes that have been identified as having crucial roles in the development of symbiotic tissues appear both to be associated with transposable elements, and also to lack orthologues in other organisms. Both these observations suggest the possibility that transposable elements have played a part in the genetic innovation underlying a move from saprotrophic to biotrophic nutrition.

Further Reading

General Works on Methodology

Grigoriev, I., 2013. Fungal genomics for energy and environment. In: Horwitz, B.A., Mukherjee, P.K., Mukherjee, M., Kubicek, C.P. (Eds.), Genomics of Soil- and Plant-Associated Fungi. Springer, Berlin Heidelberg, pp. 11–27.

Hirsch, P.R., Mauchline, T.H., Clark, I.M., 2010. Culture-independent molecular techniques for soil microbial ecology. Soil Biol. Biochem. 42, 878–887.

Lindahl, B.D., Nilsson, R.H., Tedersoo, L., Abarenkov, K., Carlsen, T., Kjøller, R., Kõljalg, U., Pennanen, T., Rosendahl, S., Stenlid, J., Kauserud, H., 2013. Fungal community analysis by high-throughput sequencing of amplified markers – a user's guide. New Phytol. 199, 288–299.

Lindahl, B.D., Kuske, C.R., 2014. Metagenomics for Study of Fungal Ecology. In: The Ecological Genomics of Fungi. John Wiley & Sons Inc, Hoboken, pp. 279–303.

Macleod, D., 2006. Principles of Gene Manipulation and Genomics, SB Primrose & RM Twyman. Blackwell Publishing, Oxford.

Martin, F., Cullen, D., Hibbett, D., Pisabarro, A., Spatafora, J.W., Baker, S.E., Grigoriev, I.V., 2011. Sequencing the fungal tree of life. New Phytol. 190, 818–821.

Neufeld, J.D., Wagner, M., Murrell, J.C., 2007. Who eats what, where and when? Isotope-labelling experiments are coming of age. ISME J. 1, 103–110.

Wellington, E.M.H., Berry, A., Krsek, M., 2003. Resolving functional diversity in relation to microbial community structure in soil: exploiting genomics and stable isotope probing. Curr. Opin. Microbiol. 6, 295–301.

Examples of Molecular Approaches to Investigating the Diversity and Ecology of Fungi

Baldrian, P., Kolařík, M., Štursová, M., Kopecký, J., Valášková, V., Větrovský, T., Žifčáková, L., Šnajdr, J., Rídl, J., Vlček, Č., 2011. Active and total microbial communities in forest soil are largely different and highly stratified during decomposition. ISME J. 6, 248–258.

Buée, M., Reich, M., Murat, C., Morin, E., Nilsson, R.H., Uroz, S., Martin, F., 2009. 454 Pyrosequencing analyses of forest soils reveal an unexpectedly high fungal diversity. New Phytol. 184, 449–456.

Butt, T.M., Greenfield, B.P., Greig, C., Maffeis, T.G., Taylor, J.W., Piasecka, J., Dudley, E., Abdulla, A., Dubovskiy, I.M., Garrido-Jurado, I., 2013. Metarhizium anisopliae pathogenesis of mosquito larvae: a verdict of accidental death. PloS One 8, e81686.

Clemmensen, K.E., Bahr, A., Ovaskainen, O., Dahlberg, A., Ekblad, A., Wallander, H., Stenlid, J., Finlay, R.D., Wardle, D.A., Lindahl, B.D., 2013. Roots and associated fungi drive long-term carbon sequestration in boreal forest. Science 339, 1615–1618.

Cornell, M., 2007. Comparative genome analysis across a kingdom of eukaryotic organisms: specialization and diversification of the fungi. Genome Res. 17, 1809–1822.

Eastwood, D.C., et al., 2011. The Plant Cell Wall–Decomposing Machinery Underlies the Functional Diversity of Forest Fungi. Science 333, 762–765.

Fierer, N., Breitbart, M., Nulton, J., Salamon, P., Lozupone, C., Jones, R., Robeson, M., Edwards, R.A., Felts, B., Rayhawk, S., Knight, R., Rohwer, F., Jackson, R.B., 2007. Metagenomic and small-subunit rRNA analyses reveal the genetic diversity of bacteria, archaea, fungi, and viruses in soil. Appl. Environ. Microbiol. 73, 7059–7066.

Hess, J., Skrede, I., Wolfe, B., LaButti, K., Ohm, R.A., Grigoriev, I.V., Pringle, A., 2014. Transposable element dynamics among asymbiotic and ectomycorrhizal Amanita fungi. Genome Biol. Evol. 6 (7), 1564–1578.

Hibbett, D.S., Ohman, A., Glotzer, D., Nuhn, M., Kirk, P., Nilsson, R.H., 2011. Progress in molecular and morphological taxon discovery in Fungi and options for formal classification of environmental sequences. Fungal Biol. Rev. 25, 38–47.

Hibbett, D.S., Donoghue, M.J., 2001. Analysis of character correlations among wood decay mechanisms, mating systems, and substrate ranges in homobasidiomycetes. Syst. Biol. 50, 215–242.

Hibbett, D., Thorn, R., 2001. Basidiomycota: homobasidiomycetes. In: McLaughlin, D.J., McLaughlin, E.G., Lemke, P.A. (Eds.), The Mycota VII, Part B: Systematics and Evolution. Springer, Berlin, pp. 121–168.

James, T.Y., et al., 2006. Reconstructing the early evolution of fungi using a six-gene phylogeny. Nature 443, 818–822.

Johnson, J., Evans, C., Brown, N., Skeates, S., Watkinson, S., Bass, D., 2014. Molecular analysis shows that soil fungi from ancient semi-natural woodland exist in sites converted to non-native conifer plantations. Forestry 87, 705–717.

Kershaw, M.J., Talbot, N.J., 2009. Genome-wide functional analysis reveals that infection-associated fungal autophagy is necessary for rice blast disease. Proc. Natl. Acad. Sci. U.S.A. 106 (37), 15967–15972.

Kohler, A., Kuo, A., Nagy, L.G., Morin, E., Barry, K.W., Buscot, F., Canbäck, B., Choi, C., Cichocki, N., Clum, A., 2015. Convergent losses of decay mechanisms and rapid turnover of symbiosis genes in mycorrhizal mutualists. Nat. Genet. 47, 410–415.

Lindahl, B.D., Ihrmark, K., Boberg, J., Trumbore, S.E., Högberg, P., Stenlid, J., Finlay, R.D., 2007. Spatial separation of litter decomposition and mycorrhizal nitrogen uptake in a boreal forest. New Phytol. 173, 611–620.

Martin, F., et al., 2008. The genome of *Laccaria bicolor* provides insights into mycorrhizal symbiosis. Nature 452, 88–92.

Martinez, D., Challacombe, J., Morgenstern, I., Hibbett, D., Schmoll, M., Kubicek, C.P., Ferreira, P., Ruiz-Duenas, F.J., Martinez, A.T., Kersten, P., 2009. Genome, transcriptome, and secretome analysis of wood decay fungus Postia placenta supports unique mechanisms of lignocellulose conversion. Proc. Natl. Acad. Sci. 106, 1954–1959.

Martinez, D., Larrondo, L.F., Putnam, N., Gelpke, M.D.S., Huang, K., Chapman, J., Helfenbein, K.G., Ramaiya, P., Detter, J.C., Larimer, F., 2004. Genome sequence of the lignocellulose degrading fungus Phanerochaete chrysosporium strain RP78. Nat. Biotechnol. 22, 695–700.

Pickles, B.J., Genney, D.R., Potts, J.M., Lennon, J.J., Anderson, I.C., Alexander, I.J., 2010. Spatial and temporal ecology of Scots pine ectomycorrhizas. New Phytol. 186, 755–768.

Richards, T.A., 2011. Genome evolution: horizontal movements in the fungi. Curr. Biol. 21, R166–R168.

Richards, T.A., Talbot, N.J., 2013. Horizontal gene transfer in osmotrophs: playing with public goods. Nat. Rev. Microbiol. 11, 720–727.

Riley, R., et al., 2014. Extensive sampling of basidiomycete genomes demonstrates inadequacy of the white-rot/brown-rot paradigm for wood decay fungi. Proc. Natl. Acad. Sci. U.S.A. 111, 9923–9928.

Rosling, A., Timling, I., Taylor, D.L., 2013. Archaeorhizomycetes: patterns of distribution and abundance in soil. In: Horwitz, B.A., Mukherjee, P.K., Mukherjee, M., Kubicek, C.P. (Eds.), Genomics of Soil- and Plant-Associated Fungi. Springer, Berlin Heidelberg, pp. 333–349.

Skrede, I., Engh, I., Binder, M., Carlsen, T., Kauserud, H., Bendiksby, M., 2011. Evolutionary history of Serpulaceae (Basidiomycota): molecular phylogeny, historical biogeography and evidence for a single transition of nutritional mode. BMC Evol. Biol. 11, 230.

Slot, J.C., Hibbett, D.S., 2007. Horizontal transfer of a nitrate assimilation gene cluster and ecological transitions in fungi: a phylogenetic study. PLoS One 2. e1097.

Talbot, J. M., Martin, F., Kohler, A., Henrissat, B., and Peay, K. G., 2015. Functional guild classification predicts the enzymatic role of fungi in litter and soil biogeochemistry. Soil Biol. Biochem. 88, 441– 456.

Tunlid, A., Rineau, F., Smits, M., Shah, F., Nicolas, C., Johansson, T., Persson, P., Martin, F., 2013. Genomics and spectroscopy provide novel insights into the mechanisms of litter decomposition and nitrogen assimilation by ectomycorrhizal fungi. Genom. Soil Plant Assoc. Fungi 191, 441–456.

Vandenkoornhuyse, P., Quaiser, A., Duhamel, M., Le Van, A., Dufresne, A., 2015. The importance of the microbiome of the plant holobiont. New Phytol. 206, 1196–1206.

van der Heijden, M.G.A., Martin, F.M., Selosse, M.-A., Sanders, I.R., 2015. Mycorrhizal ecology and evolution: the past, the present, and the future. New Phytol. 205, 1406–1423.

CHAPTER

7

Mutualistic Symbiosis Between Fungi and Autotrophs

Sarah C. Watkinson

University of Oxford, Oxford, UK

Introduction

All plants are hosts to fungi that are specialised to establish close contact with living plant cells and feed on the products of photosynthesis. This mode of fungal nutrition is termed **biotrophy**. Some biotrophic fungi are parasites, including the pathogenic rusts and smuts described in Chapter 8. Others confer advantage on their hosts and are said to be **mutualistic**, and these are the subject of this chapter. We first consider the **mycorrhizal** fungi, which sustain all terrestrial ecosystems by mutualism with roots. In **lichens**, multicellular tissues of the fungal partner accommodate unicellular green algae and/or photosynthetic prokaryotes that provide them with carbon and energy. Lichens can therefore grow on bare illuminated surfaces and include species that are the main primary producers in environments too harsh for vascular plants, on mountains, in deserts, and the Arctic and Antarctic. Fungi that grow microscopically inside the tissues of plants are termed **endophytes**. Toxic secondary metabolites of some endophytes can protect the host from insect attack. These molecules are of pharmacological interest because they target specific enzymes common to human and animal metabolism.

MYCORRHIZA

The mycorrhizal mode of nutrition, in which the development and physiology of the fungus are integrated with plant roots to form a joint nutrient-absorbing structure, has evolved separately in different clades of the fungal kingdom. In all cases, the mutualism is based on the exchange of photosynthate from the plant with nutrients scavenged from the soil by the fungal mycelium. The interface for nutrient exchange forms during development of the mycelium in the root. In the following sections we consider arbuscular mycorrhizal fungi (AMF) formed by the monophyletic Glomeromycota (Chapter 1, p.25); Ectomycorrhiza (ECM) formed mainly between forest trees and basidiomycetes but a few ascomycetes; and ericoid mycorrhiza (ERM) formed by ascomycetes with Ericaceae and other plants of acid

The Fungi
http://dx.doi.org/10.1016/B978-0-12-382034-1.00007-4

nutrient-poor soils. In all cases the plant supplies carbon and energy to the fungus, and the fungus assists plant nutrition by scavenging mineral nutrients from the soil and from organic remains.

Figures 7.1 and 7.2 show the structures of AMF, ECM, and ERM mycorrhiza. Ectomycorrhiza and ERM, being formed mainly by basidiomycetes and ascomycetes, respectively, produce mycelium capable of persistent growth and tissue formation. They can give rise to persistent and extensive networks of **extraradical mycelium** – mycelium extending into the soil beyond the root. They may form multicellular fruiting bodies and decompose substantial plant remains. The extraradical mycelium of AMF, formed by the relatively ephemeral Glomeromycota, extends only short distances from the root and forms no multicellular tissues.

Arbuscular Mycorrhizal Fungi

All AMF are monophyletic, belonging to the Glomeromycota, a basal clade of the Kingdom Fungi. They are the most universally distributed of mycorrhizal fungi, both geographically and in host range, the simplest in form, and the most ancient. They are present in over 70% of land plant species including liverworts, ferns, gymnosperms, and angiosperms. Important food and forest crop species, such as wheat (*Triticum aestivum*), rice (*Oryza sativa*), maize

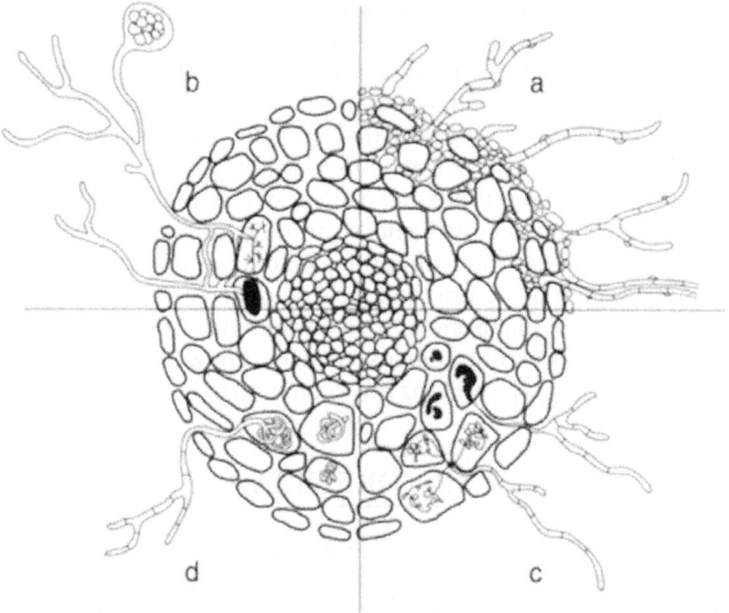

FIGURE 7.1 Types of interaction between plant roots and different kinds of mycorrhizal fungi: (a) ectomycorrhiza, with fungal mantle outside the root epidermis connected to mycelium in the soil, and to intercellular tissue of the Hartig net in the root cortex; (b) arbuscular mycorrhiza, infecting the root from a large spore in the soil, growing intercellularly in the cortex and forming intracellular vesicles and branched arbuscules in cells of the inner root cortex; (c) orchid mycorrhiza, forming intracellular hyphae which live briefly in each invaded cortical cell before being digested; (d) ericoid mycorrhiza forming bundles of hyphae in root cells, which are connected to hyphae in the soil. *Source: Carlile, M.J., Watkinson S.C., Gooday, G.W., The Fungi, second edition: Figure 7.12b.*

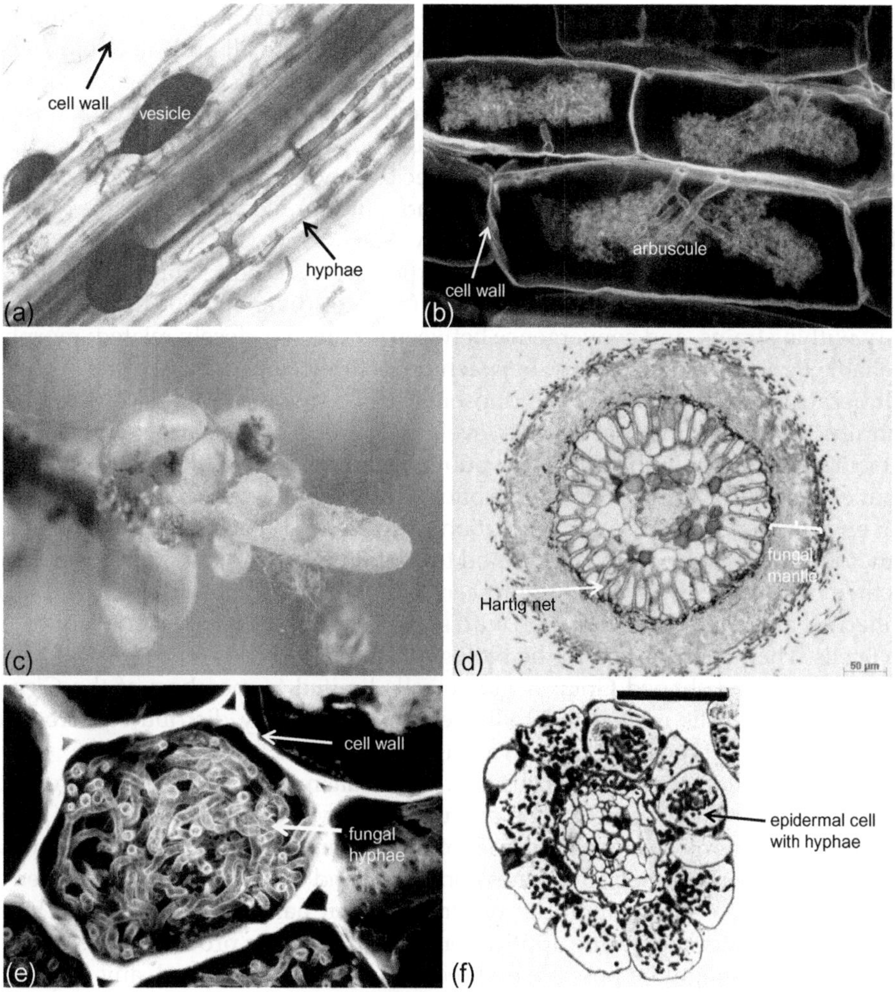

FIGURE 7.2 Typical structures of arbuscular mycorrhizas (a, b), ectomycorrhizas (c, d), and ericoid mycorrhizas (e, f). *Source: van der Heijden et al. (2015).* (See the colour plate.)

(*Zea mays*), poplar (*Populus* spp.), and soybean (*Glycine max*) are mycorrhizal, and symbiosis with AM fungi is now recognised as the default mineral nutrient acquisition strategy of land plants. Fossilised intracellular fungi identifiable as AMF have been found in the rhizomes of petrified early land plants in rocks from the Ordovician era laid down 460 Mya. Arbuscular mycorrhiza thus pre-date vascular plants with roots and would have associated with the first green plants to colonise the land. Colonisation of plants by AM fungi can result in a 20% net increase in photosynthesis, a major contribution to the global carbon cycling budget of ecosystems. Any soil where plants grow will contain spores of AMF, the largest of which, at up to 0.5 mm diameter, are visible to the naked eye and can be separated from soil by sieving and filtration of soil suspensions. The spores have thickened and resistant walls coloured white, yellow, orange, or brown with carotenoids. They contain reserves which on germination support the outgrowth of a short hypha. Because AMF fungi are all obligately biotrophic, they cannot continue to grow after exhausting their reserves without colonising the roots of a host plant, so culture collections of AMF fungi have to be maintained on pot-grown hosts.

Five genera, *Acaulospora*, *Entrophospora*, *Gigaspora*, *Glomus*, and *Scutellospora*, are distinguished on morphotype. About 230 morphospecies can be distinguished by experts on the basis of limited morphological variations. The advent of molecular techniques has revealed a much greater diversity of AMF than was known even a few years ago. Molecular methods for identification are much needed in AMF studies because their simple morphology gives insufficient information to distinguish the full range of diversity. For the usual reasons of plasticity within strains, and convergence between strains not closely related, morphology is not a reliable indicator of relatedness between types. Molecular studies of field diversity are difficult because the ITS region of the ribosomal DNA small subunit is too variable within AMF strains to identify them at species level, and multiple nuclear genotypes are usually present in a single individual. However, a public database, Maarj*AM* (http://www.maarjam.botany.ut.ee) now allows grouping into taxonomic hierarchies equivalent to families and genera, and environmental OTU's with resolution equivalent to species, are recognised.

Glomeromycota have recently been found to contain intracellular Gram-positive endobacteria in their cytoplasm, revealed by fluorescent *in situ* hybridization. These bacteria are vertically inherited and globally distributed, with sequence similarity to bacterial endosymbionts of insect cells. Their contribution to the biology of the fungus has yet to be discovered. Two kinds are known: rod-shaped Gram-negative cells limited to members of the *Gigasporaceae* family, and coccoid cells related to Mollicutes that are widely distributed across different lineages of AMF. Both types may exist in a single fungal cell.

The life cycle of Glomeromycota is obscure. No sexual morphology has been reported, and until recently they were assumed to be asexual. However, hyphae of different individuals will fuse to form heterokaryons with nuclei of more than one genotype, with segregation of genetic markers. Meiosis-associated genes conserved and expressed in four different *Glomus* species suggest that recombination can occur, though it has never been seen. This has important implications for any future application of AMF for crop productivity, since inoculant strain improvement might be possible by breeding. On the other hand, any improved inoculum might lose its superiority by interbreeding with the pre-existing soil AMF population.

The association begins when a spore germinates close to a root (Figure 7.3) in the rhizosphere (the region of soil in the vicinity of a root that is influenced by root exudates and abraded cells that support a distinctive microbiota). When the outgrowing hypha meets the root close behind the advancing root tip, it forms pads termed **hyphopodia**, that adhere like the appressoria of pathogenic fungi (Chapter 8). From the hyphopodium, an infection peg enters the root, either between two rhizodermal cells, or through a rhizodermal cell. Once in the cortex of the root, hyphae spread in the apoplast along its longitudinal axis (Figure 7.1b). AMF vary in the development and structure of the interface between fungal and plant cells. In some, hyphae grow mainly as intracellular coils inside the host's root cells, while in others extensive intercellular hyphae give rise to finely branched intracellular **arbuscules** that provide an interface for developmental and metabolic interactions between host and fungus (Figure 7.2b). The common genus *Rhizophagus* develops storage vesicles (Figures 7.1b and 7.2) as well as arbuscules within host tissue. In a root cell invaded by a hypha, the nucleus moves from the periphery to the centre of the cell, the vacuole becomes fragmented and plastid

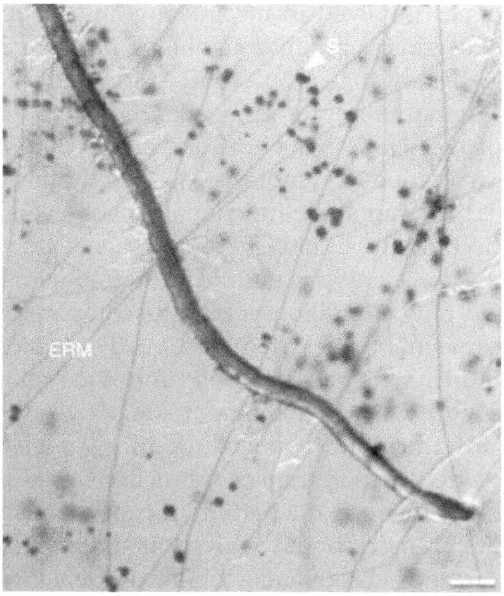

FIGURE 7.3 Growing hyphae and spores of *Glomus* sp. colonising chick pea roots *in vitro*. Extraradical mycelium, ERM spores, S. Bar 300 micrometres. *Source: Lanfranco and Young (2012).*

organisation alters. Both plant and fungal cell membranes remain intact, with the plant cell membrane expanding and invaginating around the branching hypha and retaining a thin layer of plant cell wall materials between fungal and plant cell. The fungus connects to the soil outside the root via intercellular hyphae that traverse the plant cortex and epidermis and extend short branches from the root surface into the rhizosphere.

Each AMF mycelium in a root results from a separate colonisation by a spore present in the soil (Figure 7.3). As the root tip grows, the root surface behind becomes thickened and inaccessible for live-cell colonisation. Thus, colonisation is dynamic and continuous, and maintenance of the association depends on abundant inoculum pre-existing in the soil. As each root becomes colonised, it develops a weft of outgrowing hyphae. Roots can extend at the rate of several hundred millimetres a day, and every millimetre of new root growth can support over a thousand millimetres of hyphae. The combined root-fungus structure thus expands rapidly through the soil, although AMF do not penetrate deeply and are found mostly in the top 20 cm of the soil horizon.

Development of the symbiosis depends on recognition between partners. Recognition between AMF and plant host begins with an exchange of chemical signals in the soil. Plant roots exude strigolactones which initiate the interaction by inducing tropic growth in AM germ tubes. Concentrations as low as 10^{-13} M can induce this response. Strigolactones also stimulate germination of the seeds of parasitic plants, suggesting that these chemicals are a reliable and sensitive indicator of the presence of plant roots. Germinating AMF fungi release diffusible low-molecular-weight chitin derivatives, lipo-oligosaccharides and chito-oligosaccharides, which act as the '*myc* factors' that induce symbiotic development in root cells. Genomic investigations are revealing the reciprocal molecular responses of AMF and host during mycorrhiza development. The plant genes involved in recognising and responding to the presence of AMF prove to have ancient origins. A putative signalling pathway, deduced from the successive up-regulation of the genes during colonisation, starts with recognition of fungal chito-oligosaccharides by **LysM** domain receptor kinases of the plant. This elicits calcium spiking in the root cell nucleus, which activates a calcium- and calmodulin-dependent protein kinase, leading to transcriptional activation of symbiosis-associated plant genes. Some of the genes involved in AMF symbiont recognition are functionally conserved across all vascular plants, and have homologues in Bryophytes and in Charales (green algae). Others, found only in leguminous plants, are involved in symbiotic nodule formation with Rhizobium bacteria. Since leguminous plants diverged only 60 million years ago, compared with the 400 million year history of AMF, legumes presumably recruited these AMF symbiosis genes for the development of nitrogen-fixing nodules with bacteria. Entry into the root alters the patterns of gene expression in the fungus. Figure 7.4 shows a comparison between the sets of genes expressed in extraradical (ERM) and intradical (IRM) mycelium during development of the association. In the fungus growing outside the root, transport functions predominate, but once nutrient exchange is established within the root, the pattern of expression changes in favour of genes associated with metabolism. Transcriptional analysis at successive phases of spore germination, rhizosphere interaction, and subsequent symbiotic hyphal growth of *Rhizophagus irregulare* inside the root of *Medicago truncatula* have shown different sets of genes are preferentially expressed at each stage. In some cases the proteins they encode have been visualised *in vivo*. Once inside the host tissue, AMF hyphae exude multiple small secreted proteins – termed mycorrhizal small secreted proteins (MiSSP) – into the apoplastic space. This is the space inside the root but accessible to the soil solution, which includes both plant cell walls and intercellular spaces, but which lies outside the plant cell membranes. One such protein, SP7, inhibits a plant transcription factor responsible for up-regulating plant host defence genes. Transgenic expression of *SP7* in a transformed host plant resulted in greater mycorrhizal colonisation, confirming the role of this gene in suppressing host resistance. The SP7 protein enters the plant cell from the apoplast and localises to the plant cell nucleus where it binds to an ethylene-responsive transcription factor regulating the expression of several defence-related genes. This protein shares homology with **MiSSPs** from ectomycorrhizal fungi and also with the protein effector molecules that enable plant pathogenic fungi to overcome their hosts' resistance (Chapter 8, p.255). Thus, mutualistic symbiotic fungi share fundamental mechanisms with parasitic biotrophs, using similar components of signal and response pathways to achieve entry into the plant cell.

Once an interface is established, nutrient exchange occurs between plant and fungus. Fungal sugar transporters that facilitate carbon supply to the fungus are among the first host genes to show increased expression, a sequence of events which is presumably adaptive, because the obligate AMF requires a carbon supply once the spore reserves have been exhausted. One of these up-regulated transporters, **MST2**, has affinity and specificity to xylose. Since no invertase was found in the AMF genome, this suggests that symbiosis may depend on apoplastic plant invertases. Plant cell wall formation is inhibited in the part of the plant plasmalemma forming the arbuscule, but wall precursor sugars may still be produced in this region and assimilated by the fungus instead of being used for host cell wall synthesis.

When symbiosis has been established, further physiological changes occur that are associated with the role of the fungal partner in host mineral nutrient uptake. From physiological

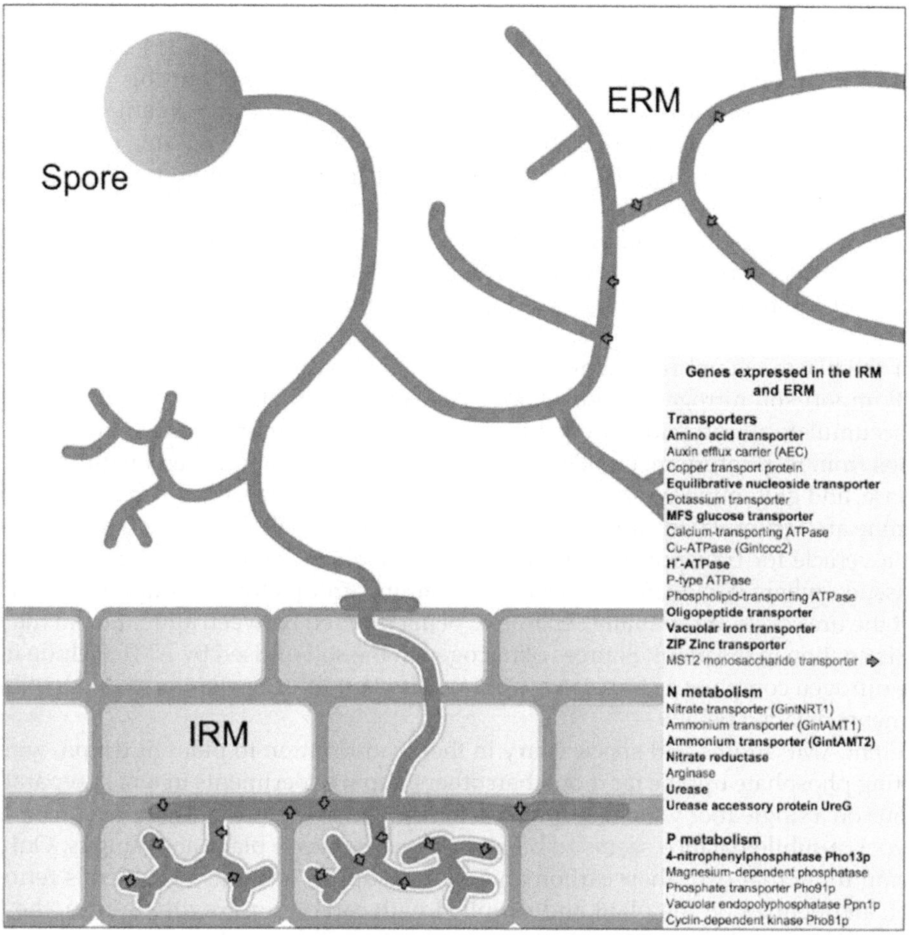

FIGURE 7.4 Gene expression during different life stages of AMF. Comparison of genes expressed in extraradical (ERM) and intraradical IRM mycelium, shows a switch from the expression of genes encoding proteins with transport functions to those involved in metabolism as the host–fungus nutrient exchange interfaces become established. *Source: Lanfranco and Young (2012).* (See the colour plate.)

experiments, uptake by the plant of both phosphate and nitrogen is enhanced following AMF colonisation. Cell membrane transporters, including those involved in transport of inorganic phosphate, nitrate, and cations including iron and zinc, are all up-regulated. A unique clade of plant phosphate transporters (**Pht1**, subfamily I), includes members expressed exclusively in the AM symbiosis, for example, **MtPT4** of *Medicago trunculata*, which is essential not only for the acquisition of phosphate from the fungus, but also for development of the fungal symbiotic structures within the root. Other genes expressed at this stage include proteins involved in the interconversion of metabolites shuttled between partners, and those for phosphate acquisition, such as secreted phosphatases able to act on phosphate esters. Free soluble phosphate is low in soil because soil cations – calcium, iron, and aluminium – form poorly-soluble salts. Roots become surrounded by a diffusion-limited zone depleted in phosphate, nitrogen, nitrate, and other nutrients. AMF can relieve this limitation as they can grow beyond the depletion zone, but this is not their only contribution to phosphorus acquisition. Not only do fungi take up soluble soil phosphate more efficiently than unaided roots, they can also dissolve poorly-soluble phosphate esters present in acid soils, as well as dissolving phosphate from rock minerals (Chapter 5, p.181). Phosphate taken up by extraradical hyphae is accumulated as polyphosphate, in which form it is translocated in vacuoles towards the interfacial apoplast separating host and fungal cells. Several proteins involved in this process have been identified from the transcriptome of the fungus growing inside the root. Lipid metabolism is up-regulated here, presumably to facilitate export of carbon from lipid vesicles of AMF and to provide precursors for the plasma membrane synthesis demanded by arbuscule formation.

AMF import soil nitrogen into the root by means of spatially differentiated metabolism. Fungi accumulate amino acids by combining soil-acquired nitrogen compounds with carbon skeletons from respiratory metabolism, in amination reactions catalysed by glutamine dehydrogenase, and glutamine synthase/glutamine aminotransferase. Base-rich amino acids such as arginine are accumulated, stored, and transported in vacuoles (Chapter 5, p.155). Arginine acts as a vehicle for transporting amino nitrogen by translocation through hyphae into the plant tissues, where it is deaminated to release amino groups for plant uptake. In effect, the steps of the urea cycle are spatially separated, being divided between amination in the fungus and deamination in the plant. Sources of nitrogen in the soil utilised by AMF include not only soluble nitrogen compounds but also nitrogen-rich plant litter, as demonstrated by ^{15}N tracer experiments detailed below.

Different AMF strains and species vary in their contribution to plant nutrition, with some promoting phosphate uptake far more than others. From experiments in which separate AMF infections on a single root were manipulated to supply varied amounts of phosphate or none to the root, a subtle balance seems to be maintained between plant and fungus. Only those colonising fungi that earn their carbon and energy supply, in terms of minerals returned to the root, are retained on the plant and supplied with carbon for growth. In turn, the fungus provides more phosphorus to the plant in response to an increased sugar supply from it.

High-throughput transcriptional analysis reveals that the effect of AMF symbiosis on the plant is not limited to the root, but modifies the entire plant transcriptome systemically to an extent that was not previously appreciated. Experimental AMF colonisation of the root was found to affect the expression of hundreds of plant genes in rice, *Oryza sativa*, and in tomato, *Solanum lycopersicum*, including many that confer important phenotypic characteristics, such as lycopene synthesis in tomatoes. The results point to the potential of AMF inoculation in crop nutrition and protection.

AMF in Ecosystems

In natural and semi-natural ecosystems, AMF fungi play vital, but still poorly-understood parts in plant productivity and community diversity (Figure 7.5) as well as making essential contributions to soil structure and carbon sequestration. The ability to connect to an AMF network can affect nutrient capture by roots. Plants colonised by AMF may be more efficient at extracting soil phosphate and/or nitrogen than plants grown in sterile soil. Roots typically form mycorrhizal associations under nutrient limitation. Plant roots containing abundant

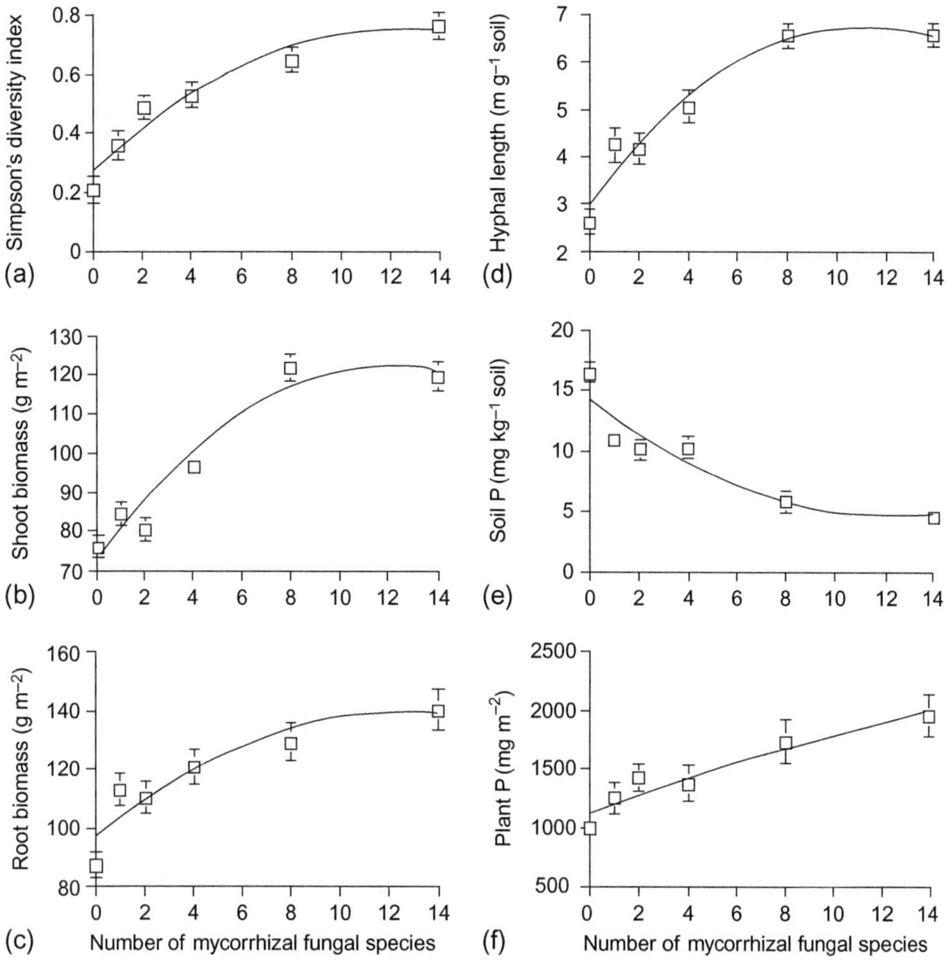

FIGURE 7.5 The effect of AMF species-richness on plant and fungal growth. Sets of replicate trays of sterilised soil, placed out of doors in a field site, were inoculated with soil containing 1, 2, 4, 8, or 14 different strains of species of AMF. The inoculated trays were then sown with 100 seeds from a mixture of 15 plant species. After one growing season, the plants growing in each tray were assessed for diversity of species (a), shoot and root biomass (b and c, respectively) and the total length of hyphae of mycorrhizal fungi in the soil (d). The levels of phosphate were also measured in the soil (e) and in plants (f). Plants grew better, and the plant community was more diverse, when 8 or 14 different AMF were present. There were accompanying increases in hyphal growth and phosphate uptake from soil to plants. *Source: van der Heijden et al. (1998).*

phosphate inhibit the formation of AMF arbuscules, the plant thereby avoiding an unnecessary drain on its carbon resources. Soil minerals accumulated by the extraradical hyphae are partitioned between the fungus and the plant.

AMF hyphae translocate nutrients not only from soil to root, but also between roots of adjacent plants. Using growth chambers divided by partitions either permeable or non-permeable by growing hyphae, it was shown that a plant can acquire nitrogen from dead leaf litter via its AMF hyphae, when these were allowed to colonise plant litter enriched with the stable isotope ^{15}N (Figure 7.6). Nitrogen import through AMF hyphal connections enhanced host plant growth as well as supplying the fungus. Some aspects of the mechanism by which AMF acquire soil nitrogen remain unclear. Readily-available forms of nitrogen are scarce in natural soils, in particular those of boreal forest, where most nitrogen is in the form of complex organic compounds resistant to microbial attack. Mycorrhizal fungi are probably the main route by which trees acquire nitrogen from these compounds in the soil. The ability of AM fungi to take up both labile and recalcitrant organic nitrogen compounds under field conditions in boreal forest has been recently demonstrated, using an ingenious technique involving test substances (glycine and chitosan as labile and recalcitrant nitrogen compounds, respectively) bound to the surface of nanoparticles called 'quantum dots'.

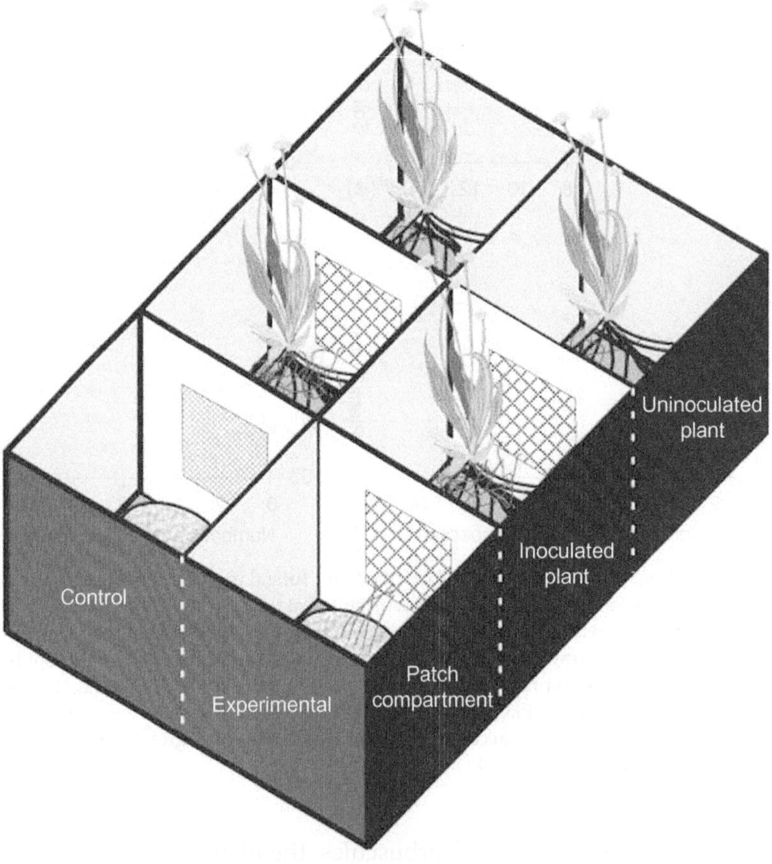

FIGURE 7.6 Microcosm used to demonstrate the ability of AMF mycelium to translocate nitrogen from plant litter into roots of living host plants. Hyphae from experimentally AMF-inoculated plants in the centre compartments of the experimental row grew into the patch of ^{15}N isotope-labelled grass litter in the 'Patch' compartment. In the control row, intervening mesh prevented hyphae growing between compartments but allowed solute diffusion. It was found that AMF-colonised plants allowed to access the grass litter acquired three times as much nitrogen as those where AMF hyphae were prevented from reaching the litter. *Source: Hodge et al. (2001).* (See the colour plate.)

Natural and semi-natural plant communities host a much wider diversity of AMF fungi than arable soils. Ploughing, fertilising, and monoculture reduced AMF diversity in a UK arable wheat field compared to that in an adjacent woodland on the same type of soil. In field experiments, the diversity of AMF fungi added to sterile soil influenced the productivity and diversity of a plant community subsequently developing from a standard seed mix sown in the soil (Figure 7.5). The presence of AMF relaxes competition between plant species, allowing different species to flourish alongside each other. Several independent parameters of plant growth and phosphate uptake were enhanced in the mixed plant community when a variety of AMF types were present, suggesting that different plants might require different types of mycorrhiza. AMF composition may thus affect the capacity of a soil to support a diverse plant community or a particular type of plant. The invasiveness of introduced exotic plant species may also be influenced by specific interactions with local AMF. Those AMF strains or species which are found widely may have a wide range of habitat tolerances, while it is likely that those strains or species more specialised to host or habitat are the ones most under-represented in collections and databases. We still know little about the extent of host specificity, and whether AMF include rare, keystone or threatened species.

AMF have potential as inoculants to agricultural soil to improve crop productivity. An advantage of AMF is that they are biologically targeted to roots, compared with fertiliser that is quickly leached out by rain. Other potential benefits include enhanced nutrient uptake, drought resistance and protection against pathogens and parasites. A 2-year trial with two bioinoculants, *Glomus fasciculatum*, together with the mycorrhizal helper bacterium *Pseudomonas monteilii*, demonstrated significant protection of the medicinal crop plant *Coleus forskohlii* against attack by the soil fungus pathogen *Fusarium chlamydosporum*.

Although it is not yet clear that AMF could be developed for sustainable management in industrialised agriculture, it has been suggested that, rather than attempt inoculation, simply reducing tillage could preserve the integrity of AMF networks already present in all soils, and thus enhance crop growth and soil carbon sequestration. AMF not only aid nutrient uptake by plants but also channel carbon from plant photosynthesis into the soil rapidly and in significant amounts, with estimates ranging from 50 to 900 $kg\,ha^{-1}$. Fungi may add carbon to soil in the form of extraradical hyphae or as exudates such as **glomalin**, an extracellular mycelium-bound substance, believed to be a glycoprotein, which is sticky and hydrophilic. Glomalin contributes to the moist, cohesive character of soils under natural vegetation compared with the friable crumb structure of cultivated arable soils. In a 16-year-long field monitoring study, treatments that increased soil AMF hyphae also increased glomalin-related soil protein (**GRSP**) pools and water-stable macro-aggregates, while all three were reduced by fungicide application. Conserving the resident AMF population could thus be seen as a valid goal in soil management for carbon sequestration as well as plant productivity.

Ectomycorrhizal Fungi

Ectomycorrhiza (ECM) are so-called because the fungus is extracellular throughout the association, forming no intracellular structures. These fungi associate mainly with woody perennial plants, but some shrubs that occur in early stages of plant successions are ectomycorrhizal, for example, *Salix herbacea* and *Dryas octopetala*. About 3% of plant species form ECM with thousands of fungi, mainly basidiomycetes although some common types are formed by ascomycetes. Although relatively few plant species are involved, the importance of ECM fungi on a global scale cannot be overestimated, because they partner with tree hosts that form the most important terrestrial carbon sinks on the planet: Pinaceae, the vast northern conifer forests of the world, Fagaceae, temperate deciduous woodlands, and Dipterocarpaceae, the South East Asian rainforests. In ECM the fungal partner receives carbon and energy as photosynthate from the host tree via an extracellular interface within the root cortex described in detail below. However, many ectomycorrhizal species of fungi can also feed on plant remains (Figure 7.9), an ability they retain from saprotrophic ancestry (Figure 7.10).

The main clades of ectomycorrhizal basidiomycetes are the Agaricomycetes *Amanita*, *Boletus*, *Cortinarius*, *Laccaria*, *Lactarius*, *Russula*, and *Suillus*, the wood decomposing basidiomycetes *Tomentella* and *Thelephora* in Polyporales, the ascomycete *Coenococcum*, and the Sebacinales which have relatively poorly differentiated sporophores and the capacity to form symbioses with a wide variety of plants including ericaceous species and non-woody orchids. Molecular analysis is revealing much greater infraspecific variation in ECM fungi than was once supposed. For example, *Pisolithus tinctorius*, used as an inoculant in forestry, is represented worldwide by many separate species each with different host plant associations. Because the diversity of natural populations is correlated with resilience under disturbance, infraspecific diversity in mycorrhizal fungi is likely to be an important factor in the resilience of forest ecosystems under climate change, and is being intensively investigated now that sequence data representative of natural ecosystems are accessible.

Morphology and Anatomy

In the surface soil horizons of most temperate forests, fine ectomycorrhizal rootlets (Figure 7.7) will be found running through surface layers of soil rich in decomposing plant litter. Their short lateral branches, about a millimetre wide or less, differ in colour and surface characteristics from their parent axis, and are typically thicker. Their distinctive appearance is due to their surface coating of fungal tissue, the ectomycorrhizal mantle or sheath. Within the root, living root cells interface with fungal hyphae that penetrate between them forming the Hartig net, named after the German forest biologist Robert Hartig.

The mantle (Figure 7.2d) is the living tissue barrier between absorbing rootlet and soil. It develops continuously as the root grows, suppressing root hair development and forming a sock-like sheath several cells thick encasing the entire root tip. As new roots arise they are colonised by hyphae which may originate from ECM already existing on the root, from spores, or from mycelium already growing on other roots or plant litter. Some species readily colonise from spores, others require an already-established base on a living root in order to establish associations with further roots. In the initial stages of root infection, hyphae grow between epidermal cells to establish the nutrient exchange interface. Angiosperm and gymnosperm ECM differ in the radial extent of root tissue colonised. In angiosperms, only the epidermal layer is colonised, but in gymnosperms hyphae penetrate between the outer cortical cells as well. In all cases hyphae are limited to the apoplastic space outside host cell plasma membrane and thus remain outside the root endodermis. The characteristics of the mantle are the result of alterations in growth form of both partners as the association develops. The plant may be induced by colonisation to form new lateral rootlets, and the pattern of these is typical of the fungal species, so that different fungi form different characteristic ECM morphotypes, even on the same species of tree (Figure 7.7). Fungal association may stimulate radial elongation

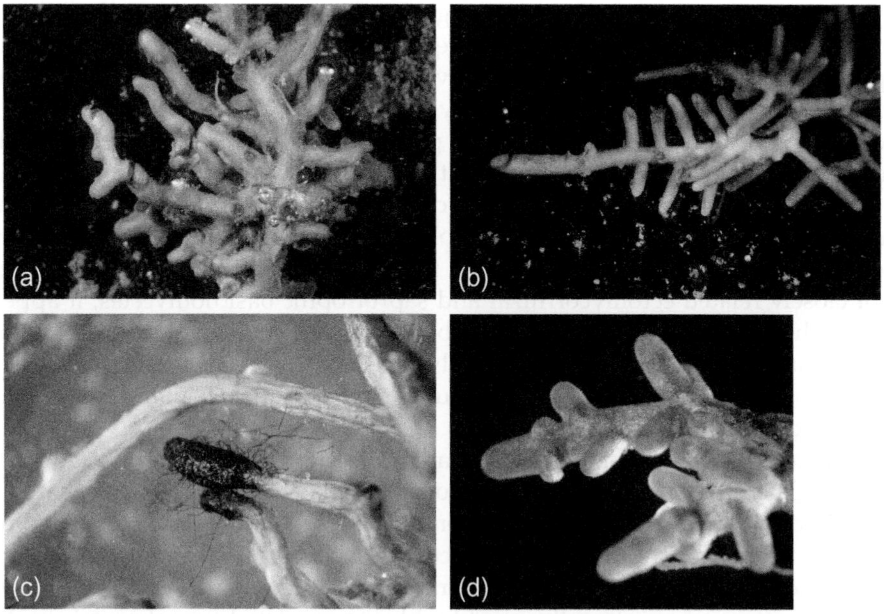

FIGURE 7.7 Ectomycorrhizal morphotypes from beech (*Fagus sylvatica*) woodland soil; (a) Laccaria sp., (b) Lactarius sp., (c) Coenococcum sp., and (d) Russula sp. *Source: John Baker.* (See the colour plate.)

of the cortical cells, resulting in a greater potential surface contact area for nutrient exchange and producing the root thickening characteristic of some ECM morphotypes. Fungal mantle tissue forms either by extensive interweaving of hyphal filaments, or by cell division to produce isodiametric cells. Differentiated cell types characteristic of the species may arise, such as latex-producing cells in *Lactarius*. Pigments give each morphotype its characteristic colour: purple, black, white, and shades between dark brown to pale yellow. Morphotypes also vary in whether they produce mycelial outgrowths from the mantle into the soil, and the extent of such outgrowth. Species vary widely in the nature and extent of their exploratory mycelium. Some, like members of the Boletales genus *Suillus,* form extensive networks, with translocating systems of mycelial cords (Chapter 5, p.176), while others, like the common ECM genus *Russula,* simply encase the rootlet in a smooth sheath with no visible emanating mycelium. The widespread black ECM formed by the ascomycete *Coenococcum geophilum* is recognisable by its hairy-looking surface bristling with melanized hyphae.

The life span of ECM rootlets is of interest in the context of forest soil carbon sequestration via ectomycorrhiza, but difficult to measure. The residence time of carbon (from ^{14}C data) in ECM of several species (identified from morphotype and RFLPs) was investigated in woodland of mixed pine and juniper. The turnover time of 4-5 years in the carbon of ECM rootlets suggested that they were relatively long-lived compared with unassociated mycelium and with mushrooms. Many variables are likely to affect longevity of ectomycorrhizal fungi on rootlets, including the species of fungus, the growth rate of roots, grazing of fungi by soil invertebrates and soil nutrient status. The contribution of dead ectomycorrhizal fungi to soil carbon is significant and is further discussed below.

The Hartig net is the interface between fungal hyphae and plant root cells, where the development and physiology of both partners becomes integrated into a functional unit. Here, fine hyphae ramify as finger-like processes over the plant cell surfaces and through the intercellular middle lamellae. Only in relatively few types of association, or in senescence of the mycorrhiza, do these hyphae invade the plant cell with intracellular growths. Instead, exchange occurs across both hyphal and plant cell walls, though plant cell walls are thinner at the Hartig net interface. Fungal growth is confined to the apoplastic space between root cells, freely accessible to the soil solution. Hyphae do not cross the endodermis, the cylinder of tissue that contains the central conducting tissue of the tree. From comparative genomic studies of closely related ectomycorrhizal and saprotrophic basidiomycetes, it appears that evolution of the ECM habit is accompanied by expansion and refinement of gene families of pectinases and glucanases active in hydrolysing components of the intracellular middle lamella, suggesting that enzymic lysis by the fungus may facilitate contact between plant and fungal cells in the Hartig net.

The sequence of root-fungus interactions leading to ectomycorrhiza formation starts when the fungus is chemotropically attracted to grow towards a nearby root, stimulated by root exudates including flavonoids and strigolactones. The exchange of signals between host and fungus during establishment and maintenance of ectomycorrhizal symbiosis has been dissected at the molecular level in the model interaction between *Laccaria bicolor* and *Populus* sp. As the partners come into contact, the most highly up-regulated fungal protein is, **MiSSP7**, one of a constellation of small secreted proteins (**MiSSP**'s) exuded by the fungus. This protein can be visualised *in situ* in the region of the developing Hartig net by immuno-localisation (Figure 7.8). **MiSSP7** enters the plant cell via active phosphatidylinositol 3-phosphate-mediated endocytosis, and is targeted to the nucleus, where it acts as a transcription factor affecting the expression of genes involved in defence, cell wall remodelling and signalling. **MiSSP7** shares some features with the protein **SP7** of AMF, described above, which prepares the root for symbiosis even before physical contact between partners. Both are effector proteins similar to the widely conserved effectors of plant pathogens (Chapter 8, p.252), and share the same role of suppressing resistance and setting up biotrophic interaction between cells of plant and fungus. *Laccaria bicolor* transformants with reduced expression of **MiSSP7** do not enter into symbiosis with poplar roots.

Following the initial burst of fungal **MiSSP** release, protein markers for plant resistance, elicited on first contact, decrease, and the plant's auxin responses change, leading to lateral root induction forming the typical ECM morphotype. As the root accepts fungal colonisation and starts to form joint tissues with the fungus, transport proteins involved in nutrient exchange start to be up-regulated. The genome of *Laccaria bicolor* has 15 genes characterised as high affinity H+/glucose transporters, which are expressed as mycorrhiza develop and are probably involved in the uptake of glucose from host cells. As fungal glucose metabolism accelerates, fed by sugars synthesised by the leaves and translocated to the Hartig nets of the roots, the ECM root system becomes a strong carbohydrate sink for the tree. Between 10% and 20% of the plant's photosynthate is estimated to be allocated to the ECM root system.

Genomic studies are revealing additional functional groups of genes up-regulated in ECM symbiosis that are considered to have roles in setting up and maintaining the plant-fungal interface. Cell wall remodelling enzymes, including pectinases, glucanases, and peptidases, presumably assist the hyphae to penetrate between cells in the apoplastic middle lamellar

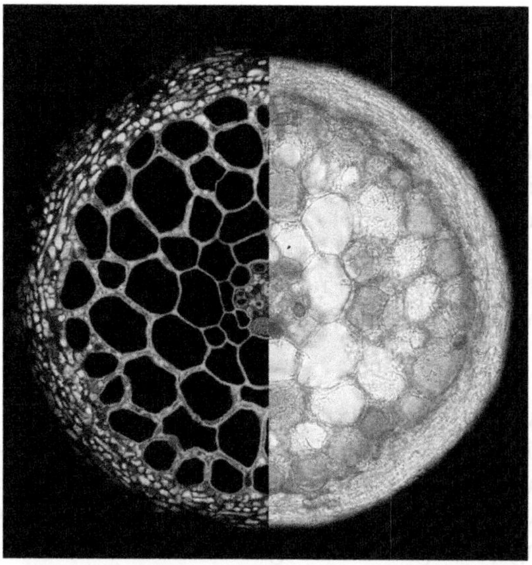

FIGURE 7.8 Immuno-localisation of a highly expressed fungal effector-like protein in a *Populus trichocarpa–Laccaria bicolor* ectomycorrhizal root tip. A transverse cross section of a poplar root colonised by the symbiotic ectomycorrhizal fungus *L. bicolor*. The green signal is an immuno-localisation of the fungal effector protein MiSSP7 highly expressed in the hyphae of *L. bicolor* while staining with propidium iodide highlights the cell walls of the root cells. *Source: Jonathan Plett and Francis Martin.* (See the colour plate.)

spaces of the root cortex. Symbiosis-related acidic polypeptides released by the fungus during ECM formation share features with adhesins of fungal pathogens of animals and plants so might help attachment of the hypha to the host cell. A third group of up-regulated proteins are hydrophobins, secreted proteins that form an amphipathic monolayer on surfaces with one hydrophilic and one hydrophobic side. These could be involved in covering the mantle (and extraradical cords), making the mycorrhizal root tip a sealed system. Genomic research such as the Agaricomycete sequencing programme of the Joint Genome Institute can be expected to reveal a diversity of function among the ECM fungi that populate forest soils.

The association between ECM partners is facilitated by soil bacteria of genera including *Pseudomonas*, *Rhodococcus*, *Streptomycetes*, *Burkholderia*, and *Bacillus*, known as **mycorrhiza helper bacteria**. These colonise the surface of fungal hyphae, stimulate hyphal growth, and may even live within hyphae as endobacteria. They assist ECM nutrient acquisition, protect their host plants from pathogens, and are probably important for tree growth in forests and nurseries, but we know little about the mechanism by which they promote the formation and function of mycorrhiza. The first genome sequence of a helper bacterium, *Pseudomonas fluorescens*, a Gram-negative rod-shaped bacterium isolated from a sporocarp of the ectomycorrhizal fungus *Laccaria bicolor*, was published in 2014.

Basidiomycete mycelium growing from long-lasting food sources such as dead wood and tree roots can persist for decades in the forest floor (unlike the ephemeral mycelia of AMF that turn over in a few days). Some species of ectomycorrhizal fungi can develop extensive and persistent foraging networks fuelled by tree hosts that in turn scavenge mineral nutrients for the host, from the soil, leaf litter, and even from pollen (Figure 7.9). The lignin/humus degrading capacity of ectomycorrhizal fungi varies between genera. Comparative genomics suggests that *Laccaria* has lost much of its wood decomposing ability, *Paxillus* decomposes complex organic matter in a way similar to brown rotters, and *Cortinarius* species which are typically found in well-established woodland have a full set of Mn-peroxidase genes and are likely to act as mycorrhizal white-rotters. Interesting parallels have been discovered between the evolution of saprotrophic brown rot and ectomycorrhizal nutritional modes in basidiomycetes. The clade Boletales, for example, includes both the wood-destroying dry rot fungus *Serpula lacrymans* and its closest known relative, the ectomycorrhizal *Austropaxillus*, both of which have lost many wood degrading enzymes in the course of evolution.

Evolution

Unlike AMF, ECM symbiosis is polyphyletic, having arisen in at least ten clades of fungi (Figure 7.10). Phylogenies of ECM basidiomycetes, such as species in the common ECM genus *Laccaria*, show many short branches indicating recent radiations. Molecular clock data, calibrated with fossil evidence, point to the earliest occurrence of ECM symbiosis around a hundred million years ago, alongside the origin and diversification of angiosperm and gymnosperm plants (Figure 7.11). Subsequently, ECM fungi appear, from present patterns of biogeographical distribution, to have accompanied their tree hosts, perhaps even enabling them to spread into new regions. Pine ECM fungi probably accompanied the earliest pine ancestors on the former southern hemisphere continent of Gondwanaland, and radiated along with the spread of pine forests across the northern hemisphere following the break-up of the older continents.

ECM associations continue to evolve dynamically today, with repeated acquisitions and some possible losses of symbiosis evident in basidiomycete phylogeny. This is not a stable mutualism like that of the *Neotyphodium* endophytes of grasses where the fungus has lost independent sexuality and is effectively part of the host (below, p. 235). Both partners retain the capacity for ecologically independent life. Sexual reproduction in each species occurs independently of the symbiotic partnership, retaining the potential for independent variation and evolution by natural selection. This is consistent with phylogenies that show both losses and acquisitions of the ectomycorrhizal mode of nutrition. While ECM symbionts are generally nested within mainly saprotrophic clades, there is some evidence, notably in Boletales, that the reverse step may have occurred, with loss of ECM-forming capability and reversion to saprotrophy.

ECM in Ecosytems

Mycelium of both ECM and plant litter saprotrophs mingles in the organic soil horizons of forests, and their interactions with each other and with other soil biota are of major significance in ecosystem nutrient dynamics. Basidiomycete mycelium dominates the microbiota of forest soils, and can amount to several tonnes per hectare. Living mycelium accumulates, stores, and redistributes carbon, nitrogen, phosphorus, and other nutrients. Mycorrhizal fungi not only cycle soil nutrients, but may deposit large amounts of recently fixed carbon in soils, building large pools of carbon in the form of complex molecules that contribute to long-term ecosystem carbon sequestration. Until recently it was assumed that soil carbon came mainly from plant remains accumulated aboveground and gradually incorporated into the upper soil

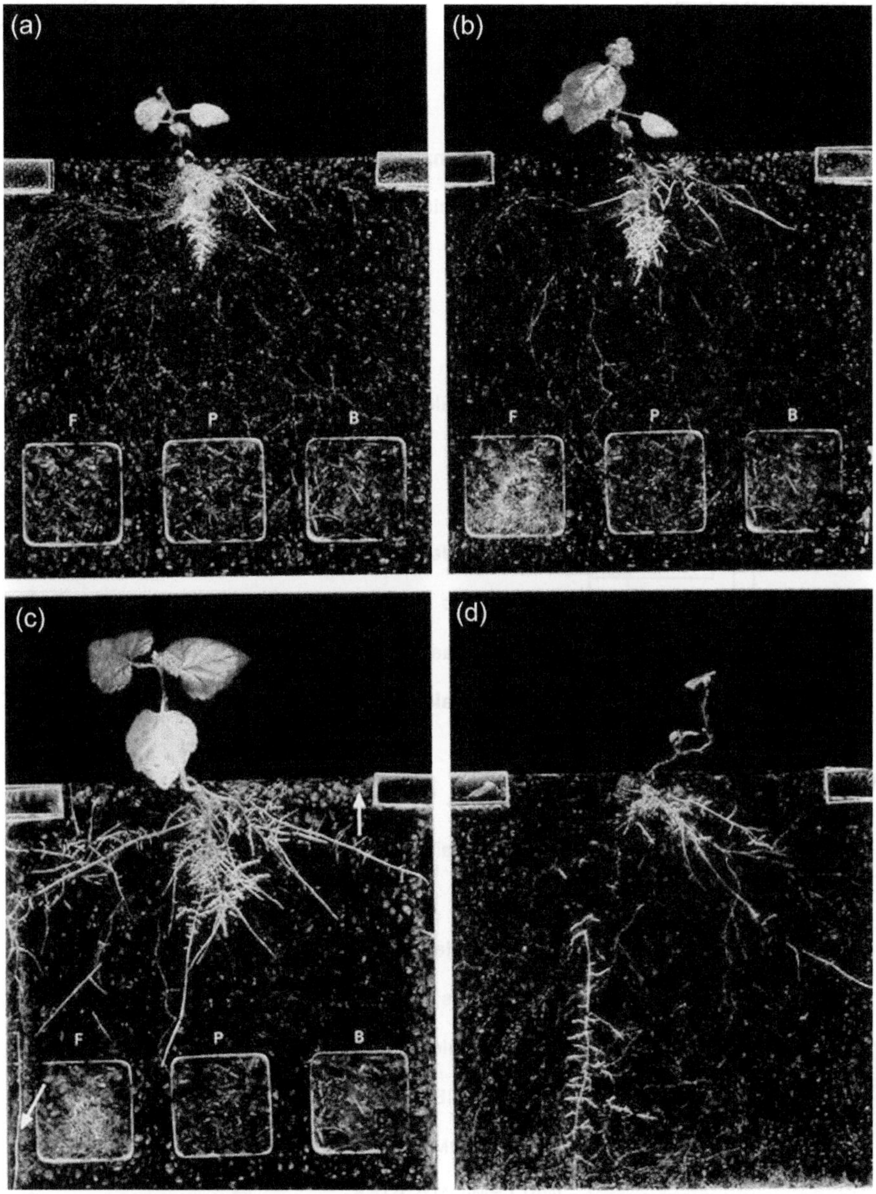

FIGURE 7.9 Photographs (a)–(c) show sequential development of ectomycorrhizal seedlings of birch, *Betula pendula*, associated with *Paxillus involutus* ectomycorrhiza in observation chambers containing trays of litter of beech (F), pine (P), and birch (B) at 8, 35, and 90 days after litter placement. Initial colonisation of litter is followed by increased plant growth, compared with d, a mycorrhizal control plant without litter addition. *Source: Perez-Moreno and Read (2000).* (See the colour plate.)

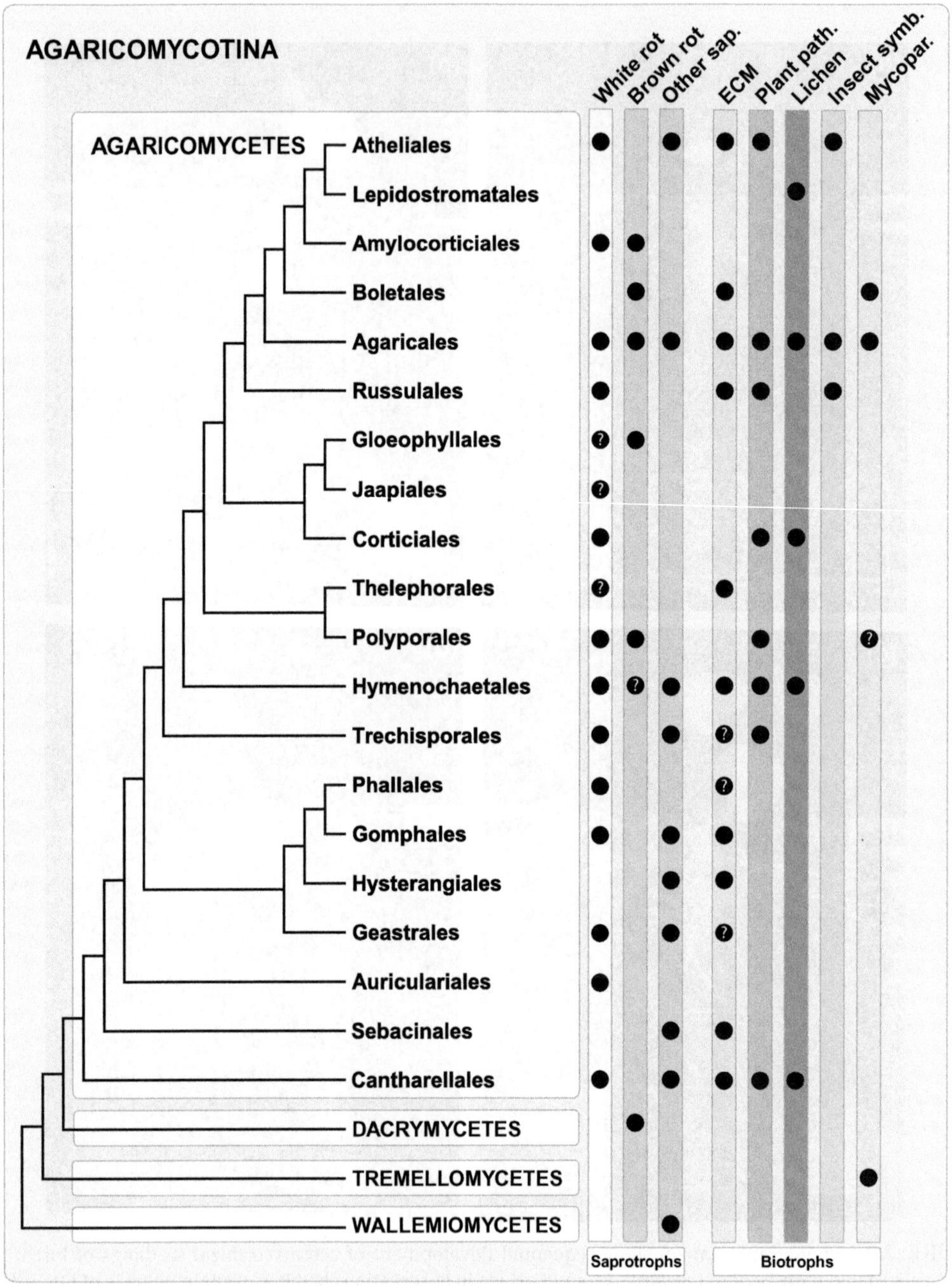

FIGURE 7.10 Phylogenetic distribution of major nutritional modes across the Agaricomycotina. The tree summarises recent phylogenomic and multi-gene phylogenetic studies. Saprotrophs include white rot and brown rot wood decay fungi, and a broad category of 'other' saprotrophs, such as litter, dung, and keratin-degrading fungi. White rot is very widespread and is probably the ancestral condition of the Agaricomycetes, but not the Agaricomycotina as a whole (note that it is absent from Dacrymycetes, Tremellomycetes, and Wallemiomycetes). Brown rot has evolved independently in at least five orders of Agaricomycetes, as well as Dacrymycetes. Biotrophs include ectomycorrhizal symbionts (ECM), plant pathogens, lichen-forming basidiomycetes, insect symbionts and mycoparasites, all of which are ultimately derived from saprotrophic ancestors. Agaricomycotina also include other biotrophs that are not shown, including endophytes, nematode-trapping fungi, bacteriovores, parasites of algae and bryophytes, and animal pathogens. Question marks indicate uncertainty. *Source: See James et al. (2006). Figure David Hibbett.* (See the colour plate.)

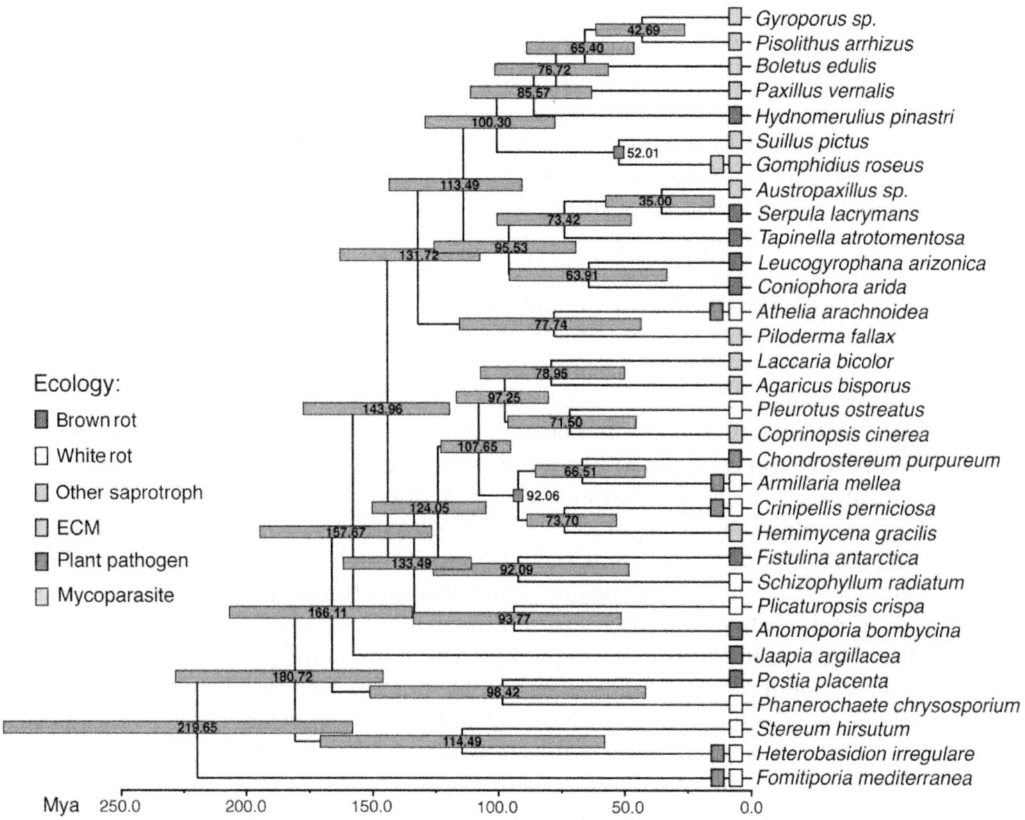

FIGURE 7.11 Molecular phylogeny and the evolution of Agaricomycete ecology. The chronogram, inferred from a combined six-gene data set by molecular clock analysis, illustrates the divergences of nutritional mode in Agaricomycetes in relation to the likely time of divergences in angiosperm and gymnosperm plants. The estimated times of divergence are shown as blue bars, with the mean node ages in the bars. Calibration points with fossil ages are shown in red (dark gray in the print version). *Source: D. Floudas, in Eastwood et al. (2011) Chapter 5, Further Reading.* (See the colour plate.)

horizons as litter fragments and humus. However, recent evidence suggests that organic soil layers may also grow from below, through continuous additions of carbon compounds from roots and their mycorrhiza. A chronosequence of soils under boreal forest that had developed over periods between centuries and millennia was investigated on a cluster of Scandinavian lake islands. The smaller the island, the older the soil carbon, because smaller islands are less frequently burned from lightning strikes, being statistically less likely to be struck. On these small islands, unburned for over 2000 years, the deeper soil layers harbouring predominantly mycorrhizal fungi and roots had accumulated proportionately larger amounts of persistent organic compounds of root and fungal origin, which was associated with tightly-bound nitrogen, leaving little available nitrogen to support plant growth. In other studies, forest fungi have been found to have a range of enzymes able to liberate nitrogen and phosphorus from such complexes. However, because of the variety of chemical bonds, they cannot be easily targeted by most soil microbes and so decompose only slowly.

Extraradical mycelium facilitates acquisition of nutrients from poorly-soluble minerals. Organic acid secretion and hyphal intrusion into rock enables ectomycorrhizal trees growing on nutrient-poor rocky ground to acquire mineral nutrients via mycelium, from underlying rock and from insoluble mineral particles in the soil, as described in Chapter 5, p.181. This is a key process in the establishment and maintenance of boreal forest, since ultimately these inorganic materials are the only source of mineral nutrients for the entire overground biota. Several common ectomycorrhizal species of *Suillus*, *Lactarius*, and *Piloderma* penetrate at least half a metre down into mineral horizons, with some species present exclusively in these layers. Typically the acid, nutrient-poor soils of these forests are strongly layered **podsols**, with a dark organic horizon overlying a pale, highly-leached eluvial horizon containing mineral particles such as the phosphorus-containing mineral apatite. Other minerals utilised by mycorrhiza include silicates containing calcium, magnesium, and potassium. Hyphae have been found growing into apatite particles, using both hyphal pressure and exudation of citric and oxalate acids, which act both by protonation and chelation to release soluble phosphate. In this way, ECM roots have direct access to phosphorus, bypassing competing biota in the soil and avoiding toxic metal ions such as aluminium that are often present in the acid soil solution of podsols. Other essential cations acquired by ECM from rock include iron, captured by exuded iron-chelating siderophore molecules (Chapter 5, p.159) such as ferricrocin, secreted by the common symbionts *Coenococcum geophilum* and *Hebeloma crustuliniforme*. Mineral solubilisation is not confined to mineral particles in soil. Rock surfaces can also be dissolved by fungal **bioweathering**, and the ability of fungal root symbionts to utilise solid rock as a nutrient source presumably underlies the ability of some pine trees to root directly into the bedrock.

The processes by which a diverse ECM community becomes established is of interest because of the interaction between plant community diversity and that of mycorrhizal fungi. A fungal foray through woodland will typically find scores of sporophores of species belonging to mycorrhizal genera such as *Amanita*, *Boletus*, *Laccaria*, *Lactarius*, *Cortinarius*, *Tricholoma*, and more. Excavating tree rootlets will reveal a similar number of ectomycorrhizal morphotypes, while molecular sampling will reveal the presence of even more species. Most of the diversity can be ascribed to a few common taxa, while intensive sampling will reveal an almost inexhaustible tally of rarities. How do these fungi arrive?

In temperate climates *Laccaria amethystina*, with distinctive purple sporophores and purple-tipped pale ectomycorrhizal roots, are among the first colonisers of new tree seedlings. They occur on a variety of host trees and colonise readily from easily-germinated spores, so behaving as widespread ruderal members of the ECM community. Others, including, for example, the *Cortinarius* species typical of ancient woodland, are later colonists of roots. They colonise roots more readily by mycelial growth from already-established ECM, and their spores are slow to germinate.

Arrival and colonisation is limited by the dispersal capacity of fungal species. While airborne spores are produced in staggering abundance by many ECM species, their concentration falls away exponentially with distance from the source, and root colonisation is dose-dependent. Analysis of species and infraspecies diversity across landscape and geographical scales shows that ECM communities are dispersal-limited. It is not true of mycorrhizal fungi that ‘everything is everywhere’ and ECM fungi do not show cosmopolitan distributions. Large geographical areas such as neotropical and palaeotropical rainforest have largely endemic ECM populations. Within a species range, airborne spores serve to facilitate gene flow throughout the population.

Ericoid Mycorrhizal Fungi

The Ericaceae are dominant plants of acid heathland and upland soils including the genera *Calluna, Erica, Vaccinium, Azalea, Rhododendron*, and the Epacrids of Australasia which grow in dry sandy soils. Ericaceae are enabled to grow in acid and upland soils too poor in mineral nutrients for other plants, by forming ERM, mainly with Helotiales (ascomycetes). The plants provide the fungi with sugars from photosynthesis, while partnership with fungi enables plants to acquire mineral nutrients from insoluble organic residues, through enzymes and siderophores deployed for nutrient acquisition by the fungus. The Helotiales appear to have arisen and diversified during the Cretaceous era in Gondwanaland, which might therefore be a site of origin of the ericoid mycorrhizal association.

Ericaceous plants have distinctive root morphology. Instead of the single-celled root hairs of other plants, they have very fine multicellular roots only 100–750 μm in diameter, termed hair roots, in which the outermost epidermal layer immediately behind the advancing root tip meristem consists of cells packed with fungal hyphae (Figures 7.1d and 7.2e & f). The central vascular strand and cortex are highly reduced so that as much as 80% of the hair root volume is composed of intracellular fungus. Each epidermal cell is separately colonised from the soil by hyphae which invade by penetrating the cell wall and proliferating inside, with the plant cell membrane invaginated around the hyphae. This symbiotic epidermal tissue of the hair root lasts only a few weeks until overtaken by root secondary thickening. An endemic population of suitable soil fungi is thus essential to maintain the continual root recolonisation which is a feature of ERM. Electron micrographs of established partnerships show a clear matrix between the host and symbiont membranes, with surrounding plant cytoplasm enriched in rough endoplasmic reticulum and mitochondria, indicating active physiology likely to be associated with uptake of fungus-acquired nutrients. Helotiales include many well described species of readily-cultured saprotrophic soil ascomycetes identifiable by morphology of ascocarps and spores. Those isolated from roots of Ericaceae grow slowly in culture producing dark mycelium that very rarely sporulates.

Molecular phylogeny indicates the existence of a much wider taxonomic group of ericoid mycorrhizal fungi than previously known from direct observation and culture of fungi on colonised roots. *Rhizoscyphus (=Hymenoscyphus) ericae* consists not of a single species but of a broad clade of Helotiales. Culture experiments have suggested a functional as well as genetic distinction between symbiotic and saprotrophic members of the clade. Many other soil fungi are emerging from environmental genomic analyses as root symbionts, although to be classed as true ERM fungi it is necessary to determine that they form the characteristic morphological association described above. The previously overlooked basidiomycete group Sebacinales appear to be ubiquitous as root symbionts and include members that form ERM, as well as ECM, mycoheterotrophic, and saprotrophic fungi. Members of the Chaetothyriales (black yeasts) are particularly important formers of ERM. They are the source of most fungal DNA extracted from ericoid roots, and in axenic culture they have been shown to associate with roots to form the diagnostic intracellular coils.

The enzymes of ericoid mycorrhizal fungi acquire nitrogen and phosphorus for their plant hosts by breaking down organic residues of plants and animals in which these elements are bound. In the cold, acid conditions of ecosystems where ERM predominate, microbial decomposition is slower than photosynthetic production. Organic material, such as woody plant and fungal remains and insect exoskeletons, accumulate as deep layers of brown peat forming the surface horizons of **mor** soils. Phenolic lignin residues bind proteins and form complex and long-lasting compounds with nitrogen, making it unavailable to roots. Phosphorus similarly becomes immobilised by covalent linkage with polysaccharides in the form of phosphomonoesters (**phytate**). Mycelium emanating from ERM-colonised hair roots can decompose these materials with secreted and wall-bound enzymes, including phytases and phosphodiesterases that respectively hydrolyse phytate and nucleic acids, liberating phosphate. Microcosm experiments demonstrate the capacity of the ERM fungus to accumulate and import liberated mineral nutrients into its host plant.

Mycoheterotrophic Associations

Mycoheterotrophy is a form of plant nutrition in which a plant species that has lost its chlorophyll in the course of evolution depends on the mycelium of a mycorrhizal fungus to supplement or replace photosynthesis as a source of carbon/energy. In effect, a mycoheterotrophic plant is a parasite on the mycorrhizal symbiosis, cheating it of the carbon resources shared in the mycorrhizal mutualism. Around 400 plant species in 87 genera and 10 families have lost all chlorophyll and receive all their carbon from green plant hosts via fungal connections. There are also about 20,000 species of partially mycoheterotrophic plants (**mixotrophs**), most of which depend on fungi only during seedling establishment. Mycoheterotrophic plants are commonly found in forest understorey habitats where low light limits photosynthesis. AMF mycoheterorophs include representatives from many different plant phyla including mosses, liverworts and ferns, and many non-green flowering plants. Mycoheterotrophic plants exploiting ECM fungi include representatives of several plant families, but many are Orchidaceae, all of which require a fungal symbiont to germinate, some being only transiently mycoheterotrophic, and others achlorophyllous and wholly dependent on fungal supply. There is even an orchid which is subterranean throughout its life cycle, the Australian species *Rhizanthella gardneri*. Of the 400 fully mycoheterotrophic angiosperm plant species, 35% are orchids. The dust-like seeds of orchids carry insufficient food reserves for germination and the embryo must be colonised by hyphae of the appropriate fungus which imports sugars and stimulates development of root and shoot primordia. Orchid seeds placed in forest soils were found to need colonisation by mycelium of local ectomycorrhizal fungi. For example, seeds of a *Neottia nidus-avis* orchid would only germinate within 5 m of a beech tree, *Fagus sylvatica*, where they were colonised and supplied with nutrients by the beech ectomycorrhizal basidiomycete *Sebacina* sp. Mycoheterophic plants show a remarkable degree of specificity for their fungal partners compared with ECM or ERM plants. The conservation of such plant species thus depends upon the conservation of habitats where their specific mycorrhizal fungus is present in the soil.

Common Mycorrhizal Networks

Mycelium of mycorrhizal fungi that can extend into the soil beyond the host root may colonise adjacent host plants. Separate individual plants may thus become linked into an underground mycorrhizal network – the so-called wood-wide web. Nutrient exchange can occur between plants of different species linked by a mycelial network where plant and fungal partners are compatible. Fungi that form mycelial cords have been found to redistribute significant amounts of nutrients between plants. In nature, some plant species regularly occur together in some habitats, prompting speculation that they might be mutually supportive in this way. In the north of Scotland, pine (*Pinus*) trees commonly grow with an understorey of cowberry, *Vaccinium vitis-idaea*. Molecular analysis identified a fungal ECM of pine roots as *Meliniomyces*, in the ascomycete *Hymenoscyphus ericae* clade, which also formed ERM with cowberry roots. The fungus formed mycelial connections between plants in microcosms, and was found to mediate reciprocal carbon and nitrogen nutrient exchange between pine and cowberry, indicating that *Meliniomyces* mycelium may also channel a mutualistic exchange of resources in the field. The ERM association presumably helps the partners acquire nitrogen form the peaty, N-poor soil, while the ECM-pine connection supplies the shaded cowberry with photosynthate from the sunlit pine canopy.

The physical extent of fungal webs in soil must, at least in part, depend on specificity between plant and fungus, and on genetic compatibility between fungi. Where all the plants are compatible with predominant mycorrhizal fungi, and the fungi themselves are compatible and capable of fusing with adjacent mycelia, a mycelial web encompassing many plants is theoretically possible. A community composed of different plant species may be connected via a single genetic type of mycelium provided that the fungus is sufficiently generalist in its plant specificity. There may, however, be greater specificity at sub-specific taxonomic levels than has been recognised. Interactions between mycorrhizal fungi and plants in communities have been investigated using network analysis to discern pattern in multiple interactions. Instead of conventional pair-wise analysis, network theory deals with patterns of multiple interactions, such as those between the whole fungal population of a habitat and the community of plants which potentially interact with them. The topology of interaction networks can predict previously undetected host specificity. 'Nested' network structures, where specialists interact with symbionts that also interact with generalists, are common in mutualistic associations. For example, when Glomeromycota sequences from 450 plants in 100 square metres of woodland were analysed, the pattern of interaction predicted the existence of both generalist and specialist AMF fungi. This had been expected from previous findings, for example, the fact that natural plant communities host a more diverse AMF population than agricultural monoculture. However, network analysis has the additional power of potentially identifying the interacting species via their sequences, providing a basis for *in situ* investigation of the taxa of interest.

Habitat factors can be as influential as phylogeny in the composition of mycorrhizal assemblages. An intriguing result was obtained in a network analysis of 430 orchid plants and their mycorrhizal fungi on the island of Reunion in the Indian Ocean. The orchids are either epiphytic, living in the branches of trees, or terrestrial, with roots in the soil. The fungi all belong to the basidiomycete anamorphic group *Rhizoctonia*. However, there was little phylogenetic overlap between fungal populations in terrestrial and epiphytic orchids. The two different fungal populations had apparently assembled from varied phylogenetic origins through shared ecology. The analysis suggested a difference in fungal niche adaptation. Both terrestrial and epiphytic plants utilise fungi to supplement the tiny reserves carried in their powder-like seeds, but the terrestrial orchids continue to employ fungi to acquire carbon resources from soil, and so it might be predicted that these would be selected for an additional set of attributes such as cellulolytic ability.

Common mycorrhizal networks pre-existing in a habitat can help newly-arrived plants to become established by nurturing their seedlings, providing photosynthate from established mature plants. For example, mycorrhizal networks play a key role in the gradual establishment of vegetation on the volcanic desert slopes of Mount Fuji. Here, the first vegetation consists of scattered clumps of willow, whose ectomycorrhiza, acquired from airborne fungal spores, then fosters the growth of other plants which are compatible with the mycorrhizal fungus. This 'nurse' function of common mycorrhizal networks is exploited in forestry to regenerate forest ecosystems by planting pre-colonised saplings which host suitable ECM fungi. For example, nursery plants of *Arbutus menziesii*, which hosts a wide range of mycorrhizal taxa that develop extramatrical networks into surrounding soil, is planted to help re-establish mixed evergreen forests in Oregon.

In mature woodland, the pattern of tree colonisation by ectomycorrhizal fungal individuals was mapped in relation to host trees of various sizes. Different fungal species, *Rhizopogon vinicolor*, and *Rhizopogon vesiculosus*, each formed 13–14 genets (see Chapter 4, p.100) within a 30 m × 30 m plot of Douglas fir (*Picea abies*) forest, each genet colonising up to 19 trees. *Rhizopogon vesiculosus* genets were larger and connected more trees than *Rhizopogon vinicolor*. Large trees provided dominant nodes in spatial networks, forming centres of mycelial systems which extended over an area of several metres, containing other trees of various ages which were also colonised.

Fungal connections in common mycorrhizal networks are dynamic and variable. The spatial extent and duration of connectivity can be expected to vary as hyphal connections are continually made and broken under the influence of physical changes and biotic interactions. Mapping the area of underground distribution of ECM fungi of several common species in woodland showed that the relative size of the area occupied by each species changes. Some, but not all, ECM fungal species are patchily distributed, and the size of patches differs between species and seasons.

LICHENS

Lichens are fungi that have evolved to house a population of unicellular or filamentous photosynthetic cells (**photobionts**) that provide the fungus host (**mycobiont**) with carbon compounds. Unlike the mycorrhizal associations described above, in lichens it is the fungal partner that forms the main structural component. This takes the form of a differentiated multicellular body termed a **thallus**, with defined tissues including a specialised layer housing the photobiont cells. The association is traditionally considered to be mutualistic because of its stability, with the captive photobiont persisting unharmed, closely integrated in form and function with its fungal host. Having its own internal photosynthetic carbon supply frees a lichenised fungus from the need to grow through soil and organic material to forage for carbon and energy sources. Instead, lichens grow aboveground, exposed to the light, attached to solid surfaces: stable soil, rocks, trees, and even manmade surfaces, including concrete, rubber, and plastic. Lichens show a variety of phenotypes (Figure 7.12). Some form leafy (foliose) or branching (fruticose) three-dimensional structures that grow from defined areas at the base or margin. In humid environments these can grow into gelatinous plates or tangles of branches, tens of centimetres across, although, having no internal long-distance water conduction they cannot grow tall enough to compete with vascular plants for light. Others make hard, dry crusts so closely applied to rock (crustose), and so slow-growing, that they may look more like paint marks than anything living, and others produce a pebble-like growth close to the

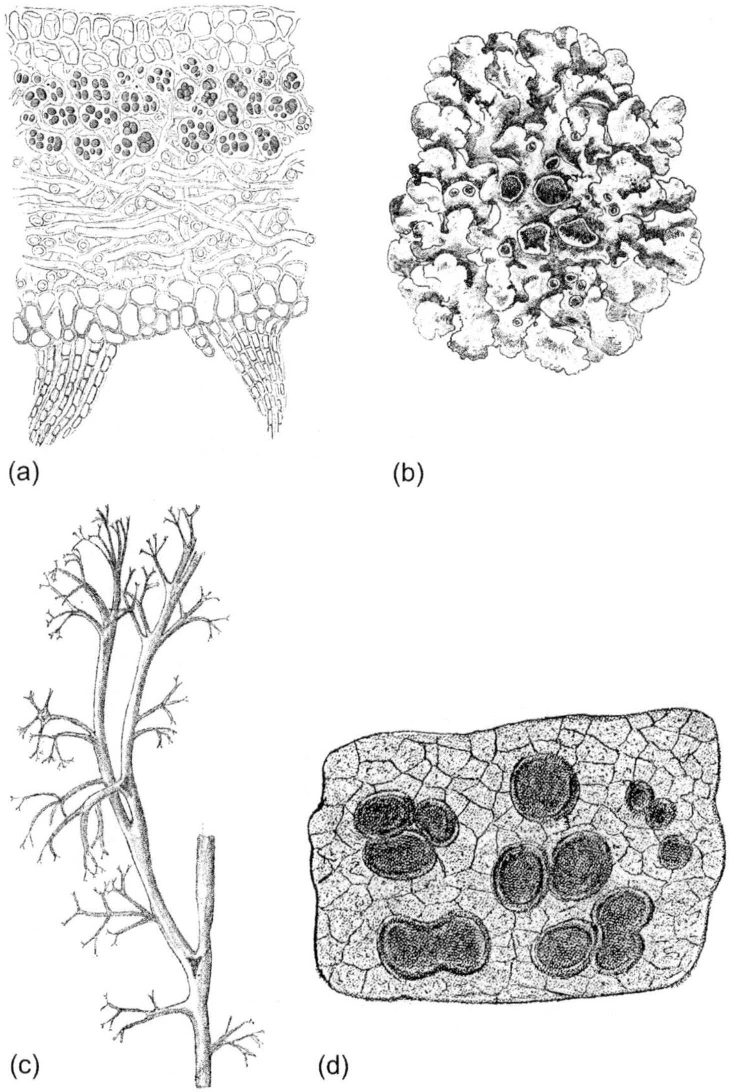

FIGURE 7.12 Anatomy and range of form in lichens: (a) vertical section of *Sticta fuliginosa*, a foliose lichen, showing the position of photobionts in the upper cortical layer of the thallus; (b) Foliose form, *Parmelia acetabulum;* (c) fruticose form, *Cladonia rangiferina*, (d) crustose form, *Lecidea confluens*. *Source: The Bodleian Library, University of Oxford, from Engler and Prantl (1907).*

substrate. Some are beautifully coloured by secondary metabolites produced by the combined metabolism of both partners. Form is determined by the fungal species, with the photobiont retaining its free-living form. Grown in isolation by culturing spores of the fungal partner, a lichen-forming fungus typically produces only slow and weak hyphal growth which is not ecologically viable. Some lichenised fungi form associations with two, or rarely, three different photobionts, producing different phenotypes with each. A lichen is identified as a partnership between a named mycobiont and named photobiont(s). In spite of the intimacy and complementarity of function between the mycobiont and photobiont partners, sexual reproduction is confined to the ascospores (rarely, basidiospores) produced by the mycobiont. New individuals may be produced when clumps of combined myco- and photobiont tissues (**soredia**) are dispersed, enabling vegetative spread of the association. Sexual reproduction involves recombination of only the fungal partner, via ascospores. To form a new lichen individual, these must encounter and combine with suitable photobionts, for example, when an ascospore lands on a damp surface already inhabited by green algae or cyanobacteria.

Classification and Nomenclature

Even though lichen symbiotic phenotypes are identified and referred to as species, the biological species concept does not apply and lichens are classified by phylogeny of the mycobiont. The species name refers to the fungal partner. There are between 17,500 and 20,000 described species, of which 99% are Ascomycetes, with over 40% of known Ascomycete species lichenised, all in Pezizomycotina. A relatively very small number of lichens, around 150, belong to four families of Basidiomycota. In contrast, fewer than 150 lichen photobionts are known, mostly only by genera. Few have been assigned to species, but it appears that photobiont species diversity is much less than that of lichen fungi, with an estimated 85% belonging to the green algae, mainly in the genus *Trebouxia*. There appears to be little selectivity for species within the **Trebouxia** genus. For example, the lichen *Lecanora rupicola* is a crustose species found on siliceous rocks worldwide. *Trebouxia* samples of *L. rupicola* from geographically widespread sites were found to belong to numerous distantly related lineages. About 10% of lichens have cyanobacterial rather than green algal photobionts, and around 4% of lichens have both. In many lichens the Cyanobacterium is partitioned within structures called **cephalodia**, where microaerophilic conditions are maintained, promoting the activity of nitrogenase which catalyses reduction of molecular nitrogen to ammonia. Cyanobacterial photobionts contribute combined nitrogen as well as carbon to the symbiosis, by virtue of their nitrogen-fixing capability. New lichen photobionts continue to be discovered, particularly in simpler forms and less-studied habitats. No differences have yet been discovered between the photosynthetic green algae and cyanobacteria isolated from lichen associations and those of the same taxonomic group found as free-living microorganisms.

An enormous diversity of small organisms lives in or on lichens. It has been said that a lichen is not an individual, but rather a consortium with an unknown number of participants. The metagenome of the lichen *Lobaria pulmonaria*, a foliose epiphytic lichen found in European montane woodland, has recently revealed an extensive microbiome including hundreds of bacterial types with the genetic potential to contribute to the viability and productivity of the symbiosis. Rhizobia are the most abundant, and grow both on the surface and within the thallus.

Endophytic fungi, not directly involved in the symbiotic partnership, are represented by more types than in any other habitat. They associate preferentially with the green algal partner and it has been hypothesised that lichen endophytes might have been evolutionary breeding grounds for fungal parasites of vascular plants. Other lichen inhabitants include bacteria, parasitic fungi, and lichen-eating invertebrates including specialised mites.

Physiology and Adaptation

Lichens are **poikilohydric**, their water potential equilibrating with that of their environment, like mosses. The lichen grows as a fungal tissue within which photosynthesising photobiont cells occupy a relatively small volume in particular positions within the thallus. Many lichen species can grow, though slowly, under conditions of intermittent water supply. The unique cellular structure and physiology that makes this possible has been analysed in detail in the large structurally complex lichen *Sticta sylvatica* (Figure 7.13) where the hyphae of the outer top and bottom layers (upper and lower cortex) are more or less isodiametric, embedded and stuck together in a hydrophilic matrix. When wet, the upper cortex transmits light to the

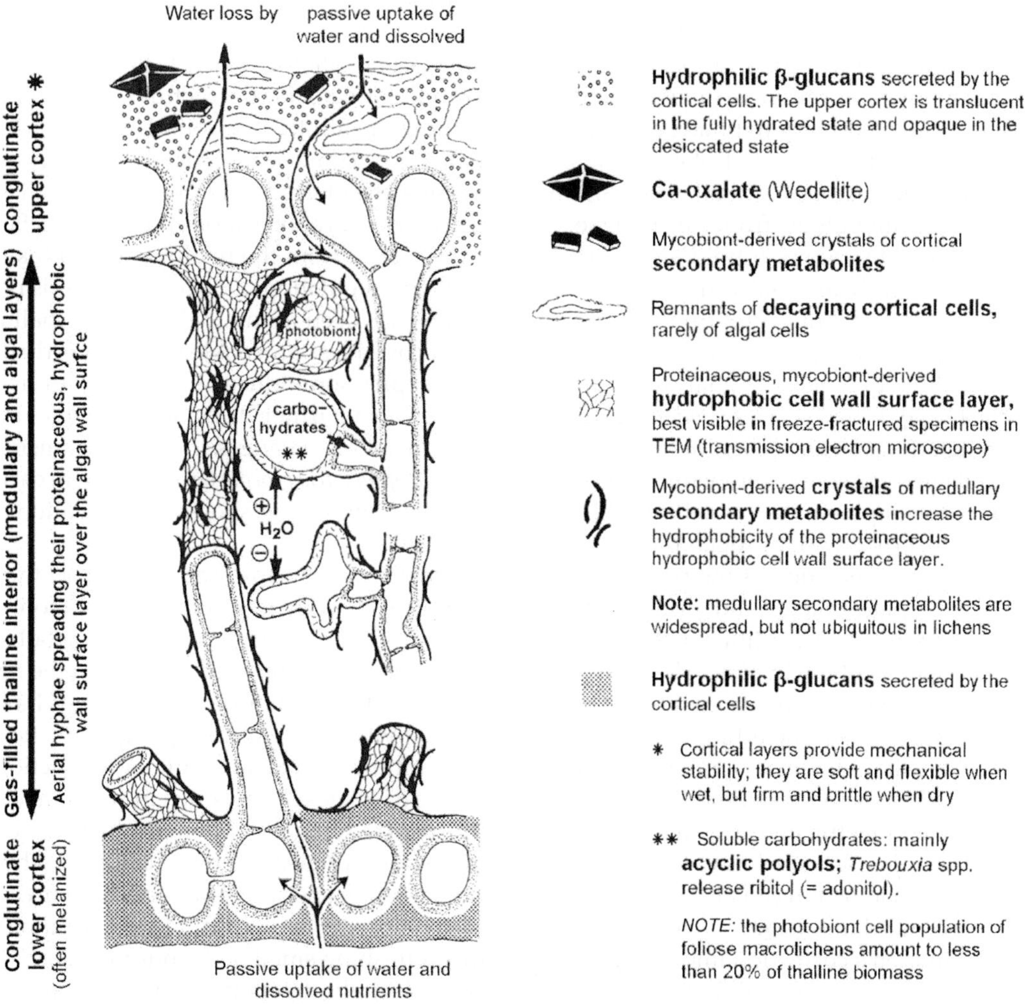

FIGURE 7.13 The functional anatomy of internally stratified thalli of lichenized ascomycetes. *Source: Honegger (2009).*

underlying photobiont cells, which form a layer at its lower side. The central part (medulla) consists of loosely interwoven filamentous hyphae with gas-filled spaces between them, which are prevented from becoming waterlogged by layers of **hydrophobin** (Chapter 2, p.53) coating their walls. Differentiation of the lichen thallus involves a remarkable localisation of hydrophilic and hydrophobic cells and tissues. Mass flow of water over short distances occurs in the tissues (thalli) of some large lichens such as *Peltigera* spp., through vein-like ribs of thickened tissue. These contain cells with highly hydrophilic walls, insulated by outer hydrophobic layers, which channel passive but rapid capillary flows of water and solutes. In dry weather the cortical layers can lose so much water that they shrink and become brittle. Remarkably, cell damage is minimal even though air spaces can occur within fungal cells, and on rewetting the thallus rehydrates and cell structure is regained. There is, however, a cost in photosynthetic assimilation on each drying episode, because on rehydration photosynthesis takes longer to resume than respiration.

The interface for nutrient exchange between myco- and photobiont in lichens is unlike that in many biotrophic associations, in that no intracellular haustorial structures are formed. Instead, the medullary hyphae grow into the gelatinous sheath that surrounds the photobiont cells and the associated photobiont and hyphal cells become sealed together within the hydrophobic coating material. The carbon compounds transferred from the photobiont to the fungus are polyols from green algae, and glucose from cyanobacteria. When the lichen is moistened by rain or humid air, photosynthate is released from the photobiont and taken up by the closely-associated hyphae. The mechanism and regulation of carbon transfer is not well understood, but C flux from photobiont to fungus has been found to depend on thallus water content. The interaction between partners has proved less amenable than mycorrhizal partnerships to molecular analysis. Changes in gene expression induced by mixing cultures of the model lichen species *Cladonia grayi* with its green alga partner *Asterochloris* sp. include up-regulation of cell recognition in the fungus and metabolic changes in both organisms. A similar experiment in the cyanobacterial lichen *Pseudocyphellaria crocata* showed that in early thallus development the cyanobacterial symbiont, *Nostoc punctiforme*, shows up-regulation of genes concerned with nitrogen fixation in the heterocyst.

The close interaction of the partners in the lichen symbiosis is expressed not only in their integrated morphology and primary metabolism but also in their secondary metabolites (Chapter 5, p.161), including the 'lichen acids' that are produced by the fungus and crystallise on the surfaces of the hyphae. However, in pure culture the quantities and sometimes the nature of the products differ from those in the lichenized state. Over a thousand compounds have been characterised, including some with biological activities that include photoprotection, as well as allelochemical, antibacterial, anti-tumour, anti-herbivore, and antioxidant action. Usnic acid has activity against bacteria, including clinical isolates of vancomycin-resistant enterococci and methicillin-resistant *Staphylococcus aureus*. Traditionally lichen products have a variety of folk uses. They have been used to produce a range of pleasing muted colours when used to dye cloth, and can also be used in the identification of species. Vulpinic acid, a mycotoxin produced in *Letharia vulpine*, has been used as a poison for wolves and foxes. Atranorin is used in men's cosmetics, imbuing products with a refreshing smell of the outdoors.

Many lichens have a remarkable ability to survive drought, freezing, high temperatures, and scarcity of key nutrients, and dominate terrestrial ecosystems too harsh for vascular plants, including Arctic and Antarctic, high alpine, desert, and steppe. Ecosystems covering more than 12% of Earth's land mass are lichen-dominated. Endolithic and epilithic lichens that grow in or on rocks are important agents in eroding and solubilising minerals at both cell and landscape scales. Crust-forming lichens that bind the surface soil in arid zones play an important part in stabilising soils against wind erosion and preventing desertification. In the boreal Arctic, poikilohydry, and the nitrogen-fixing capacities of lichens such as *Stereocaulon* and *Peltigera*, enable lichens to colonise bare and dry ground. Lichens that tolerate freezing have been shown to do so by virtue of ice nucleation sites in wall surfaces that ensure that ice crystallises in the intercellular spaces, and not within cells. The photobiont partner may affect stress tolerance, for example, different clades of the alga *Trebouxia* have been found in Arctic lichens, with one clade preferentially associated with lichens of extremely cold habitats.

In boreal Arctic regions the vegetation can be dominated by species of *Cladonia, Cetraria, Stereocaulon,* and *Alectoria*, which form closed mats loosely attached to the soil (Chapter 12, pp. 383–386). These lichen mats can provide as much as 60% of the winter food of caribou and reindeer. Disastrously, they have been affected by radioactive pollution. Lichens bind metal cations and accumulate metal-rich particles within the thallus, absorbing them over the surface. Radionuclides from fallout from H-bomb testing and accidents such as those at Chernobyl and Fukushima accumulate in lichens and constitute a health hazard for Scandinavian Sami people and North American Eskimo, who depend on meat of the caribou and reindeer that graze lichen and accumulate high concentrations of radionuclides in their bodies.

Lichens are valuable indicators of several forms of atmospheric pollution. Assay of lichen-bound lead has provided longitudinal data on global atmospheric lead pollution, and showed a decline in levels following the introduction of catalytic converters on car exhausts. They are sensitive to sulphur dioxide and nitrogen pollution (e.g. ammonia, oxides of nitrogen and elevated deposition in rainfall) but different species are affected to varying degrees, so that lichen diversity can be used to monitor air pollution. Changes in the physiology of arctic lichens can indicate atmospheric pollution from nitrogen oxides originating from fossil fuel combustion in distant industrialised regions. When their environment becomes nitrogen-sufficient, lichen metabolic profiles indicate a change from nitrogen-limited to phosphorus-limited metabolism.

Evolution

The earliest fossils that have been identified as undeniably lichens, with photobiont partners and stratified tissues comparable to extant lichen species, have been found in 415 million year old rock from the Lower Devonian. Internal anatomy was intact and could be examined by scanning electron microscopy. It appeared to consist of septate hyphae forming tissue containing cyanobacterial and a unicellular, presumably green algal photobiont, and asexual spores in a pycnidium typical of Pezizomycotina (Ascomycota).

Molecular genetics has revolutionised our understanding of the processes that have led to lichenisation in fungi. We now know that the lichen symbiosis is polyphyletic, having evolved convergently in separate fungal lineages, and in some, including those of well-known and widespread saprotrophs such as *Aspergillus* and *Penicillium* in the Eurotiales, there is evidence that a formerly lichenised mode of nutrition has been lost. This is of particular interest because a lichenised past may have endowed some of these saprotrophs with important biological attributes such as their pathways for the synthesis of bioactive secondary metabolites including penicillin.

ENDOPHYTES

Microscopic fungi of remarkable diversity inhabit the intercellular and apoplastic spaces of plant tissues, and are collectively termed **endophytes**. Plants provide the habitat for thousands of endophytic fungi, as well as bacteria and viruses. Some cause host disease and produce reproductive structures at the hosts' expense. However, many endophyte fungi produce no visible structures on the plant surface and may only be found when searched for by microscopic inspection, isolation into culture or from the presence of their DNA. Environmental genomic sampling is now revealing microbiomes associated with every larger organism examined, and plants are no exception. Endophytic fungi are ubiquitous in plants throughout natural ecosystems. Some, described in detail below, are known to confer fitness benefits on their hosts, but tantalisingly little is known of the biology and host interactions of the majority. Discerning the effects of this hyperdiverse normal microbiota on plant physiology and ecology is a current challenge. Most are Ascomycota, the majority belonging to Hypocreales. They fall into four broad functional types according to the range of hosts colonised, the extent of colonisation, mode of transmission between hosts, taxonomic diversity and known effects on host fitness (Table 7.1).

Clavicipitaceous endophytes Fungi belonging to the Ascomycete group Hypocreales (Clavicipitaceae) occur as endophytes in Gramineae (grasses, including cereal crops) (Figure 7.14). They show a range of plant interactions from parasitism to mutualism. Mutualistic endophytes confer selective advantage on their hosts by producing anti-insect and anti-vertebrate toxins that protect their host plants from insect attack and herbivore grazing. The genus *Epichloe* has given rise to an exclusively asexual form, known as *Neotyphodium*, which differs in life cycle on the host grass (Figure 7.15). *Epichloë* species reproduce both asexually by conidia, and also through a sexual cycle where ascospores are formed on a stroma produced in place of the inflorescence. In cereal crops this is known as 'Choke' disease. Because both sexually produced ascospores and asexual conidiospores are released, the fungus can spread horizontally through the cereal population from one individual plant to another. However, asymptomatic endophytes evolved from *Epichloë* have lost the capacity for release of ascospores, and grow exclusively as the asexual anamorphic form within host tissues (Figure 7.15), and

TABLE 7.1 Criteria Used to Characterise Classes of Fungal Endophytes

	Clavicipitaceous	Nonclavicipitaceous		
Criteria	**Class 1**	**Class 2**	**Class 3**	**Class 4**
Host range	Narrow	Broad	Broad	Broad
Tissue(s) colonised	Shoot and rhizome	Shoot, root, and rhizome	Shoot	Root
In planta colonisation	Extensive	Extensive	Limited	Extensive
In planta biodiversity	Low	Low	High	Unknown
Transmission	Vertical and horizontal	Vertical and horizontal	Horizontal	Horizontal
Fitness benefits[a]	NHA	NHA and HA	NHA	NHA

[a]*Nonhabitat-adapted (NHA) benefits such as drought tolerance and growth enhancement are common among endophytes regardless of the habitat of origin. Habitat-adapted (HA) benefits result from habitat-specific selective pressures such as pH, temperature, and salinity. Adapted from Rodriguez et al. (2009).*

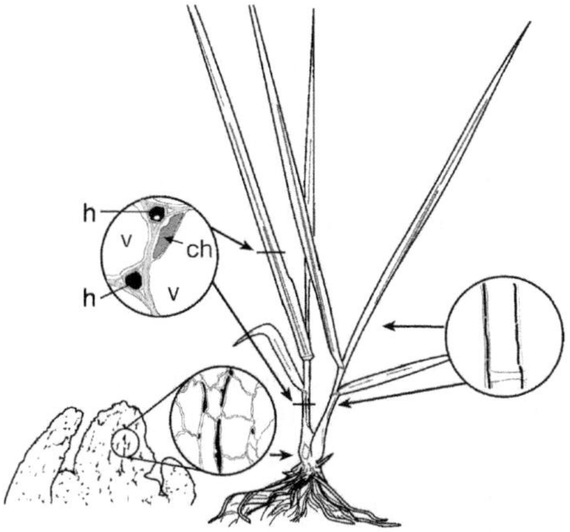

FIGURE 7.14 Endophyte growth in a grass plant. Bottom left: fungal growth in the stem and leaf primordia. The fungal hyphae are shown darkly stained with osmium as they would appear in transmission electron microscopy. Upper left: a cross section of leaf or leaf sheath reveals hyphae (h) of the Epichloë endophyte between host cells. Also shown are a chloroplast (ch) and vacuoles (v). Right: the endophyte as it appears in a leaf epidermal peel, stained for hyphae, which are arranged mainly along the longitudinal axis of plant cells. Unstained septa separate individual fungal cells, each of which bears a single haploid nucleus (not shown). *Source: Selosse and Schardl (2007).*

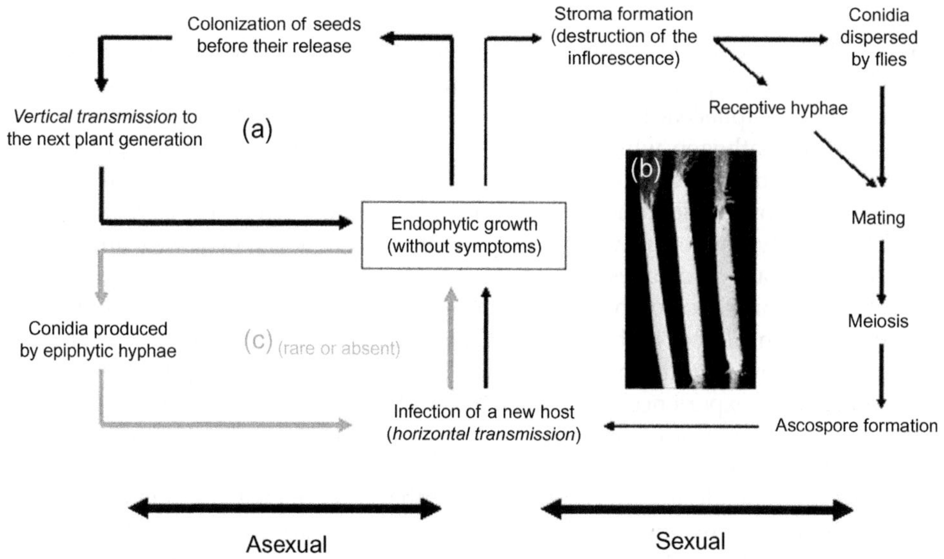

FIGURE 7.15 Life cycles of *Epichloë* endophytes of Gramineae. *Epichloë* can reproduce asexually by invading the host's seeds (a), by sexual cycles where ascospores are formed (b) on a stroma or 'choke' that destroys the inflorescence, and more rarely through asexual spores (c). Transmission is thus vertical (a) or horizontal, (b, c). Neotyphodium species, endophytes derived from Epichloë, mostly carry out asexual reproduction through host seeds (a), resulting in mainly vertical transmission. *Source: Selosse and Schardl (2007).*

these anamorphic forms are classified as *Neotyphodium*. Transmission in the host population is predominantly vertical, from one generation to the next, via seedborne mycelium. Mycelium enters the ovule from tissues of the shoot in which it is systemic, and colonises the seed itself. No independent sexual reproduction occurs and the evolutionary fate of the fungus thus becomes indissolubly linked to that of its host. From population studies of *Neotyphodium* species in grasses of natural ecosystems, it has been found that individuals consist mainly of multiple asexual hybrid genotypes, probably arising from genetic exchange between individuals within a single plant, each arising from separate conidial infections. Because the hybrids are only transmitted through host reproduction, they are likely to be selected for traits that improve host fitness. The selective advantage the mutualism confers on both partners explains the stability of the partnership. As a result, the plant carries the endophyte throughout its geographical range and most of the population is infected. The characteristics conferred by infection, such as increased vigour of vegetative growth, and the presence of secondary metabolites antagonistic to herbivores, become general features of the grass species. Many species of the common temperate grass family *Poaceae* are affected in this way by *Neotyphodium* endophytes. The toxins produced by the endophyte are alkaloids: indole-deterpenes and ergot alkaloids poisonous to vertebrates, and peramines and lolines that act mainly against insects. Alkaloids produced by *Epichloe* or *Neotyphodium* endophytes across a range of grass species were analysed. The production of anti-insect alkaloids peramine and loline was found to be greater than anti-vertebrate ergot alkaloids in the asexual **Neotyphodium**, perhaps because, having lost sexual reproduction, it no longer depends on insects to transmit male gametes in fertilisation. Mutualism based on the anti-insect properties of alkaloids appears to represent some cost to the plant. The grass *Bromus setifolius* is used by leaf cutting ants. Where the ants are abundant, for example, in Argentinian desert, 80–100% of sampled plants contained the endophyte *Neotyphodium tembladerae*, but in places with no ants, where the pressure of herbivory was relaxed, endophyte levels were down to 20–0%.

Alkaloids of Clavicipitaceae have a potent effect on humans and domestic animals. *Claviceps purpurea* is a plant pathogen which produces black sclerotia (ergots) instead of grain in the ear of cereal crops. In the past, local outbreaks of ergotism were caused by this mycotoxin which causes vasoconstriction and hallucinations. Ergot alkaloids have been well characterised in terms of biosynthetic pathways and gene clusters encoding them and are important in pharmacology. Various syndromes occur among farm animals due to endophyte alkaloid consumption. Ergot alkaloid poisoning of cattle causes vasoconstriction resulting in sloughing off of hoofs and abortion. Horses in the western United States who eat 'sleepy grass', *Achnatherum robustum* colonised by an endophyte producing lysergic acid, may sleep for days, then gradually recover, but the experience leaves its mark and they do not eat the grass again. Similar phenomena are reported from Asia (drunken horse grass) and South Africa (dronk grass).

Even endophytes that are not acutely toxic to animals can deter feeding, and a variety of turf grass cultivars are now identified as 'endophyte enhanced'. Some Class 1 endophytes confer resistance to disease, for example, *Epichloë festucae*-infected turf grass is resistant to some leaf spot diseases (e.g. dollar spot caused by *Sclerotinia homeocarpa* and red thread disease caused by *Laetisaria fuciformis*). Systemic endophytes can affect not only individual plants but also whole ecosystems. *Neotyphodium coenophialum* in tall fescue (*Lolium arundinaceum*) not only affects plant–herbivore interactions and plant productivity but also plant–plant competition, decomposition rates, and grassland species diversity.

Class 2 endophytes mostly belong to Pezizomycotina (Ascomycota) with some Basidiomycota. They are an ecologically distinct class as they colonise the roots, stems, and leaves of monocotyledonous and dicotyledonous plants, often forming extensive infections, with especially high infection frequencies (90–100%) in plants growing in stressful environments. They can be transmitted vertically via seed coats and rhizomes, and also horizontally, some (e.g. species of *Phoma* and *Arthrobotrys*) being abundant in soil, others having low abundance in soil. The latter are probably unable to compete outside the host, whereas those abundant in soil may have several lifestyles – symbiotic and saprotrophic. As with other endophytes, they colonise plant tissues by direct penetration or using infection structures such as appressoria, and grow mainly between plant cells, upon which they have little or no obvious impact. In healthy plants, sporulation or appressorial formation is low, but the fungi rapidly emerge and sporulate when plants senesce. Species of *Alternaria*, *Cladosporium*, *Epicoccum*, and *Phaeosphaeria* are dominant endophytes of *Dactylis glomerata* and other grasses and sporulate over a significant area of the aerial plant when it senesces.

In most cases Class 2 endophytes are found by culture and/or molecular sampling, rather than through altered host characteristics, and their effect on plant fitness is not known. However, a few species have been demonstrated to have a remarkable effect in allowing their hosts to grow in habitats that are too stressful for either plant or fungus alone (Rodriguez et al., 2009). For example, *Curvularia protuberata*, which colonises all tissues of the geothermal plant *Dichanthelium lanuginosum*, appears to be confer thermotolerance on its host. The effect is reciprocal since neither host nor endophyte can tolerate temperatures above 40 °C when growing alone, but the colonised plant can grow in places where the temperature is up to 65 °C. Similarly, the coastal dune grass *Leymus mollis* can only grow at the salinity of sea water when colonised by its endophyte *Fusarium culmorum*. *Colletotrichum* spp. confer disease resistance on their hosts. These habitat-adapted fitness benefits are specific to the fungal isolates concerned, but can be conferred on a taxonomically wide range of host plants. Clearly these fungi are of great agronomic interest and the physiological mechanisms of their effects await investigation.

Class 3 endophytes are characterised by ubiquity in plants, and by their hyperdiversity. They occur in the aboveground herbaceous and woody tissues of an extremely wide range of plants and can be extremely diverse even within a single plant, for example, over 80 species have been isolated from juniper (*Juniperus communis*) and from oak (*Quercus petraea*). Diversity is related to latitude, with much higher diversity in the tropics than in boreal forest and arctic tundra, though local abiotic conditions can increase endophyte diversity, for example, wet microclimates in temperate regions. Plants with long-lived evergreen foliage may harbour a greater diversity of endophytes than shorter-lived foliage. Unlike fungi in Classes 1 and 2, they form highly localised infections, involve a wide range of species in Ascomycota (especially Pezizomycotina but also Saccharomycotina), and some Basidiomycota belonging to Agaricomycotina, Pucciniomycotina, and Ustilaginomycotina. Unlike Class 1 and 2 endophytes, Class 3 endophytes are transmitted horizontally via spores and/or hyphal fragments and are major contributors to the air spora at leaf fall in temperate regions (Chapter 3, p. 92). Sterile seedlings and newly-emerged leaves become colonised rapidly in the field. Over 80% of leaves of endophyte-free cocoa tree (*Theobroma cacao*) seedlings contained endophytes within 2 weeks of leaf emergence in a tropical forest, in the early wet season.

Hyperdiversity makes it almost impossible to distinguish single ecological roles for Type 3 endophytes. Some protect host from parasites, for example, bark endophytes may have a role in protection against Dutch elm disease and some endophyte assemblages decrease lesion formation and leaf death by *Phytophthora* in cocoa trees (*Theobroma cacao*), and might be thus exploited as inoculants in biological control of disease. Endophyte infection can damage host plants, for example, by making seedlings wilt faster under drought. The endophyte flora includes latent saprotrophs and parasites that become active once tissues start to senesce. Early stages in the decomposition of fallen leaves are dominated by endophytes such as *Alternaria* that live asymptomatically in and on healthy leaves in summer, but start to decompose cellulose and sporulate when the leaf falls.

Wood decay fungi (typically Basidiomycota and xylariaceae in the Ascomycota), present as latent endophytes in the functional sapwood of angiosperm trees, can start to grow once loss of the xylem water column results in aerobic conditions required for hyphal growth. Decay columns of single fungal genotypes develop very rapidly – sometimes in less than a single growing season, probably by the growth of genetically identical mycelia from many different foci. These xylem-inhabiting endophytes are broadly species-specific in the development of decay columns, even though a wide range of species is revealed to be latent in most tree species. For example, in the UK, *Daldinia concentrica* fruit bodies are usually found on ash (*Fraxinus excelsior*) and occasionally on beech (*Fagus sylvatica*) in the south and on birch (*Betula* spp.) in the north; *Eutypa spinosa* and *Hypoxylon fragiforme* are usually found on beech and *Fomes fomentarius* on birch (*Betula* spp.) *Hypoxylon fuscum* and *Stereum rugosum* on hazel (*Corylus avellana*), and *Stereum gausapatum* and *Vuillemina comedens* on oak (*Quercus* spp.) and beech. However, all of these fungal species and many more are found by molecular sampling in a wide range of asymptomatic angiosperm tree xylem, suggesting a narrower range of conditions for mycelial development than for endophytic survival.

Class 4 endophytes consist of a poorly-understood group of fungal root associates characterised by melanised hyphae and melanised septa, and known as Dark Septate Endophytes (DSE). They have been found in over 600 plant species in more than 100 plant families. They appear to have evolved repeatedly in Ascomycota, and are found in the genera *Cadophora*, *Microdochium*, *Trichocladium*, *Phialophora*, *Leptodontidium*, and *Phialocephala*. Root colonisation by *Phialocephala fortinii* begins by superficial colonisation of the root surface by a loose network of hyphae. Subsequently, individual hyphae grow along the main axis of the root, between root cortex cells and within depressions between epidermal cells, ultimately penetrating some root cells. Some DSE form a structure similar to the Hartig net of ectomycorrhiza, and may be mycorrhizal. *Cadophora* includes both DSE and ectomycorrhizal taxa, and some DSE form ericoid mycorrhiza *in vitro*. Ubiquitous worldwide, they are particularly prevalent in environments with high abiotic stress, including arid sandy soils, the Arctic and Antarctic, and heavy metal contaminated sites. They may confer lead tolerance on plants, believed to be mediated by their intracellular antioxidant systems, including the melanin in their mycelium. DSE can precede plants in early succession because they can grow and form propagules in the absence of a host plant, and have been sampled from the barren forefront of glaciers where they are assumed to arrive as airborne propagules. Their transmission is likely to be horizontal, via conidia or microsclerotia. They probably have low host specificity. Groups of DSE from arid sandy plains in Hungary were characterised by sequence and tested for root association by inoculation, and were found to be generalists able to colonise both native and invasive host species.

Bioactive Compounds Produced by Endophytes

Interaction with other organisms is believed to accelerate selection of microbial secondary metabolites for bioactivity (Chapter 5). Endophytes, with their long-term association with hosts and multitude of interactions with other plant inhabitants including fungi, bacteria, nematodes, and insects, provide a fertile area for exploring fungal biosynthetic potential. Since the discovery that taxol is produced by endophytes of the yew tree (*Taxus baccata*), there has been an explosion of endophyte screening studies, and over 4000 biologically active secondary metabolites have already been found with varied structural groups. These are investigated for potential antibacterial, anti-viral, antifungal, anti-cancer, anti-inflammatory activity, as insulin receptor activators, acetylcholinesterase inhibitors, β-glucuronidase inhibitors, eosinophil inhibitors, insecticides, and root growth accelerators. Novel compounds from endophytes include new lactones with potential as leads for anti-malarial drugs, (+)-ascochin and (+)-ascodiketone, apiosporic acid, chaetocyclinones, colletotrichic acid, cyclopentanoids, enalin derivative, isofusidienols, myrocin A, naphthoquinone, pestalotheols A–D, phomopsilactone, and spiroketals. As well as producing novel bioactive chemicals, endophytes can affect biotransformations of chemicals that can be used in drug modifications, and offer the potential for conversions relevant to biofuel production. **Bioprospecting**, the search for useful genes in environmental sequences, can identify gene clusters encoding biosynthetic pathways in unculturable or poorly culturable fungi such as lichens and endophytes, and promising genes can be transformed into more tractable heterologous hosts for expression.

With the advent of metagenomics, including expression analyses linked to taxa, new knowledge is certain to elucidate the huge taxonomic and functional diversity of fungal symbionts that populate plant tissues and microenvironments. Symbiosis, once considered a specialised way of life, is being revealed as an engine of plant evolution that enabled plants to colonise the land and continues to maintain the productivity and diversity of all Earth's terrestrial biomes.

Further Reading

Books

Aroca, R., 2013. Symbiotic Epiphytes. Springer, Berlin Heidelberg.

Brodo, I.M., Sharnoff, S.D., Sharnoff, S., 2001. Lichens of North America. Yale University Press, USA.

Cairney, J.W.G., Chambers, S.M. (Eds.), 1999. Ectomycorrhizal Fungi: Key Genera in Profile. Springer, Berlin Heideberg.

Cheplick, G.P., Faeth, S., 2009. Ecology and Evolution of the Grass-Endophyte Symbiosis. Oxford University Press, Oxford.

Hock, B. (Ed.), 2012. The Mycota IX. Fungal Associations. Springer, Berlin.

Lugtenberg, B., 2015. Principles of Plant Microbe Interactions. Springer International Publishing, Switzerland.

Nash, T.H., 2008. Lichen Biology, second ed. Cambridge University Press, Cambridge.

Peterson, R.L., Massicotte, H.B., Melville, L.H., 2004. Mycorrhizas: Anatomy and Cell Biology. CABI, Wallingford.

Purvis, W., 2009. Lichens. Natural History Museum, London.

Schulz, B.J.E., Boyle, C.J.C., Sieber, T.N., 2010. Microbial Root Endophytes. Springer-Verlag, Berlin Heidelberg.

Smith, S.E., Read, D.J., 2010. Mycorrhizal Symbiosis, third ed. Academic press, London.

Southworth, D. (Ed.), 2011. Biocomplexity of Plant-Fungal Interactions. Wiley-Blackwell, Oxford.

Strobel, G., 2012. Genetic diversity of microbial endophytes and their biotechnical applications. In: Nelson, K.E., Jones-Nelson, B. (Eds.), Genomics Applications for the Developing World. Springer, New York, pp. 249–262.

Varma, A. (Ed.), 2008. Mycorrhiza: Genetics and Molecular Biology, Eco-function, Biotechnology, Eco-physiology, Structure and Systematics. third ed.. Springer, Berlin.

White, J.F., Bacon, C.W., Hywel-Jones, N.L., Spatafora, J., 2003. Clavicipitalean Fungi. CRC Press, Raton, Boca.

Zambonelli, A., Bonito, G.M., 2012. Edible Mycorrhizal Mushrooms. Springer, Berlin.

Journal Articles and Reviews

Mycorrhiza and Mycoheterotrophs

Bago, B., Pfeffer, P., Shachar-Hill, Y., 2001. Could the urea cycle be translocating nitrogen in the arbuscular mycorrhizal symbiosis? New Phytol. 149, 4–8.

Bidartondo, M.I., 2005. The evolutionary ecology of myco-heterotrophy. New Phytol. 167, 335–352.

Bodeker, I.T.M., Nygren, C.M.R., Taylor, A.F.S., Olson, A., Lindahl, B.D., 2009. ClassII peroxidase-encoding genes are present in a phylogenetically wide range of ectomycorrhizal fungi. ISME J. 3, 1387–1395.

Cairney, J.G., 2011. Ectomycorrhizal fungi: the symbiotic route to the root for phosphorus in forest soils. Plant Soil 344, 51–71.

Clemmensen, K.E., Bahr, A., Ovaskainen, O., Dahlberg, A., Ekblad, A., Wallander, H., Stenlid, J., Finlay, R.D., Wardle, D.A., Lindahl, B.D., 2013. Roots and associated fungi drive long-term carbon sequestration in boreal forest. Science 339, 1615–1618.

Deveau, A., Gross, H., Morin, E., Karpinets, T., Utturkar, S., Mehnaz, S., Martin, F., Frey-Klett, P., Labbé, J., 2014. Genome sequence of the mycorrhizal helper bacterium Pseudomonas fluorescens BBc6R8. Genome Announcements 2.

Dunham, S.M., Mujic, A.B., Spatafora, J.W., Kretzer, A.M., 2013. Within-population genetic structure differs between two sympatric sister-species of ectomycorrhizal fungi, Rhizopogon vinicolor and R. vesiculosus. Mycologia 105 (4), 814–826.

Fitter, A.H., Helgason, T., Hodge, A., 2011. Nutritional exchanges in the arbuscular mycorrhizal symbiosis: implications for sustainable agriculture. Fungal Biol. Rev. 25, 68–72.

Govindarajulu, M., Pfeffer, P.E., Jin, H., Abubaker, J., Douds, D.D., Allen, J.W., Bucking, H., Lammers, P.J., Shachar-Hill, Y., 2005. Nitrogen transfer in the arbuscular mycorrhizal symbiosis. Nature 435, 819–823.

Grelet, G.-A., Johnson, D., Paterson, E., Anderson, I.C., Alexander, I.J., 2009. Reciprocal carbon and nitrogen transfer between an ericaceous dwarf shrub and fungi isolated from Piceirhiza bicolorata ectomycorrhizas. N. Phytol. 182, 359–366.

Gutjahr, C., Parniske, M., 2013. Cell and developmental biology of arbuscular mycorrhiza symbiosis. Annu. Rev. Cell Dev. Biol. 29, 593–617.

Hart, M.M., Aleklett, K., Chagnon, P.-L., Egan, C., Ghignone, S., Helgason, T., Lekberg, Y., Öpik, M., Pickles, B.J., Waller, L., 2015. Navigating the labyrinth: a guide to sequence-based, community ecology of arbuscular mycorrhizal fungi. New Phytol. 207 (1), 235–247.

Hodge, A., Campbell, C.D., Fitter, A.H., 2001. An arbuscular mycorrhizal fungus accelerates decomposition and acquires nitrogen directly from organic material. Nature 413, 297–299.

Hodge, A., Fitter, A.H., 2013. Microbial mediation of plant competition and community structure. Funct. Ecol. 27, 865–875.

James, T.Y., et al., 2006. Reconstructing the early evolution of fungi using a six-gene phylogeny. Nature 443, 818–822.

Jin, H., Liu, J., Liu, J., Huang, X., 2012. Forms of nitrogen uptake, translocation, and transfer via arbuscular mycorrhizal fungi: a review. Sci. China Life Sci. 55, 474–482.

Johnson, D., Martin, F., Cairney, J.W.G., Anderson, I.C., 2012. The importance of individuals: intraspecific diversity of mycorrhizal plants and fungi in ecosystems. N. Phytol. 194, 614–628.

Kiers, E.T., Duhamel, M., Beesetty, Y., Mensah, J.A., Franken, O., Verbruggen, E., Fellbaum, C.R., Kowalchuk, G.A., Hart, M.M., Bago, A., Palmer, T.M., West, S.A., Vandenkoornhuyse, P., Jansa, J., Bücking, H., 2011. Reciprocal rewards stabilize cooperation in the mycorrhizal symbiosis. Science 333, 880–882.

Kuo, A., Kohler, A., Martin, F.M., Grigoriev, I.V., 2014. Expanding Genomics of Mycorrhizal Symbiosis. Front. Microbiol. (5) Published online 2014 Nov 4. http://dx.doi.org/10.3389/fmicb.2014.00582

Lanfranco, L., Young, J.P.W., 2012. Genetic and genomic glimpses of the elusive arbuscular mycorrhizal fungi. Curr. Opin. Plant Biol. 15, 454–461.

Leake, J.R., Cameron, D.D., 2012. Untangling above- and belowground mycorrhizal fungal networks in tropical orchids. Mol. Ecol. 21, 4921–4924.

Lindahl, B.D., Tunlid, A., 2015. Ectomycorrhizal fungi – potential organic matter decomposers, yet not saprotrophs. New Phytol. 205, 1443–1447.

Martin, F., Aerts, A., Ahren, D., Brun, A., Danchin, E.G.J., Duchaussoy, F., Gibon, J., Kohler, A., Lindquist, E., Pereda, V., Salamov, A., Shapiro, H.J., Wuyts, J., Blaudez, D., Buee, M., Brokstein, P., Canback, B., Cohen, D., Courty, P.E., Coutinho, P.M., Delaruelle, C., Detter, J.C., Deveau, A., DiFazio, S., Duplessis, S., Fraissinet-Tachet, L., Lucic, E., Frey-Klett, P., Fourrey, C., Feussner, I., Gay, G., Grimwood, J., Hoegger, P.J., Jain, P., Kilaru, S., Labbe, J., Lin, Y.C., Legue, V., Le Tacon, F., Marmeisse, R., Melayah, D., Montanini, B., Muratet, M., Nehls, U., Niculita-Hirzel, H., Secq, M.P.O.-L., Peter, M., Quesneville, H., Rajashekar, B., Reich, M., Rouhier, N., Schmutz, J., Yin, T., Chalot, M., Henrissat, B., Kues, U., Lucas, S., Van de Peer, Y., Podila, G.K., Polle, A., Pukkila, P.J., Richardson, P.M., Rouze, P., Sanders, I.R., Stajich, J.E., Tunlid, A., Tuskan, G. and Grigoriev, I.V., 2008. The genome *of Laccaria bicolor* provides insights into mycorrhizal symbiosis. *Nature 452*, 88–92.

Massicotte, H.B., Peterson, R.L., Melville, L.H., Luoma, D.L., 2012. Biology of mycoheterotrophic and mixotrophic plants. In: Biocomplexity of Plant–Fungal Interactions. Wiley-Blackwell, Hoboken, pp. 109–130.

Nara, K., 2006. Ectomycorrhizal networks and seedling establishment during early primary succession. N. Phytol. 169, 169–178.

Naumann, M., Schusler, A., Bonfante, P., 2010. The obligate endobacteria of arbuscular mycorrhizal fungi are ancient heritable components related to the Mollicutes. ISME J. 4, 862–871.

Nehls, U., Göhringer, F., Wittulsky, S., Dietz, S., 2010. Fungal carbohydrate support in the ectomycorrhizal symbiosis: a review. Plant Biol. 12, 292–301.

Oldroyd, G.E.D., 2013. Speak, friend, and enter: signalling systems that promote beneficial symbiotic associations in plants. *Nat. Rev. Micro. 11*, 252–263.

Öpik, M., Davison, J., Moora, M., Zobel, M., 2013. DNA-based detection and identification of Glomeromycota: the virtual taxonomy of environmental sequences. Botany 1, 1–13.

Peay, K., Bidartondo, M., Arnold, A., 2010. Not every fungus is everywhere: scaling to the biogeography of fungal-plant interactions across roots, shoots and ecosystems. N. Phytol. 185, 878–882.

Perez-Moreno, J., Read, D.J., 2000. Mobilization and transfer of nutrients from litter to tree seedlings via the vegetative mycelium of ectomycorrhizal plants. *New Phytol. 145*, 301–309.

Pickles, B.J., Genney, D.R., Potts, J.M., Lennon, J.J., Anderson, I.C., Alexander, I.J., 2010. Spatial and temporal ecology of Scots pine ectomycorrhizas. N. Phytol. 186, 755–768.

Plett, J.M., Martin, F., 2011. Blurred boundaries: lifestyle lessons from ectomycorrhizal fungal genomes. Trends Genetics 27, 14–22.

Rineau, F., Roth, D., Shah, F., Smits, M., Johansson, T., Canbäck, B., Olsen, P.B., Persson, P., Grell, M.N., Lindquist, E., Grigoriev, I.V., Lange, L., Tunlid, A., 2012. The ectomycorrhizal fungus Paxillus involutus converts organic matter in plant litter using a trimmed brown-rot mechanism involving Fenton chemistry. Environ. Microbiol. 14, 1477–1487.

Salvioli, A., Bonfante, P., 2013. Systems biology and "omics" tools: a cooperation for next-generation mycorrhizal studies. Plant Sci. 203–204, 107–114.

Selosse, M., Le Tacon, F., 1998. The land flora: a phototroph-fungus partnership? Trends Ecol. Evol. 13, 15–20.

Simard, S.W., Beiler, K.J., Bingham, M.A., Deslippe, J.R., Philip, L.J., Teste, F.P., 2012. Mycorrhizal networks: mechanisms, ecology and modelling. Fungal Biol. Rev. 26, 39–60.

Simard, S.W., Perry, D.A., Jones, M.D., Myrold, D.D., Durall, D.M., Molina, R., 1997. Net transfer of carbon between ectomycorrhizal tree species in the field. Nature 388, 579–582.

Singh, R., Soni, S., Kalra, A., 2013. Synergy between Glomus fasciculatum and a beneficial Pseudomonas in reducing root diseases and improving yield and forskolin content in Coleus forskohlii Briq. under organic field conditions. Mycorrhiza 23, 35–44.

Talbot, J.M., Treseder, K.K., 2010. Controls over mycorrhizal uptake of organic nitrogen. Pedobiologia 53, 169–179.

Tisserant, E., Kohler, A., Dozolme-Seddas, P., Balestrini, R., Benabdellah, K., Colard, A., Croll, D., Da Silva, C., Gomez, S.K., Koul, R., Ferrol, N., Fiorilli, V., Formey, D., Franken, P., Helber, N., Hijri, M., Lanfranco, L., Lindquist, E.,

Liu, Y., Malbreil, M., Morin, E., Poulain, J., Shapiro, H., van Tuinen, D., Waschke, A., Azcón-Aguilar, C., Bécard, G., Bonfante, P., Harrison, M.J., Küster, H., Lammers, P., Paszkowski, U., Requena, N., Rensing, S.A., Roux, C., Sanders, I.R., Shachar-Hill, Y., Tuskan, G., Young, J.P.W., Gianinazzi-Pearson, V., Martin, F., 2012. The transcriptome of the arbuscular mycorrhizal fungus Glomus intraradices (DAOM 197198) reveals functional tradeoffs in an obligate symbiont. New Phytol. 193, 755–769.

van der Heijden, M.G.A., Klironomos, J.N., Ursic, M., Moutoglis, P., Streitwolf-Engel, R., Boller, T., Wiemken, A., Sanders, I.R., 1998. Mycorrhizal fungal diversity determines plant biodiversity, ecosystem variability and productivity. Nature 396, 69–72.

van der Heijden, M.G.A., Martin, F.M., Selosse, M.-A., Sanders, I.R., 2015. Mycorrhizal ecology and evolution: the past, the present, and the future. New Phytol. 205, 1406–1423.

Wagg, C., Jansa, J., Stadler, M., Schmid, B., van der Heijden, M.G.A., 2011. Mycorrhizal fungal identity and diversity relaxes plant–plant competition. Ecology 92, 1303–1313.

Weigt, R., Raidl, S., Verma, R., Agerer, R., 2012. Exploration type-specific standard values of extramatrical mycelium – a step towards quantifying ectomycorrhizal space occupation and biomass in natural soil. Mycol. Prog. 11, 287–297.

Whiteside, M.D., Digman, M.A., Gratton, E., Treseder, K.K., 2012. Organic nitrogen uptake by arbuscular mycorrhizal fungi in a boreal forest. Soil Biol. Biochem. 55, 7–13.

Zhang, Q., Blaylock, L.A., Harrison, M.J., 2010. Two Medicago truncatula half-ABC transporters are essential for arbuscule development in arbuscular mycorrhizal symbiosis. Plant Cell Online 22, 1483–1497.

Lichens

Arnold, A.E., Miadlikowska, J., Higgins, K.L., Sarvate, S.D., Gugger, P., Way, A., Hofstetter, V., Kauff, F., Lutzoni, F., 2009. A phylogenetic estimation of trophic transition networks for ascomycetous fungi: are lichens cradles of symbiotrophic fungal diversification? Syst. Biol. 58, 283–297.

Blaha, J., Baloch, E., Grube, M., 2006. High photobiont diversity associated with the euryoecious lichen-forming ascomycete Lecanora rupicola (Lecanoraceae, Ascomycota). Biol. J. Linn. Soc. 88, 283–293.

Chua, J.P.S., Wallace, E.J.S., Yardley, J.A., Duncan, E.J., Dearden, P.K., Summerfield, T.C., 2012. Gene expression indicates a zone of heterocyst differentiation within the thallus of the cyanolichen Pseudocyphellaria crocata. N. Phytol. 196, 862–872.

Engler, A., Prantl, A., 1887–1915. Die Natürlichen Pflanzenfamilien nebst ihren Gattungen und wichtigeren Arten, insbesondere den Nutzpflanzen, unter Mitwirkung zahlreicher hervorragender Fachgelehrten Abt. 1*, 1907; Euthallophyta (Abt. II): Eumycetes: Lichenes.

Freitag, S., Feldmann, J., Raab, A., Crittenden, P.D., Hogan, E.J., Squier, A.H., Boyd, K.G., Thain, S., 2012. Metabolite profile shifts in the heathland lichen Cladonia portentosa in response to N deposition reveal novel biomarkers. Physiol. Plant. 146, 160–172.

Grube, M., Cernava, T., Soh, J., Fuchs, S., Aschenbrenner, I., Lassek, C., Wegner, U., Becher, D., Riedel, K., Sensen, C.W., Berg, G., 2015. Exploring functional contexts of symbiotic sustain within lichen-associated bacteria by comparative omics. ISME J. 9, 412–424.

Hawksworth, D., 2015. Lichenization: the origins of a fungal life-style. In: Upreti, D.K., Divakar, P.K., Shukla, V., Bajpai, R. (Eds.), Recent Advances in Lichenology. Springer, India, pp. 1–10.

Hill, D., 2009. Asymmetric co-evolution in the lichen symbiosis caused by a limited capacity for adaptation in the photobiont. Bot. Rev. 75, 326–338.

Honegger, R., 2009. Lichen-forming fungi and their photobionts. In: Deising, H.B. (Ed.), Plant Relationships. Springer, Berlin Heidelberg, pp. 307–333.

Honegger, R., 2012. The symbiotic phenotype of lichen-forming ascomycetes and their endo-and epibionts. In: Hock, B. (Ed.), Fungal Associations. Springer, Berlin Heidelberg, pp. 287–339.

Joneson, S., Armaleo, D., Lutzoni, F., 2011. Fungal and algal gene expression in early developmental stages of lichen-symbiosis. Mycologia 103, 291–306.

Molnár, K., Farkas, E., 2010. Current results on biological activities of lichen secondary metabolites: a review. Z. Naturforsch. C 65, 157–173.

Schoch, C.L., Sung, G.-H., López-Giráldez, F., Townsend, J.P., Miadlikowska, J., Hofstetter, V., Robbertse, B., Matheny, P.B., Kauff, F., Wang, Z., Gueidan, C., Andrie, R.M., Trippe, K., Ciufetti, L.M., Wynns, A., Fraker, E., Hodkinson, B.P., Bonito, G., Groenewald, J.Z., Arzanlou, M., Sybren de Hoog, G., Crous, P.W., Hewitt, D., Pfister, D.H., Peterson, K., Gryzenhout, M., Wingfield, M.J., Aptroot, A., Suh, S.-O., Blackwell, M., Hillis, D.M., Griffith,

G.W., Castlebury, L.A., Rossman, A.Y., Lumbsch, H.T., Lücking, R., Büdel, B., Rauhut, A., Diederich, P., Ertz, D., Geiser, D.M., Hosaka, K., Inderbitzin, P., Kohlmeyer, J., Volkmann-Kohlmeyer, B., Mostert, L., O'Donnell, K., Sipman, H., Rogers, J.D., Shoemaker, R.A., Sugiyama, J., Summerbell, R.C., Untereiner, W., Johnston, P.R., Stenroos, S., Zuccaro, A., Dyer, P.S., Crittenden, P.D., Cole, M.S., Hansen, K., Trappe, J.M., Yahr, R., Lutzoni, F., Spatafora, J.W., 2009. The ascomycota tree of life: a phylum-wide phylogeny clarifies the origin and evolution of fundamental reproductive and ecological traits. Syst. Biol. 58, 224–239.

Viles, H.A., 2012. Microbial geomorphology: a neglected link between life and landscape. Geomorphology 157–158, 6–16.

Endophytes

Alvin, A., Miller, K.I., Neilan, B.A., 2014. Exploring the potential of endophytes from medicinal plants as sources of antimycobacterial compounds. Microbiol. Res. 169 (7-8), 483–495.

Arnold, A.E., Lamit, L.J., Gehring, C.A., Bidartondo, M.I., Callahan, H., 2010. Interwoven branches of the plant and fungal trees of life. N. Phytol. 185, 874–878.

Freitag, S., Feldmann, J., Raab, A., Crittenden, P.D., Hogan, E.J., Squier, A.H., Boyd, K.G., Thain, S., 2012. Metabolite profile shifts in the heathland lichen Cladonia portentosa in response to N deposition reveal novel biomarkers. Physiol. Plant. 146, 160–172.

Mejía, L.C., Rojas, E.I., Maynard, Z., Bael, S.V., Arnold, A.E., Hebbar, P., Samuels, G.J., Robbins, N., Herre, E.A., 2008. Endophytic fungi as biocontrol agents of Theobroma cacao pathogens. Biol. Control 46, 4–14.

Nagabhyru, P., Dinkins, R., Wood, C., Bacon, C., Schardl, C., 2013. Tall fescue endophyte effects on tolerance to water-deficit stress. BMC Plant Biol. 13, 127.

Panaccione, D.G., Beaulieu, W.T., Cook, D., 2014. Bioactive alkaloids in vertically transmitted fungal endophytes. Funct. Ecol. 28, 299–314.

Parfitt, D., Hunt, J., Dockrell, D., Rogers, H.J., Boddy, L., 2010. Do all trees carry the seeds of their own destruction? PCR reveals numerous wood decay fungi latently present in sapwood of a wide range of angiosperm trees. Fungal Ecol. 3, 338–346.

Rodriguez, R.J., White, Jr, J.F., Arnold, A.E., Redman, R.S., 2009. Fungal endophytes: diversity and functional roles. *New Phytol. 182*, 314–330.

Selosse, M.A., Schardl, C.L., 2007. Fungal endophytes of grasses: hybrids rescued by vertical transmission? An evolutionary perspective. New Phytol. 173, 452–458.

Strobel, G.A., 2014. Methods of discovery and techniques to study endophytic fungi producing fuel-related hydrocarbons. Nat. Prod. Rep. 31, 259–272.

Wilberforce, E.M., Boddy, L., Griffiths, R., Griffith, G.W., 2003. Agricultural management affects communities of culturable root-endophytic fungi in temperate grasslands. Soil Biol. Biochem. 35, 1143–1154.

CHAPTER

8

Pathogens of Autotrophs

Lynne Boddy
Cardiff University, Cardiff, UK

There are currently about 7.0 billion people on planet Earth, and this will increase to 9 billion by 2050, with 80% in the developing world. Feeding this growing population depends on agricultural crops. Plant pests and diseases can cause yield losses of over 50% for major crops, and up to 80% for some crops (e.g. cotton). In the first few years of the twenty-first century, average losses of rice were 37.4%. When you consider that currently, rice feeds about 25 people per hectare, and this will have to increase to about 45 per hectare by 2050, the importance of dramatically reducing losses from pest and disease is clear. Crop losses worldwide cost US$550 billion each year. While about 60% of losses are caused by invertebrate pests, weeds and weather – frosts, drought, floods, etc., 40% are due to disease, of which fungi cause about two-thirds. Fungal infections currently destroy over 125 million tonnes of rice, wheat, maize, potatoes, and soybeans each year. Rice blast disease, caused by *Magnaporthe oryzae* (pp. 265–267), is at present the most destructive disease of rice worldwide with losses of over 50 million tonnes per annum. Ninety percent of wheat varieties grown worldwide are susceptible to a recently emerged lineage of the stem (black) rust pathogen *Puccinia graminis* f.sp. *tritici.* race Ug99. These examples highlight the importance of mitigating disease spread by fungal pathogens to safeguard world food security. Fungus and fungus-like diseases of plants can have terrible consequences to the human population, as attested to by the Irish Potato famine of the mid 1800s, where the crop was devastated by the fungus-like *Phytophthora infestans* (oomycete) (pp. 277–279). This led to death by starvation, of 1.5 million people, and emigration of over a million, largely to North America. The necessity to understand fungal diseases of plants is evident!

In this chapter we will first consider the variety of fungi that causes diseases of plants and then the differences in susceptibility of plants to pathogens, and how plants defend themselves. We will next examine the main events in the disease cycle of pathogens, from arrival, attachment, and entry into the plant, through the different ways that pathogens establish in and exploit the plants, finally to exit of the pathogen from the plant, and survival until it finds another suitable host. These general concepts will be illustrated with case studies of a range of types of disease, mainly of crop plants. Of course, plants in natural environments also suffer

The Fungi
http://dx.doi.org/10.1016/B978-0-12-382034-1.00008-6

from fungal diseases, as do other autotrophs such as lichens and seaweeds, and these are mentioned in separate sections. Finally we look at newly emerging diseases and the potential threats they pose to the security of our food supply.

SPECTRUM OF INTERACTIONS OF FUNGI WITH PLANTS

Fungi that attack living plants are called pathogens, or phytopathogens – to distinguish them from fungi that cause disease in other organisms. This is a catchall term that includes fungi that destroy living cells and feed on their contents, as well as fungi that absorb nutrients from living cells without killing them, but as a consequence, considerably reduce plant fitness. Plant pathologists commonly use the terms **necrotrophs** and **biotrophs** to describe these two categories of phytopathogen, respectively. The situation is complicated further by the fact that some fungi can initiate plant infection as biotrophs and switch later to the wholesale destruction of tissues as necrotrophs, a relationship termed **hemi-biotrophic**. Even dividing into three categories is not sufficient to cover the range of different characteristics and the mechanisms operated by plant pathogens, and does not get full support from genomic studies. Nonetheless we will use these categories here for simplicity. 'Parasite' is also used to describe pathogenic fungi. Some mycologists reserve this term for biotrophs, regarding all biotrophic pathogens as parasites; others treat parasite and pathogen as synonymous, using parasitism to describe plant infections by biotrophs and necrotrophs. Given the potential for ambiguity, we have avoided the term parasite in this chapter.

Necrotrophic and biotrophic pathogens behave differently in many ways, including the way in which they gain entry into the host, the way in which a host responds defensively to invasion, and in the signalling that occurs (see below; Table 8.1). Fungi also vary in the degree to which they obtain their nutrition as pathogens. Some are obligate whereas others are facultative. Biotrophic pathogens tend to be obligate, often having specialised structures for nutrient absorption, whereas necrotrophs and some hemibiotrophs are often facultative, being able to survive to some extent saprotrophically. Some fungal pathogens have a broad host range whereas others, especially biotrophic pathogens, have a very narrow host range, attacking only a few plant species. Some pathogen species are divided into specialised groups (*formae speciales*, f.sp.) which only cause disease of a few plant species; some *formae speciales* even have races that only colonise a few varieties of a species (e.g. *Fusarium oxysporum* has many *formae speciales* each of which is pathogenic to a narrow range of plant species [see the section on 'Case Studies', pp. 265–281]).

From the plant's perspective it is important to realise that though necrotrophic fungi kill host cells that does not necessarily imply that the whole plant will be affected. On the other hand, although biotrophic pathogens feed off living host cells, and often there are few visible signs of early infection, they can considerably reduce host plant fitness/productivity. It is also important to recall that although biotrophic pathogens reduce host fitness, many fungi, though feeding biotrophically, are mutualists that increase plant fitness, such as mycorrhizal, lichen, and many endophytic fungi (Chapter 7). This highlights the difficulty of determining the nature of some mycorrhizal symbioses in which it is unclear whether the fitness of both partners is enhanced by the relationship.

TABLE 8.1 A Comparison of Biotrophic and Necrotrophic Plant Pathogen Characteristics

Characteristic	Biotrophs	Necrotrophs
Obligacy	Specialised, obligate	Relatively unspecialised, facultative
Host range	Narrow	Broad
State of host on entry	All ages; in possession of host defence	Often immature, overmature or damaged
Axenic culturability	Not easy and sometimes not yet possible	Easy
Entry into host	Specialised, e.g. direct penetration of cell walls, e.g. powdery mildews; or via natural openings, e.g. *Puccinia hordei* (Fig. 8.3)	Unspecialised: via wounds or natural openings
Production of appressoria and haustoria	Appressoria or Appressoria-like structures evident; haustoria generally present (Figs. 8.5, 8.6, 8.7)	Not usually
Damage to host tissues	Little in compatible host	Rapid cell death
Production of lytic enzymes	Localised to hyphae, and limited quantity	Depending on mode of killing: often copious doing massive damage
Production of toxins	Not usually produced	Depending on mode of killing: often produced acting relatively locally or spread extensively in xylem. Some produce host-specific toxins (HSTs)
Survival following host death	Little or no saprotrophic ability; often survives as dormant spores	Can often grow saprotrophically
Control by resistance/susceptibility genes	PRR proteins; *R* gene products	PRR proteins; host specific toxin (HST) binding susceptibility gene products
Host defence pathways	NPR1; salicylic acid (SA)	COI1/EIN2; jasmonic acid (JA)/ethylene (ET)

DISTRIBUTION OF PATHOGENS AMONGST FUNGAL GROUPS

Most fungal phyla contain species that are plant pathogens, but most pathogens are found in a limited number of taxonomic orders, and within these orders often cause similar types of disease (Table 8.2). Ascomycota contains several orders with many or all species that are plant pathogens (e.g. Taphrinales). The Erisyphales are important biotrophic pathogens causing 'powdery mildews'; the Helotiales, Pleosporales, and Hypocreales contain many necrotrophic pathogens; several other orders also contain necrotrophic pathogens. In Basidiomycota, the majority of pathogens are the biotrophic Uredinales (smuts) and Ustilaginales (rusts); a few others are important root and butt pathogens of trees. There are many fungus-like oomycete pathogens of plants, within the Peronosporales; the Saprolegniales have few terrestrial pathogenic representatives, though some *Aphanomyces* spp. cause root disease in several crop plants. Other phyla contain relatively few pathogens.

TABLE 8.2 Distribution of Pathogens Among Fungal Groups

Phylum and order	Fungus species	Pathogenic mode	Disease	Symptoms	Host	Comments[a]
Mucoromycotina: mucorales	*Rhizopus* spp.	Necrotroph	Soft rot	Rotting of fruit and fleshy organs	Many fruits and vegetables	
Chytridiomycota	*Synchytrium endobioticum*	Obligate	Potato wart disease	Cauliflower-like warts on the tubers	Potato	
Ascomycota: Capnodiales	*Mycosphaerella graminicola*		Leaf spot	Leaves: water-soaked lesions, black pycnidia later	Cereal grasses, mainly wheat	
Ascomycota: Diaporthales	*Cryphonectria parasitica*	Necrotroph	Blight	Sunken cankers and split bark	Chestnut (*Castanea*) trees	Destroyed about 4 billion American chestnut trees
Ascomycota: Erisyphales	*Blumeria* (*Erisyphe*) *graminis*	Biotroph	Powdery mildew	Leaves: chlorotic or necrotic Leaves, stems, heads: covered with spores and mycelium giving a powdery/'woolly' appearance	Cereal grasses	
Ascomycota: Eurotiales	*Penicillium* spp.	Necrotroph	Soft rot	Tissues with high water content mostly affected, including fruit	Many fruits and vegetables, including citrus and apple	
Ascomycota: Glomerellales	*Colletotrichum*		Anthracnose	Stems and fruits: dark sunken regions Can cause rot of fruit	Beans, fruits, rye, some fruit and forest trees	
Ascomycota: Helotiales	*Botrytis* spp., including *B. cinerea*	Necrotroph	Blight and rots	Blossoms; Water-soaked rot blossoms; Leaves, stems, fruit: powdery lesions Fruit: rot	Fruit, ornamental and fruit trees	
Ascomycota: Helotiales	*Monilinia* spp.		Brown rot	Brown rotting of fruit	Stone fruits	
Ascomycota: Helotiales	*Sclerotinia sclerotiorum*	Necrotroph	Soft rot and wilt	Tissues with high water content mostly affected. Chlorosis, wilting, leaf drop; colonisation of fruit toughing the ground and in storage	Wide range; over 400 species	

Ascomycota: Hypocreales	*Claviceps purpurea*	Necrotroph		Dark coloured sclerotia in ears	Rye and related grasses	Consumption of the sclerotia causes ergotism in humans and other mammals
Ascomycota: Hypocreales	*Fusarium* spp., including *F. graminearum*	Necrotroph	Head blight	Shrivelled grain, premature bleaching of spikelets	Cereal grasses	
Ascomycota: Hypocreales	*Fusarium oxysporum*	Necrotroph	Wilt	Wilting, chlorosis, necrosis, leaf drop, stunted growth, damping off	Many, including most crop plants	
Ascomycota: *Incertae sedis*	*Gaeumannomyces graminis* var. *tritici*	Necrotroph	Take-all disease	Blackening of roots and a black crust around the stem base. Fungus invades phloem and xylem, preventing uptake and movement of water and assimilates	Wheat and barley	The most widespread cereal root disease in the world
Ascomycota: *Incertae sedis*	*Verticillium dahlia*, *V. albo-atrum*	Necrotroph	Wilt	Wilting, stunting and yellowing of older leaves	Many crop plants and trees	
Ascomycota: Magnaporthales	*Magnaporthe oryzae*	Hemibiotroph	Rice blast	Leaves; lesions whitish/grey when older, which may coalesce to kill entire leaf (Fig. 8.5)		Estimated to destroy annually 10–30% of rice crops, enough to feed 210–740 million people
Ascomycota: Microascales	*Ceratocystis fagacearum*	Necrotroph	Oak wilt	Wilting, browning of leaves; defoliation Sapwood beneath bark: brown discolouration	Oaks	Caused considerable damage in American mid-west; originated in Russia
Ascomycota: Ophiostomatales	*Ophiostom ulmi*, *O. novo-ulmi*	Necrotroph	Dutch elm disease	Wilting, yellowing/browning of leaves Sapwood beneath bark: brown/green streaks (Fig. 8.10)	Elm	Devastated elms across Europe and North America; identified first in the Netherlands
Ascomycota: pleosporales	*Alternaria* spp.	Necrotroph	Leaf spot	Dark blemishes on leaves, stem lesions and on fruit	Wide range of crops including tomato, potato, cucurbits, cotton, brassica	

(*Continued*)

TABLE 8.2 Distribution of Pathogens Among Fungal Groups—cont'd

Phylum and order	Fungus species	Pathogenic mode	Disease	Symptoms	Host	Comments[a]
Ascomycota: Pleosporales	*Cochliobolus sativus*	Necrotroph	Blight and root rot	Diseases of root, stem, leaf, and head tissues	Cereals and other grasses	
Ascomycota: Pleosporales	*Venturia inaequalis*	Necrotroph	Scab	Fruit and leaves: dark lesions	Apple	
Ascomycota: Taphrinales	*Taphrina* spp.	Biotroph	Leaf curl and blisters	Leaf curls, blisters, fruit deformities and stem dieback, depending on the plant species	Fruit trees	
Basidiomycota: Agaricales	*Armillaria mellea*	Necrotroph	Root rot	Above ground symptoms are similar to other root rotters: browning of foliage; decline of vigour; stem dieback. In main roots and at trunk base sapwood is water-soaked with creamy fan-shaped sheets of mycelium growing over the surface, and black zone lines within. Rhizomorphs often present (pp. 58–60)	Many species, especially woody	
Basidiomycota: Cantharellales	*Rhizoctonia solani* (=*Thanatephorus cucumeris*)	Necrotroph	Damping-off disease	Seeds are colonised and fail to germinate or seedlings are killed	Wide host range	
Basidiomycota: Russulales	*Heterobasiodion annosum*	Necrotroph	Root and butt rot	Above ground symptoms: abnormal needle growth; bark becomes paler; crown thinning; decline in vigour; eventual death. Below ground: white-rotted woody roots (Fig. 8.11)	Conifer and angiosperm trees	Most economically important forest pathogen in the northern hemisphere
Basidiomycota: Tilletiales	*Tilletia* spp.		Smut/bunt	Grain: replaced by black spore masses	Wheat	
Basidiomycota: Uredinales	*Hemileia vastatrix*	Biotroph	Coffee rust	Leaves: rust-coloured, oval lesions on lower surface; leaves eventually drop	Coffee	Devastating in coffee plantations

Basidiomycota: Uredinales	*Puccinia graminis*	Biotroph	Black stem rust	Leaves, stems and heads: rust-coloured, diamond shaped, raised lesions; black teliospores when mature	Cereal grasses	Causes wheat losses of 10–70%, enough to feed 200–1410 people per annum
Basidiomycota: Uredinales	*Uromyces viciae-fabae*	Biotroph	Rust	Rust-coloured pustules surrounded by a yellow halo	Bean rust	
Basidiomycota: Ustilaginales	*Ustilago hordei*	Biotroph	Smut	Grain: replaced by black spore masses	Oats	
Basidiomycota: Ustilaginales	*Ustilago maydis*	Biotroph	Smut	Grain: galls are formed (Fig 8.8)	Maize	Formerly was a major problem. Now causes mean losses worldwide of 2–20%, enough to feed 26–260 million people. Losses in individual fields can approach 100%. Eaten as a delicacy in central America called huitlacoche
Oomycota: Peronosporales	*Bremia lactucae*	Biotroph	Downy mildew	Pale yellow angular patches on leaves delimited by veins. Downy white masses on underside of leaves beneath these patches, also appearing on upper surface later	Lettuce (*Lactuca sativa*)	
Oomycota: Peronosporales	*Phytophthora* spp.	Necrotroph	Root rots	Wilting; rotting roots; death when severe	Trees, shrubs, vegetables	Many emerging diseases
Oomycota: Peronosporales	*Phytophthora infestans*	Hemibiotroph	Late blight	Leaves: water-soaked lesions, followed by dead brown areas on lower leaves; white woolly growth on lower surface. Tubers: watery, dark rotted tissue	Potato	Cause of the Irish potato famine in nineteenth century. Causes 50–78% crop loss, enough to feed 80–1270 million people
Oomycota: Pythiales	*Pythium* spp.	Necrotroph	Damping off	Rotting of seeds and roots, and seedling death		
Phytomyxea	*Plasmodiophora brassicae*	Necrotroph	Clubroot	Wilting Roots: distortion, swelling, rotting when severe	Cabbage and other Brassicas	

[a]*Estimates of number of people who could be fed if crops were not destroyed were made by Sarah J Gurr.*

Examples of some common and economically significant fungal and fungus-like pathogens and diseases of plants.

SUSCEPTIBILITY TO AND DEFENCE AGAINST FUNGAL PATHOGENS

Despite the large scale of some crop losses, most plants are resistant to most pathogens, most of the time. Three factors must interact to enable the development of disease: susceptibility of host, favourability of environment, and virulence/abundance of the pathogen. The likelihood of plants becoming diseased is often visualised by considering a disease triangle, in which these three factors are indicated on its sides (Figure 8.1a). If any of the three factors are zero, disease will not develop. The pathogen may be of a more or less virulent race or the inoculum size of the pathogen may vary (Figure 8.1b). Equally, the pathogen must be in an infective stage of its lifecycle and not in a dormant state for successful disease development to occur. With regard to the host, different plant species and cultivars often vary in their susceptibility to different pathogens, depending on their age, vigour, and their defence mechanisms. Plants may also be genetically uniform (monocultures) which will influence the rate of disease development by a given pathogen. With regard to environmental conditions, temperature, soil moisture, atmospheric humidity, wind, insolation, pH, nutrient status, and other aspects of soil quality, all have a major impact on disease development and severity, via their effects on fungal inoculum production, spread, growth and colonisation, and via their effects on plant vigour.

When a fungus encounters a plant, it is rarely able to be pathogenic, because plants have preformed physical and chemical defence barriers (described in more detail below p. 254), and even if these are overcome, the presence of a fungus will induce additional plant defences (explained in more detail on pp. 254–256). Most plants are, therefore, immune to most fungi (termed **non-host resistance**), and plants have the ability to reduce the severity of any potential fungal infection (termed **basal resistance**). Resistance initially involves the recognition of 'non-self' cues derived from the fungus – originally referred to as **elicitors** and more recently termed **microbe-associated molecular patterns** (MAMPs; or PAMPs in the case of pathogens). MAMPs are recognised when they bind to transmembrane proteins called pattern recognition receptors (PRRs), which go on to trigger plant defence responses termed PTI (pattern triggered immunity or plant innate immunity) (p. 254). This response is often swift and transient.

However, successful biotrophic pathogens (and mutualistic fungi), which grow in healthy hosts, can overcome the plants innate immunity by evading detection and/or by interfering with host defence responses. They do this by producing **effector** molecules (p. 255), that are induced during infection and are targeted to the plant cell apoplast, and in some cases to the plant cytoplasm, where they are able to suppress the plants innate immunity response resulting in effector triggered susceptibility (ETS). Of course the plant responds, and recognition of these fungal-derived virulence factors results in a second level of resistance, mediated by phytohormones and gene cascades culminating in effector triggered immunity (ETI), which is generally accompanied by the hypersensitive response (HR, pp. 255–256), programmed cell death, and restriction of pathogen ingress. So at the first level of plant immunity (PTI) recognises and responds to non-self molecules common to many microbes, including non-pathogens, and the second level (ETI) responds to pathogen effectors (virulence factors), either directly or indirectly, via their effects on host targets.

As mentioned above, these aspects of pathogenicity and plant immunity are attributed to fungi whose interests are to keep the invaded host plant cell alive. But what about the immunity elicited by plants in response to the group of fungi which benefit from the induction

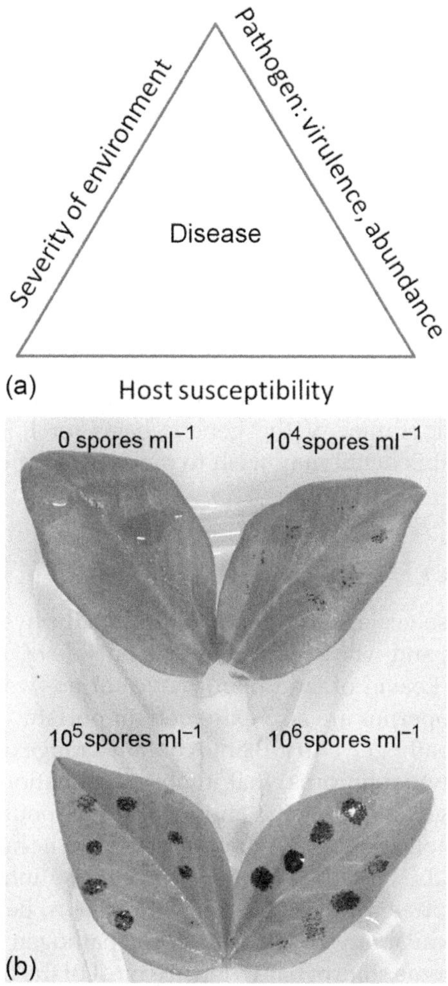

FIGURE 8.1 (a) The disease triangle. Each side of the triangle represents one of the factors involved in disease development. The length of each side can be set as proportional to the total of all aspects of each component that favour disease. The area, therefore, represents the amount of disease. If any of these factors are zero, disease will not develop. For example, if the plants are resistant then the host side and hence amount of disease would be small. In contrast, if the hosts were susceptible and planted densely, the host side would be large and so would the extent of disease, provided that the other sides were not zero. Likewise an extremely abundant, virulent pathogen would make a long pathogen side, and provided that environment was favourable and the host susceptible there would be a large amount of disease, and so on. (b) The inoculum potential (i.e. the amount of pathogen available) can influence success of colonisation, as illustrated here with different spore loads of the necrotrophic pathogen *Botrytis cinerea* added to the surface at eight points on broad bean (*Vicia faba*) leaves. *Source: (b) © Peter Spencer-Phillips.*

of plant cell death – the necrotrophs? Necrotrophic fungi can be divided into two groups, the broad host range necrotrophs and the host-specific necrotrophs. Notwithstanding that most necrotrophic fungi are unable to overcome the innate immune response of vigorous plants, broad host range necrotrophic infections on susceptible plant hosts can elicit host cell death (HCD) and other defense responses. However, unlike with biotrophic pathogens, HCD does not confine fungal ingress but rather is an indicator of successful infection, in many cases enhancing pathogen colonisation. There are a small group of host-specific necrotrophic fungal pathogens which produce 'effector-like' host-selective toxins (HSTs), which are toxic only to the specific host of the disease and are ineffective against the vast majority of other plants; these HSTs mediate susceptibility of the host to the host-specific necrotrophic pathogen. When a necrotrophic fungus produces a HST and the specific host plant has the corresponding dominant susceptibility gene, then ETS is established. A growing amount of evidence is now emerging in favour of there being common signalling pathways associated with both necrotroph susceptibility (inducing ETS) and biotroph resistance (inducing ETI). It is feasible that HST pathogens can 'deliberately' activate host ETI responses directed against biotrophic pathogens in order to establish ETS and thereby thrive in environments that would not be conducive with biotrophic lifestyles.

The following three subsections provide more detail about preformed and induced defences, and a simple consideration of the genetic basis for host/pathogen compatibility. Readers who do not want this detail may wish to move directly on to key events in the disease cycle (p. 257).

Preformed (Constitutive) Defence

Plants can defend themselves against fungal attack with physical and chemical barriers. Plant cells have thick walls, and when they are on the outside of plants, they are usually covered with cutin or suberin. Leaves of land plants have surface waxes. Bark is a physical protection, but its chemical properties are also extremely important. Size, shape, and location of stomata can also influence pathogen entry. Plants contain a range of low-molecular-weight antimicrobial compounds (**phytoanticipins**) that inhibit germination and/or growth, including phenols and quinones, long-chain aliphatic and olefinic compounds, aldehydes, cyanogenic glucosides, saponins, terpenoids, stilbenes, glucosinolates, cyclic hydroxamic acids and related benzoxazolinone compounds, tannins, lysozyme, proteinase inhibitors, polygalacturonase-inhibiting proteins (PGIPs), defensins, anti-microbial proteins, peptides, and low-molecular-weight compounds. Low availability of nutrients to the pathogen also contributes to defence. Further, most plants are in some state of basal defence most of the time, as the innate immunity of plants, though rapid and transient, is constantly being activated, since plants are continuously encountering bacteria, fungi, and damage in the rhizosphere and phyllosphere.

Active (Induced) Defence

As well as preformed defence mechanisms, plants have evolved at least two lines of active defence. As already mentioned, the first provides basal defence (akin to innate immunity) against all non-self-organisms and is based on the recognition of structurally conserved MAMPs by PRRs that then activate a pattern triggered immunity (PTI) preventing further colonisation of the host tissue. MAMPs are typically evolutionarily conserved molecules required for growth and/or development of the fungus, and include chitin found in fungal cell walls, and β-glucan found in fungal and oomycete cell walls, both of which have corresponding plant PRRs. However, fungal pathogens continue to evolve mechanisms to overcome preformed defences, to evade MAMP detection and/or produce effectors that can suppress the MAMP triggered immunity (PTI) thereby allowing the fungus to invade host tissue. Effectors are a group of mainly small secreted proteins (with very little homology to each other or existing proteins within databases). Fungal effector genes are typically induced upon host colonisation. Effector proteins are mainly produced in the ER and are secreted through the Golgi and are delivered to the apoplast where some are then subsequently transported into the host cytoplasm. Most are small secreted proteins with very little homology among them or to other proteins within fungal databases. They function either in the apoplast or within the cytoplasm of the host cell; yet how they are transported has not been established. Fungal apoplastic effectors typically include cell wall-degrading enzymes and necrosis- and ethylene-inducing (NEP1)-like proteins termed NLP's; however, the function of only a very few cytoplasmic effectors has been determined.

Plants have evolved a second active defence system, based on the recognition of such effectors or effector mediated changes in host components, by R proteins (which can be race-specific or race non-specific types) and the subsequent activation of ETI, which leads to a more rapid and enhanced defence response which is more robust than PTI. These responses are mediated by phytohormones such as jasmonic acid (JA), salicylic acid (SA), ethylene (ET), and abscisic acid (ABA). The JA and SA pathways are, in general, mutually antagonistic, with the JA/ET dependent pathway regulated by COI1/EIN2 genes and controlling responses to necrotrophs and chewing insects, and the SA pathway regulated by NPR1-like genes and controlling programmed cell death, which curbs the spread of biotrophs and hemibiotrophs as well as inducing systemic-acquired resistance (SAR). SAR is a resistance response of the whole plant at a distance from a site of localised infection by a pathogen. Effectively, it immunises the plant against future attack. The SAR response is mediated by a salicyclic acid-dependent process. MAP (mitogen-activated protein) kinase signal transduction cascades are involved in regulation of salicyclic acid accumulation. PRPs are formed and accumulate in SAR. Induced systemic resistance (ISR), unlike SAR, is mediated by a jasmonic acid/ethylene-dependent process, at least in some systems, and PRPs do not accumulate. Abscisic acid (ABA) plays a role in abiotic stress responses (drought, salinity, etc.) as well as influencing plant immunity through cross talk between the SA dependent pathway.

Plant defence responses vary in their time of activation, some occurring rapidly and others more slowly. The main rapid responses are an oxidative burst, generation of nitrous oxide (NO), cross-linking of cell wall proteins, and synthesis and deposition of callose concurrent with the HR reaction. HR is genetically programmed suicide of cells in the immediate vicinity of the invading fungus, a response which is particularly effective against biotrophs as they need living plant cells for growth and abstraction of nutrients. O_2^- and H_2O_2 are rapidly generated in many interactions between plants and fungal pathogens, and in conjunction with NO have antimicrobial activity. A weak response occurs in both compatible and incompatible interactions, but lasts longer (3–6 h) in the latter. Cross-linking of cell wall proteins occurs rapidly, making cell walls more resistant to hyphal penetration; its effect has been likened to self-sealing car tyres. Production of callose – a β-1,3-linked glucan – occurs rapidly as it does not require transcription or protein synthesis, just activation of the enzyme which is present in an inactive form. It is deposited as papillae localised in the plant cell wall, which again resists hyphal penetration.

Resistance responses that occur more slowly are: production of phytoalexins; lignification; suberisation; production of hydroxyproline-rich glycoproteins (HGRPs); production of pathogenesis-related proteins (PRPs); systemic-acquired resistance (SAR); and ISR. These responses all require gene transcription, hence their slower occurrence. However, they can be primed by the presence of some beneficial microbes, including root colonisation by AM fungi (pp. 206–212). **Phytoalexins** are low-molecular weight compounds, with antimicrobial properties against a wide range of fungi and other organisms, which are synthesised by plants and accumulate in them after exposure to a potential pathogen. Pathogens differ widely in their sensitivity, and some have detoxification mechanisms. **Lignin** and **suberin** are both found in the walls of healthy plant cells, but synthesis and deposition increases during attempted infection, increasing resistance to penetration and decomposition. Likewise, HRGPs are present in healthy cell walls, but increase during infection; their function in defence is not entirely clear but they play a role in cross-linking proteins and provide a template for lignin deposition in papillae. PRPs include a range of acidic protease-resistant proteins with different functions, for example, glucanases and chitinases break down components of fungal cell walls, thionins increase permeability of fungal cell membranes, defensins affect fungal membrane transport, and others inactivate fungal ribosomes.

Genetic Compatibility/Incompatibility of Host and Fungus

Whether fungi have the ability to establish within living plant cells is determined at the genetic level (Table 8.3). The gene-for-gene hypothesis proposed by Harold Flor, based on his pioneering work on flax (*Linum usitatissimum*) and the flax rust fungus *Melampsora lini*, states that for every dominant avirulence (*Avr*) gene in the pathogen, there is a cognate resistance (*R*) gene in the host and that the interaction of the *Avr*/*R* gene products results in the activation of host defence mechanisms, such as the hypersensitive response (HR) that causes localised cell death and halts the ingress of biotrophic fungal pathogens. The first fungal *Avr* gene was cloned in 1991 by van Kan and colleagues, and the first oomycete *Avr* gene in 2004 by Shan and colleagues. Over the last 20 years a multitude of *Avr* and cognate *R* genes have been identified which has significantly enhanced our molecular understanding of host pathogen interactions. Attempts have been made to study binding of *Avr*/*R* gene products; however, this has proved difficult and resulted in the proposal of an alternative (guard) model, whereby R proteins may not directly bind to their corresponding *Avr* gene product, but instead monitor the state of host components that are targeted by these molecules. Consequently, the term

TABLE 8.3 Gene-for-gene interaction resulting in resistance or susceptibility of a plant to a potential fungal pathogen

	Plant genotype			
Fungus genotype	R_1R_2	R_1r_2	r_1R_2	r_1r_2
A_1A_2	Resistant	Resistant	Resistant	Susceptible
A_1a_2	Resistant	Resistant	Susceptible	Susceptible
a_1A_2	Resistant	Susceptible	Resistant	Susceptible
a_1a_2	Susceptible	Susceptible	Susceptible	Susceptible

A symbolises avirulence and *a* virulence genes in the fungus, as avirulence is dominant. *R* symbolises resistance and *r* susceptibility genes in the host plant.

'effectors' was adopted to describe pathogen-derived gene products that alter (either positively or negatively) the interaction between plant and pathogen. These include the previously termed avirulence genes of the biotrophs and the host-specific toxin (HST) genes of the necrotrophs. Effectors are induced by transcriptional regulators when the host is colonised, some pathogens tailoring expression of these in individual host tissues. Effectors are targeted to the apoplast or cytoplasm and are delivered to the host through invading hyphal tips or across specialised biotrophic structures called haustoria (Figures 8.6, 8.9).

The mode of nutrition of necrotrophs does not require them to maintain a healthy viable host and it was, therefore, thought that necrotrophic fungal pathogens do not follow the gene for gene model; however, a small group of necrotrophic pathogens, including *Alternaria*, *Stagonospora*, and *Cochliobolus*, produce HST's that can be categorised as effectors. These are small peptides that are recognised by plant susceptibility genes. These HST effectors promote disease in a particular host species (and sometimes only within specific genotypes of the host) when the host expresses a specific dominant susceptibility gene. In effect, the pathogen produces an effector capable of promoting disease and the host plant produces a receptor that is required for susceptibility. This is a mirror image of the gene for gene hypothesis of biotrophic fungi where dominant host resistance and pathogen avirulence gene products are required for resistance.

KEY EVENTS IN THE DISEASE CYCLE OF PATHOGENS

A series of reasonably distinct events occurs during the development and spread of a disease in a plant and plant population (Figure 8.2). These events are termed the disease cycle and usually correspond closely to the lifecycle of the pathogen, though some pathogens (e.g. rust fungi) have very complex lifecycles. The different phases of the disease cycle are considered below.

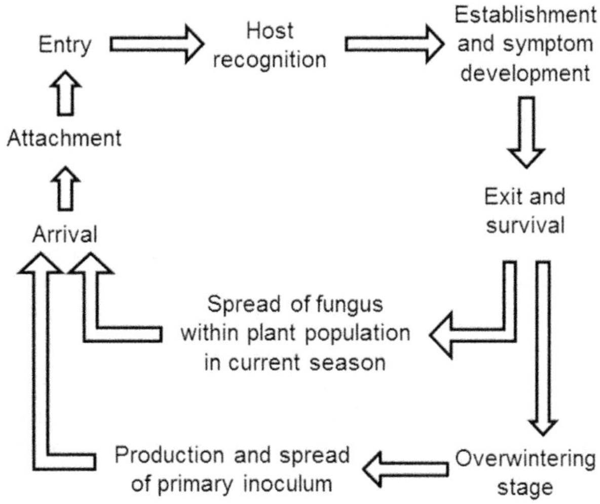

FIGURE 8.2 Events in the development of a disease cycle.

Arrival, Attachment, and Entry

The first key event in the infection cycle is for the fungus to come into contact with a potential host plant. For some fungi this is in soil, for others in the aerial environment, and yet others encounter plants in both. In soil, most plant pathogenic fungi are present as resting spores or sclerotia (pp. 60–61). Even when the microclimatic environment is suitable, most only germinate when mycostasis (p. 356) is overcome by the presence of nutrients exuded from living roots. Fungi which survive as resting spores have to await the arrival of roots in their vicinity, while others have motile zoospores that are attracted to their host roots but not to those of other species. For example, the fungus-like *Phytophthora cinnamomi* and *Phytophthora citrophthora* are attracted respectively to avocado (*Persea americana*) and citrus, but not vice versa.

Some fungi can reach roots as mycelium, particularly when an infected root contacts an uninfected one. This is a major means of spread of the tree root pathogen *Heterobasidion annosum* (Figure 8.11; pp. 276–277). The necrotroph *Rhizoctonia solani* (one cause of 'damping-off' disease of seedlings) grows freely through soil and utilises cellulose-rich substrata. It is as combative (pp. 349–351) as many obligate saprotrophs, aided by production of luxuriant mycelium and the ability to mycoparasitise (pp. 341–349) some fungi. Necrotrophic *Armillaria* spp. can grow between hosts as rhizomorphs (pp. 58–60). Biotrophs, hemibiotrophs and most necrotrophs are, however, unable to grow through soil.

Aerial plant surfaces are reached by spores transported by wind, rain splash or animal vectors (Chapter 3). There are many different types of spore (Chapter 3); some species produce several types and rust fungi produce up to five types during their life cycle (Table 8.6). Spores of some species germinate immediately in water or a humid atmosphere, rather than lying dormant. Others do not germinate in the absence of stimulants from the host, have endogenous self-inhibitors which prevent germination at high spore densities, and germination synchronises with dark periods, resulting in less chance of desiccation of hyphae. Spores of many facultative pathogens germinate more rapidly and extensively in the presence of nutrients derived by exudation, wounding or decay of the host.

Before fungi can parasitise plants they must breach their physical and chemical defences (see above, pp. 254–256). Entry occurs either through wounds, natural openings (e.g. stomata), or through an intact surface. Many obligate pathogens enter via stomata, though the powdery mildews (obligate) and many facultative pathogens enter directly through the intact plant surface. The site and mechanism of entry varies not only between species but can also vary between different types of spores of the same species.

Prior to penetration, germ tubes grow across plant surfaces and, in the case of many rusts, grow at right angles to veins, maximising the chance of encountering stomatal pores. Rust germ tubes respond to topographical and other stimuli (Figure 8.3). For the next stage of infection, adhesion to the plant surface must occur. An appressorium (Figures 8.5c–8.7, 8.9) forms on the plant surface or over openings, depending on the point where a particular species enters. These appressoria range from simple swollen cells, through lobed structures to complex infection cushions, that anchor the fungus prior to and during penetration. A thin penetration hypha (peg), which usually emanates from the centre of the appressorium adjacent to the plant, then effects entry into underlying tissues by wall-bound enzymes, including

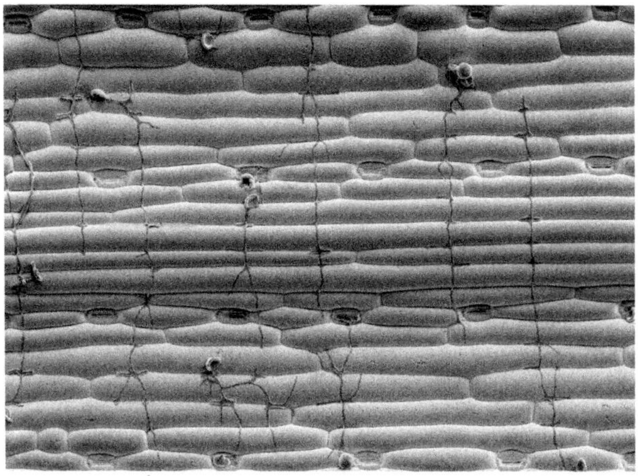

FIGURE 8.3 Hyphae of *Puccinia hordei*, cause of barley (*Hordeum vulgare*) brown rust, growing across the surface of a leaf. The hyphae are thigmotropic, and use the topography of the epidermal cells to orient there growth direction. *Source:* © *Nick Read.*

cutinases, and physical force. The physical force generated by turgor pressure in the penetration peg, which is usually insulated with melanin, can in some species reach 17 μN μm^{-2}; if a force of this magnitude were exerted over the palm of a single hand, it would be sufficient to lift an 8000 kg bus! Changes in fungal cell biology that occur during these early stages are detailed in some of the case studies.

Following penetration of the outer cuticle, subsequent growth and exploitation differs considerably depending on the host–fungus combination, but the vast majority grow between and/or into host cells; the powdery mildews only grow over the external surface and penetrate epidermal cells with haustoria (p. 272). Different fungi acquire nutrients from plants in different ways, and have different effects on them, leading to different symptoms (Table 8.4). At the two extremes, necrotrophs and biotrophs operate very differently and have different effects on hosts (Table 8.1).

Establishment and Exploitation: Necrotrophic Pathogens

In susceptible plants, necrotrophs typically produce extracellular hydrolytic enzymes, necrosis-related proteins and/or toxins which kill host cells and make them available as nutrient resources. Hyphae penetrate between cells but cell death often occurs in advance of hyphae. When necrotrophs kill with extracellular enzymes, copious quantities of a wide variety of pectolytic enzymes are secreted from hyphal tips. Though plants only contain a small quantity of pectic substances relative to cellulose and hemicelluloses, they are vital to the middle lamella, and when broken down, cells leak and lose turgor, and tissues become soft and 'watery'. This occurs, for example, in 'damping-off' (hence the name) diseases of seedlings and soft rots of fruit (Table 8.2). After using cell contents and pectic substances, cell wall components are broken down by cellulases and hemicellulases.

TABLE 8.4 Plant Disease Symptoms Caused by Fungal and Fungus-Like Pathogens

Name of symptom	Description of symptom
Abnormal growth	Including enlarged gall-like or wart-like swellings on above ground tissues, leaf curls, profuse branching to form witches brooms
Anthracnose	Sunken, dead blemishes on above ground plant parts
Blast	Destructive lesions on leaves and flowers
Blight	Browning of above ground plant parts
Canker	Localised wounds on woody stems; often sunken tissue
Damping off	Rapid death and collapse of seedlings, especially in warm, wet conditions
Decline	Loss of vigour
Dieback	Death of twigs starting at tips and advancing distally
Mildew	Chlorotic or necrotic tissue covered with mycelium and spores, giving a woolly or powdery appearance
Rots	Disintegration of plant tissues
Rusts	Small rust-coloured lesions on stems or leaves
Scab	Localised, scab-like lesions
Spots	Localised, dark-coloured lesions on leaves
Wilts	Wilting, discolouration and death of leaves, shoots and sometimes whole plants

Fungi produce a wide range of toxins with a variety of modes of action (Table 8.5). Some are host-selective/specific (i.e. toxic to host plants but not to non-host plants), whereas others are non-selective/non-specific (i.e. not related to the host range of the fungus). Plant resistance to a certain necrotrophic pathogen often results from insensitivity to the toxin that the fungus produces or from the ability to degrade the toxin rapidly. When toxins kill host cells, tissues often remain dry, but discoloured, and lesions remain discrete and spread slowly (e.g. *Alternaria* leaf spot diseases), though in vascular wilts they are spread over large distances in xylem (see below). Many toxins can, however, affect membrane permeability resulting in leakage of solutes, as with pectinases. Following killing by toxins, the fungi produce extracellular enzymes and utilise host tissues for their nutrition.

Vascular wilt fungi grow and proliferate in the vascular elements of plants, where they utilise nutrients in xylem sap. They produce both low and high molecular weight toxins in the xylem, the former being transported to leaves where they impair functioning of stomata, causing loss of transpirational control and leaf death. High-molecular-weight toxins increase the viscosity of sap, which also contributes to leaf death. In addition, vascular wilt fungi form extremely small spores which are transported in the xylem and often become lodged on and block the perforated end walls of the xylem vessels. Here they germinate, the germ tube grows through to the next vessel, spores are immediately produced and the process continues (see the section on 'Vascular Wilts; pp. 272–274). The plant's resistance mechanisms may also exacerbate the problem by forming gums and tyloses that block xylem vessels. Plants die rapidly as a result of these activities, herbaceous plants often being killed within a few days and large trees within a few weeks or months.

TABLE 8.5 Examples of Host-Specific and Non-Host-Specific Toxins Produced by Necrotrophic Pathogens of Plants

Toxin	Fungus	Disease	Effects and other properties
Host specific			
ACT-toxin, ACR-toxin, AF-toxin, AK-toxin, AM-toxin,	*Alternaria alternate fornae specialis*	Respectively leaf spots of tangerine, rough lemon, strawberry Japanese pear, apple,	Some (AF-, ACT-) act at the plasma membrane; others (ACR-) affect mitochondria
HC-toxin	*Cochliobolus carbonum* race 1	Northern leaf spot and ear rot of maize	Deacetylates chromatin, hence affects gene expression
HS-toxin	*Bipolaris sacchari*	Eye spot disease of sugar cane	The toxin is a terpenoid
Peritoxins (PC-toxin)	*Periconia circinata*	Sorghum root rot	Inhibition of mitosis and of growth of primary roots; eleectrolyte leakage
Ptr Tox A, Ptr Tox B	*Pyrenophora tritici-repentis*	Tan spot of wheat	Tan necrosis and extensive chlorosis
T-toxin	*Cochliobolus heterostrophus* race T	Southern corn (maize) leaf blight	Male sterility
Victorin	*Cochliobolus victoriae*	Victoria blight of oats	Leaves become a bronze colour. Membrane depolarisation; ion leakage; inhibition of protein synthesis
Nonhost specific			
Cercosporin	*Cercospra* spp.	Various	Generates active oxygen species; it is photactivated
Cerato-ulmin	*Ophiostoma novo-ulmi*	Dutch elm disease	A hydrophobin
Enniatins	*Fusarium* spp.	Dry rot of potato	Ionophores
Fomannoxin	*Heterobasidion annosum*	Root and butt rot of conifers and also some broadleaves	Inhibits plant cell growth and protein synthesis
Fumonisins	*Fusarium moniliforme* (=*Giberella fujikuroi*)	Ear rot of maize	Inhibits sphingolipd biosynthesis
Fusicoccin	*Fusicoccum amygdali*	Canker of almonds and peaches and wilt	Ireversible opening of stomata
Solanopyrones	*Alternaria solani*, *Ascochyta rabiei* (=*Didmella rabeiei*)	Chickpea blight	Inhibitor of DNA repair
Tentoxin	*Alternaria* spp.	Various	Chlorosis
Trichothecenes	*Fusarium* species	Head blight, root rot, and seedling blight of wheat	Inhibit protein synthesis

Information largely from Strange (2003).

Even at early stages of disease, or when pathogens remain localised, physiological processes are influenced throughout the plant. Host metabolism and biosynthesis increase, leading to rapid use of energy, increased respiration, and rapid depletion of carbohydrate reserves. Photosynthesis can be affected by damage to chloroplasts, rapid ageing and premature senescence of leaf tissues. Changes in plant water relations result from decreased uptake from soil if roots are damaged, changes in rate of transpiration, changes in water potential gradients, and resistance to flow through the plant, as well as from wilt diseases. Carbohydrate metabolism and movement around the plant are altered indirectly by disruption of water transport and directly by reducing the amount of carbohydrate leaving infected leaves. Plant-wide symptoms often result from hormone changes as a result of production by the fungus or modification of those produced by the plant. For example, epinasty (downward curving of leaves) is a symptom of over production or increased sensitivity to ethylene. Ethylene also has a role in chlorosis and necrosis. Decrease in indole acetic acid (IAA) can cause premature abscission resulting in leaf drop.

Establishment and Exploitation: Biotrophic Pathogens

In contrast to necrotrophs, there is a more balanced relationship between plant and pathogen, at least for much of the time of the association. Biotrophic pathogens abstract nutrients from host cells, but rather than producing vast quantities of extracellular enzymes or toxins, their production is controlled and probably only 'switched on' when required. Some enzymes are probably wall-bound having only localised effects. Further, it has been suggested that host enzymes may work for the fungus; invertases may cleave sucrose into glucose and fructose, (e.g. in broadbean, *Vicia faba*, infected with *Uromyces fabae*), when carbohydrate source tissues become sinks. As well as producing extensive intercellular hyphae, biotrophic fungi penetrate host cells with a narrow hyphal branch which then expands to form variously shaped structures – termed haustoria – which provide a large surface (Figures 8.6, 8.9). Haustoria, do not, however, penetrate the plasmalemma (see Broad bean rust case study for further details). Nutrient acquisition from the host can occur both via intercellular hyphae and haustoria. In the case of *Uromyces fabae*, amino acid uptake is via both, but sugar uptake is solely by haustoria. Haustoria also perform some biosynthetic activities, including synthesis of vitamin B1.

Whilst biotrophic pathogens do not kill host cells, they affect host functioning considerably. Their success depends largely on maintaining a balance between extracting nutrients for growth plus reproduction and impairing the ability of the host plant to produce further nutrients. The fungus acts as a sink for nutrients. Metabolites commonly accumulate at infection sites, both as a result of increased rate of transport to these regions and a reduction in nutrient movement away from them (e.g. rusted primary leaves of broadbean import up to forty times as much assimilate as uninfected leaves). Movement of nutrients towards infected regions occurs partly because the fungus acts as a sink, causing diversion of nutrients from other plant parts, and as a result of the fungus producing cytokinins – plant hormones. Localised increase in cytokinin concentration gives rise to 'green islands' typical of powdery mildew and rust infections of many plants. These result from chlorophyll retention in the vicinity of infections for longer than in other parts of leaves, and in retention of the capacity for protein synthesis. Effectively, the plant cells are maintained in a juvenile condition.

Biotrophs also have other effects on plant development via effects on hormones, either by synthesising or degrading plant hormones, or by altering the plant's metabolism of hormones and by interfering with the plant tissue response to hormones. Alterations in the quantity and type of auxins, cytokinins and gibberellins can result in hypertrophy (increase in the size of tissues/organs), hyperplasia (cell proliferation), and abnormal differentiation of organs. The witch's brooms caused by *Moniliophthora perniciosa* on cocoa, for example, form when lateral buds are released from apical dominance.

Exit and Survival

A successful pathogen produces dispersal structures once it has become established. These can be asexual or sexual spores, or mycelium, depending on the type of pathogenesis, fungus species and time of year (see the section on 'Case Studies'). During the growing season of the host, dispersal structures aid spread between potential hosts, whereas following death of whole plants or parts of plants, pressure on the pathogen is for survival. These different requirements are reflected in many leaf and stem pathogens by the production of asexual spores which spread rapidly during the growing season, and sexual spores that lie dormant throughout the adverse season for host growth. Interestingly, in black stem rust of wheat caused by *Puccinia graminis* f.sp. *tritici*, pustules which had been forming uredospores switch to produce thick-walled teliospores (p. 18) as autumn approaches. Switching from vegetative growth to spore formation, and between formation of different spore types relates to changes in weather, light, and the host. Unspecialised, necrotrophs often kill their hosts rapidly and produce resting stages after a short period of growth: the fungus-like *Pythium* and *Phytophthora* species form oospores on young colonies, while *Fusarium* spp. and *Rhizoctonia* spp., respectively, produce chlamydospores and sclerotia in response to nutrient depletion. Obligate biotrophs (e.g. those species that cause downy mildew, powdery mildew, and rusts) are dependent on spores for dispersal as they are unable to grow in dead tissues.

Spores can be dispersed by wind over long distance – across continents and even between continents (Figure 8.4, see also Chapter 3). This is particularly important for establishing emerging diseases in new areas and for re-establishing diseases in areas where host plants have been absent for several seasons. Single-step invasions of disease from one continent to another by airborne spores are rare. However, examples include sugarcane rust (*Puccinia melanocephala*) which was introduced into the Dominican Republic and thence more widely in America from Cameroon (West Africa) as uredospores moved across the Atlantic Ocean in June 1978 by cyclonic winds. Coffee leaf rust (*Hemileia vastatrix*) probably moved in the same way from Angola to Brazil in 1970. Other diseases have been moved between continents first in infected material and then by wind-blown spores across the continent to which they have been introduced. Potato late blight (see the section on 'Case Studies') was introduced to Europe in infected tubers and then spread as spores across Europe. Yellow rust (=stripe rust) of wheat moved to Australia in 1979 as spores of *Puccinia striiformis* f.sp. *tritici* on clothing and then across the continent and to New Zealand as wind-blown uredospores. Many biotrophs have cycles of extinction after host death followed by recolonisation of the new crop. This can occur over large distances (500–2000 km) and is in some ways analogous to the annual migration of some insects and birds, for example, the downy mildew *Peronospora tabacina* on tobacco in the eastern

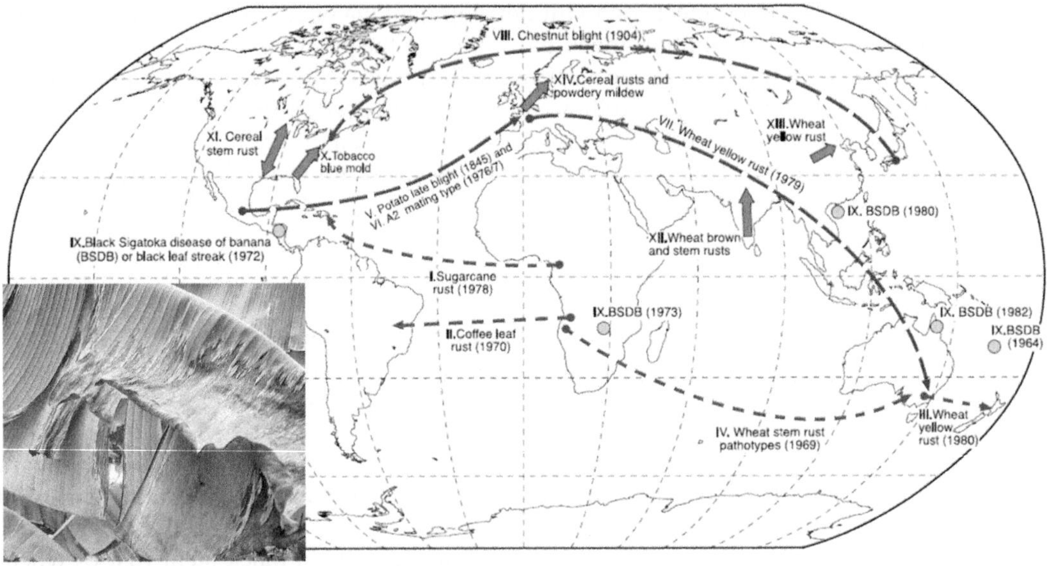

FIGURE 8.4 Plant pathogenic fungi can be transported extremely long distances by wind dispersal of spores. Though travel between continents is a rare event, unusual atmospheric conditions do sometimes occur, leading to dispersal and establishment of disease far from its long established origins. Red (black in the print version) arrows (on lines with short dashes) indicate disease spread by direct movement of airborne spores while blue (dark grey in the print version) arrows (on lines with long dashes) show where pathogens were first spread to new regions in infected plant material or by man and then onwards as airborne spores. Orange circles (grey in the print version) show the global spread of black Sigatoka disease of banana (inset) caused by *Mycospaerella fijiensis*, the first recorded outbreak on each continent is indicated by IX. Green arrows show five examples of diseases which periodically migrate via airborne spores from one region to another in extinction (due to absence of crops over one or more seasons)-recolonisation cycles (X, XI, XII, XIII, and XIV). The inset shows the symptoms of black Sigatoka disease of banana caused by *Mycosphaerella fijiensis* © John Lucas. *Source: Brown and Hovmøller (2002).* (See the colour plate.)

United States, rust pathogens of wheat in North America, and wheat yellow rust which reestablishes each autumn in northern China as there are no wheat plants during summer (Figure 8.4).

Exit and/or survival as somatic structures is largely confined to necrotrophs, since most biotrophic pathogens have highly specialised physiological relationships with their hosts, and little or no saprotrophic ability. However, some species of Taphrinales and Ustilaginales can grow saprotrophically in a yeast form on the surface of leaves, and this has a survival role.

Initially, necrotrophs utilise tissues which they have killed, unhindered by other fungi. However, competition from antagonistic obligate saprotrophs (Chapter 10) will increase with time and the initial advantage lost. Four different types of behaviour occur amongst necrotrophs in response to competition. Firstly, some necrotrophs simply survive in a dormant vegetative form or grow and utilise/decompose organic matter very slowly. This is typical of some stem pathogens. *Oculimacula yallundae* (=*Cercosporella herpotrichoides*), the cause of eyespot lodging in wheat and barley, only has limited ability to decompose cellulose and hence grows slowly, but it is able to form pseudoparenchyma (p. 63) within dead host tissue and can remain viable for up to 3 years in this form despite the presence of competitive saprotrophs.

Secondly, some necrotrophs actively colonise tissues that were uninvaded prior to host death (e.g. *Cochliobolus sativus* in cereal roots and *Gaeumannomyces graminis* [cause of take-all disease] in cereals). If sufficient resources have been secured before combative saprotrophs invade, then survival is ensured, although resistant spores may also assist. Thirdly, some necrotrophs actively grow into soil. *Rhizoctonia solani* (one cause of 'damping-off' of seedlings) grows freely through soil producing luxuriant mycelium, utilising cellulose-rich substrata, and is as combative as many obligate saprotrophs, being mycoparasitic on some fungi. It also produces resistant sclerotia (pp. 60–61). Lastly, some necrotrophs extend into soil but are not combative whilst migrating to new resources. For example, *Armillaria* species pathogenic on trees grow as rhizomorphs, insulated from their surroundings and supplied with nutrients from host residues.

CASE STUDIES

Rice Blast Disease

Rice blast disease, caused by *Magnaporthe oryzae* (Ascomycota), occurs in about 80 countries on all continents where rice is grown, in both paddy fields and upland cultivation. The extent of damage caused depends on environmental factors, but worldwide it is one of the most devastating cereal diseases, resulting in losses of 10–30% of the global yield of rice. In rice seedlings, small necrotic regions appear initially, which become larger and coalesce, and have chlorotic margins (Figure 8.5). In older rice plants, disease symptoms can occur in leaves, collar – junction of the leaf blade and leaf sheath, nodes, neck, and panicle (Figure 8.5). Neck rot and panicle blast are particularly devastating causing up to 80% yield losses in severe epidemics. Triangular, purple-coloured lesions form on the neck node which elongate on both sides, seriously impairing grain development. The panicles become white when young neck nodes are invaded; infection later in plant growth results in incomplete grain filling.

The disease cycle begins when a conidium lands on a rice plant and becomes attached to the host surface through production of spore tip mucilage. This is followed by a series of developmental steps: germination, germ tube growth, formation of an appressorium, emergence of a penetration peg from the appressorium, and subsequently invasive growth in the host (described in Figure 8.5). Appropriate development of infection structures on the leaf surface involves sensing plant-derived cues, transduction of signals through G-protein-coupled receptors; activation of signalling cascades mediated by cAMP, mitogen-activated protein kinase (MAPK; proteins that are evolutionarily conserved and function as key signal transduction components in fungi, and also plants and mammals), phospholipase- and calmodulin-dependent pathways and resultant up-regulation and down-regulation of genes. The genome sequence for strain 7–15 was published in 2005 and today there are over 30 genomes available from strains around the globe. There are up to 1500 secreted proteins in the genome, the function of many being unknown, but some are certainly effector proteins. Many novel genes/metabolic clusters are being found in different isolates of the fungus, so large scale sequencing will be revealing. Over 20% (2154) genes are differentially expressed, mostly up-regulated during the processes prior to invasion. Temporal analysis of the transcriptome during infection, together with gene replacement experiments, identifies virulence-related genes required for successful invasion. For example, avirulence in some strains of

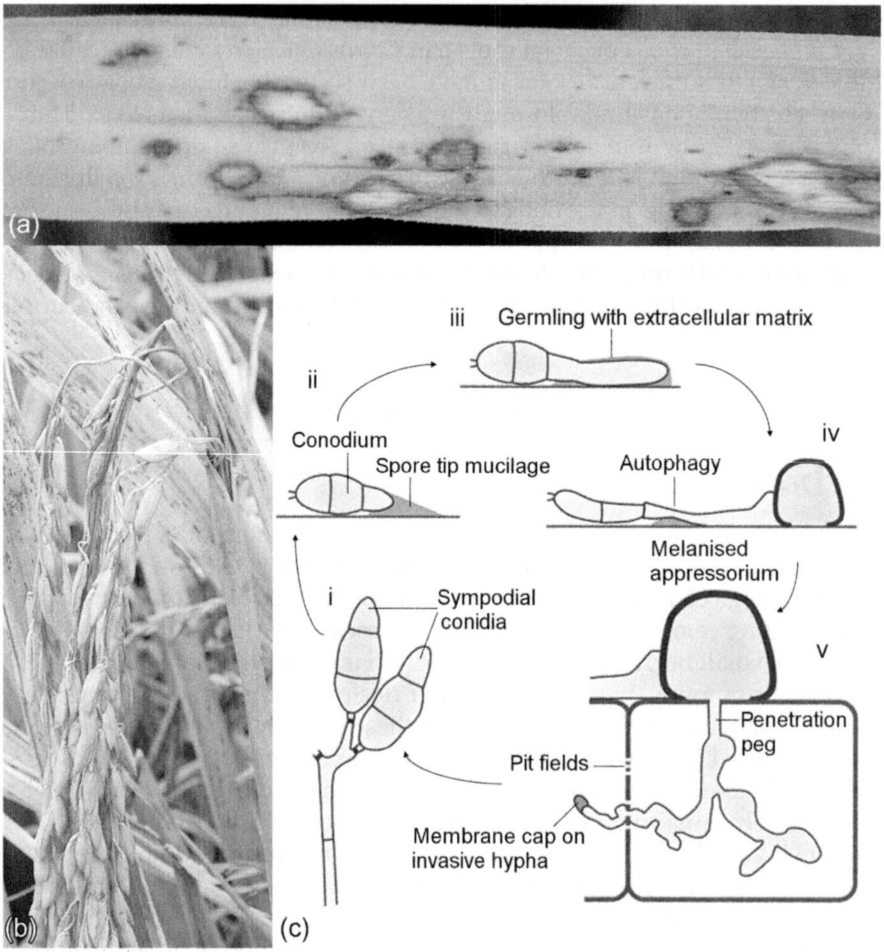

FIGURE 8.5 Rice blast disease caused by *Magnaporthe oryzae.* (a) necrotic lesions on rice seedlings. (b) Neck and panicle blast. (c) The disease cycle: (i) The teardrop shaped conidia have three cells each with a single nucleus. (ii) Germination proceeds commonly from the tapering end and the nucleus passes into the germ tube, where mitosis occurs. One daughter nucleus moves back into the spore and the other moves into the swollen tip of the germ tube. Appressorium initiation is controlled by progression of the latter nucleus into the S-phase of the cell cycle, which is within 2–4 h on a hydrophobic surface. Appressorium differentiation is controlled at the G2-M transition. At an environmental level, appressorium formation depends on plant-based and non-plant cues, including physical cues, such as hydrophobicity and hardness (appressoria will form on artificial surfaces), plant cutin monomers and nitrogen depletion. (iii) Autophagy of the spore and germ tube occurs and glycerol synthesised from the lipid and glycogen moves into the appressorium. Autophagy is essential for virulence, since the appressorium forms at the expense of the conidial contents, as shown by targeted mutation of the autophagy gene *MgATG8* which resulted in an avirulent strain. (iv) Up to 8 MPa turgor pressure (i.e. 40 times higher than a car tyre) is generated, by accumulation of glycerol, within the single-celled, dome-shaped appressorium, which is insulated by a layer of melanin in the wall. This turgor pressure, supplied in a very controlled manner, allows the penetration peg to force its way through the cuticle and cell wall, but the plasma membrane remains intact and becomes invaginated around the penetrating hypha. This is a pivotal recognition point for fungus and host. The plant responds by production of reactive oxygen species (ROS) around the infection site, but the fungus has evolved a rapid ROS detoxification mechanism. Growth into the next cell is through clusters of plasmodesmata, termed pit fields. The hyphae that penetrate into adjacent cells have a membrane cap comprising membrane lamellae, which probably secrete protein into the cytoplasm of the living host cell. (v) Emergence of conidiophores 4–5 days after infection. *Source: (a, b) From Galhano and Talbot (2011); (c) Ebbole (2007).*

Magnaporthe oryzae has been traced to loss of function in several specific genes: two of these encode proteins in signalling pathways linked to G-protein-coupled receptors, both of which are required for appressorium formation.

In susceptible hosts, the fungus ramifies between and within cells, probably spreading from one plant cell to another via plasmodesmata. The fungus is hemibiotrophic, initially feeding on living cells but subsequently causing death. Lesions on the plant surface start to form when host cell death begins 4–5 days after infection. Melanised macroconidia are then produced on conidiophores that protrude from lesions, with up to about 20,000 spores produced from one lesion on a leaf, and 60,000 on a single spikelet in one night. Spore release is triggered by a 1–2h period of darkness. The fungus can sporulate repeatedly for around 20 days. The disease is polycyclic (i.e. multiple infection cycles in a growing season) with 7 days between spore germination and production of conidia.

Because rice provides almost 25% of the calories consumed by humans, understanding the biology of this pathogen is particularly important for developing disease control strategies. Moreover, the pathogen has now jumped hosts to wheat, and is now an urgent problem in South America. Due to its agricultural impact and tractability, *Magnaporthe oryzae* has become a model fungus for studying host pathogen interactions, at both the cell biology and molecular level. Rice blast populations comprise sets of discrete clonal lineages with a limited spectrum of virulence. The most cost-effective, and hence preferred, way of managing rice blast disease is by growing high yielding cultivars with single, dominant resistance genes; over 70 blast resistance (*R*) genes have been identified. However, the problem with single-locus resistance is that it is often short-lived, only lasting for a few years. To lessen the chance of *R* gene breakdown broad-spectrum *R* genes can be deployed, which provide resistance to different strains of the fungus (e.g. resistance gene *Pi*2 confers resistance to over 450 isolates). Another proposed strategy is to stack several *R* genes, having different resistance spectra, into a single rice cultivar; this has also been proposed for control of cereal rusts. The use of antagonistic phylloplane bacteria and fungi is also being tested. Fungicides are available but these have attendant problems of cost, development of fungal resistance and damage to ecosystems.

Rust

Black stem rust of wheat, caused by *Puccinia graminis* f.sp. *tritici* (Basidiomycota), is a major concern for the world's food security, causing crop losses of up to 70%, enough to feed several hundred million people (Table 8.2). Not surprisingly, this rust fungus has been extensively studied. Its lifecycle has already been described in Chapter 1 (pp. 18–19), so here we consider broad bean rust which, though less important overall for global food security, has also been studied in detail and has long been used as a model for studying cytology, physiology, biochemistry, and molecular aspects of rust fungus biology. *Uromyces fabae* causes disease of many legumes, including broad bean (*Vicia faba*), with up to 50% losses of yield, and over 50 other *Vicia* spp., 20 *Lathyrus* spp., lentil (*Lens culinaris*), and pea (*Pisum sativum*). Like other rusts, it is an obligate biotroph.

Uromyces fabae is macrocyclic (i.e. multiple disease cycles during each growing season), and produces all five spore forms found in the Uredinales. Its lifecycle and the role of the different spore types are shown in Figure 8.6 and Table 8.6. **Telia** form on stems and leaves

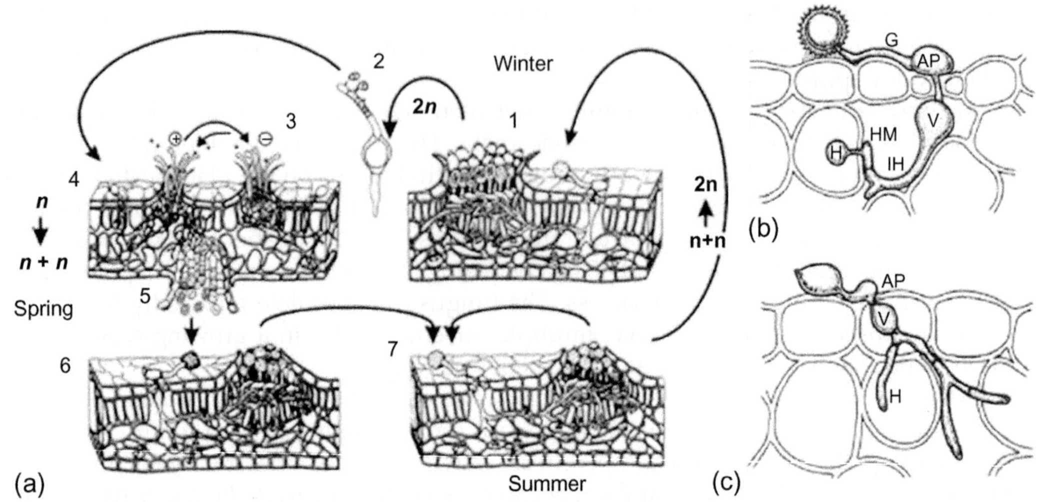

FIGURE 8.6 The life cycle of *Uromyces fabae*, cause of broad bean (*Vicia faba*) rust, is extremely complex. (a)(1) The fungus overwinters as telia, which produce teliospores (2*n*), surviving in crop residues in the field and also on seed. The teliospores are spread by wind. (2) They germinate on a plant in spring to form a metabasidium with four haploid basidiospores (with two mating types [p. 115]). (3) These germinate and infect the plant (b) – see below for detail. (4) Pycnia are formed, on the upper surface of leaves, which contain haploid pycniospores. The latter are exchanged between pycnia of different mating type (+, −). (5) Spermatisation (p. 112) occurs, and then aeica, containing dikaryotic ($n + n$) aeciospores, are produced on the lower surface of the leaf. The aeciospores spread the infection within the crop and to other nearby bean crops. (6) Uredea are then produced on the stems and leaves; they contain dikaryotic uredospores, which (7) also spread the fungus throughout the crop canopy. Late in the season, the uredia differentiate into telia and remain in crop debris ready for the infection cycle to start again on next season's crop. (b) When uredospores land on a plant surface they have an irregular shape, because they are completely dehydrated. They rehydrate rapidly and become ellipsoidal. Their surface is hydrophobic, which facilitates initial adhesion to plant surfaces. An extracellular matrix of carbohydrates and glycosylated polypeptides form, followed by an adhesion pad. Spores only germinate when fully hydrated, at temperatures between 5 and 26 °C (optimum 20 °C), and after at least 40 min of darkness, but will do so on almost any surface. When the germ tube (G) grows, cytoplasm moves into it from the spore. An appressorium (AP) forms when appropriate signals concerning leaf surface topography are received. Cytoplasm moves from the germ tube into the appressorium and a septum separates the two structures. A penetration peg is forced, by turgor pressure, into the stomatal cavity. A vesicle (V) forms and from this an infection hypha (IH) develops. On contact with a mesophyll cell a haustorial mother cell (HM), with a thick multi-layered wall, is formed and attaches to the plant cell wall. All of the cytoplasm moves into this, and is walled off by a septum. A penetration hypha breaches the mesophyll cell wall by localised production of enzymes, including pectin esterases, pectin methylesterases, and cellulases, and a haustorium (H) is formed causing the host plasma membrane to invaginate – so the haustorium is not truly intracellular. A matrix forms around the haustorium, and this is the site of nutrient uptake and information exchange. (c) Unlike uredospores, teliospores have smooth, thin walls. The infection structures that develop from them – appressoria (AP), vesicles (V) and haustoria (H) – are much less differentiated. Further, penetration pegs enter directly into plant cells rather than via stomata. *Source: Modified from Voegele (2006).*

late in the season, remaining in crop debris after harvest, the fungus over-wintering as **teliospores**. In spring teliospores germinate to form a **metabasium** that produces **basidiospores**, that themselves germinate to form monokaryotic (p. 113) infection structures. During the rest of the growing season infection mainly occurs following germination of **uredospores**, which with **aeiciospores** spread the fungus throughout the canopy and between plants.

TABLE 8.6 *Uromyces fabae* spore types

Spore type	Ploidy	Produced by/in	Infect plant tissues	Position in lifecycle
Teliospores	Diploid	Telia	No	Nuclei fuse during teliospore production. Overwintering stage in plant remains in the field and on seed. Germinate to produce a metabasidium
Basidiospores	Haploid	Four produced by each metabasiudium following meisois. Two mating types	Yes	Smooth, thin-walled. Germinate on host in spring and produce infection structures
Pycniospores (spermatia)	Haploid	Pycnia	No	They act as sperm and enter a receptive hypha of a pycnium of a different mating type. Dikaryotisation occurs in aecial primordial
Aeciospores	Dikaryotic	Aecia	Yes	Aecial spores germinate and produce an infection structure from which uredia are produced
Uredospores	Dikaryotic	Uredia	Yes	Thick-walled with spines and darkly pigmented. The main asexual spore produced in vast quantities and dispersed aerially. Repeatedly infect host plants during the summer. Uredia differentiate into telia during the fall

The events following germination are well characterised for basidiospores and uredospores. The infection processes are rather different. Both produce appressoria with a penetration peg beneath, followed by a vesicle when entry into the plant has been effected, and subsequently a haustorium within a plant mesophyll cell, but these structures are less differentiated when they are produced following germination of basidiospores (Figure 8.6). Also, the penetration peg produced following basidiospore germination enters an epidermal cell directly and then forms a vesicle, whereas that from an uredospore enters via a stoma and produces the vesicle in the stomatal cavity. As mentioned previously (p. 258, Figure 8.3), the germ tubes produced by rust spores can follow topographical features of leaves to locate natural openings. *Uromyces appendiculatus* can recognise the height of the lip of the guard cells (0.5 μm) on broad bean leaves. Rust fungi also detect chemical signals (e.g. leaf alcohols).

The haustorium is the site of uptake of carbohydrate and amino acids, though the latter can also be taken up via intercellular hyphae, and also synthesises other nutrients *de novo*. The haustorium plays a major role in primary metabolism. *Uromyces fabae* has evolved several mechanisms to avoid host recognition, including masking the chitin in infection structures by acidic cellulases and protease action, conversion of chitin to chitosan, release of mannitol that suppresses host defence responses involving reactive oxygen species (ROS), production of protein effectors that take control of host metabolism. *Vicia faba* gene expression patterns not only alter in the vicinity of infection, but also in remote organs (e.g. stems and roots).

Control is by use of resistant races, fungicide application and/or horticultural practices such as appropriate rotation with non-host crops and removal of plant debris containing the fungus.

Colletotrichum Anthracnose and Blights

Almost all crops worldwide are susceptible to one or more *Colletotrichum* spp. (Ascomycota), causing anthracnose (sunken dead spots) and blights (tissue browning) of aerial tissues. *Colletotrichum* can also be latently present causing post-harvest rots, infecting tissues pre-harvest but not developing overtly until after harvest. They cause major economic loss of fruits and vegetable crops, including staples in developing countries (e.g. banana, cassava, and sorghum). Spores germinate and enter the plant via a fine penetration peg produced beneath an appressorium (Figure 8.7). *Colletotrichum* is hemibiotrophic, initially establishing itself biotrophically within the plant. First an intracellular vesicle is formed, and from this a few large intracellular primary hyphae develop and extend into only a few cells (Figure 8.7). These hyphae and the vescicle are surrounded by a matrix which is the interface with the plant apoplast. Specific genes are expressed during the biotrophic phase, including *C1H1* which encodes a glycoprotein, and *CgDN3* which is thought to maintain the biotrophic phase of development. Subsequently, the fungus switches to a necrotrophic phase in which narrower hyphae ramify through host tissue. These hyphae secrete endopolygalacturonases and other cell wall-degrading enzymes. *Colletotrichum gloeosporiodes* secretes a *pelB-encoded*

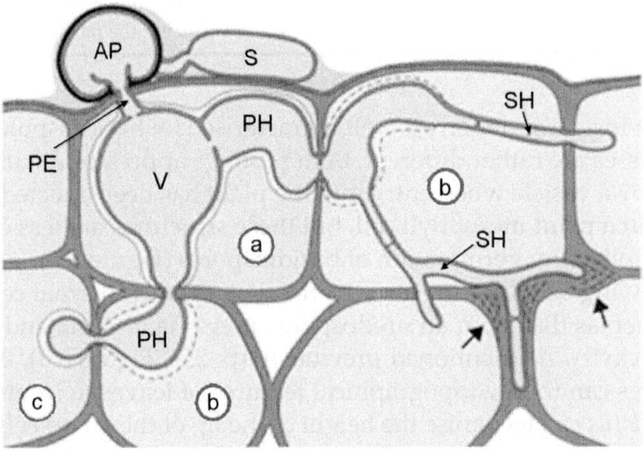

FIGURE 8.7 *Colletotrichum* anthracnose. Most species are hemibiotrophic as seen in this diagram of infection by *Colletotrichum lindemuthianum.* A spore (S) attaches to the surface of the host. When it germinates, it produces a short germ tube, which differentiates into an appressorium (A), from the underside of which develops a penetration peg (PE) which pierces the cuticle and wall of the epidermal cell. The hypha swells to form a vesicle (V) from which develop broad primary hyphae (PH) surrounded by plant plasma membrane. This is the biotrophic stage (a); the plant cell remains alive, and the host and fungal protoplasts remain separated by an interfacial matrix (indicated by yellow (light grey in the print version) colouring). After 1 or 2 days the plant plasma membrane begins to disintegrate and the host cell dies (b). A sequence of colonisation of plant cells by new primary hyphae occurs (c) with subsequent death after a few days. The biotrophic phase ends when narrow secondary hyphae (SH) develop from the primary hyphae. The secondary hyphae are not surrounded by host membrane/interfacial matrix, and secrete plant cell wall-degrading enzymes (indicated by arrows) in this necrotrophic phase. *Source: Mendgen and Hahn (2002).* (See the colour plate.)

pectate lyase which not only breaks down cell wall components but also reduces host defence responses which are triggered by released oligogalacturonides.

The fungus survives between cropping seasons within crop residues where it can grow saprotrophically. The disease is spread by asexual spores via water splash, wind, and invertebrates. Germination and infection require high (near 100%) humidity, and pre-harvest disease is most serious at warm (25–20 °C) temperatures. Post-harvest disease, however, can occur in much drier conditions, when tissues are damaged or through ageing, as the fungus is already latently present.

Corn Smut

Smuts are caused by members of the Ustilaginales (Basidiomycota) and occur worldwide, with over 1200 species. Most attack the ovaries of Gramineae, and develop in them and the grain. Until the twentieth century, smuts, along with rusts, were the major cause of grain loss. They are now controlled using resistant varieties and seed treatment to kill teliospores and mycelium.

Corn smut, caused by *Ustilago maydis*, affects corn wherever it is grown. Its complex life-cycle has been introduced earlier (pp. 16–18). It overwinters as teliospores (diploid) in soil and on crop debris, and these resistant spores can remain viable for several years (Figure 8.8).

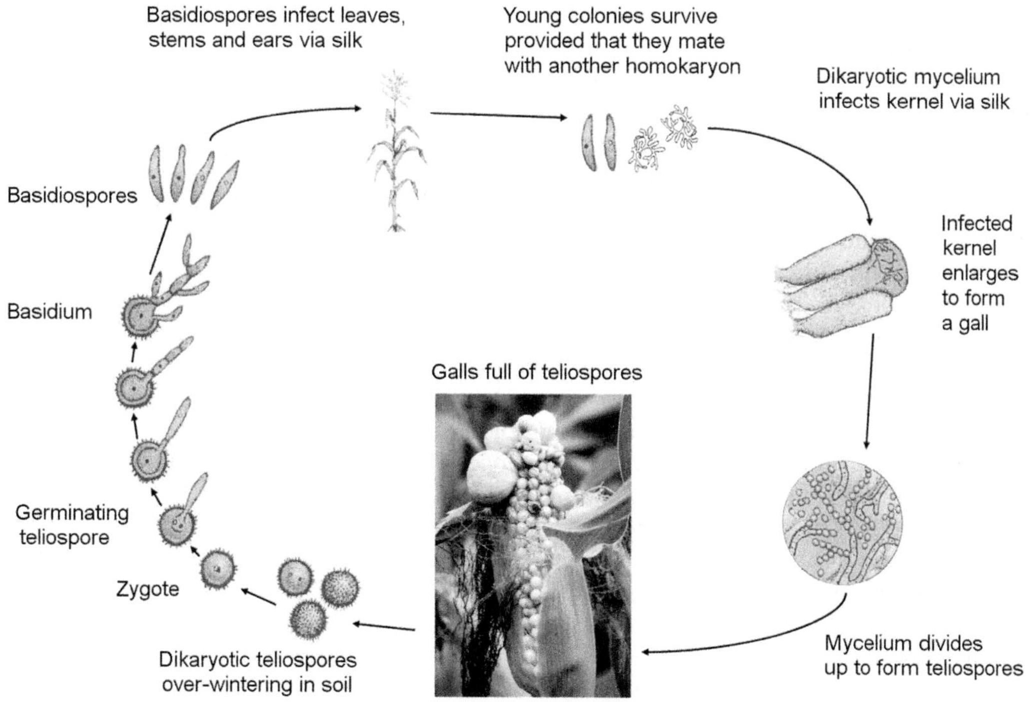

FIGURE 8.8 The disease cycle of Corn (*Zea mays*) smut, caused by *Ustilago maydis*. *Source: Modified from Agrios (2005); image © Albert Brand.*

The teliospores germinate to form a **promycelium**, on which a basidium and basidiospores (sporidia) form, following meiosis. The basidiospores, spread onto plants by rain splash and air currents. They bud in a yeast-like way, and are a saprotrophic phase. Under nutrient limiting conditions they form fine hyphae. If fusion with a compatible homokaryotic mycelium occurs, the dikaryotic hyphae penetrate directly into epidermal cells and spread intercellulary within plant tissues, but if successful mating does not occur the hyphae often die. Cells in the vicinity of living hyphae enlarge and divide forming galls, with the hyphae remaining intercellular for most of gall formation. In seedlings, systemic infections occasionally occur; in older plants infections are local, many of the galls remaining too small to be visible, only a few forming the characteristic large galls (Figure 8.8). The fungus invades galls prior to sporulation, uses plant cell contents and then converts the dikaryotic mycelium to teliospores. Released teliospores that land on corn meristematic tissue may cause new infections, but the majority are survival structures until the next season.

Powdery Mildews

Powdery mildews are extremely common and widespread, and economically one of the most important groups of diseases infecting many plant taxa, though not gymnosperms, the most severely infected crops including cereals and cucurbits. They are caused by many species within the family Erisyphaceae (Ascomycota) (e.g. *Blumeria graminis* on cereals and other grasses, and *Erisiphe cichoracearum* on *Arabidopsis*). They are obligate biotrophic pathogens and are mostly unculturable, though artificial growth media have now been developed for *Blumeria graminis*. Powdery mildew is most common on the upper surfaces of leaves, and to a lesser extent on lower surfaces of leaves and other organs. The mycelium grows only on the surface of the plant and does not invade tissues, nutrients being obtained via haustoria in the plant epidermal cells. Chains of round, ovoid, or rectangular conidia (p. 67) are formed on condiophores on the surface, giving the powdery appearance. These spores are dispersed by air. Sexual **cleistocthecia** (p. 20), containing one or a few asci are produced when conditions become unfavourable. Powdery mildews rarely kill their hosts but their pathogenesis causes considerable loss of productivity (as much as 20–40%) due to nutrient removal, reduced photosynthesis, increased respiration and transpiration and impaired growth.

The infection process has largely been studied in *Erisyphe pisi* on pea (*Pisum sativum*), *Blumeria graminis* on barley (*Hordeum vulgare*) and *Erisyphe cichoracearum* on *Arabidopsis*. Conidia are released, germinate and infect epidermal cells in the absence of a water film provided that air relative humidity is high. Once infection has begun, surface colonisation continues irrespective of atmospheric humidity, causing severe disease in warm, dry climates as well as cooler and more humid regions. When conidia land on a susceptible host they germinate and within 24 h produce an appressorium from which a penetration peg develops and penetrates the epidermal cuticle of the host, forming a haustorium within an epidermal cell, as described in Figure 8.9. The haustorium is separated from the host cytoplasm by an extrahaustorial membrane and gel-like amorphous matrix derived from the host – though the pathogen may contribute components. The haustorium obtains water and nutrients from the host. As soon as the haustorium is completely formed, another hypha emerges from the spore, usually on the opposite side to the germ tube, fed by nutrients from the plant. By 5 days conidiophores start to form.

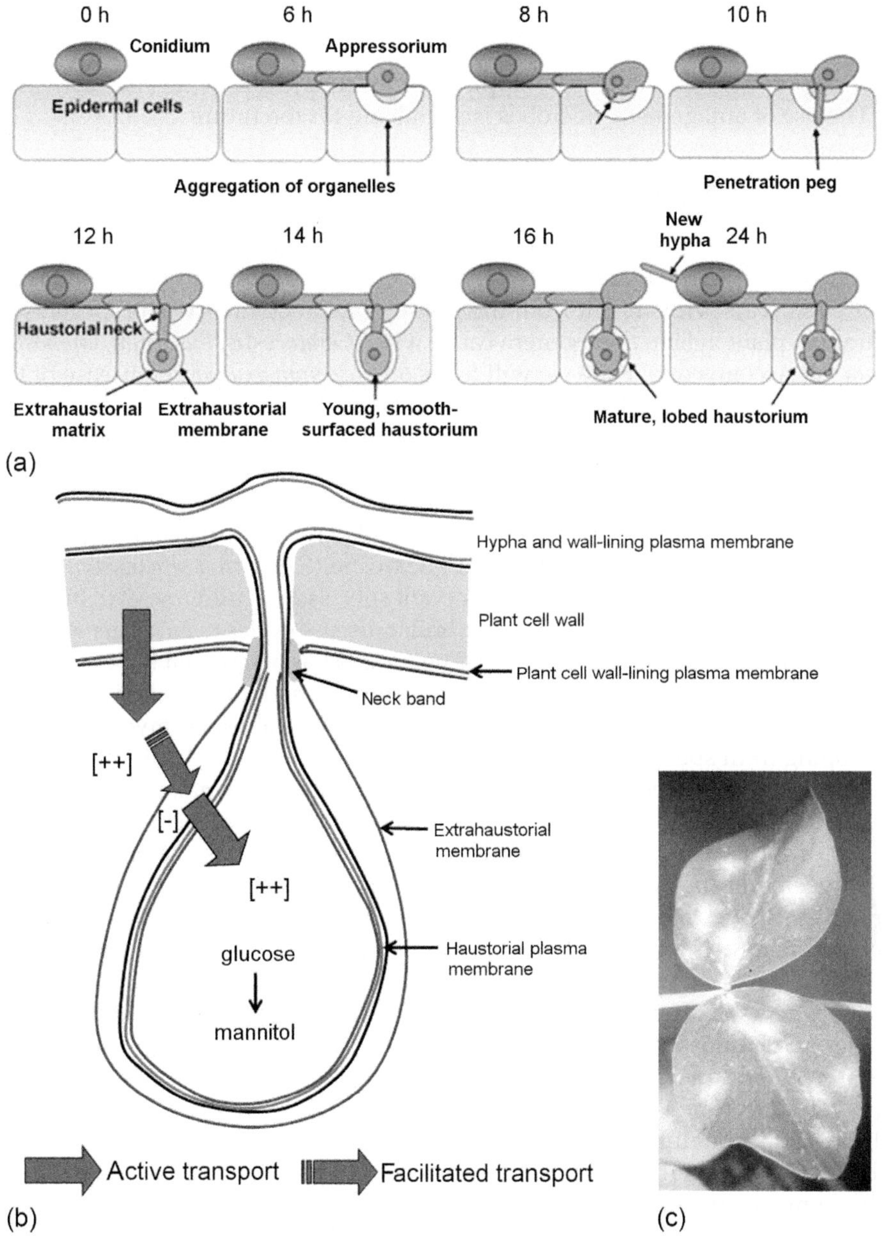

FIGURE 8.9 Powdery mildew. (a) The infection processes is illustrated diagrammatically for colonisation of susceptible *Arabidopsis* by *Erisyphe cichoracearum.* Within about 6 h following inoculation, conidia germinate and form an appressorium at the tip of the germ tube. The host cell (whether susceptible or not), and even the adjacent one if the appressorium is close to it, responds by accumulation of numerous organelles and vesicles in the cytoplasm beneath the appressorium (stage 1). They are probably responsible for production of papillae–callose-rich deposits in the plant cell wall, which is a defence response. During the next 6–10 h (stage 2) a penetration peg develops beneath the appressorium and penetrates the plant epidermal cell. The hastorium then begins to develop (stage 3; 10–14 h) as a swollen, smooth-surfaced, elongated sac at the tip of the penetration peg, the nucleus migrates from the appressorium into the haustorium, and a septum separates them. By 24 h (stage 4) the lobed haustorium, with its extrahaustorial membrane and matrix, is fully formed. (b) Uptake of sugars from the host plant cell, illustrated here for powdery mildew of pea, is via the haustorium. Glucose from the host is converted to mannitol in the haustorium, and then moves out to allow further growth of mycelium. (c) Powdery mildew on pea (Pisum sativum). *Source: (a) Modified from Koh et al. (2005); (b, c) © Peter Spencer-Phillips.*

Control of powdery mildews is by application of fungicides or defence activating compounds. The use of antagonistic microbes is promising for the future.

Vascular Wilts: *Fusarium oxysporum*

Vascular wilts are widespread, extremely destructive diseases which can cause rapid wilting and death of entire plants within a few weeks, though death of perennials can take months or even years. Wilts result from the presence and activities of the specialised necrotrophic fungi in plant xylem. Four genera cause wilts: *Ceratocystis*, *Fusarium*, *Ophiostoma*, and *Verticillium* (all Ascomycota). Here we will focus on *Fusarium oxysporum* disease of tomatoes, and then Dutch elm disease (*Ophiostoma ulmi* and *Ophiostoma novo-ulmi*).

Fusarium oxysporum is common in soil and rhizosphere communities throughout the world. All strains are able to grow saprotrophically and survive in dead organic matter. Many strains can penetrate roots but not all cause disease. Those which do cause disease can result in severe losses of most vegetables, flowers and several field crops, e.g. banana (*Musa* spp.), coffee (*Coffea* spp.), plantain (*Musa paradisiaca*) and sugarcane (*Saccharum officinarum*). Pathogenic fusaria have a high level of host specificity for plant species and cultivars, with over 120 *formae speciales* and races. Most *Fusarium* wilts have similar disease cycles to *Fusarium oxysporum* f.sp. *lycopersici* on tomato (*Solanum lycopersicum*). The fungus has an asexual life cycle. It produces three types of asexual spores: **microconidia**, **macroconidia**, and thick-walled **chlamydospores** on older mycelium or within macroconidia. Hyphae from established mycelia, and the germ tube developing from spores, perceive signals from root exudates. The hyphae secrete a battery of cell wall-degrading enzymes, including pectate lyases, polygalacturonases, proteases, and xylanases, which assist entry into roots through wounds, at branching points or directly through root tips. It has multiple mechanisms for overcoming host defence. The mycelium spreads between root cortex cells until it reaches xylem vessels, which it enters through pits. As described earlier (p. 260), from there the fungus travels through the plant, mostly upwards by microconidia in the sapstream. Later spread into adjacent vessels can occur by penetrating pits. The xylem becomes clogged by mycelium and spores, and by plant-produced gels, gums and tyloses. Water transport to the leaves fails, and the plant wilts and dies. The fungus then invades all plant tissues and obtains nutrition by decomposing them. When the fungus reaches the plant surface it sporulates profusely. Control of the disease is difficult, because the fungus can survive as a saprotroph for a long time in the absence of the host. Greenhouse soil can be sterilised, but this is impracticable in the field. Use of resistant varieties is the main approach, combined with fungicides. Biological control is a future possibility, with promising results from use of antagonistic bacteria (e.g. *Burkholderia cepacia* strains) and fungi (e.g. *Gliocladium* spp., *Trichoderma* spp.) and non-pathogenic *Fusarium oxysporum* strains.

Vascular Wilts: Dutch elm Disease

Dutch elm disease (DED) is a devastating wilt disease of elm (*Ulmus*) trees. In the last century there were two extremely destructive pandemics of DED, which spread across Europe and North America (Figure 8.10a and b). The first, caused by *Ophiostoma ulmi* (Ascomycota), started in about 1910 and had died down by the 1940s after killing 10–40% of elms. The second epidemic, which appeared around the 1940s was caused by

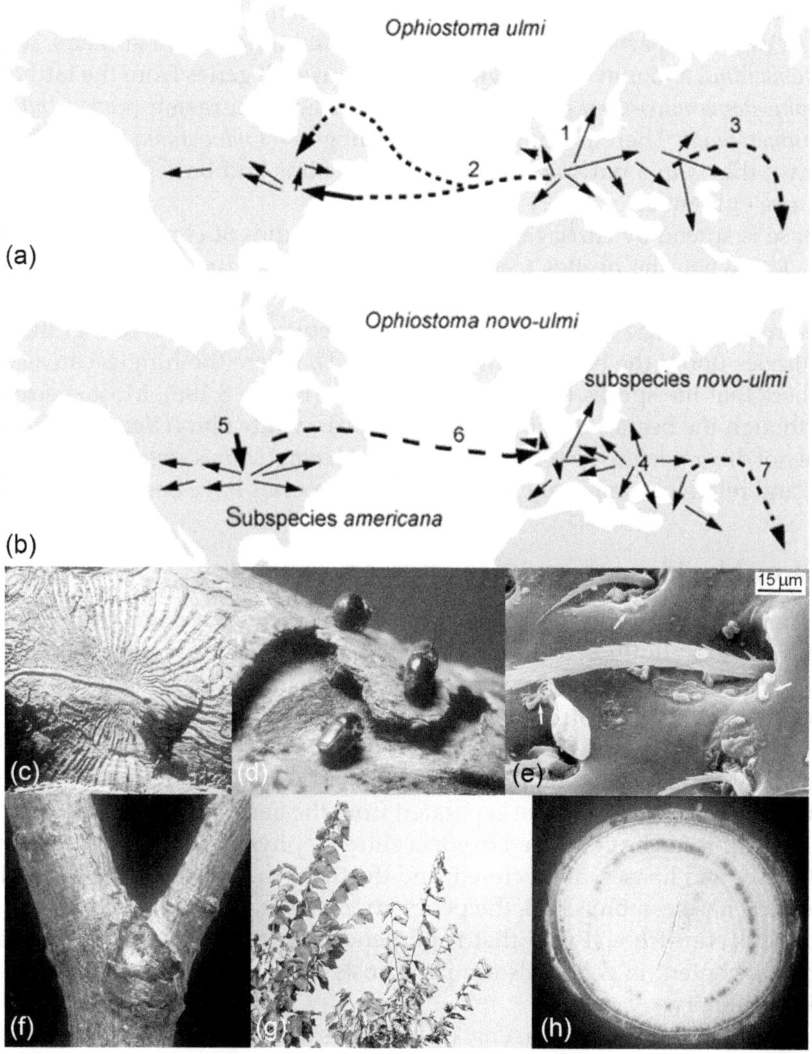

FIGURE 8.10 Dutch elm disease. (a) The first outbreak of *Ophiostoma ulmi* occurred in northwest Europe (1), from where it spread westwards to the UK and North America (2), and eastwards into central Asia (3). (b) The *Ophiostoma novo-ulmi* epidemics began simultaneously in (5) the southern Great Lakes region of North America (subspecies *americana*), from where it spread across the continent and to Europe in the 1960s (6), and in (4) Moldova-Romania (subspecies *novo-ulmi*) from where it spread westward across Europe, and eastward into the Tashkent area (7) in the 1970s. Where the distribution of subspecies *americana* and subspecies *novo-ulmi* overlaps, they are hybridising, and a new form may emerge in the future. The solid arrows indicate natural migrations from likely sites of initial introduction. The dashed arrows, show subsequent spread following additional importation events. Natural vectoring of spores is on the bodies of adult bark beetles while in the galleries that the larvae create beneath elm bark (c and d). Conidia of the pathogen are carried to new host trees directly on the exoskeleton surface of the beetles, not in a protective mycangial cavity (p. 333), seen here (arrowed) lodged in setal pits of *Scolytus scolytus* (e). Unsurprisingly, after flight, often less than half of the beetles still carry a viable spore load. The beetles feed in twig crotches (f), providing a route of entry for the fungus into the tree. The pathogen causes trees to wilt, typically with yellowing of the leaves (g). Cross sections of diseased branches have characteristic dark spots (h). *Source: (a, b) From Brasier and Buck (2001); (c, d, f, g, h) © Forestry Commission Picture Library; (e) courtesy of Joan Webber.*

Ophiostoma novo-ulmi, a much more aggressive pathogen. These two species evolved independently in different parts of the world. As *Ophiostoma novo-ulmi* spreads, it is replacing the *Ophiostoma ulmi*, and may well have picked up 'useful' genes from the latter during this process. *Ophiostoma novo-ulmi* exists as two subspecies – Eurasian (*novo-ulmi*) and North American (*americana*). There is another DED pathogen – *Ophiostoma himal-ulmi* – found in the Himalaya; this is in a natural balance with the elms and bark beetles in the area, and does not cause epidemics.

The disease is spread by carriage of spores on the bodies of elm bark beetles. The spores enter the xylem when the beetles feed in twig crotches (Figure 8.10f). The vessels become blocked by gums and tyloses and fungal material (Figure 8.10h), and foliage wilts, ultimately resulting in tree death. The fungus can also spread between trees via root grafts. The beetles breed in galleries under the bark (Figure 8.10c and d), where the fungus can also grow, and it is from there that the spores lodge on their bodies (Figure 8.10e). Most species of elm are affected, although the beetles do not favour European white elm (*Ulmus laevis*). Though the disease will not die out in the near future, some Wych elm (*Ulmus glabra*) in Scotland has been unaffected, and relatively resistant cultivars of other species have been bred.

Botrytis Grey Mould

Botrytis cinerea (Ascomycota) infects over 200 plant species, causing grey mould, evident on the surface as grey fluffy mycelium. Worldwide, it causes annual losses of $10 billion to $100 billion. It is able to counteract a broad range of plant defence chemicals. It is one of the most extensively studied necrotrophic plant pathogens.

Botrytis cinerea produces vast quantities of asexual spores which, when they land on a plant surface, germinate and form an appressorium and penetration peg that breaches the plant cuticle. Since the appressorium is not separated from the germ tube by a septum it is unlikely that sufficient turgor can be generated to effect entry by physical pressure alone. Enzymes, including cutinases and lipases, are secreted, and the tip of the penetration peg produces H_2O_2. When the cuticle has been breached, the penetration peg reaches an epidermal cell and often grows into the pectin-rich cell wall that is perpendicular to the plant surface. Plant species with low pectin content in cell walls are poor hosts for *Botrytis cinerea*, which has effective pectinolytic machinery.

Botrytis cinerea produces a wide arsenal of chemicals that cause host death, including a spectrum of low-molecular weight metabolites (e.g. botrydial, oxalic acid, and HSTs). During cuticle penetration and formation of primary lesions, *Botrytis cinerea* triggers an oxidative burst from the plant, accumulation of free radicals and hypersensitive cell death in the plant cells. While this confers resistance against biotrophic pathogens (p. 252), programmed plant cell death is beneficial to necrotrophs, including *Botrytis cinerea*, since they feed on dead cells. The fungus is also able to suppress host immunity by producing small RNA (sRNA) molecules which cause gene silencing (p. 134). As well as the aforementioned pectinases, *Botrytis cinerea* produces cellulases and hemicellulases to decompose plant cell walls to obtain nutrition.

Grey mould can be partially controlled in the field by combinations of fungicides. Biological control of grey mould on flowers and fruits using antagonistic microbes has future potential. Removal of infected plant material and reduction of humidity during storage of fruits and bulbs, and reduction of humidity in glasshouses are important control measures.

Annosum Root and Butt Rot

Root and butt rot caused by *Herobasidion annosum* (Basidiomycota) and other *Heterobasidion* species (Figure 8.11) are one of the most economically important diseases of conifers in the northern hemisphere, especially Europe, causing annual losses of over US$1030 million. The fungus was, for a long time, thought to be a single species – *Heterobasidion annosum*. Mating experiments, however, revealed host-specialised intersterile groups, with three species in Europe, two in North America and others elsewhere, though not necessarily pathogenic (Table 8.7).

FIGURE 8.11 Annosum root and butt rot is caused by several species of *Heterobasidion* (Table 8.7). (a) Basidiospores infect freshly felled stumps at thinning and final harvest (1 and 3). Mycelium colonises the stumps and spreads into the woody roots from where it spreads as mycelium to roots of mature trees (1) and seedlings (4) by grafts and root contact. Spread from colonised trees (2) can occur in the same ways. (b) Jeffrey pine at the edge of a large and still expanding disease centre caused by *Heterobasidion irregulare* in the Sierra Nevada. (c) Fruit body of *Heterobasidion annosum*. (d) Severe white pocket rot decay in sitka spruce (*Picea sitchensis*) caused by *Heterobasidion annosum*. (e) Felled Norway spruce (*Picea abies*) showing heart rot due to *Heterobasidion annosum*. *Source: (a, b) Modified from Asiegbu et al. (2005); (c) © Alan Outen; (b, d, e) © Stephen Woodward.* (See the colour plate.)

TABLE 8.7 Species of *Heterobasidion* and Their Hosts

Species	Inter-sterility group	Host	Location
H. annosum	P	Scots pine (*Pinus sylvestris*), but also other conifers and broadleaved trees	Europe up to 66°N
H. abietinum	F	Silver fir (*Abies alba*), but also other hosts	Europe, where *A. Alba* grows naturally
H. parviporum	S	Relatively strictly on Norway spruce (*Picea abies*)	Europe, where *P. abies* grows naturally. In China two separate populations exist
H. irregulare	P	*Pinus*	North America, in eastern (Quebec to Florida) and western (British Columbia to Mexico) forests, but less common in central parts
H. occidentale	S	*Abies, Tsuga, Picea, Pseudotsuga,* and *Sequoiadendron*	Western North America, from Alaska to California
H. insulare			Eastern and southern Asia, from Russia and Japan in the north to Borneo and New Guinea in the south, and India and Nepal in the west
H araucariae		A saprotroph on dead ewood of *Auracaria, Cunninghamia,* and *Pinus*	Australia, New Zealand, New Guinea, and the Fiji islands

Basidiospores infect freshly felled stump surfaces or stem and root wounds, between 5 and 35 °C. Conidia are also produced but their role is unclear. Basidiospores germinate and colonise stumps including the roots. The mycelium produced can live in stumps for a long time without causing disease in a living tree. The fungus spreads from stumps to healthy trees by mycelial growth through root grafts or contacts (Figure 8.11). Only occasionally does colonisation occur through fine roots. The fungus spreads necrotrophically in the sapwood in living trees, but later grows in the heartwood of most tree species despite the presence of fungistatic compounds. It produces a wide range of wood decay enzymes and several toxins: fommanoxin, fommanosin, fommanoxin acid, oosponol, and oospoglycol. Several morphological and chemical responses are activated in the host at late stages of infection, including the production of a range of phenolic compounds which may damage fungal membranes, lignification which prevents diffusion of toxins and enzymes, suberisation and papillae formation also occurs, and volatile and non-volatile terpenes are produced, and resins are mechanical barriers. Depending on the time of year, moisture content of the sapwood, and the age and vitality of the tree, *Heterobasidion annosum* can extend at rates of up to 1 to 2 m year^{-1} in stems and roots, respectively. It is not known how long an individual *Heterobasidion* genet (clone) can remain alive on the same site, but they are capable of occupying disease centres of 50 m diameter which would probably be over 100 years old; individuals have been found in stumps 62 years after felling, and in root systems of diseased trees for several decades.

Strategies for control of annosum root rot include use of chemicals, biological control agents, and silvicultural practices. Chemical and biological control methods involve applications to the stumps after felling. Urea and borates are chemicals used commercially for control, and the saprotrophic *Phlebiopsis gigantea*, is available in several commercial formulations for biocontrol. Planting species with low susceptibility can reduce root rot problems, and disease in mixed stands is less than in monocultures. Delaying thinning until trees are older can reduce the problem, as can thinning at times when basidiospores are not being dispersed.

Oomycete Diseases: Potato Blight, Downy Mildews, Pythium Damping-Off, and Rot

The oomycetes, though not fungi (p. 35), operate in many similar ways, cause a range of diseases of plants (Table 8.2), and are studied by mycologists. Pathogens include many *Phytophthora* species (Peronosporales), the downy mildews (Peronosporales) and *Pythium* species (Pythiales) (Table 8.2), mentioned below. Late blight, caused by *Phytophthora infestans*, is the most devastating disease of potatoes worldwide, especially in regions which often experience cool, damp weather. *Phytophthora infestans* is a specialised necrotroph, and also causes major problems with other members of the Solonaceae (e.g. tomato). It kills stems and foliage at any time during the growing season, and can kill whole fields of plants in less than 2 weeks under optimal cool, wet conditions. Potato tubers and tomato fruits are also attacked, rotting in the field or during storage.

In the past, the pathogen over-wintered solely as mycelium within infected potato tubers, growing through aerial parts producing sporangiophores that project through stomata. The sporangia are released into the air or can be dispersed by rain. At 12–15 °C, sporangia germinate releasing three to eight motile zoospores, but above 15 °C sporangia can germinate directly to form a germ tube. Until the 1980s, only the *A1* mating type was present in most areas of the world, and in the absence of a compatible mating type sexual reproduction could not occur (pp. 114–115). The compatible mating type *A2* has now spread from Mexico to the rest of the world, allowing sexual reproduction resulting in the formation of resistant oospores in infected tissues both above and below ground. Oospores can over winter and survive for 3 or 4 years in soil, and recombination allows emergence of more virulent strains.

On wet leaves or stems, spores germinate and the germ tube penetrates directly through the epidermis or enters through a stoma. The hyphae grow extensively between cells and penetrate cells forming long, curled haustoria. Infected cells die, but the mycelium continues to spread into living tissue, lesions enlarge and new ones develop, foliage is killed and the tuber yield considerably reduced. The pathogen is multicyclic with many asexual generations each growing season; in optimal conditions sporangia can form within 4 days from initial infection. In wet weather, when sporangia are washed from leaves into soil, the second phase of the disease develops. Zoospores emerge and enter the tubers via wounds and lenticels. Mycelium develops mostly intercellularly, haustoria again being formed within cells.

The development of epidemics depends very much on climatic conditions. Optimal conditions for production of sporangia are close to 100% humidity with temperatures between 15 and 25 °C. At over 30 °C growth ceases, though the oomycete is not killed and it can sporulate again when conditions become favourable. The disease is controlled by sanitary measures (destroying infected material and planting with disease-free tubers), planting with resistant varieties (though each variety is only resistant to some races of *Phytophthora infestans*), and appropriately timed application of chemical fungicides. The arrival of the A2 mating type, with consequent formation of resistant oospores, and the emergence of new pathogenic races (see Chapter 4) may considerably change man's ability to control this disease. One promising approach with tomatoes is the induction of systemic-acquired resistance (SAR) by infecting with the tobacco necrosis virus or application of DL-3-amino-butyric acid.

Downy mildews, another type of oomycete disease, are all caused by obligate biotrophic pathogens. For example, *Bremia lactucae* is the most important cause of disease of lettuce (*Lactuca*) worldwide; *Hyaloperonospora parasitica* causes downy mildew of *Arabidopsis*, and though not hugely destructive nor of economic significance, it has been used extensively as a model organism in molecular studies. *Hyaloperonospora parasitica* downy mildew of *Arabidopsis*, like other downy mildews of crucifers, typically occurs in cool (10–15 °C), moist conditions. The life-cycle is relatively simple (Figure 8.12); a conidium germinates and forms an appressorium either directly or on a short germ tube, within about 6 h after landing on a leaf. A penetration hypha forms beneath the appressorium and penetrates the leaf where two epidermal cells meet, or occasionally through a stoma. As the coenocytic penetration hypha grows between cells, haustoria are often inserted into adjacent epidermal and then mesophyll cells. In compatible interactions there is minimal macroscopic disruption to the host until sporulation, when the coniodiophores protrude from stomata as a downy growth (hence the disease name).

Incompatible host/pathogen combinations can result in various resistance phenotypes. Usually there is an oxidative burst and a salicylic acid (SA) dependent hypersensitive response (HR) in the epidermal cells and a few adjacent mesophyll cells, with a shift from housekeeping to defence metabolism, and about a 10% change in the transcriptome. HR involves at least one phytoalexin – camalexin. Systemic-acquired resistance (SAR) is also induced, and is associated with increased SA and systemic accumulation of PR-proteins.

A third major type of oomycete disease is caused by *Pythium* species. They are necrotrophic pathogens, causing damping-off diseases of seedlings, and seed, root and fruit soft-rot worldwide, though the species responsible vary according to abiotic environment. *Pythium ultimum*, for example, is common in soils of cool regions, while *Pythium phanidermatum* and *Pythium irregulare* are common where soil temperatures are higher. Species of *Fusarium* (Ascomycota), *Rhizoctonia* (Basidiomycota), and many *Phytopthora* species (oomycete) also cause similar damping-off diseases. The diseases kill young and over-mature plants and tissues, but mature plants are rarely killed, though lesions can develop on stems (at the soil line) and roots; roots rot and plant growth and yield can be severely reduced. In infected soils, seeds can fail to germinate, and seedlings can be attacked before or after emergence; invaded tissues become water-soaked, discoloured and soon collapse, the fungus-like organism continuing to colonise the fallen seedling. Broadleaf plants and Gramineae are especially susceptible. The disease is spread in infected plant material and via soil water.

Pythium survives in soil as thick-walled, sexual oospores and asexual sporangia. At 10–18 °C, germination of both oospores and sporangia tends to be by means of zoospores, whereas above 18 °C germ tubes tend to be produced. Germination, hyphal growth and tissue penetration can be induced by plant exudates. *Pythium* penetrates the plant directly by physical force and enzymic activity. Pectinases break down pectins in the middle lamella causing cells to part and tissues to break up. Cellulases result in plant cell wall disintergration. The oomycete is unable to advance into lignified tissue.

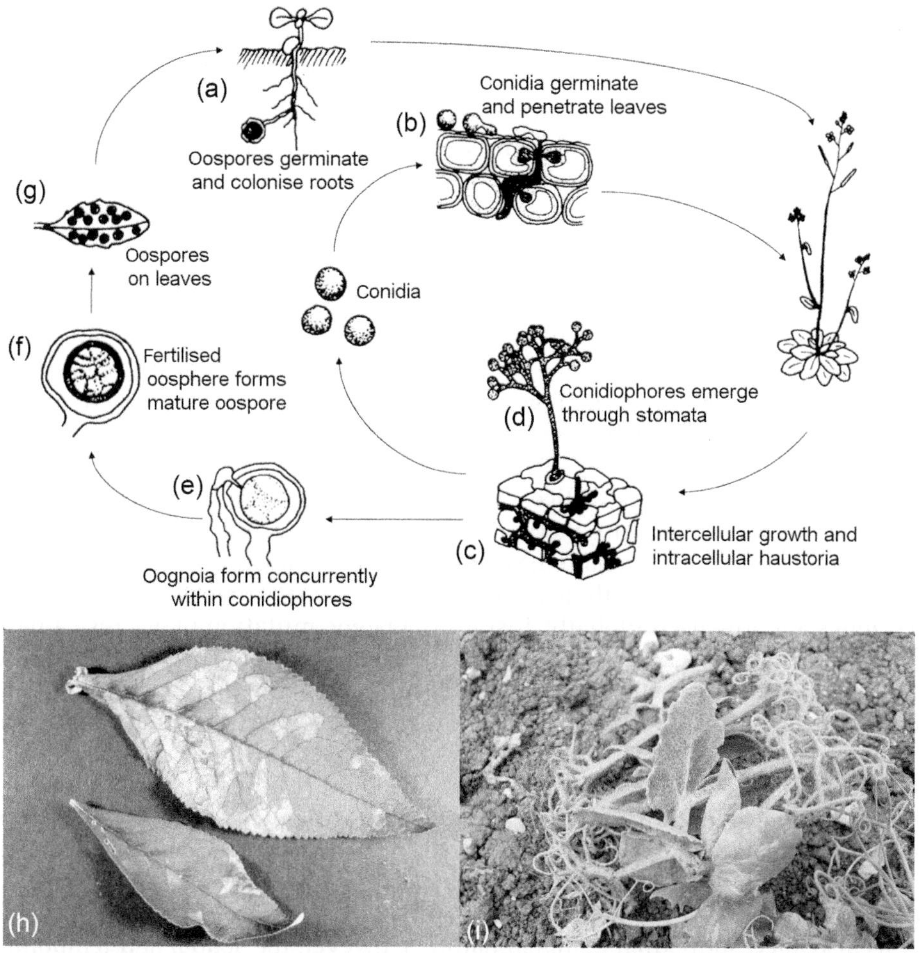

FIGURE 8.12 Downy mildews. (a–g) The life cycle of *Hyaloperonospora parasitica* causing downy mildew on *Arabidopsis thaliana*. Plants can be colonised from oospores that germinate in soil (a) and by conidia on leaves that germinate and penetrate between epidermal cells (b). (c) Coenocytic mycelium grows between cells and swells to fit spaces. Pear-shaped haustoria penetrate cells and absorb nutrients. After 1–2 weeks conidiophores, bearing asexual conidia, protrude through stomata (d) and at the same time oospores are formed (e–g). Oogonia (female sex organs) contain an oosphere which is fertilised via a fertilisation tube that grows from the male antheridium through the oosphere wall. The fertilised oosphere matures into a thick-walled oospore (f). Different components of the lifecycle are drawn at different scales (a–g). (h) *Buddleja globosa* leaves, showing characteristic downy mildew (*Peronospora hariotii*) lesions where hyphae are restricted to colonising islands delimited by larger veins. (i) Pea (*Pisum sativum*) leaves and tendrils with *Peronospora viciae* infection. *Source: (a–g) Modified from: Slusarenko and Schlaich (2003); (h, i) © Peter Spencer-Phillips.*

Plant varieties resistant to *Pythium* are virtually unknown, and control is by good horticultural practice, including sanitation, drainage, and shallow planting; soil in glasshouses can be sterilised. Variable success has been obtained with biological control agents, using bacteria, including species of *Bacillus*, *Burkholderia*, and *Pseudomonas*, fungi including *Gliocladium*, *Trichoderma*, and non-pathogenic *Fusarium oxysporum*, and non-pathogenic *Pythium*.

DEVELOPMENT OF DISEASE IN NATURAL ECOSYSTEMS AND CROPS

Domestication of crop plants started 10,000–12,000 years BP in the Fertile Crescent in the Middle East, and is the origin of many of our currently most important crop species. The development of new agricultural practices and plant breeding resulted in significant changes to plant populations, and the simultaneous emergence of new pathogens. Genetic mechanisms behind evolution of new fungal species are reviewed in Chapter 4, and specific examples of evolutionary mechanisms resulting in emergence of new strains/species of plant pathogenic fungi are presented in the following section. In current, modern agroecosystems, populations of fungal pathogens are frequently challenged by new plant genes for resistance, fungicides, and a range of management practices aimed at reducing infection of crop plants. Genotypes that allow fungi to overcome these will increase rapidly and spread through the population. Spread from one plant to another is often rapid as plants are grown at very high density.

Man's modern cultivation practices leave us wide open to crop loss on a massive scale. Most crops have limited genetic diversity and are grown in vast monocultures. Since the nineteenth century plants have been bred for desired characteristics such as large yield and resistance to pathogens. Germplasm is shared worldwide, so several resistance genes are now used globally. Just a single gene mutation may cause a pathogen to become virulent, and spread worldwide is then just a matter of time. In the tropics, for example, banana (*Musa* spp.) and coffee (*Coffea* spp.) are planted as single clones and are, respectively, prone to black Sigatoka disease (Figure 8.4 inset) caused by *Mycosphaerella fijiensis*, and leaf rust. When a pathogen infects a crop with little or no genetic diversity, spread throughout the crop is often devastating. Monocultures inevitably lead to high disease incidence because of high rates of host–pathogen encounter, and disease is positively density dependent. Competition can stress plants and reduce their resistance to infection.

Though the majority of research effort has focussed on crops of economic importance, pathogens also affect individual plants in nature. However, the spatial and temporal scales at which plant–pathogen interactions occur is very different, because of the spatially and genetically heterogeneous nature of natural populations. Fungal pathogens cause death and reduced fitness of individual plants (Table 8.8), which can result in declines of populations of some host species and shifts in plant community composition. These effects can help maintain plant species diversity and genetic diversity, and affect plant community succession. Plant population dynamics are affected directly as a result of pathogen effects on survival, growth, and fecundity. Plants killed by pathogens before reproduction occurs do not contribute to the next generation, and since disease affects growth, this influences fecundity, and decreases contribution to the next generation. Competition between plants can be affected by differential susceptibility to pathogens. For example, infection of groundsel (*Senecio vulgaris*) with the rust *Puccinia lagenophorae* reduces its growth and competitiveness against lettuce (*Lactuca sativa*) and petty spurge (*Euphorbia peplus*). Succession of plant communities on sand dunes in Europe is influenced by pathogens; Marram grass (*Ammophila arenaria*), which dominates wind-blown coastal foredunes, is debilitated by pathogenic soil fungi and nematodes, and is replaced by the resistant fescue grass (*Festuca rubra*), which dominates

TABLE 8.8 Examples of Direct Effects of Fungal Diseases on Individual Plants in Nature

	Effects on		
Plant stage	**Survival**	**Growth**	**Fecundity**
Seed decay	Rates of disease related death are high, e.g. in the tropics ranging from 10% in the invasive *Mimosa pigra* in Australia to 47% in pioneer trees in Panama. In Wyoming shrub-steppe, ranging from 0–90% mortality of 5 plant species	NA	NA
Seedling diseases	Rates of death due to damping-off disease are high. In Barro Colorado Island, Panama, it was the primary cause of death of 80% of plant species tested killing 74% of a parent tree's seedlings	?	NA
Foliar diseases	Plants are sometimes killed, especially if seedlings	Foliar diseases reduce leaf area, hence photosynthetic activity and growth. Small plants are competitively disadvantaged. In Mexican tropical rain forest, mean leaf area damage was <1% and always <20%. However, in Costa Rica, growth of the tree *Erythrochiton gymnanthus*, infected by the petiole pathogen *Phylloporia chrysita*, was reduced by 52%	Reproduction can be reduced because of reduced growth
Systemic infections	Some fungi and oomycetes can have major effects. The systemic smut *Urocystis trientalis* caused 50% mortality of *Trientalis europaea* (Primulaceae). The oomycetes *Albugo candida* and *Peronospora parasitica* caused death of up to 88% of shepherd's purse, *Capsella bursa-pastoris*, seedlings	?	?
Cankers, wilts, and dieback	There have been several widespread, dramatic epidemics resulting in rapid death of trees when cankers girdle stems or block vascular transport causing wilts. These include: *Ophiostoma ulmi* and *O. novo-ulmi* on elms (p. 274); chestnut blight caused by *Cryphonectria parasitica*; sudden oak death caused by the oomycete *Phytophthora ramorum*	With some canker diseases, if cankers remain small and localised, then death may not ensue, but growth will be impaired	?

(*Continued*)

TABLE 8.8 Examples of Direct Effects of Fungal Diseases on Individual Plants in Naturevcont'd

	Effects on		
Plant stage	**Survival**	**Growth**	**Fecundity**
Root diseases and butt rots	In native North American conifer forests, the basidiomycetes *Heterobasidion annosum* (pp. 276–277) and *Phellinus weirii* cause high mortality. Death of large dominant trees alters forest structure	Root rots do not always cause death of whole trees, but growth and reproduction can be dramtically reduced. The basidiomycete *Armillaria ostoyae* reduced Douglas fir trunk radial increase by up to 60%. *H. annosum* reduced *Pinus taeda* trunk radial increase by 36%	?
Floral infections	NA	NA	Attack of flowers and developing fruits can considerably reduce fecundity. *Exobasidium vaccinii* caused a 50% reduction in flowers of *Rhododendron calendulaceum* in the Appalachian Mountains in the eastern United States. Anther smut of *Silene* spp., vectored by pollinators, and caused by *Microbotryum violaceum*, replaces stamens and staminoids with spore-bearing structures, and hence has a major effect on plant reproductive capacity

NA – not applicable; ? – effects are likely but examples are not available.
Information from Gilbert (2002).

stabilised dunes. The root rotter of trees, *Phellinus weirii*, drives forest structure and succession in conifer forests of the Western USA; *Phellinus weirii* removes overstory trees of the extremely susceptible Douglas fir (*Pseutodtsuga menziesii*) and mountain hemlock (*Tsuga mertensiana*), resistant plants taking their place.

DISEASES OF OTHER AUTOTROPHS: LICHENS AND SEAWEEDS

Fungi are not only pathogenic to plants, but also to other photoautotrophs, including algae and lichens (Chapter 7). Some fungi – termed lichenicolous fungi – live exclusively on lichens. Though some are saprotrophs, most are either host-specific or broad-spectrum pathogens.

There are over 1800 described species currently, 95% of which are Ascomycota (in 19 orders) and 5% Basidiomycota (in 8 orders). *Athelia arachnoidea* is an extremely common, widespread perennial, destructive lichenicolous basidiomycete of numerous lichen-forming Ascomycota and their photobionts. Little is known about the modes of pathogenesis/parasitism, virulence or nutrient exchange. There are also lichenicolous lichens that colonise another lichen host. For example, *Fulgensia bracteata* colonises the lobulate *Toninia caeruleonigricans*. The infection process begins with ascospores of the *Fulgensia bracteata* fungus; small scales form and eventually its yellow thallus covers the grey thallus of *Toninia caeruleonigricans*.

Sometimes, the lichenicolous fungus or lichen acquires photobionts from the host and becomes a lichen with that photobiont species. For example, the mycobiont that eventually forms the lichen *Diploschistes muscorum*, is initially lichenicolous, parasitising squamules of *Cladonia* species. Eventually, it acquires photobiont cells from the host and forms the independent lichen *Diploschistes muscorum*.

Aquatic autotrophs are also affected by fungus and fungus-like pathogens. There are three main categories of association with algae: (1) biotrophic with few macroscopic symptoms; (2) biotrophic with severe disease symptoms in which host organelles are destroyed and the entire cell is occupied by the fungus; and (3) necrotrophic on a partially senescent host with further tissue destruction. Examples include chytrids on green algae (Chlorophyta) (e.g. *Olpidium* spp. on *Cladophora*). The chloroplasts turn brown and surround the sporangium. *Lindra* spp. (Ascomycota) grow on air vesicles of *Sargassum*, turning them into soft, dark brown, wrinkled structures giving the name 'raisin disease'. *Phycomelaina laminariae* is common on brown algae, causing 'stipe blotch' of *Laminaria*. The pathogen forms black oblong or circular patches on the host's stipe, but although infected areas are severely damaged the host is not killed.

EMERGING DISEASES AND THE BIOSECURITY THREAT

As plant communities evolve in different parts of the world, pathogenic fungi evolve in association with them locally. They often cause little damage in the regions where they co-evolved, being in natural balance. However, when they arrive in different parts of the world where native plants have little resistance and/or their natural enemies are absent and unable to control them, damaging disease episodes can occur. Many new diseases are emerging (e.g. diseases caused by *Puccinia graminis* f.sp. *tritici*, *Magnaporthe oryzae* and *Phytophthora* species,Figure 8.13). Man has been responsible for many of these emerging infectious diseases (EIDs) by moving plant material, or soil harbouring the fungus, around the world (Table 8.9). Potato late blight (pp. 277–279), for example, emerged when *Phytopththora infestans*, that co-evolved with wild potato (*Solanum tuberosum*) in the Andes, was transported to Mexico, and then to Europe in the mid-nineteenth century. Movement by man or natural agents is, however, not the only cause of EIDs. New infectious diseases emerge when pathogens have:

(1) increased in incidence, host range or geographical range;
(2) altered pathogenesis;
(3) newly evolved;
(4) been newly discovered/recognised.

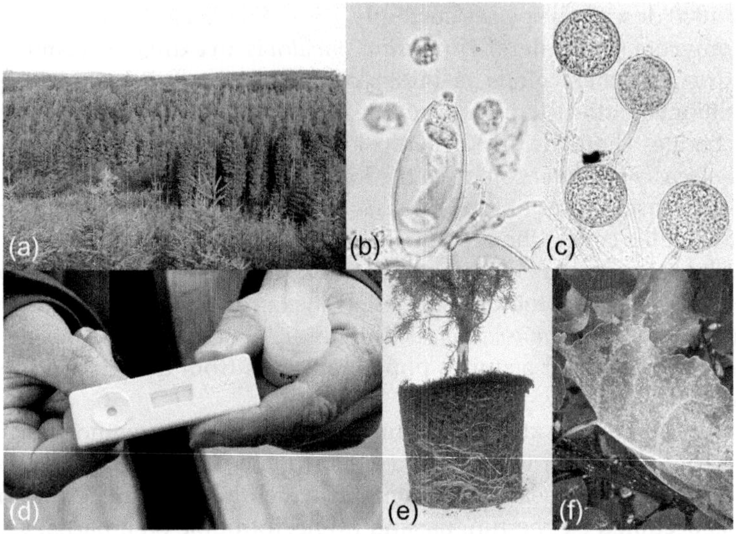

FIGURE 8.13 Emerging diseases. Many new *Phytophthora* (oomycete) species are emerging to cause disease. *Phytophthora ramorum* kills oak (sudden oak death in California, USA) and infects many other tree species, significantly rhododendron, causing non-fatal foliar dieback, which can be a source of infection for other plant species. Spores are spread by rainwater and wind. *Phytophthora ramorum* is also now killing larch (*Larix*) over large areas in the UK. (a) *Phytophthora ramorum* affecting larch in south west England. (b) *Phytophthora ramorum* sporangium with zoospore emerging. (c) *Phytophthora ramorum* chlamydospores on a leaf. (d) Test kit for *Phytophthora ramorum*. (e) Root and stem base killing of container grown *Juniperus communis suecica* by *Phytophthora* sp. (f) *Populus* rust caused by *Melampsora* sp. Uredosori and orange spore masses can be seen on the underside of a leaf. *Source: All images © Forestry Commission Picture Library.*

Some crop diseases might now be considered persistent, having caused man problems for hundreds or even thousands of years, but new races of these emerge which may cause major epidemics (e.g. *Puccinia graminis* f.sp. *tritici*; Table 8.10). The large amount of repetitive DNA and the high frequency of single-nucleotide polymorphisms (SNPs) in the genomes of some major crop pathogens, such as *Puccinia graminis* and *Phytophthora infestans* (Table 8.10), indicate that more virulent strains and resistance to fungicides are likely to evolve. A recent study of a few major ascomycete plant pathogens revealed that evolution of pathogenicity in this group is not only ancient but is both rapid and on-going. Fungal pathogens have undergone and continue to undergo extensive 'genomic tillage' where chromosome duplications, genomic rearrangements, horizontal gene acquisitions (pp. 134, 201), and large scale gene evolution caused by proliferation of transposons (pp. 132–134) and repeat-induced point mutations (p. 134), drive the evolution and expression of 'effectors' (see Chapter 4 for a discussion of these genetic mechanisms). In the dynamic 'evolutionary arms race' that exists between plant and pathogen, these mechanisms allow the pathogen to evolve novel effectors which exploit plant metabolism and evade host defence mechanisms allowing pathogens to take the upper hand and successfully infect the host plant.

Rapid evolution of new pathogens and adaptation to new environments and new hosts can arise as a result of hybridisation between resident pathogens, horizontal gene transfer (as seen with HSTs of *Stagnospora nodorum*) and host jumps or shifts (as seen in *Magnaporthe*

TABLE 8.9 Examples of Emerging Infectious Diseases (EIDs) of Plants, Including Staple Crops, Other Crops, and Plants in Natural Ecosystems

Pathogen	Disease	Hosts	Region and host origin	Time and place of emergence	Factors driving emergence
EIDs of world staples					
Phytophthora infestans (oomycete)	Late blight (pp. 277–279)	Potato *(Solanum tuberosum)*	Mexico; wild *Solanum* spp.	Mid-nineteenth century in Europe; 1990s in North America	Repeated introductions by man into non-native regions
Magnaporthae oryzae	Rice blast (pp. 265–267)	Rice (*Oryza sativa*), barley, wheat, and other grasses	China; rice	Twentieth century in all rice producing areas; 1996 United States	Worldwide spread by seed exchange
Tilletia indica	Karnal bunt	Wheat	India; wheat	1972 Mexico; 1992 USA; 2000 South Africa	Dispersal by contaminated seed
EID of other crops					
Puccinia kuehnii	Sugarcane orange rust	Sugarcane *(Saccharum officinarum)*	Australia; sugarcane	Australia	Evolution of a new strain that broke resistance
EIDs in natural ecosystems					
Ophiostoma ulmi and *Ophiostoma novo-ulmi*	Dutch elm disease (p. 274)	Elm (*Ulmus* spp.)	Elm; various places worldwide	Twentieth century; repeated pandemics in Europe, Asia, and North America	Introductions in imported timber. Hybridisation resulting in increased virulence
Cryphonectria parasitica	Chestnut blight	Chestnut (*Castanea dentata, C. sativa*)	Japanese chestnut *Castanea crenata;* Southeast Asia	Early twentieth century; Eastern USA	Imported chestnut plants
Phytophthora ramorum (oomycete)	Sudden oak death	Many woody plant species	California bay laurel-Oregon myrtle (*Umbellularia californica*) and *Rhododendron* spp.	1990s; Europe	Imported nursery stock
Phytophthora cinnamomi (oomycete)	Jarrah dieback/ Cinnamomi root diesease	>3000 trees, woody ornamentals and herbs	Wide range of plants; South-west Pacific area	1800s–1900s; Europe, North America, Australia	Imported plants

Extracted from Anderson et al. (2004) and Brasier (2008).

oryzae), often associated with newly introduced exotic pathogens (Table 8.11). *Ophiostoma novo-ulmi* (p. 274) has acquired 'useful' major genes by hybridising with *Ophiostoma ulmi*, the resident species, as it has spread across Europe and North America. *Phytophthora alni* subsp. *alni* (oomycete) has recently emerged by hybridisation, and is killing alder (*Alnus*) across Europe. Other examples include *Botrytis allii* an onion pathogen, *Melampsora columbiana* the

TABLE 8.10 Characteristics of Three Persistent Diseases of World Staple Crops, with Emerging Races

	Disease	Host range	Epidemics since	Growth/ survival temperature (°C)	Asexual and sexual life-cycle	Spore forms	Clonal	Genome size	Gene number	Repetitive DNA	SNP[a] frequency
Magnaporthe oryze (Ascomycota)	Rice blast	50 grass species	Seventeenth century	18–24 4–35	Yes	2	Yes	41.7 Mb	12 841	9.7%	1 per 2.3 Kb
Puccinia graminis (Basidiomycota)	Wheat stem rust	365 cereal grass species; barberry, *Mahonia* spp.	690 BC	15–24 −10 to 35	Yes	5	Yes	88.6 Mb	20 567	80%?	>1 per Kb
Phytophthora infestans (oomycete)	Late blight of potato	Solonaceae	1845	8–20 −5 to 28	Yes	3	Yes	228.5 Mb	18 179	74%	1 per 426 bp

[a]*SNP, single nucleotide polymorphism.*
Compiled from Gurr et al. (2011).

poplar rust, and *Verticillium longisporum* a pathogen of crucifers. Not only does man's propensity for moving plant and soil material about bring pathogens into contact with new hosts and other closely-related pathogens with potential for hybridisation, but also plant nurseries in Europe and elsewhere are infested with many species in the same genus (e.g. *Phytophthora*), and are potential breeding grounds for more aggressive hybrid pathogens.

New pathogens can emerge by adaptation to a new, closely-related plant (e.g. wild to domesticated barley (*Hordeum vulgare*); host shift) or to one that is genetically distant (e.g. different genus; host jump) (Table 8.11). Common scenarios are from: (1) wild hosts (e.g. weeds close by); (2) planting new crops in natural ecosystems (e.g. introduction of soybean, *Glycine max*) into cleared areas of Amazonian rainforest); and (3) by transporting infected plant material into new areas with naive populations of related species with a lower level of resistance

TABLE 8.11 Examples of Evolutionary Mechanisms Resulting in Emergence of Plant Pathogens in Agro-ecosystems

Evolutionary mechanism	Plant pathogen	Time scale
Domestic/host-tracking		
Co-evolution of a pathogen with its host	*Mycosphaerella graminicola* on wheat	10–12,000 BP
	Magnaporthe oryzae on rice	7000 BP
	Phytophthora infestans on potato	7000 BP
	Ustilago maydis on maize	8000 BP
Host jump/host shift		
Adaptation allows a pathogen to shift to a new closely-related host or jump to a more distantly related host species	*Magnaporthe oryzae* from Setaria millet to rice	Abrupt evolutionary change, ~7000 BP
	Rhynchosporium secalis from wild grasses to barley and rye	Abrupt evolutionary change, ~2000 BP
	Phytophthora infestans from wild *Solanum* species to potato	Abrupt evolutionary change, <500 BP
Horizontal gene transfer		
HGT: Part of a genome is transferred to another organism (species or vegetatively incompatible individual), (e.g. by a transposon)	*ToxA* from *Phaeosphaeria nodorum* into *Pyrenophora tritici-repentis*	Abrupt evolutionary change, ~60 BP
	PEP cluster in *Nectria haematococca*	Unknown
	Host-specific toxins in *Alternaria alternate*	Unknown
Hybridisation		
Mating between two closely-related species	Hybrid of *Phytophthora cambivora* and relative of *Phytophthora fragariae* on alder (Alnus)	Abrupt evolutionary change within the last century
	Hybrids of *Ophistoma ulmi* and *Ophiostoma novo-ulmi* on elm (Ulmus) trees	Abrupt evolutionary change within the last century

From Stukenbrock and McDonald (2008).

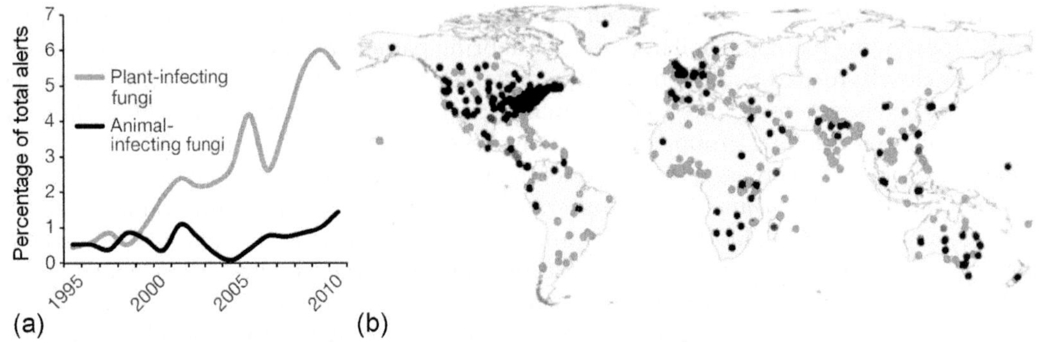

FIGURE 8.14 Increase in emerging infectious diseases (EIDs). New plant pathogenic diseases are increasing as revealed by the alerts in the ProMED (Program for Monitoring Emerging Diseases; http://www.promedmail.org) database. Alerts rose from 0.4% in 1995 to 6% in 2010. (a) Disease alerts and (b) the location of the reports. *Source: Modified from Fisher et al. (2012).*

(e.g. chestnut blight moving from imported Asian chestnuts, *Castanea crenata* to North American chestnuts, *Castanea dentata*). Over millions of years jumps can even occur between kingdoms. For example, *Claviceps purpurea* (Clavicipitaceae) is a pathogen causing ergot of rye (*Secale cereale*), but the common ancestor of the Clavicipitaceae was an animal pathogen.

EIDs are increasing (Figure 8.14). The huge impact that EIDs can have on the natural environment, and on crop losses, is illustrated by the death of 100 million elm (*Ulmus*) trees in the UK alone, due to Dutch elm disease, and 3.5 billion sweet chestnut (*Castanea sativa*) trees killed by chestnut blight in the United States. The cost in terms of human life has already been mentioned at the start of this chapter, with regard to the Irish potato famine caused by *Phytophthora infestans*. A worst case food loss scenario would be if the five main staple food crops (rice (*Oryza* spp.), wheat (*Triticum* spp.), maize (*Zea mays*), potatoes (*Solanum tuberosum*), and soya (*Glycine max*)) simultaneously succumbed to severe epidemics (of *Magnaporthe oryzae*, *Puccinia graminis*, *Ustilago maydis*, *Phytophthora infestans* and *Phakospora pachyrhizi*, respectively) leaving sufficient food for less than 40% of the world's population, though this is an unlikely scenario. The impact of EIDs goes beyond plant death: 270 megatonnes of CO_2 is predicted to be released, during 2000–2020, due to loss of western Canadian pine (*Pinus* spp.) trees due to the pathogenic association of the mountain pine beetle (*Dendroctonus ponderosae*) with blue stain fungi (*Grosmannia clavigera*).

Further Reading

General

Agrios, G.N. (Ed.), 2005. Plant Pathology, fifth ed. Acadaemic Press, California.
Lane, C.R., Beales, P.A., Hughes, K.J.D. (Eds.), 2012. Fungal Plant Pathogens. CAB International, Wallingford.
Strange, R.N. (Ed.), 2003. Introduction to Plant Pathology. John Wiley & Sons, New York.

Spectrum of Interactions of Fungi with Plants

Newton, A.C., Fitt, B.D.L., Atkins, S., Walters, D.L., Daniell, T.J., 2010. Pathogenesis, parasitism and mutualism in the trophic space of microbe-plant interactions. Trends Microbiol. 18, 365–373.
Oliver, R.P., Ipcho, S.V.S., 2004. *Arabidopsis* pathology breathes new life into the necrotrophs vs. biotrophs classification of fungal pathogens. Mol. Plant Pathol. 5, 347–352.

Susceptibility to and Defence Against Fungal Pathogens

Ballini, E., Lauter, N., Wise, R., 2013. Prospects for advancing defense to cereal rusts through genetical genomics. Front. Plant Sci. 4, 117. http://dx.doi.org/10.3389/fpls.2013.00117

Chisholm, S.T., Coaker, G., Day, B., Staskawicz, B.J., 2006. Host-microbe interactions: shaping the evolution of the plant immune response. Cell 124, 803–814.

Corrion, A., Day, B., 2015. Pathogen resistance signalling in plants. Encyclopedia of Life Sciences. John Wiley & Sons Ltd., Chichester. http://dx.doi.org/10.1002/9780470015902.a0020119

Glazebroook, J., 2005. Contrasting mechanisms of defense against biotrophic and necrotrophic pathogens. Annu. Rev. Phytopathol. 43, 205–227.

Scholthof, K.-B.G., 2007. The disease triangle: pathogens, the environment and society. Nat. Rev. Microbiol. 5, 152–156.

Jung, S.C., Martinez-Medina, A., Lopez-raez, J.A., Pozo, M.J., 2012. Mycorrhiza induced resistance and priming of plant defences. Chem. Ecol. 38, 651–664.

Yadeta, K.A., Thomma, B.P.H.J., 2013. The xylem as a battleground for plant hosts and vascular wilt pathogens. Front. Plant Sci. 4, 1–12. http://dx.doi.org/10.3389/fpls.2013.00097. Article 97.

Key Events in the Infection Cycle of Pathogens

Brown, J.K.M., Hovmøller, M.S., 2002. Aerial dispersal of pathogens on the global and contintal scales and its impact on plant disease. Science 297, 537–541.

De Wit, P.J.G.M., Mehrabi, R., Van Den Burg, H.A., Stergiopoulos, I., 2009. Fungal effector proteins: past, present and future. Mol. Plant Pathol. 10, 735–747.

Divon, H.H., Fluhr, R., 2007. Nutrition acquisition strategies during fungal infection of plants. FEMS Microbiol. Lett. 266, 65–74.

Fawke, S., Doumane, Mm, Schornack, S., 2015. Oomycete interactions with plants: infection strategies and resistance principles. Microbiol. Mol. Biol. Rev. 79, 263–280.

Friesen, T.L., Faris, J.D., Solomon, P.S., Oliver, R.P., 2012. Host-specific toxins: effectors of necrotrophic pathogenicity. Cell. Microbiol. 10, 1421–1428.

Kays, A.-M., Barkovich, K.A., 2004. Signal transduction pathways mediated by heterotrimeric G-proteins. In: Brambl, R., Marzluf, G.A. (Eds.), The Mycota III. Biochemistry and Molecular Biology, second ed. Springer-Verlag, Berlin, pp. 175–207.

Koeck, M., Hardham, A.R., Dodds, P.N., 2011. The role of effectors of biotrophic and hemibiotrophic fungi in infection. Cell. Microbiol. 13, 1849–1857.

Mendgen, K., Hahn, M., 2002. Plant infection and the establishment of fungal biotrophy. Trends Plant Sci. 7, 352–356.

Mengiste, T., 2012. Plant immunity to necrotrophs. Annu. Rev. Phytopathol. 50, 267–294.

Rafiqi, M., Ellis, J.G., Ludowici, V.A., Hardham, A.R., Dodds, P.N., 2012. Challenges and progress towards understanding the role of effectors in plant–fungal interactions. Curr. Opin. Plant Biol. 15, 477–482.

Stam, R., Mantelin, S., McLellan, H., Thilliez, G., 2014. The role of effectors in nonhost resistance to filamentous plant pathogens. Front. Plant Sci. 5, 582. http://dx.doi.org/10.3389/fpls.2014.00582

van Kan, J.A.L., 2006. Licensed to kill: the lifestyle of a necrotrophic plant pathogen. Trends Plant Sci. 11, 247–253.

Voegele, R.T., Mendgen, K., 2003. Rust haustoria: nutrient uptake and beyond. N. Phytol. 159, 93–100.

Case Studies

Asiegbu, F.O., Adomas, A., Stenlid, J., 2005. Conifer root and butt rot caused by *Heterobasidion annosum* (Fr.) Bref. s.l. Mol. Plant Pathol. 6, 395–409.

Dean, R., Van Kan, J.A.L., Pretorius, Z.A., Hammond-Kosack, K.E., Di Pietro, A., Spanu, P.D., Rudd, J.J., Dickman, M., Kahmann, R., Ellis, J., Foster, G.D., 2012. The top 10 fungal pathogens in molecular plant pathology. Mol. Plant Pathol. 13, 414–430.

Ebbole, D.J., 2007. *Magnaporthe* as a model for understanding host-pathogen interactions. Annu. Rev. Phytopathol. 45, 437–456.

Galhano, R., Talbot, N.J., 2011. The biology of blast: understanding how *Magnaporthe oryzae* invades rice plants. Fungal Biol. Rev. 25, 61–67.

Koh, S., André, A., Edwards, H., Ehrhardt, D., Somerville, S., 2005. *Arabidopsis thaliana* subcellular responses to compatible *Erisyphe cichoracearum* infections. Plant J. 44, 516–529.

Skamnioti, P., Gurr, S.J., 2009. Against the grain: safeguarding rice from rice blast disease. Trends Biotechnol. 27, 141–150.

Slusarenko, A.J., Schlaich, N.L., 2003. Downy mildew of *Arabidopsis thaliana* caused by *Hyalosperonspora parastica* (formerly *Peronospora parasitica*). Mol. Plant Pathol. 4, 159–170.

Voegele, R.T., 2006. *Uromyces fabae*: development, metabolism, and interactions with its host *Vicia faba*. FEMS Microbiol. Lett. 259, 165–173.

Development of Disease in Natural Ecosystems and Crops

Gilbert, G.S., 2002. Evolutionary ecology of plant disease in natural ecosystems. Annu. Rev. Phytopathol. 40, 13–43.

Diseases of Other Autotrophs: Lichens and Seaweeds

Lawrey, J.D., Diederich, P., 2003. Lichenicolous fungi: interactions, evolution, and biodiversity. Bryologist 106, 80–120.

Zuccaro, A., Mitchell, J.I., 2005. Fungal communities in seaweeds. In: Dighton, J., White, J.F., Oudemans, P.O. (Eds.), The Fungal Community: Its Organization and Role in the Ecosystem. Taylor & Francis, Boca Raton, pp. 533–579.

Emerging Diseases and the Biosecurity Threat

Anderson, P.K., Cunningham, A.A., Patel, N.G., Morales, F.J., Epstein, P.R., Daszak, P., 2004. Emerging infectious diseases of plants: pathogen pollution, climate change and agrotechnology drivers. Trends Ecol. Evol. 19, 535–544.

Brasier, C.M., 2008. The biosecurity threat to the UK and global environment from international trade in plants. Plant Pathol. 57, 792–808.

Brasier, C.M., Buck, K.M., 2001. Rapid evolutionary changes in a globally invading fungal pathogen (Dutch elm disease). Biol. Invas. 3, 223–233.

Fisher, M.C., Henk, D.A., Briggs, C.J., Brwonstein, J.S., Madoff, L.C., McCraw, S.L., Gurr, S.J., 2012. Emerging fungal threats to animal, plant and ecosystem health. Nature 484, 186–194.

Gurr, S., Samalova, M., Fisher, M., 2011. The rise and rise of emerging infectious fungi challenges food security and ecosystem health. Fungal Biol. Rev. 25, 181–188.

Oliver, R., 2012. Genomic tillage and the harvest of fungal phytopathogens. N. Phytol. 196, 1015–1023.

Stukenbrock, E.H., McDonald, B.A., 2008. The origins of plant pathogens in agro-ecosystems. Annu. Rev. Phytopathol. 46, 75–100.

CHAPTER

9

Interactions with Humans and Other Animals

Lynne Boddy
Cardiff University, Cardiff, UK

Interactions between fungi and animals are very different from those with plants, since both fungi and animals are heterotrophs. When you consider the broad diversity of Kingdom Fungi and their lifestyles, and the similarly broad diversity of the animal Kingdom, it comes as no surprise that interactions between the two are many and varied. Interactions can be direct or indirect, and can prove beneficial or detrimental to either or both of the interacting partners (Table 9.1). We begin by considering the interactions in which fungi use vertebrate tissues, concentrating on humans and invertebrates as a food source, causing detrimental effects to the animals. Fungi are themselves – as mycelium, fruit bodies and lichen thalli – fed on by animals. However, associations have also evolved between fungi and animals that, rather than having negative impacts, are based on mutual benefit. The benefits are commonly nutritional, but also often include other additional or even sole benefits, such as provision of a suitable environment, protection against antagonists, and carriage of fungal propagules.

FUNGI AND HUMANS: MEDICAL MYCOLOGY

A small number of fungi can directly affect vertebrates by colonising and growing on or within the skin, or inside the body. About 400 species of fungi are able to cause disease in humans. These diseases are called **mycoses**. A wider range of fungi cause humans problems by producing mycotoxins which, when ingested, give rise to **mycotoxicoses**. Also, some fungi are **allergens**. We deal first with allergens, then mycotoxicoses and finally mycoses.

Allergies

Fungal spores are extremely common in the air, with outdoor concentrations typically ranging between 200 and 10^6 spores m^{-3} (see also Chapter 3), the mean spore content outdoors being 100 to 1000 times greater than that of pollen. Outdoor spore concentrations

The Fungi
http://dx.doi.org/10.1016/B978-0-12-382034-1.00009-8

TABLE 9.1 Examples of the Wide Range of Interactions that have Evolved Between Fungi and Animals

Nature of symbiosis	Fungus	Animal	Effect on fungus	Effects on animal
Intracellular parasites	Microsporidia	Arthropods, fish and to a lesser extent other vertebrates	Nutritional	Death of cells and sometimes whole organisms
Pathogenic	Species of *Blastomyces*, *Histoplasms*, *Coccidioides*, *Paracoccidioides* (Ascomycota)	Humans[a] and other vertebrates	Nutritional	Death of cells, tissues and sometimes the whole animal
Pathogenic	*Pseudogymnoascus destructans* (Ascomycota)	Many bats, especially little brown myotis (*Myotis lucifugus*)	Nutritional	Epidermal tissues eroded and fat reserves lost, which are essential to hibernation, hence death
Pathogenic	*Batrachochytrium dendrobatidis* (Chytridiomycota)	Amphibians	Nutritional	Death; some species have become extinct
Pathogenic	Species of *Achlya*, *Saprolegnia*, *Pythium* (oomycetes)	Fish	Nutritional	A range of effects depending on pathogen: skin lesions, organ damage; blocked blood vessels and gills; mass mortality with some species
Pathogenic	*Entomophthora* (zygomycete) *Beauveria bassiana*, *Metarhizium anisopliae* (Ascomycota)	invertebrates	Nutritional	Death
Pathogenic	*Aspergillus* (Ascomycota)	Sea fan corals (*Gorgonia* spp.)	Nutritional	Depends on host immune status; lesions, galls and sometimes death
Biotrophic parasite and Mutualism	*Septobasidium* (Basidiomycota)	Scale insects	Nutritional and dispersal	Scales live within a mycelial mat on the surface of plants. Some scales are parasitized, others benefit from a buffered microclimatic environment and protection from predators
Predation	*Coprinus comatus*, *Hohenbuehelia*, *Hyphoderma* and *Pleurotus* mycelium	Nematoda and rotifers	Nutritional	Killing by toxins or trapping on adhesive on constricting ring structures; subsequent utilisation of whole body contents
Commensalism or mutualism: symbionts within invertebrate gut	*Asellariales*, *Harpellales* (Kickxellomycotina)	Freshwater, marine and terrestrial Crustacea, Insecta, Myriapoda	Obligate gut symbionts: nutritional/ environmental	Slight loss of gut nutrients but possibly aid to digestion

TABLE 9.1 Examples of the Wide Range of Interactions that have Evolved Between Fungi and Animals—cont'd

Nature of symbiosis	Fungus	Animal	Effect on fungus	Effects on animal
Mutualism: symbionts within invertebrate gut	Yeasts in Saccharomycetes (Ascomycota) and Tremellales (Basidiomycota)	Coleoptera and Insecta	Nutritional and habitat	Nutrition: enzymes for digestion; provision of essential nutrients; detoxification of plant metabolites
Mutualism: ants and higher termites cultivate the fungus	*Attamyces*, *Leucoagaricus* and *Lepiota* spp. *Termitomyces* (Basidiomycota)	Attine ants Macrotermitinae	Provision of plant resources; maintenance of favourable abiotic and biotic environment; carriage of spores to new nests	Nutrition; ingested enzymes
Mutualism: females carry asexual spores to trees; larvae develop in colonised wood	*Amylostereum* spp. (Basidiomycota) *Ophiostoma* spp. (Ascomycota) *Ophiostoma*, *Ceratocystis* spp. (Ascomycota), *Entomocorticium* spp. (Basidiomycota)	Siricid woodwasps Ambrosia beetles Bark beetles	Carriage and inoculation into a suitable environment	Softening of wood; improved nutrition; ingested enzymes
Mutualism	Phallaceae (Basidiomycota)	Diptera	Spore dispersal	Nutrition: feeding on spore masses
Commensalism/ mutualism/ mycophagy: larvae burrow in wood colonised by the fungi	Wood-rotting species, e.g. *Laetiporus sulphureus*, *Trametes versicolor* and *Coniophora puteana* (Basidiomycota)	Death watch beetle *Xestobium rufovillosum*	May benefit from nutrient input in faeces; harm may accrue from comminution	Softening of wood; improved nutrition
Mycophagy within fruit bodies	Agarics and polypores (Basidiomycota)	Gamasid mites, Insecta and Coleoptera	Decreased reproductive output; spore destruction; spore dispersal	Nutrition and breeding ground
Mycophagy of mycelium	Mycelia of many soil fungi	Collembola, woodlice, nematodes, some millipedes	Morphological and enzyme production changes; increases and decreases in hyphal coverage and biomass; alteration of outcome of inter-specific mycelia interactions	Nutrition

[a]*See Tables 9.4 and 9.5 for more examples of human pathogens.*

vary with climate, especially temperature, moisture, and wind, and hence vary daily. Daily changes in fungal spore, as well as pollen, counts are monitored in many cities, which is useful for the huge population of asthmatics and people who suffer from allergic rhinitis (hay fever), because allergy symptoms tend to increase with spore concentrations (see below). In the United States, a simple scale for fungal spore concentrations has been developed by the National Allergy Bureau: below 6500 spores m^{-3} is categorised as low; 6500–12,999 spores m^{-3} qualifies as moderate; 13,000–49,999 spores m^{-3} is high, and greater than 50,000 spores m^{-3} is very high. The Spores of ascomycetes in the genera *Alternaria, Cladosporium*, and *Epicoccum*, and basidiomycetes in the genus *Ganoderma* are examples of common allergy-causing species.

Indoor air contains spores that have entered from outdoors as well as from those fungi growing indoors, but the concentration is usually half that of the outdoor environment. The indoor concentration depends on humidity, temperature, ventilation, the presence of decomposing material, carpets, pets, and plants. Unlike other allergic sources (e.g. pollen), fungal spores, and hyphal fragments are common in the air throughout the year, though there are seasonal peaks. In a study of badly infected buildings in Denmark, the most commonly occurring species were ascomycetes, especially in the genera *Penicillium* and *Aspergillus*, and to a lesser extent *Chaetomium, Cladosporium, Ulocladium*, and *Stachybotrys*. There has been a lot of concern, particularly in the United States, about the purported toxicity of certain fungi that grow in flooded homes. The spores of some of these indoor fungi, including a black-pigmented ascomycete, *Stachybotrys chartarum*, carry toxins that can cause a range of illnesses if they are absorbed in high concentrations. However, it is not clear how often people who inhale spores of this fungus in water-damaged buildings are exposed to levels of these mycotoxins that can cause illness. Nevertheless, the inhalation of large quantities of allergenic spores in these circumstances remains a serious public health concern. Fungal growth inside a building is indicated if the concentration of spores in indoor air exceeds the measurement for outdoor air on the same day, and/or if different fungi are identified in indoor and outdoor air.

The human body defends itself with its immune system that recognises and responds to different antigens, destroying, for example, potential pathogens. However, occasionally there is an overactive immune response, known as hypersensitivity, which causes more damage than the potential pathogen. There are different types of hypersensitivity (Table 9.2). In Type 1 (immediate hypersensitivity), for example, fungal antigens (proteins on the surface of fungi) are recognised by immunoglobulin E (IgE), and then termed allergens. Binding of the IgE with the allergen triggers allergic responses including asthma, eczema, hay fever, rhinitis (inflammation of nasal mucous membranes), and urticaria (nettle rash). Susceptible individuals can become sensitised by continual low-dose exposure to allergens. A wide range of fungi cause various allergic diseases (Table 9.2), with over 80 genera inducing Type 1 allergies in humans and over 20 genera producing allergic proteins. Allergies are a serious global health problem, with an estimated 10% of the human population showing allergic sensitivity to fungal spores, 300 million people suffering from asthma, and 250,000 deaths each year attributed to the illness. Fungal spores tend to be smaller than other allergen sources and can reach the alveoli. Unlike other allergen sources, some fungi may also colonise tissues (pp. 303–309).

TABLE 9.2 Types of Fungal Allergic Reactions Based on Information in Simon-Nobbe et al. (2008)

Type	Clinical manifestation	Allergic mechanism[a]	Examples of most prominent genera of inducers[b]
Allergic rhinitis	Nasal obstructions, pleuritis, rhinorrhea, and sneezing	Type 1 allergy	Ascomycota: *Alternaria, Aspergillus, Bipolaris, Cladosporium, Curvularia* and *Penicillium*
Asthma	In children: increased bronchial activity. In adults: severe asthma and even death	Type 1 allergy	Ascomycota: *Alternaria, Aspergillus, Cladosporium, Epicoccum, Helminthosporeum* and *Penicillium*
Atopic dermatitis	Chronic skin inflammation	Type 1 allergy, associated with high levels of allergen-specific and total IgE	Ascomycete yeasts: *Malassezia furfur*
Allergic bronchopulmonary mycoses (ABPM)	Growth in bronchial lumen, which leads to persistent inflation. Bronchiectasis is induced in asthma sufferers	Type 1, 111, 1V	Ascomycota: *Aspergillus fumigatus, Candida albicans, Curvularia, Gotrichum*, and *Helminthosporeum*
Allergic sinusitis	Multiple sinuses are affected; hyphae are detectable in mucus, but no tissue invasion	Type 1, 111, 1V, specific IgE and IgG antibodies and raised levels of total IgE	Ascomycota: *Alternaria, Aspergillus, Bipolaris*, and *Curvularia*.
Hypersensitivity pneumonitis (extrinsic allergic alveolitis)	Repeated allergen inhalation may lead to irreversible lung damage. Precipitating antibodies and antigen-induced lymphocyte stimulation occurs	Type 111, 1V	Ascomycota: *Aspergillus* and *Penicillium* Basidiomycota: *Lentinula edodes, Pleurotus ostreatus*, and *Serpula lacrymans*

[a] *There are different types of allergic/hypersensitive reaction by humans. Type 1 is an immediate reaction causing an inflammatory response, as a result of immunoglobulin E (IgE) causing excessive activation of some white blood cells (mast cells and basophils). In Type 2, antibodies bind to antigens on the body's own cell surfaces. In Type 3 (immune complex) there is binding of an antibody (e.g. immunoglobulin G, IgG) to a soluble antigen. Type 4 is cell-mediated and does not involve antibodies.*

[b] *Though Ascomycota and Basidiomycota are the most allergic fungi, other groups do have allergic members, for example, zygomycete genera, Absidia, Mucor, and Rhizopus.*

Mycotoxicoses

Fungi produce an enormously wide range of metabolites, as described in detail in Chapter 5. Unsurprisingly, some of them are toxic. Not all toxic compounds produced by fungi are termed mycotoxins; those that are toxic mainly to bacteria are commonly called antibiotics, those toxic to plants are termed phytotoxins, and those that are found in mushroom fruit bodies are often referred to as mushroom toxins or poisons. The term mycotoxin is reserved for low molecular weight fungal secondary metabolites that are toxic to humans and other vertebrates in low concentrations. Not only is it challenging to define mycotoxins, it is also hard to classify them, and depends on the purpose of the classification. Clinicians tend to categorise mycotoxins by effect (e.g. neurotoxins and immunotoxins) physicians by the illness they cause (e.g. St. Anthony's fire) organic chemists by chemical structure, biochemists by their biosynthetic origins, and mycologists by the fungi that produce them. Three to four hundred mycotoxins and mushroom toxins have now been identified, and they occur in families of chemically related metabolites. Only about 20, however, are the usual causes of vertebrate health problems (Table 9.3).

Humans and other animals are most commonly exposed to mycotoxins by unwittingly consuming them in contaminated food, and to mushroom toxins by mistakenly eating poisonous mushroom species. Food can be contaminated while growing, postharvest during storage, or indirectly via the food chain (e.g. in milk from cows that ate contaminated food). Mycotoxins contaminate up to 25% of the world's food supply. Exposure is most common in places where methods of food handling and storage are poor, malnutrition is a problem, and few regulations exist to protect populations from exposure. In many countries there is

TABLE 9.3 Some Mycotoxins and Mushroom Toxins

Toxin	Example producing species	Pathological effects
Mycotoxins		
Aflatoxins	*Aspergillus flavus*	Liver damage, liver cancer
Citrinin	*Penicillium citrinum*	Kidney damage
Gliotoxin	*Aspergillus fumigatus*	Immunosuppressant
Ochratoxins	*Aspergillus ochraceus*	Kidney damage
Patulin	*Penicillium expansum*	Kidney damage
Trichothecene: T-2	*Fusarium sporotrichioides*	Alimentary toxic aleukia
Trichothecene: Vomitoxin	*Fusarium graminearum*	Vomiting, anti-feedant
Zearalenone	*Fusarium graminearum*	Gynaecological disturbances
Fumonisins	*Fusarium moniliforme*	Oesophageal cancer
Ergot alkaloids	*Claviceps purpurea*	Vasoconstriction, gangrene, convulsions
Mushroom toxins		
Amanitin	*Amanita phalloides*	Liver damage
Phalloidin	*Amanita phalloides*	Liver damage
Muscarine	*Amanita muscaria*	Sweating, vomiting
Gyromitrin	*Gyromitra esculenta*	Liver and kidney damage
Orellanine	*Cortinarius speciosissimus*	Kidney damage
Coprine	*Coprinus atramentarius*	Alcohol poisoning
Psilocybin	*Psilocybe cubensis*	Psychotropic effects
Ibotenic acid	*Amanita muscaria*	Psychotropic effects

legislation governing permitted levels of mycotoxins in food, and food samples are tested to enforce these. In the UK, a maximum of $4\,\mu g\,kg^{-1}$ of aflatoxin (produced by *Aspergillus* species) is permitted in human food and $50\,\mu g\,kg^{-1}$ of patulin (produced by *Penicillium* species) in apple juice. Controlling mycotoxin production in food largely revolves around prevention by good agricultural practice and storage under conditions not conducive for fungal growth (e.g. low humidity for grain storage). In the future, plant breeding programmes and genetic engineering may produce crop plants with enhanced antifungal genes, and biocontrol strategies may be developed.

The most notorious mycotoxins are the aflatoxins from *Aspergillus flavus* (hence 'A-flatoxin'), *Aspergillus parasiticus*, and *Aspergillus nominus*. These were first discovered in 1960 after the death, from liver disease, of over 100,000 turkeys in Norfolk, UK, followed by deaths of other farm animals. The cause turned out to be their food – ground peanut meal contaminated with *Aspergillus flavus*. There are four main aflatoxins B_1, B_2, G_1, and G_2, the first being the most important. They are not only acutely toxic but also cause chronic illness, being the most active natural carcinogenic substances known. Several *Aspergillus* species, notably *A. ochraceus*, also produce ochratoxin A (e.g. on cereals, especially barley, cocoa and coffee beans). Kidney damage is the main problem, but ochratoxin A is also a liver toxin, an immune suppressant and a carcinogen. Patulin is produced by a range of *Aspergillus* and *Penicillium* species, but the main problem is from *Penicillium expansum* causing soft rot of apples and other fruit, and is often found in unfermented apple juice. The trichothecenes are a family of over 60 sesquiterpenes, produced by a range of genera including *Fusarium*, *Phomopsis*, *Stachybotrys*, and *Trichoderma*. Trichothecene T-2, which has been the most studied, is produced by *Fusarium sporotrichoides* and *Fusarium poae*, growing on millet left in the fields under snow. It caused a terrible epidemic of alimentary toxic aleukia (involving degeneration of bone marrow, haemorrhaging, necrosis of the alimentary tract and blood abnormalities) in the former Soviet Union, in the 1940s. The zearalenone family of mycotoxins, produced by *Fusarium* species, particularly *Fusarium graminearum* and *Fusarium culmorum*, mimic oestrogen, and in some formulations can be called drugs. The ergot alkaloids, with lysergic acid as a structure common to all, are a toxic cocktail produced in the sclerotia of *Claviceps* (Ascomycota) species – pathogens of various grasses. Human ergotism (St. Antony's fire) caused by eating bread made with flour from infected cereals, especially rye, was common in the Middle Ages, with major epidemics in Russia as late as 1927 and the last reported outbreak in France in 1951. The disease takes two forms – convulsive ergotism affects the central nervous system, while the second form results in gangrene in the extremities. It is still an important animal disease.

There are relatively few toxic mushroom fruit bodies (Table 9.3). The most toxic – *Amanita phalloides* (the death cap) – produces two closely related families of bicyclic peptide toxins, the most abundant of which are α-amanitin and phalloidin. They have completely different actions: α-amanitin specifically inhibits RNA polymerase II, preventing mRNA synthesis, while phalloidin irreversibly binds to filamentous actin, disrupting cell structure. Both toxins cause liver damage, and eventually death. A single fruit body contains only a few thousandths of a gram, but this is sufficient to kill an adult human. This potency is being harnessed, and has been successful in arresting pancreatic cancer in mice, by linking the α-amanitin toxin to an antibody that attaches to cell surface EpCAM protein found on cancer cells. Other toxic examples include *Amanita muscaria* (fly agaric) and species of *Clitocybe* and *Inocybe*, which produce muscarine, an acetylcholine analogue which binds to nerve synapses causing continuous stimulation. The fly agaric also produces the amino acid ibotenic acid, whose decarboxylated derivative muscimol causes hallucinations and dizziness. *Tricholoma equestre* and *Russula subnigricans* cause destruction of muscle tissue, coma and heart failure, the toxin from the latter being cycloprop-2-ene carboxylic acid. *Gyromitra esculenta* (false morel) is also fatal when uncooked, as a result of the toxin gyromitrin, which gives rise to toxic hydrazines. Some species of *Cortinarius* produce orellanine, which causes kidney damage. *Psilocybe* species produce psilocybin, which is hallucinogenic for several hours (p. 162), and some people also experience nausea and panic attacks.

Mycoses

Fungal diseases can be categorised according to increasing severity as superficial, subcutaneous, or systemic. The latter can be further divided depending on whether the causative organism is a true pathogen able to invade tissues of an otherwise healthy host, or whether it is an opportunist able to invade tissues of a debilitated or immunocompromised host (Table 9.4). There are around half a dozen extremely unpleasant fungal diseases, including aspergillosis, candidiasis, histoplasmosis, coccidiomycosis, and blastomycosis. Ironically, with advances in some medical treatments there has been a rise in life-threatening fungal diseases, especially in the areas of transplant and chemotherapy, where the immune system is suppressed.

There are three main attributes that fungi need to cause disease in humans: (1) the ability to grow well at 37 °C; (2) the ability to utilise many different carbon and nitrogen sources, and to scavenge limiting elements (e.g. iron); and (3) the ability to recognise and adapt to the conditions within the human host, which are very different from those outside. After first briefly considering superficial and cutaneous infections, we will describe some of the main fungi that are able to invade living humans (i.e. those that exhibit these three main attributes). There will also be other fungi that share these attributes and could grow in the expanding population of immunocompromised patients, some of which are already starting to emerge as pathogens (e.g. filamentous species of *Acremonium*, *Alternaria*, *Bipolaris*, *Fusarium*, *Penicillium marneffei* and *Pseudoallesheria*, *Scedosporium prolificans*, and the yeast-like *Candida krusei*, *Rhodotorula rubrum*, and *Trichosporon* species).

Superficial Infections

There are fungi on the skin all of the time, most of which are commensal, causing no harm, though others are capable of causing disease. Many of the commensals are yeast forms, while the pathogens can be in yeast or mycelial form. Species in the genus *Malassezia* live permanently on the skin, at population densities varying between individual people and between sites, often with less than 4 cm^{-2} on hands and feet, and up to $10^4 cm^{-2}$ on chest and backs, with maximum densities occurring between late teens and early middle age. Most have an absolute growth requirement for lipids and, hence, are prevalent in areas rich in sebaceous glands (e.g. chest, back, face, and scalp). *Malassezia furfur* is the cause of dandruff. *Malassezia* species are also able to cause skin complaints, including pityriasis versicolor and seborrhoeic dermatitis. Pityriasis versicolor is scaly pigmented lesions, typically on the upper trunk, containing both yeast and mycelial forms, and often the most common fungal infection in hot climates, as warmth and humidity favour development. Seborrhoeic dermatitis is also seen as scaly lesions in around 3% of the immune competent population but about 80% of HIV-positive patients.

TABLE 9.4 Categories of Fungal Infections of Humans Based on Severity of Effects

Infection/mycosis	Definition and general description	Examples of disease	Causal fungus	Phylum
Superficial	Superficial infection of skin or hair shaft; no invasion of living tissue	Seborrhoeic dermatitis, dandruff, folliculitis pityriasis	*Malassezia furfur* (lipophilic yeast)	Basidiomycota
Cutaneous	Superficial infections of the hair, skin, or nails. No living tissue is invaded, but a variety of allergic or inflammatory response occurs in the host due to the fungus and its metabolic products	Candidiasis of skin, mucous membranes and nails/Thrush Tineas/ringworm	*Candida albicans* *Epidermophyton, Microsporum, Trichophyton*	Ascomycota Ascomycota
Subcutaneous	Chronic, localised infections of skin and subcutaneous tissue following accidental implantation of the fungus, mostly saprotrophs from soil or plant material	Chromoblastomycosis Entomophthoromycosis Mycotic mycetoma Sporotrichosis	*Phialophora, Cladosporium* *Basidiobolus ranarum* *Exophiala* and others *Sporothrix schenckii*	Ascomycota zygomycete Ascomycota Ascomycota
Systemic – Dimorphic/True pathogen	Able to invade and develop in tissues of an otherwise healthy host with no recognisable predisposing factor (i.e. can overcome the physiological and cellular defences of the human host). Primary site of infection is usually pulmonary. The morphology outside of the host differs from that inside the host	Blastomycosis Coccidioidomycosis Histoplasmosis Paracoccidioidomycosis	*Blastomyces dermatitidis* *Coccidioides immitis* *Histoplasma capsulatum* *Paracoccidioides brasiliensis*	Ascomycota Ascomycota Ascomycota Ascomycota
Systemic – Opportunistic	Infections occur almost exclusively in immunocompromised patients, (e.g. AIDS, advanced cancer, post-organ-transplant, following steroid/antibiotic/chemo-therapy). Incidence is rising	Aspergillosis Candidiasis (candidosis) Cryptococcosis Hyalohyphomycosis Phaeohyphomycosis Pneumocytosis Penicilliosis Zygomycosis	*Aspergillus fumigatus* *Candida albicans* *Cryptococcus neoformans* Non-pigmented, conidial fungi (e.g. *Fusarium* spp.) darkly pigmented conidial fungi (e.g. *Cladosporium, Curvularia*) *Pneumocystis jirovecii* *Penicillium marneffei* *Rhizopus, Mucor, Absidia*	Ascomycota Ascomycota Basidiomycota Ascomycota Ascomycota zygomycete

Fungal pathogens have emerged independently in different phyla. They have also emerged independently many times within phyla, for example in Ascomycota ranging amongst diverse species such as *Pneumocystis jirovecii* (archiascomycete), *Candida* spp. (hemiascomycetes), and *Aspergillus* spp. (euascomycetes). They emerged in three different ways: (1) from commensals to pathogens (e.g. *Candida* spp. on skin and mucosal surfaces) which can be transmitted to other humans; (2) fungi that are not part of the normal human mycobiota, but can be transmitted from person to person (e.g. lung-colonising *Pneumocystis* spp.) transmitted as aerosols and inhalation, and *Malassezia* spp. and dermatophytes; and (3) opportunistic fungi with no human to human transmission, infection being from the natural environment, though how they can be so well adapted to a human host, yet lacking direct transmission is unclear.

Cutaneous Infections

About 20 species in three genera – *Epidermophyton, Microsporum, Trichophyton* – all with the ability to utilise keratin, grow in the non-living tissues of hair, nails, and skin, in the region above the layers where keratin is deposited. They cause a complex of diseases known clinically as tinea (ringworm) in humans and other vertebrates, and are spread in a keratin-tissue fragment containing viable fungus. Some are **anthropophilic**, largely growing on humans but occasionally other animals (e.g. *Trichophyton rubrum* and *Trichophyton tonsurans*). Others are **zoophilic**, primarily found on other mammals, but can be transmitted to humans via direct contact (e.g. *Microsporum canis*, in cats and dogs). A third group are **geophilic**, decomposing keratin rich tissues in soil, but can also form infections in humans (e.g. *Microsporum gypseum*). The lesions caused by ringworm vary considerably in appearance, but often there is inflammation, swelling, and vesicles. Spreading ring-like lesions, from which the disease gets its name, are found on face, scalp, limbs, and body (Figure 9.1a). On the scalp, skin becomes scaly and hair is lost. When nails are infected they become discoloured, raised, thickened, and crumbly. In temperate regions, 75% of all tinea diseases are foot ringworm (athlete's foot). In the UK, 10–15% of the population have foot ringworm and 5% nail diseases. It is more prevalent in men than women, is higher in sufferers of diabetes and the immunosuppressed, and increases with age to about 25% in the elderly.

There are around 200 *Candida* species ubiquitous in the natural environment, commonly associated with plants and animals, but only a dozen or so associated with human disease, the most common of which are *Candida albicans, Candida glabrata, Candida parapsilosis*, and *Candida tropicalis*. They are carried innocuously by many people on the skin, in the mouth, vagina and gastro-intestinal tract. *Candida albicans* and other *Candida* species can cause cutaneous infections at many sites on the body, especially those that are moist, such as folds of flesh and armpits (Figure 9.1b). Infection of the mouth and vagina is commonly called thrush, because of the white yeast plaques that it forms on the surface of mucous membranes. Oral infections are most common in babies and the elderly; about 75% of women will have a vaginal infection at some time. Mostly, *Candida* is in balance with other skin microbes, but the balance can be shifted by antibiotic therapy and immune suppression, occurring in almost all AIDS sufferers. These cutaneous infections are relatively simple to treat. However, *Candida* can cause a different and serious disease if the cells enter and spread within the body (see Systemic opportunistic infections, below); such diseases are not a consequence of cutaneous infection.

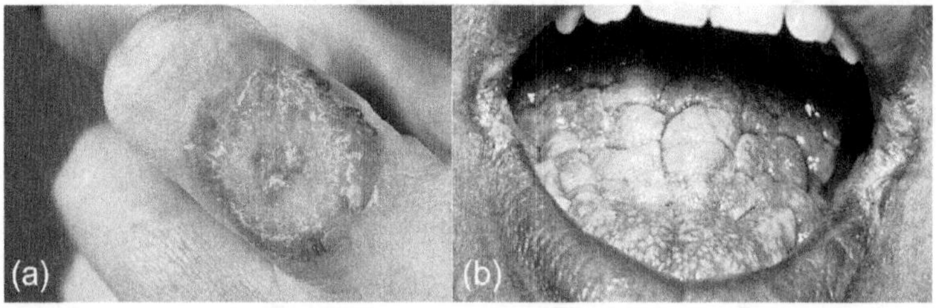

FIGURE 9.1 Human pathogens: cutaneous infections. (a) Ringworm is caused by *Tinea* species infecting the skin, seen here on a finger. (b) Candidiasis of the tongue and mouth corners of an immune deficient human adult. *Source: (a) © Roy Watling; (b) courtesy of www.doctorfungus.org © 2007*

Subcutaneous Infections

Subcutaneous infections mainly occur in the tropics and subtropics, as a result of a saprotrophic fungus being implanted via a wound, and the majority of infections occur in people who walk barefoot, farmers, gardeners, florists, and miners. There have also been increases in recent history as a result of blast injuries from war-settings introducing soils, and fungal infection, into human tissue. Chromoblastomycosis, mycetoma, and sporotrichosis are the most common diseases, and are caused by ascomycetes, but some Entomophthoromycotina (entomopathogenic fungi; pp. 314–320) (e.g. *Basidiobolus ranarum* and *Conidiobolus coronatus*) also form subcutaneous infections (Table 9.5). An outbreak of 3000 cases of sporotrichosis occurred in a South African gold mine in the 1940s, contracted from infected pit props.

Systemic–Dimorphic/True Pathogens

Systemic mycoses are usually acquired by inhalation, starting in the lungs and subsequently affecting the whole body. The four main examples are blastomycosis, histoplasmosis, coccidioidomycosis, and paracoccidioidomycosis. The fungi involved are all dimorphic, switching between yeast and mycelial phases. In each case the yeast is the pathogenic state and the filamentous form is saprotrophic. The dimorphism is regulated by temperature, with mycelial growth in nature at 25–30 °C, and yeast growth in tissue or enriched media at 37 °C (body temperature). Some of these diseases are confined to certain geographical regions.

Blastomycosis, caused by *Blastomyces dermatitidis*, is found in the United States in the Mississippi and Ohio River valleys, and in Canada in states that border the St. Lawrence Seaway and the Great Lakes (Figure 9.2). Outbreaks of the disease are associated with activities around water courses, since the fungus grows in moist soil with rotting plant material. It is, however, difficult to isolate, hence its ecology in nature is unclear. Even in areas where it is prevalent, only one or two people in 100,000 get the disease. Infection occurs when the spores become airborne, entering the lungs and multiplying there, causing an acute disease resembling pneumonia, chronic tuberculosis, or lung cancer, and can result in acute respiratory distress. The fungus can also spread to other organs via blood and lymph. The disease can be fatal, particularly in patients with compromised immune systems. **Histoplasmosis** occurs in the same region (Figure 9.2), and is caused by *Histoplasma capsulatum* var. *capsulatum*, which grows well in nitrogen-rich wild bird, chicken, and bat guano. The disease is similar to blastomycosis, and again in most cases no obvious symptoms are produced. In Central and West Africa, the disease takes a different form, causing deep mycoses in the skin and bones. It is caused by a different variety of the fungus – *Histoplasma capsulatum* var. *duboisii* – likely a different species.

Coccidioidomycosis, commonly known as 'valley fever', is a disease found in southwest United States, Central America, northern South America and Argentina (Figure 9.2). The ascomycetes responsible – *Coccidioides immitis* and *Coccidioides posadasii* – thrive in dry, salty soils typical of desert areas (though it is not a problem in the deserts of Africa or Asia). *Coccidioides immitis* is endemic to the southern deserts and central valley of California, and probably Baja California, while *Coccidioides posadasii* is endemic to southern Arizona, New Mexico, northern Mexico, western Texas, and some parts of South America. In the soil, these fungi are associated with heteromyid (a family of rodents) burrows, and grow as septate mycelium that produces arthroconidia (p. 67), which can be inhaled by humans when the spores rise in dust storms. In the lungs, the arthroconidia enlarge to form large multinucleate spherules (80 μm), which form many small (2–5 μm) uninucleate endospores, which can spread the infection

TABLE 9.5 Subcutaneous Fungal Infections

Disease	Example causative species	Type of fungus	Natural habitat	Disease symptoms	Prognosis and therapy
Chromo-blastomycosis	*Cladophialophora carrionii, Fonsecaea compacta, F. pedrosoi, Phialophora verrucosa*	Filamentous Ascomycota	Soil and woody plant matter	Localised crusted, verrucoid, ulcerated lesions form. Satellite lesions can spread through lymph. Sometimes there is dissemination to the brain	Infections are usually localised. Early stages are treated by topical anti-fungals or surgical removal. Advanced infections may require long systemic treatments with itraconazole or terbinafine
Entomophthoromycosis	*Basidiobolus ranarum, Conidiobolus coronatus*	Zygomycete (Entomophthorales)	Soil and plant litter	*B. ranarum* causes gradually enlarging granulomas in arms and trunk; *C. coronatus* typically colonises nose tissues	*B. ranarum* has been treated with amphotericin B, potassium iodide, and itraconazole; surgery is often necessary
Mycetoma	*Acremonium falciforme, A. redifei, Aspergillus nidulans, Exophiala jeanselmei, leptosphaeria senegalensis, Madurella mycetomatis, M. grisea*	Filamentous Ascomycota	Ubiquitous, mainly soil	Localised infections of cutaneous and subcutaneous tissues. Lesions are locally invasive tumour-like abscesses. Lesions rupture resulting in ulcers, swelling, and distortion of the infected part of the body	Mycetomas are resistant to chemotherapy, often leaving surgery as the only option
Sporotrichosis	*Sporothrix schenckii*	Ascomycota, dimorphic, filamentous and conidia-forming at 25 °C or less, switching to cigar-shaped yeast cells at 37 °C	Globally in soil	Usually localised lesions of cutaneous and subcutaneous tissues, typically following lymphatic pathways. Sometimes sporotrichosis can cause osteoarthritis, pulmonary infections and meningitis. Internal dissemination is by yeast cells. Peru has highest incidence of infections	Prolonged (3–6 months for local lesions; at least 12 months for osteoarticular disease) treatment is needed, but usually responds. Local hypothermia, and oral azoles and is effective

Information from a variety of sources, including http://www.doctorfungus.org/ (accessed 30 Nov 2011).

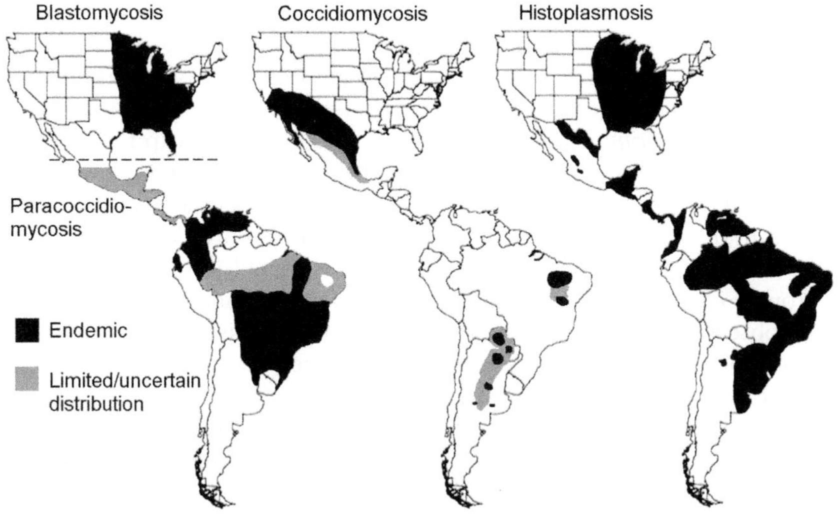

FIGURE 9.2 Approximate distribution of some systemic dimorphic pathogens endemic in North and South America. *Source: Information from Hector and Laniado-Laborin (2005) and Colombo et al. (2011).*

in the surrounding tissues. Millions of people have been infected, but about 60% of infections are asymptomatic; 30% experience a range of symptoms, including coughs, for several months but are ultimately self-limiting; less than 10% require medical intervention. In severe cases, the fungus can spread to adrenal glands, bones, central nervous system, joints, lymph nodes and skin, and the disease can be fatal. There has been a steady increase in reported cases from about 1500 in 1998 to over 20,000 in 2013.

Paracoccidiomycosis – caused by *Paracoccidioides brasiliensis* – is largely confined to South and Central America (Figure 9.2). Infection occurs by inhalation of spores, though their origin is uncertain, as the fungus has rarely been isolated from soil. There is a rare acute form that affects juveniles of both sexes, which sometimes leads to death from fungal growth in the liver, spleen or bone marrow. Skin testing shows that a high proportion of the population of both sexes has come into contact with the fungus by the age of 20 years. However, most cases of pulmonary infections in adults are males (often >50:1 males:females) between 30 and 50 years old. This remarkable difference in the frequency of infections between males and females results from inhibition of the dimorphic switch from mycelium/conidium to yeast by physiological concentrations of the female sex hormone oestradiol; growth of yeast cells when formed is unaffected. The cytosol of yeast and hyphae contains a high-affinity binding protein with similar properties to mammalian receptor proteins responding to steroid hormones, and so there may be a signal transduction pathway operating in the fungus similar to those of mammals. Also, there are major differences in yeast and mycelial cell walls: in hyphae, β-(1-3)-glucan is the major component, whereas it is α-(1-3)-glucan in yeast. The latter may play a role in evasion of host defences.

Systemic–Opportunistic Infections

Since the 1980s mortality from fungal diseases has been increasing, largely reflecting the increase in immunocompromised patients, but also due to increased travel to tropical regions and an increased clinical awareness and improved diagnosis. The most common systemic opportunist fungal pathogens in immunocompromised patients are species of *Candida* and saprotrophic (i.e. act as saprotrophs in the natural environment outside of human bodies) *Aspergillus* and zygomycete species. Also, though relatively infrequent in western countries, the basidiomycete yeast *Cryptococcus neoformans* causes very high (up to 100%) mortality in AIDS sufferers in regions where HAART (highly active antiretroviral therapy) is not available (i.e. around 400,000 per annum worldwide). Over 40 million people in the world suffer from HIV/AIDS, but in developed countries AIDS has become a chronic illness rather than rapidly terminal, thanks to the introduction of HAART. Nonetheless, globally around 3 million people die from it each year. *Pneumocystis jirovecii* (=*carinii*) and *Penicillium marneffei* have emerged from obscurity to being major pathogens of AIDS patients. All of these fungi can be thought of as opportunists, growing inside humans if they happen to find themselves there, but are usually found living in the natural environment.

Aspergillosis covers a broad spectrum of diseases caused by species of *Aspergillus*. Of over 200 *Aspergillus* species, less than 20 cause disease in humans, other mammals and birds, the most common being *Aspergillus fumigatus*, and to a lesser extent *Aspergillus flavus*, *Aspergillus nidulans*, *Aspergillus niger*, and *Aspergillus terreus*. These are all saprotrophs, which are extremely common in nature. They produce large quantities of small (2–4 μm), airborne conidiospores, to which we are all exposed, yet mostly do not succumb to disease. Until the mid-twentieth century, disease was usually associated with occupations in which people were exposed to abnormally high spore loads, such as agricultural workers frequently handling hay and agricultural produce stored in confined spaces, giving rise to disease commonly called farmer's lung. Reduction of spore load, due to changes in harvesting methods and improved storage protocols, reduced the incidence of the disease. It is, however, now a major concern because of its lethal invasion of immunocompromised patients, and is one of the most frequently acquired nosocomial (i.e. hospital-acquired) infections following immunosuppressive therapy, with around 700,000 deaths world-wide each year caused by *Aspergillus* infections.

There are three main types of disease caused by *Aspergillus* species: allergic bronchopulmonary aspergillosis, pulmonary aspergilloma, and invasive aspergillosis. Allergic bronchopulmonary aspergillosis is a hypersensitive reaction to spore and hyphal surface antigens, leading to asthmatic reactions, and is particularly common in asthma and cystic fibrosis patients. With pulmonary aspergilloma, *Aspergillus* forms balls of hyphae plus host cells, tissue debris, and other substances in cavities within the lungs. In this form of disease, the fungus does not usually invade surrounding lung tissue; in about 10% of cases it resolves itself without treatment, and only rarely does it enlarge. Invasive aspergillosis starts from the primary focus of infection, which is usually the lower respiratory tract, resulting from inhaled spores, though less commonly invasion can be via the sinuses or skin through catheter insertion sites etc. From the primary focus, *Aspergillus* invades blood vessels and is transported to other organs, particularly the brain. Unlike several other human pathogenic fungi, *Aspergillus* species do not have a yeast-mycelial dimorphism, but grow strictly in filamentous form. Though *Aspergillus fumigatus* is the major cause of invasive aspergillosis, cases resulting from *Aspergillus terreus* are increasing, and *Aspergillus lentulus* is emerging; *Aspergillus terreus* and *Aspergillus lentulus* both have low susceptibility to currently available antifungal drugs.

Candidiasis. While *Candida albicans* and a few other species of *Candida* usually cause, at worst, cutaneous infections (see above), sometimes they penetrate through the skin or mucosal surfaces, eventually reaching the bloodstream. From there they can be disseminated to organs, including the brain, liver, and kidneys, eventually leading to death. How the yeast cells enter the bloodstream is not entirely clear, but routes include damage to the intestinal tract by chemotherapy and surgery, and direct entry via catheters and intravenous lines. The immune system can eradicate low numbers of yeast, though not if the immune system is compromised. It is the fourth most common nosocomially acquired bloodstream infection. It is difficult to attribute mortality rates to *Candida* infection, because almost all sufferers of invasive candidiasis have an underlying illness, but it probably ranges between 15% and 50%, much higher than other systemic infections (e.g. MRSA, methicillin-resistant *Staphylococcus aureus*). The annual financial cost in the United States alone in 2002 was estimated at US $1.7 billion. *Candida albicans* is the most common cause of candidiasis, followed by *Candida glabrata* and *Candida parapsilosis*. *Candida glabrata* rapidly develops resistance to antifungal drugs, which may select for this species. *Candida parapsilosis* commonly grows from biofilm on plastic surfaces, and probably enters via catheters and intravenous lines. In Asia, *Candida tropicalis* is a particularly common cause of candidiasis.

Several traits of *Candida albicans* are putative virulence factors. (1) The ability to adhere to host tissue, so as not to be dislodged by the bloodstream or host secretions (e.g. sweat and saliva). *Candida albicans* produces **adhesins** that bind to a range of host proteins, including fibronectin, and carbohydrate moieties of membrane glycoproteins and glycolipids. Als (agglutin-like sequence) proteins and Hwp (hyphal wall proteins) are also adhesins. (2) *Candida albicans* produces a wide range of extracellular enzymes that break down host proteins, including antibodies, lipases, and phospholipases. (3) The ability to switch between phenotypes is significant not only in *Candida* spp. but also in other invasive pathogenic fungi. Though it is frequently referred to as the yeast/mycelial dimorphism, *Candida albicans* is polymorphic also growing as pseudohyphae (Figure 9.1). Yeast forms are typical outside the body. Both yeast and hyphae are found in tissues, and the different forms may be important at different stages or in different types of infections; some mutants unable to switch forms have a reduced ability to cause disease. The switch from yeast to filamentous morphology is critical for invading host tissue. It can be induced *in vitro* with blood serum and nutrients, but the very complex interacting signal networks, found by genetic analysis, indicate that morphogenesis in *Candida* can respond to many different chemical and physical cues (p. xxx). This reflects a mode of development highly adapted to the multiple and changeable niches presented by host tissues.

***Cryptococcus* infection** in humans is acquired by inhalation of spores or yeast cells from the natural environment, especially soil containing pigeon guano (*Cryptococcus neoformans* var. *neoformans* and *Cryptococcus neoformans* var. *grubii*), and where eucalyptus trees and decaying wood are present (*Cryptococcus neoformans* var. *grubii* and *Cryptococcus gattii*). *Cryptococcus neoformans* var. *neoformans* and *Cryptococcus neoformans* var. *grubii* have a worldwide distribution in the natural environment, and cause the vast majority of cryptococcal infections in humans with underlying immunosuppression. On the other hand, *Cryptococcus gattii* causes the majority of infections reported in immunocompetent hosts. *Cryptococcus* species are not obligate human pathogens, and also occur in other organisms, including domesticated and wild animals, insects, and amoebae, but there is no evidence of direct transmission between animals and humans, nor between humans. *Cryptococcus* is the most important life-threatening fungal infection of AIDS patients. It can colonise the host respiratory tract without causing disease, and can be cleared or enter a latent phase, which may subsequently be reactivated and disseminated in the blood to cause systemic infection – cryptococcosis. It most commonly infects the brain and central nervous system, causing meningoencephalitis, but can cause localised infection in any organ. *Cryptococcus neoformans* is usually isolated, from both patients and the environment, as budding yeast. In its filamentous form it can undergo monokaryotic fruiting or mating, the latter involving fusion of two haploid cells with different mating type alleles, a and α (Chapter 4). Over 98% of both clinical and environmental isolates are the α-mating type, which is more virulent than the a-mating type. *Cryptococcus neoformans* can also form thick-walled chlamydospores (p. 67.) that could act as long-term survival structures in the natural environment. The production of polyphenoloxidase, which converts phenolic substrates into melanin, is a virulence factor. Melanin and the thick, acidic, mucopolysaccharide capsule of the yeast inhibit phagocytosis.

Penicillium marneffei is endemic to Southeast Asia and has emerged as a significant mycosis in humans since the 1980s with the rise of AIDS. In the north of Thailand, about 25% of AIDS patients are infected with it. It causes systemic infections resulting in considerable mortality. Bamboo rats (*Rhyzomys*) are a possible source of infection, as surveys have shown a high prevalence in these animals. However, the fungus is present, and may grow, in soil in the endemic region, and soil is currently assumed to be its main environmental reservoir. Of over 200 species in the genus *Penicillium*, *Penicillium marneffei* is the only species that is highly pathogenic, and the only species that has a temperature-dependent dimorphism, growing intracellularly as a fission yeast.

Pneumocytosis caused by *Pneumocystis jirovecii*, is one of the major opportunistic pathogens of immunocompromised patients. Though previously mistaken as a protozoan, *Pneumocystis jirovecii* is an ascomycete (Taphrinomycotina), but a rather unusual one. It has cholesterol, not ergosterol, in its cell membranes, which negates the use of amphotericin B and azole antifungal drugs. Its yeast-like cells cluster together in host tissue; its vegetative form is probably haploid, replicating asexually. Following fusion, a diploid zygote is formed, and meiosis occurs in a cell termed the precyst (terminology from the time when it was thought to be a protozoan), giving rise to the early cyst. Eight spores develop within the cyst, which should be called an ascus, and the spores ascospores. The mature ascus ruptures, and the ascospores are released to germinate into feeding forms. This lifecycle information comes from growth in animals. It is currently not possible to grow it in artificial culture, so little is known of its ecology and physiology, nor what infective agent is released into the environment. Though its DNA has been detected in air and water samples, it is probably not viable and close host proximity (such as vertical transmission between mother and child) is likely necessary for transmission to occur. Strains from humans, rodents, rabbits and other animals are genetically distinct, with high host specificity, as shown by cross-infection experiments. It appears to be passed from human to human by breathing in spores, which germinate and invade the alveoli, resulting in extensive damage to the alveolar epithelium. It only causes problems in premature babies, malnourished infants, and immunocompromised patients, where it causes pneumonia and, in a few cases (<3%), lesions in the lymph nodes, liver, spleen or bone marrow.

Zygomycosis, though hitherto rare, zygomycoses are increasing, especially in immunocompromised patients, and are often lethal due to resistance to many common antifungal drugs. With *Mucor circinelloides*, sporangiospore size dimorphism is linked to virulence.

The larger sporangiospores are virulent, germinating inside the host (demonstrated in wax moth (*Achroia*)) and lysing macrophages, whereas the smaller spores are not.

Dimorphisms are common in human pathogenic fungi. *Cryptococcus neoformans* infections occur in the yeast mode, the hyphal mode being present in nature and the phase when sexual reproduction takes place. *Cryptococcus neoformans* forms giant cells (up to 50 μm) in the lungs of infected hosts, formation being enhanced in the presence of opposite mating types, as a result of signalling by a mating pheromone, somewhat analogous to quorum sensing. Cell signalling circuits may govern both virulence and mating. With *Candida albicans*, the yeast phase is critical for spread within the host, whereas the hyphal form allows the fungus to escape from macrophages following phagocytosis, and also to form biofilms on catheters, etc. As mentioned earlier, *Candida* spp. exist as yeasts, pseudohyphae and filamentous forms (Figure 5.15). Switching between morphological forms is obviously also crucial in the pathogens categorised as systemic/dimorphic true pathogens (e.g. *Paracoccidioides brasiliensis*). On the other hand, other pathogens (e.g. *Aspergillus fumigatus*) are strictly filamentous. Clearly, there is no simple rule that one morphological form is pathogenic and others are not.

Antifungal Agents for Treatment of Mycoses

The main treatment of fungal infection is chemotherapy, with drugs whose main actions are (1) inhibition of plasma membrane synthesis or disruption of plasma membrane integrity; (2) disruption or inhibition of cell wall biosynthesis; and (3) inhibition of metabolism and disruption of mitosis (Table 9.6). These antifungal agents also differ in structure, solubility, spectrum of activity, extent of fungistatic/fungicidal activity, and ability to induce resistance. Superficial and cutaneous fungal diseases of skin usually respond well to topical antifungal creams, including various azoles, terbinafine, and amorolfine. Fluconazole or Amphotericin are taken internally in severe *Candida* and other systemic fungal infections.

PATHOGENS OF OTHER VERTEBRATES

Though much of our attention inevitably falls on humans, all other mammals have fungal pathogens, some closely allied to those affecting humans. Just one rather different example is given below – the emerging bat white nose syndrome. Birds too are subject to fungal diseases, again often similar to those affecting mammals (e.g. aspergillosis and candidiasis are the most common). Reptiles also have fungal diseases, the integumentary system being most affected, with lesions often containing soil saprotrophs; systemic mycoses are rare. Neither birds nor reptiles are considered further here. With regard to amphibians, the devastating emerging chytridiomycosis disease overshadows all others. Fish, too, are affected by fungal diseases, especially caused by members of the fungus-like oomycetes. Both chytridiomycosis of amphibians and some oomycete diseases are described below.

Bat White-Nose Syndrome

An emerging fungal disease of bats – White-nose syndrome – is having devastating effects on bat populations. The recently (2006) discovered psychrophilic *Pseudogymnoascus destructans* (Ascomycota: Helotiales; previously known as *Geomyces destructans*) is causing

TABLE 9.6 Commonly Used Antifungal Drugs, Their Mode of Action, and Susceptible Fungi

Type of drug	Mechanism of action	Susceptible fungi
Drugs active against plasma membrane, synthesis or integrity		
Allylamines (naftifine and terbinafine) and thiocarbamates (tolnaftate)	Inhibits activity and formation of lanosterol	*Asperillus, Acremonium, Arthrographis, Fusarium, Penicillium, Trichoderma*
Azoles (imidazoles and triazoles) and echinocandins (cilofungin)	Inhibits ergosterol synthesis	*Aspergillus, Candida, Cryptococcus*
Folimycin (concanamycin A)	Inhibits V-type proton-ATPase	Various fungal species
Hydroxypyridones	Inhibition of ATP-synthesis and uptake of essential components	Various fungal species
Octenidine and pirtenidine	Inhibition of ergosterol biosynthesis	*C. albicans, Saccharomyces cerevisiae*
Polyenes (Amb and nystatin)	Auto oxidation of ergosterol; formation of free radicals which damage the plasma membrane	Species of *Aspergillus, Candida, Coccidioides, Cryptococcus, Histoplasma, Saccharomyces*
Sphingofungin	Interrupts sphingolipid synthesis	Various fungal species
Active against cell-wall components		
Aureobasidin-nikkomycin polyoxins	Inhibition of chitin synthesis and assembly	*Candida* and *Cryptococcus* spp.
Benanomycin A-pradimicin A	Membrane disruption causing leakage of intracellular potassium; calcium-dependent complexing with saccharides of mannoprotein	*Aspergillus* spp., *Candida* spp., and *Cryptococcus neoformans*
Echinocandins (caspofungin, micafungin, anidulafungin)	Inhibition of cell-wall glucan synthesis	*Candida* spp.
Active against cellular anabolism		
5-Fluorocytosine	Inhibition of pyrimidine metabolism	Species of *Asperillus, Candida,* and *Cryptococcus*
Sordarin (sordaricin methyl ester)	Disrupts placement of tRNA from A site to P site; disrupts movement of ribosones along the mRNA thread	*C. albicans* and *S. cerevisiae*

From Abu-Elteen and Hamad (2011).

massive mortality in hibernating bats in eastern North America. Many bat species are affected, but the little brown myotis (*Myotis lucifugus*), which was formerly among the most common bat species, is now facing extirpation in that region. *Pseudogymnoascus destructans* hyphae replace hair follicles and associated sweat and sebaceous glands; the very obvious white mycelial growth on nose, ears and wing membranes is reflected in the name of the disease. The fungus erodes the epidermal tissues, and infected bats have no fat reserves, which are crucial for surviving hibernation. *Pseudogymnoascus destructans* is psychrophilic, growing optimally at 5–10 °C, with little growth above 15 °C; the temperatures in infected bat hibernacula range between 2 and 14 °C all year round, allowing continual fungal growth.

Chytridiomycosis of Amphibians

Perhaps the most devastating of all vertebrate pathogens is the chytrid *Batrachochytrium dendrobatidis* (shortened to *Bd*), which first came to notice in 1987 when the golden frog (*Atelopus zeteki*) was extirpated in Costa Rica. It causes a disease of amphibians that has resulted in serious declines of over 200 species, and the extinction of at least three species, the Panamanian golden frog, the Australian gastric brooding frogs (*Rheobatrachus* sp.), the sharp-snouted day frog (*Taudactylus acutirostris*), and probably many more. The disease has spread worldwide, but Southeast Asia has a much lower incidence, and the island of Madagascar, the most amphibian-rich place on the planet, appears currently to be free from the disease (Figure 9.3a). It spreads rapidly, as seen following extirpation of amphibians in Costa Rica in 1987, moving through Central America, and reaching Panama by 2008 (Figure 9.3b). This movement is too rapid to be spread by frogs and toads themselves, so this must be by another agent, perhaps on the feet of birds. Global spread is due to the human trade in amphibians, but the origin of this emerging disease is still uncertain. Patterns of genetic diversity of *Bd* isolates from infected populations indicate that it is a recently emerged lineage (*Bd*GPL) that evolved in the twentieth century, as the product of recombination in a single mating between closely related strains; isolates from worldwide locations are extremely closely related, and the two other strains that have been discovered (*Bd*CAPE and *Bd*CH) are much less virulent.

The motile flagellate zoospores of *Bd* seek the amphibians and penetrate the skin, where they form sporangia (Figure 9.3c). Symptoms of colonisation are not usually obvious to the naked eye, though sometimes there are skin lesions. There is epidermal hyperplasia (abnormal increase in the number of normal cells) and hyperkeratosis (thickening of the skin), and possibly increased skin shedding. The mechanism by which *Bd* causes death is related to disruption of electrolyte transport across the epidermis, that is critical in maintaining homeostasis. Not all species succumb to the disease and some species and populations in the wild have survived initial declines. This variation is likely to be due to differences in virulence between *Bd* strains, environmental conditions, host behavioural characteristics and immune responses. Rate of production and load of zoospores is probably an important determinant of the rate of disease development and mortality, as mortality occurs above a threshold of infections. Possible treatments, on a small scale, include application of antifungal drugs (e.g. itraconazole) to all of the tadpoles in infected populations, and application of 'probiotic' bacteria that produce antifungal compounds that kill *Bd* on amphibians' skin.

Recently, a new species of chytrid that is pathogenic to amphibians, *Batrachochytrium salamandrivorans*, was discovered causing dieoff in Dutch and Belgium fire salamanders. This chytrid is suspected to have been introduced to Europe from another region of the world, and suggests that there is probably a large number of species of these chytrids waiting to be discovered, as well as to emerge as pathogens.

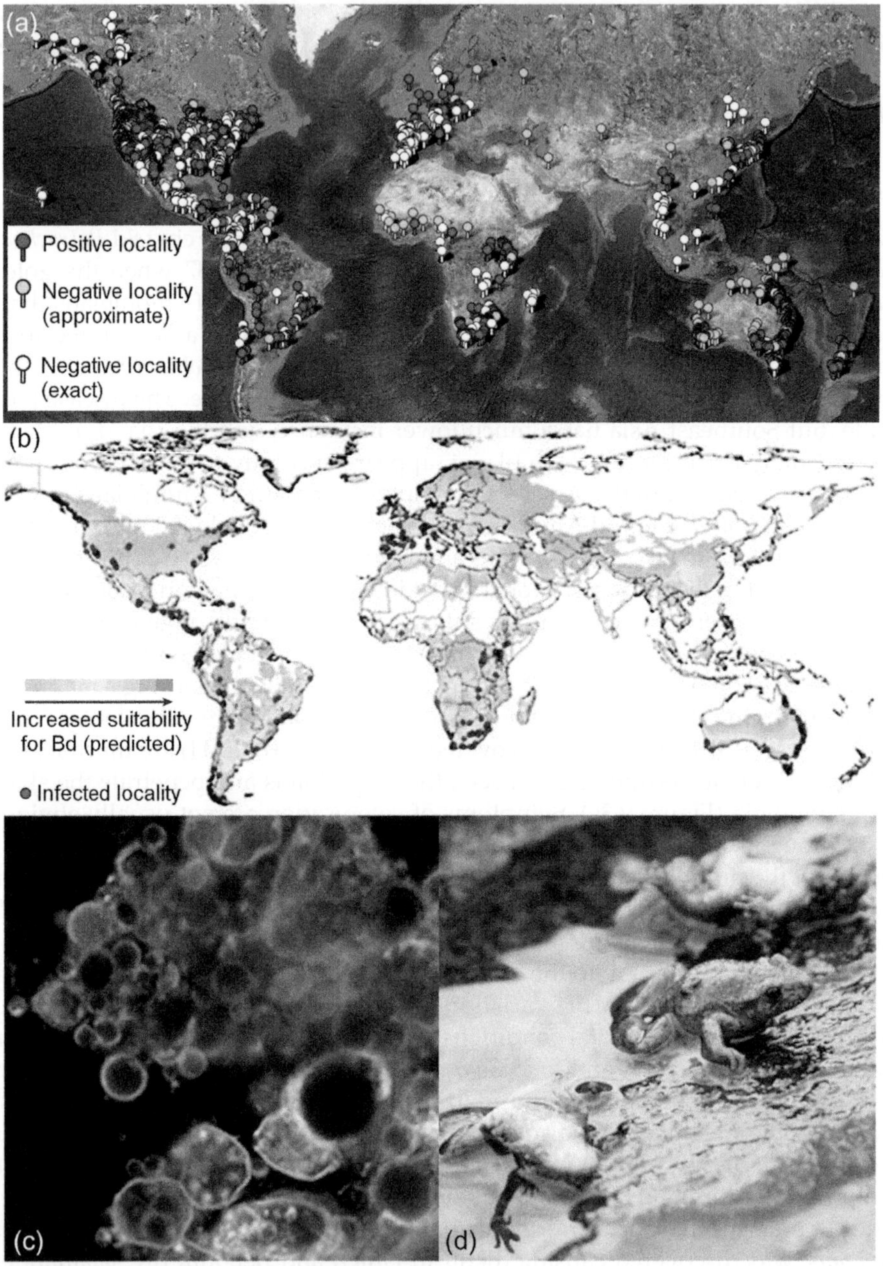

FIGURE 9.3 (a) Chytridiomycosis of amphibians caused by *Batrachochytrium dendrobatidis* (Bd) is now widespread across much of the globe, though Southeast Asia is still largely free of the disease, as seen in this screen shot from the Global Mapping Project (http://bd-maps.net). (b) The disease is predicted to spread even more widely by the Climate Envelope Model developed by D. Rödder, J. Kielgast, J.B. Schmidtlein, et al. (unpublished data). (c) Laser-scanning confocal micrograph of Bd in culture. Blue stained structures are metabolically active sporangia. (d) Pyreneen Midwife toads *Alytes obstetricians* sufferening from lethal chytridiomycosis. *Source: Panel (b) from Fisher et al. (2009), (c) from Fisher et al. (2009), (d) © Mat Fisher.* (See the colour plate.)

Pathogens of Fish

Oomycetes (Kingdom Stramenopila) are perhaps the most widespread 'fungal' disease of fish, especially species of *Achlya* and *Saprolegnia* (Saprolegniales), but also Saprolegniales species in the genera *Aphanomyces, Calyptralegnia, Dictyuchus, Leptolegnia, Pythiopis* and *Thraustotheca, Pythium* (Peronosporales), and *Leptomitus* (Leptomitales). Saprolegniasis is a disease of the epidermis of fish (Figure 9.4). It typically starts on the fins or head and often spreads over the entire body, being visible as white or grey mycelial patches. Spores commonly enter the fish body via damaged gills. In Salmonids, saprolegniasis is associated with stress. *Saprolegnia* species can also infect fish eggs, swimming from dead to live eggs via positive chemotaxis. Ulcerative mycosis and epizootic ulcerative syndrome, which can cause mass mortality, are attributed to *Aphanomyces invadans*. Distinct skin lesions, which appear as red-spots, black marks, or red-centred, white-rimmed deep ulcers, contain hyphae that can sometimes penetrate deeply into the fish beyond the muscles, damaging the brain, vertebrae, and other organs. Branchiomycosis is a widespread disease, especially in warmer climes, and can cause major problems in carp farms. The disease obstructs the blood vessels in the gills, and appears initially as flecks on the gills. The gills later become grey-white, and can even drop off exposing underlying cartilage. *Branchiomyces sanguinis* is associated with carp (*Cyprinus*), tench (*Tinca*), and sticklebacks (*Gasterosteidae*), and *Branchiomyces demigrans* is common on pike (*Esox*) and tench.

Ichthyophonosis is one of the most well-known diseases of fish, especially marine, killing over 80 species. Unlike most fish pathogens that are facultative, the causal agents *Ichthyophonus hoferi* and *Ichthyophonus gasterophilum* are obligate pathogens. Their phylogenetic position is not, however, clear. Fish are usually infected via the digestive tract. Fins disintegrate and can fall off; organs including liver, kidneys and spleen can be colonised, causing swelling, distension of the body and accumulation of exudates; eyes bulge and erode when colonised.

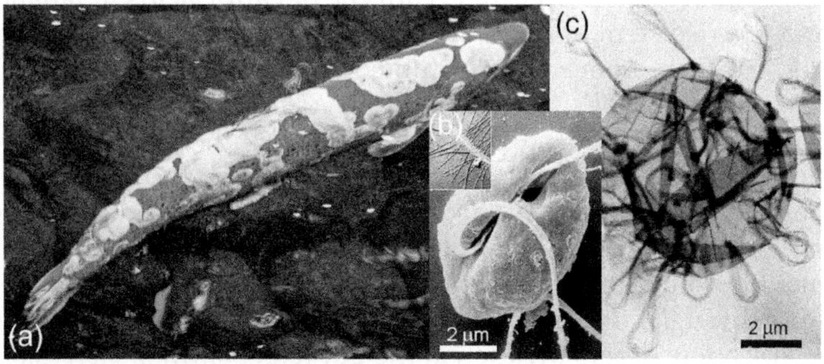

FIGURE 9.4 Saprolegniosis – oomycete (Kingdom Stramenopila) pathogens of fish and their eggs. These fungus-like organisms cause serious losses to fish both in commercial hatcheries and fish farms and can threaten wild stocks of salmonids when they return to their spawning grounds. (a) Mature brown trout (*Salmo trutta*) showing characteristic white lesions of *Saprolegnia parasitica*. (b) Secondary zoospore of *Saprolegnia parasitica*, with ventral groove from which flagella emerge. The hairs which decorate the anterior flagellum are shown in this electron micrograph inset. (c) A secondary cyst case of *Saprolegnia parasitica*, showing bundles of hooped spines that characterise isolates of the fish pathogenic species. *Source: © Gordon Beakes.*

KILLERS OF INVERTEBRATES

There are probably fungal pathogens of all invertebrate species, but those of insects have been most studied. We will first describe these and then mention the intriguing nematode-killing fungi, the recently emerging aspergillosis of coral, and crayfish plague.

Entomopathogens

There are estimated to be in excess of 1000 species of fungi parasitic on insects, in about 90 genera. Insect pathogenesis as a way for fungi to obtain nutrition has arisen *de novo* in all of the major fungal groups (Table 9.7). There are only a few Chytridiomycota that are parasitic on soil- or aerial-inhabiting invertebrates, partly because they are largely dependent on free water for dispersal. The zygomycete order Entomophthorales, and the ascomycete Hypocreales comprise a vast number of entomopathogens, but there are relatively few basidiomycete entomopathogens. The most studied fungal entomopathogens are *Beauveria bassiana* and *Metarhizium anisopliae* (Hypocreales).

Fungal entomopathogens are dispersed as spores, which must land on the cuticle of an insect host and remain there until they can germinate. Spores are usually adhesive, for instance *Entomophthora* spp. have mucilaginous coats, *Verticillium* spp. have slime drops, the zoospores of *Coelomomyces psorophora* produce host specific secretions that attach them to susceptible mosquitoes, and conidia of *Metarhizium anisopliae* produce an adhesion-like protein, MAD1. Spores of most fungal entomopathogens require nutrients to be available on the surface of the cuticle before they germinate, and the fungus' lipolytic activity helps. Different spore types (e.g. aerial conidia, submerged conidia, blastospores) have different adhesion properties and different cell wall surface carbohydrates, which affect pathogenesis as a result of differences in insect immune system recognition.

Germination is influenced by temperature, humidity, UV light, nutritional, and chemical environment. Fungal pathogens also have to be able to tolerate the toxic compounds present in the cuticles of some insects (e.g. caprylic or capric acids in Japanese silkworm, *Bombyx mori* and the rice stem borer, *Chilo suppressalis*). Following germination, hyphae must then enter the host. Some invade immediately upon germination, while others grow extensively over the host's surface first (e.g. *Metarhizium anisopliae* on wireworms, *Hyalius pales*). Extensive surface growth is correlated with hard host cuticle, and may allow the pathogen to detect thinner cuticle areas and build up inoculum potential. As with some plant pathogens, some fungal entomopathogens produce an appressorium, at the tip of the germ tube, prior to penetration (Figure 9.5). The site of penetration is variable and includes the arthrodial membranes between joints and between segments, direct penetration through cuticle, via the more vulnerable ventral surface, sense organs, and spiracles. Only a few others invade via the gut (e.g. *Smittium morbosum* in mosquito larvae), as digestive enzymes make this an inhospitable environment. Cuticle thickness is a major determinant of where entry occurs; *Entomophthora* spp. penetrate any part of the thin cuticle of flies, mosquitoes and aphids, but enter larger insects with thicker cuticle via arthrodial membranes. Penetration is brought about by a dual mechanical and enzymatic process.

When the fungus has succeeded in entering the insect, mycelial growth may be localised around the point of entry in the epidermis, but ultimately the body is usually extensively

TABLE 9.7 Fungal Pathogens of Insects

Kingdom/Phylum/Subphylum	Examples of pathogenic fungal species	Host
Fungus-like oomycetes	*Lagenidium giganteum*[a]	Mosquito larvae
	Aphanomyces labis	Mosquito larvae
	Pythium flevoense	Mosquito larvae
Chytridiomycota	*Myiophagus* spp.	Scale insects
Blastocladiomycota	*Catenaria* spp.	Small flies (but mostly nematodes)
	Coelomycidium spp.	Scale insects, beetle larvae, dipteran pupae
	Coelomyces spp.	Obligate pathogens requiring two aquatic hosts, mosquito larvae and crustaceans (e.g. copepods bugs) at different life cycle stages
Zygomycetes: Mucoromycotina	*Sporodiniella umbellata*	Pathogens of weak insects
Zygomycetes: DKH clade	*Smittium morbosum*	Mosquito larvae
	Some *Harpellales*	Adult black flies (Simuliidae)
Zoopagomycotina	*Zoopagaceae*	Trap hosts with adhesives and produce a restricted haustorium
	Cochlonemataceae	Rotifers, amoebae, rhizopods
Entomophthoromycotina	*Basidiobolus* and *Conidiobolus* spp.	Range of insects (and vertebrates)
	Entomophthora, Pandora, Zoophthora spp.	Range of insects
Ascomycota: Hypocreales	*Beauveria* spp.[a]	Broad host range
	Metarhizium spp.[a]	Broad host range
	Hypocrella spp.	Broad host range
	Cordyceps	Broad host range
	Ophiocordyceps	Broad host range
	Laboulbeniales	Obligate haustorial ectoparasites of insects and a few other arthropods
	Lecanicillium[a]	
Basidiomycota (only Septobasidiales)	*Auriculoscypha, Ordonia, Septobasidium, Uredinella* spp.	Scale insects

[a]*Registered biocontrol agent.*
Information extracted largely from Vega et al. (2012).

colonised in the **exploitation phase** (Figure 9.5a). Most fungal entomopathogens produce dispersal structures, such as blastospores or hyphal fragments, which circulate in the haemolymph before forming mycelium. Successful entry into the host does not ensure colonisation since insects have immune systems that can respond to invasion by eliminating or confining the pathogen. Exploitation is usually **necrotrophic**, with host death resulting from

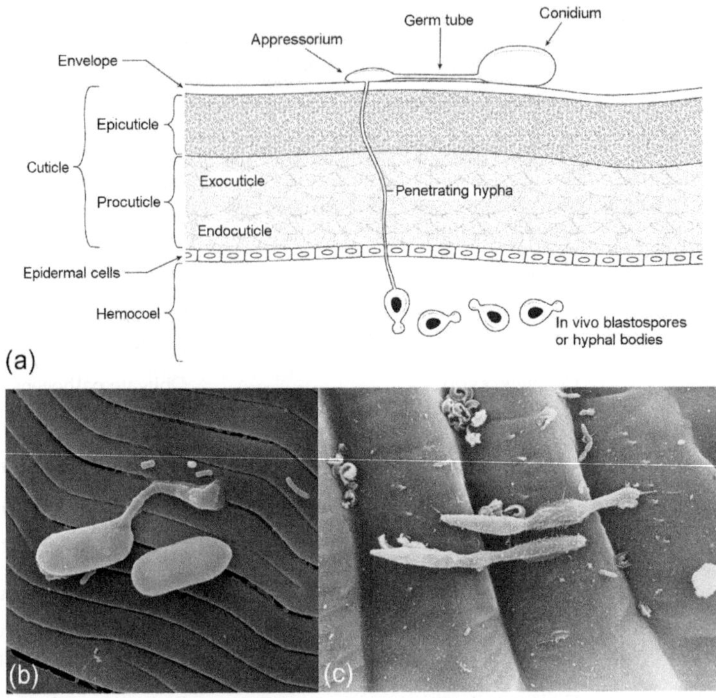

FIGURE 9.5 (a) Fungal pathogen infection of insects. A spore lands on the insect cuticle, adheres and then germinates, forming a germ tube and an appressorium. From the appressorium, a penetrating hypha enters. An appressorium forms, and from its under surface a fine hypha penetrates through the insect cuticle. Hyphae sometimes branch within the procuticle. The hypha breaches the epidermis and reaches the haemocoel, where it produces blastospores, which spread in the haemolymph before forming mycelium. (b) Two conidia of *Aschersonia* on the cuticle of whitefly. The upper conidium has germinated and produced an appressorium. (c) Germinated conidia of *Metarhizium anisopliae* on a tick. *Source: (a) From Vega et al. (2012), (b, c) © Tariq Butt.*

fungal toxins (e.g. destruxins), toxic proteases, or chronic disruption of host physiology following extensive mycelial development. Zygomycetes tend to colonise as mycelia whereas ascomycetes form budding, blastospores that colonise the haemocoel (body cavity), and then other tissues by mycelial spread. Some pathogens only colonise certain tissues, for example, *Entomophthora erupta* is confined to the abdomen of the green apple bug (*Lygus communis*), where it utilises the internal contents and posterior leg and wing muscles. Because the muscles of the pre- and mesothorax remain uncolonised, the insect is still able to move around even with ballistospores discharging from the ruptured abdomen, reminiscent of the sci-fi movie, *Alien*.

The parasitic relationship is sometimes **hemibiotrophic**, being initially biotrophic before becoming necrotrophic, or completely **biotrophic**. *Septobasidium* spp. (Basidiomycota) are epiphytic on tree bark, forming colonies with a tough outer layer of interwoven hyphae. Within the lower layers of mycelium armoured scale insects (Diaspididae) form a labyrinth of tunnels and chambers, in which they live and move about. A proportion of the scale insects are infected with the fungus; hyphae form coiled structures, analogous to haustoria, within the haemocoel of the host. Hyphae emanate from the natural orifices of the host, interconnecting with the external mycelia network, but this does not appear to interfere with the mobility of the insect. The lifespan of the scale insects is often lengthened, the insects remaining in a juvenile condition, resembling the effects of biotrophic parasites on host plants (p. 262). Nutritional advantages to the fungus are obvious, but since scale insects are provided with an environment buffered from external climate and protected from predators, and only some are colonised, the relationship could be considered mutualistic to the population.

Most fungi that are biotrophic parasites of insects and other arthropods have a small body size. They are commonly restricted to the host surface, with haustoria penetrating the host cuticle. Many have lost structures usually found in related taxa, even certain lifecycle states (e.g. sexual or occasionally asexual states). The Laboulbeniomycetes (Ascomycota) is the only group of insect ectoparasites that has diverged into many clades (over 2000 species, with 80% parasitizing beetles (Coleoptera)), appearing to be evolutionarily very successful. Unlike most fungi, Laboulbeniomycetes have a thallus with determinate growth (Figure 9.6). Some are dioecious; in some species the male comprising as few as three cells though the thalli of others can have several thousand cells. When a potential host makes contact with a mature thallus, the fungus releases sticky spores that attach to the host. The thallus that develops is attached to the host by an enlarged basal cell (foot cell), and nutrients are abstracted from the host through a peg-like or root-like haustorium below the surface of the cuticle. Spores are produced by the thallus and spread to other parts of the host's body, as well as to new hosts. There is usually no obvious major damage to the host. Like most obligate fungal biotrophic parasites of plants and animals, the Laboulbeniomycetes are largely highly host specific, with some even being specific to a particular sex of host, and others specific to certain parts of the body. For example, *Stigmatomyces baeri* usually develops on the upper side of the female fly host, but on the ventral surface of the male host.

The final stages of the life cycle of fungal entomopathogens are **exit** and **survival**, enabling the fungus to spread to a new host food source (Figure 9.7). Prior to death, the behaviour of parasitized insects is often altered by the fungus in a way that benefits the fungus, which is sometimes referred to as a zombie fungus. With fungi that cause 'summit disease', the infected insects (e.g. grasshoppers) congregate at the top of plants during late afternoon (Figure 9.7e and h), because of interference with the nervous system and demands for oxygen which is depleted as spiracles are blocked by hyphae. Fungal entomopathogens of above-ground insects commonly anchor the host in an exposed aerial position, using hyphae which grow into the plant. Sap-feeding insects are further anchored by their stylets, and many insects clasp plant tissues and each other when rigour mortis sets in (Figure 9.7). On the other hand, the zombie fungus *Ophiocordyceps unilateralis* causes the infected canopy nesting ants to move to the forest floor, where humidity and temperature are conducive for fungal growth; the ants affix themselves with their mandibles to the undersides of leaves. With some entomopathogens, volatile attractants lure potential hosts to the vicinity of the fungus; healthy green apple bugs (*Lygus communis*) are attracted to, and insert there stylets into, the spore masses of *Entomophthora erupta* on the dorsal surface of infected insects.

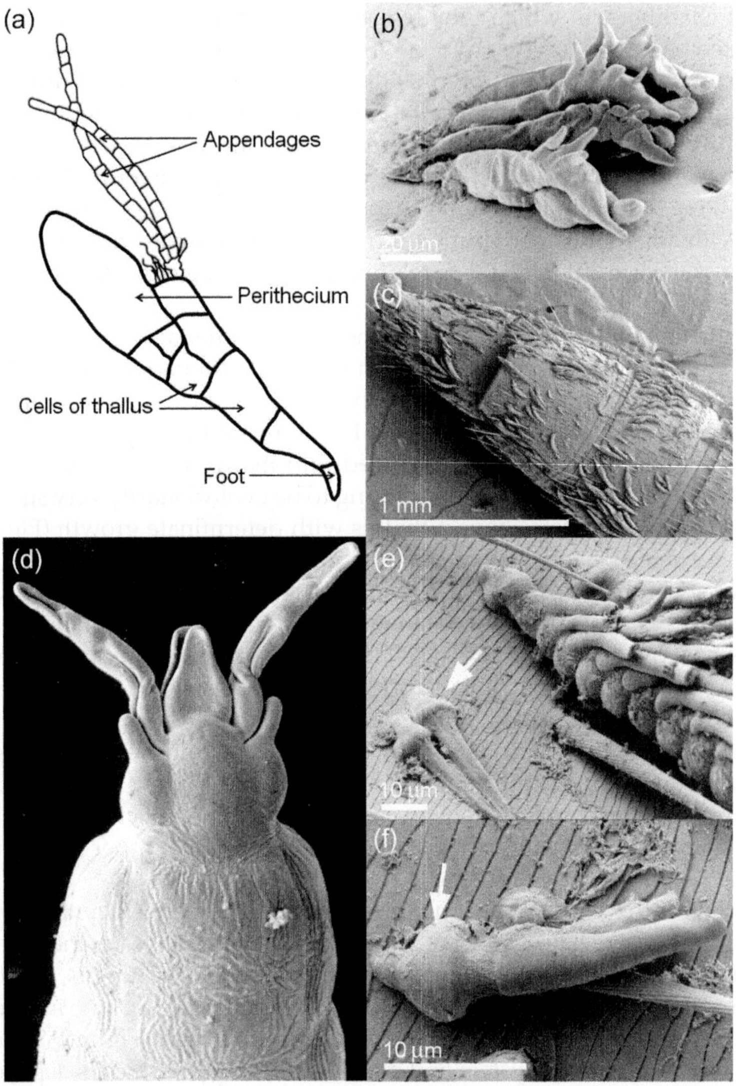

FIGURE 9.6 Laboulbeniales are biotrophic parasites on invertebrate hosts. Their simple, determinate thalli are very different from other fungi. Each ascospore has a foot cell – a swollen structure at the base – that attaches the spore to the insect cuticle. The foot cell acts like an appressorium and a fine penetration peg arises from it. The penetration peg grows in a very limited way into the insect cuticle – but presumably sufficiently to enable the fungus to tap into nutrients from the host. (a) Diagram showing main features. Scanning electron micrographs of: (b) young thalli of *Hesperomyces* sp. on a two spot ladybird (*Adalia bipunctata*); (c) many thalli of *Rhacomyces philonthinus* on the cuticle of a rove beetle (*Philonthus* sp.); (d) the apex of a *Hesperomyces* sp.; (e) *Rhacomyces philonthinus* ascospores (arrowed) and thalli; (f) *Rhacomyces philonthinus* ascospores (arrowed). The horn-like appendages are elaborate in some species, and are thought to play a role in ascospore discharge. *Source: All images © Alex Weir and Gordon Beakes.*

FIGURE 9.7 Pathogens of invertebrates have different ways in which they survive and spread in time and space. (a) *Metarhizium anisopliae*, the green muscardine fungus, is a common insect pathogen, and has been used extremely effectively as a bioinsecticide. Here the white mycelium is seen emerging from the body sutures of a chrysomelid beetle, with subsequent formation of green conidia. (b) Fruit bodies of *Ophiocordyceps amazonica* emerging from a tropical forest grasshopper that it has killed. (c) *Beauveria bassiana*, a bioinsecticide, has mummified this caterpillar. (d) The asexual stages of several different *Hypocrella* spp. emerging from whitefly (Aleyrodidae hemiptera) colonies that they have killed, on bamboo leaves. (e) Spores of an *Erynia* sp. bursting through the abdominal sutures of an onion fly (*Delia*), which has climbed to an elevated position on vegetation – classic symptoms of a 'summit' disease. (f) Not only do fungi kill insects but also arachnids. Here a *Cordyceps* sp. is emerging from a trapdoor spider. (g) *Lecanicillium lecanii* causing mass mortality of coffee green scale pest (*Coccus viridis*), another fungus effectively used as biopesticide. (h) *Ophiocordyceps nymphoides* emerging from the neck region of a ground dwelling ponerine ant that has climbed a tropical forest shrub and died clinging to it. *Source: All images © Harry Evans.*

Pathogen dispersal is usually via spores, often with different spore types fulfilling different roles such as spread within the insect population during the active season, and survival from one season to the next. Pathogens in the Entomophthorales have a range of spore types; ballistospores are shot off, by sudden pressure release, but if they fail to hit a suitable target, sticky secondary spores emerge from the ballistospores. Other spore types are also produced for survival, and in some for dispersal in water. Spore release is often timed to coincide with times when potential hosts are abundant. Species of *Erynia*, for example, that parasitize the biting blackflies (simuliids) of water courses, release ballistospores from corpses anchored to rock surfaces above water at precise times in the late afternoon, coinciding with arrival of healthy flies. Some Ascomycota (e.g. *Hypocrella* spp.) also produce different spore types, with ascospores being shot off and behaving similarly to ballistospores of the Entomophthorales. If these ascospores fail to land on a suitable host, they produce sticky spores on needle-like projections. Resting/survival spores are often produced by pathogens within the host, and released when the cadaver disintegrates. Others (e.g. *Cordyceps* and *Hirsutella* spp.) form pseudosclerotia (p. 61). The type of infection propagule can affect insect mortality. For example, mortality of tobacco budworm (*Heliothis virescens*) following infection by *Beauveria bassiana* is greater and earlier when infected by blastospores than by other condia.

The success of entomopathogenic fungi in nature has led to the development of their use as biopesticides (p. 423). Understanding their biology and ecology is important to the successful application and formulation of such products. It would be wrong to assume that all fungal entomopathogens are obligate, and additional roles have recently been discovered for some. Several, including *Beauveria* species, could be endophytes (pp. 234–239). Others, including *Beauveria bassiana*, are antagonistic to plant pathogens, and *Lecanicillium* species are parasitic on fungal pathogens of plants. Species of *Beauveria*, *Isaria*, and *Metarhizium* are common members of the soil mycobiota and may grow in the rhizosphere on exuded carbon sources.

Nematophagous and Rotifer-Trapping Fungi

There are over 300 species of fungi from Ascomycota, Basidiomycota, Chytridiomycota, Mucoromycotina and also fungus-like oomycetes that obtain their nutrition by predation or parasitism of nematode adults, instars, eggs or cysts (Table 9.8). These fungi have attracted considerable attention for over 100 years, perhaps, in part, because of the fascinating and dramatic mechanisms which some of them adopt for capturing their prey, but also because of their potential as biocontrol agents of plant-parasitic nematodes. They are found worldwide from the tropics to the Arctic and Antarctica, as saprotrophs in soil, mosses, dung, and decomposing wood and leaf litter. There are at least three different categories of **predatory fungi**: (1) those which form trapping structures; (2) endoparasites which infect nematodes as spores whose saprotrophic phase is predominantly within the nematode body; and (3) parasites of cyst nematodes that almost exclusively infect the females, eggs or larvae (Table 9.8).

Those in the first category form extensive mycelia that produce a variety of trapping structures depending on species, including adhesive pegs, knobs, rings, and three-dimensional networks of loops, and non-adhesive constricting rings, which attract and capture nematodes (Figure 9.8) and rotifers. The constricting ring traps are particularly spectacular; nematodes are captured when their bodies pass through the three-celled noose. When triggered by the touch of a nematode's body within the noose, rapid flow of water into the cells causes them to expand instantaneously, securing the nematode. Nematode-trappers are found in diverse

TABLE 9.8 Examples of Nematophagous and Rotifer-Trapping Fungi and Fungus-Like Organisms, and the Ways in which they Obtain their Prey

Predacious mechanism	Fungal species	Fungal phylum	Ecological characteristics
Spontaneously produced adhesive knobs, branches, and non-constricting rings	*Dactylaria candida*, *Dactylaria gracilis*, *Monacrosporium cionopagum*	Ascomycota	Slow growing soil saprotrophs; great predacious ability
Adhesive networks inducible by the presence of nematodes	*Arthrobotrys conoides*, *Arthrobotrys oligospora*	Ascomycota	Fast growing soil saprotrophs; weak predacious ability
Constricting rings	*Arthrobotrys dactyloides*	Ascomycota	Soil saprotroph
Adhesive projections	*Hyphoderma* spp.	Basidiomycota	Adhesive projections – stephanocysts – on basal hyphae of fruit bodies
	Pleurotus spp.		Hour glass-like projections, with a viscous adhesive and immobilising toxin
	Hohenbuehelia spp.		Similar structure to *Pleurotus* All are wood decay fungi
Endoparasites; conidia are either ingested or adhere to cuticle	*Meristacrum* spp.	Zygomycete	Adhesive conidia on conidiophores that protrude from the nematode
	Drechmeria coniospora	Ascomycota	Adheres to mouthparts
	Harposporium spp.	Ascomycota	Hook onto mouthparts. Lodge in oesophagus
	Meria coniospora	Ascomycota	Endoparasites are mostly obligate
	Catenaria anguillulae	Chytridiomycota	Zoospores colonise at natural openings
Endoparasites of cyst nematodes	*Dactylella oviparasitica*	Ascomycota	Egg parasite
	Nematophthora gynophila	Oomycete	On females and cysts of *Heterodera* spp.
Endoparasites; infective gun cells	*Haptoglossa* spp.	Oomycete	Obligate parasites of bactiverous nematodes and rotifers
Toxin production	*Arthrobotrys* spp. *Pleurotus ostreatus*	Ascomycota Basidiomycota	Many species that produce trapping structures also produce toxins
Rotifer predators	*Cephaliophora navicularis*	Ascomycota	Adhesive knobs, often attached to mouthparts of bdelloid rotifers
	Sommerstorffia spinosa	oomycete	Adhesive pegs catch hard-bodied rotifers

fungal groups (Table 9.8), but the majority belong to a monophyletic group within the family Orbiliales (Ascomycota). In the Orbiliales, the trapping mechanisms have evolved along two main lineages, one which produces adhesive trapping structures and the other constricting rings. Some fungi produce traps spontaneously, whereas others require environmental triggers. In some (e.g. *Arthrobotrys oligospora*) traps tend to form when nitrogen is limiting.

There is a sophisticated chemical 'dialogue' between the fungal predator and its nematode prey. The presence of nematodes stimulates trap formation, the triggers being small peptides, with a high proportion of non-polar and aromatic residues. On the other hand, the mycelium of many, perhaps all, species often produces chemical attractants to nematodes, and trapping organs can provide further attractants. The initial event in capture by fungi is mediated by lectins on the trap and carbohydrates on the surface of the nematodes. Different fungal species

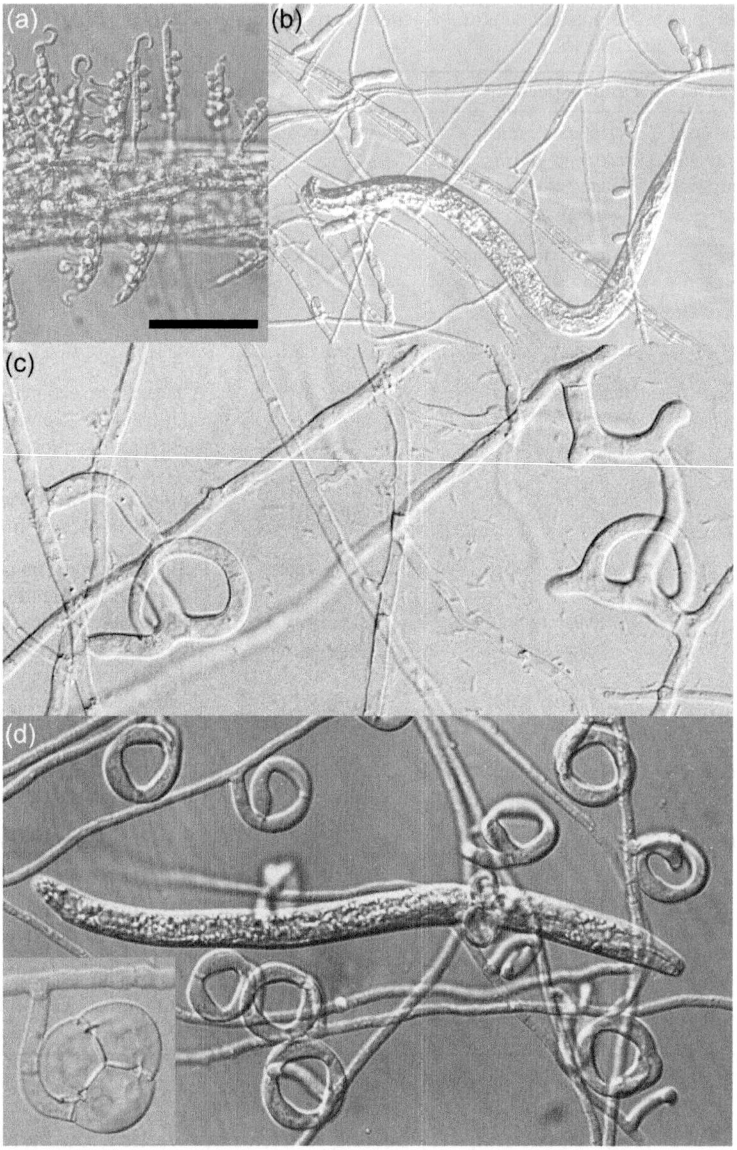

FIGURE 9.8 Nematode trapping fungi. A range of fungi are able to capture and utilise the contents of nematodes, using a range of different mechanisms. Some employ (a) conidia that lodge in the mouth or gut of the nematodes, germinate, grow, and eventually emerge producing more conidia, such as the hook-shaped conidia of *Harposporium anguillulae*. Others form adhesive structures, such as (b) small projections on hyphae of *Monacrosporium cionopagum* or (c) adhesive reticulate networks of *Monacrosporium* sp. (d) The constricting ring traps, such as those of *Drechmeria coniospora*, are a more complex mechanism. They comprise three celled loops, the cells of which rapidly expand when triggered (d and inset – *Monacrosporium doedycoides*) by the passage of a nematode. In all cases, when the nematode has been captured, hyphae penetrate the cuticle and digest the body content of the nematode. Scale bar = 50 μm (a); 100 μm (b, d); 20 μm (c). *Source: © John Webster.*

have different carbohydrate-binding proteins with different specificities for carbohydrates – 2-deoxyglucose in the case of *Dactylaria candida* and *N*-acetylglucosamine for *Arthrobotrys oligospora*. After contact with the nematodes, the prey is immobilised within a few hours by production of a nematotoxin (e.g. serine protease PII by *Arthrobotrys oligospora*). The wood decay basidiomycete *Pleurotus ostreatus* produces hour-glass shaped projections on its hyphae that exude an adhesive substance, and a toxin – ostreatin (2-decenedioic acid) – that immobilises prey within a minute. Other wood decay fungi can grow into nematodes through body orifices (e.g. *Amylostereum* species grow into the vulva of *Deladenus*, Figure 9.9a).

When the nematodes are attached to the fungal surface, a penetration peg then pierces the cuticle using mechanical pressure and hydrolytic enzymes. Inside the nematode's body, the penetration tube swells to form an infection bulb from which hyphae develop and spread, producing extracellular enzymes that rapidly digest the nematode. They spread by producing non-infective conidia.

In contrast to trap formers, most **endoparasites** form limited mycelium outside of their hosts. Rather, they produce infective zoospores or conidia that are either sticky and adhere to the host's cuticle or mouthparts, or are non-adhesive and lodge in the oesophagus when ingested. The spores germinate on or in the host nematode or rotifer, mycelium permeates and digests the body, and infective conidia are produced on hyphae that emerge (Figure 9.8a). The infective zoospores of the Chytridiomycota and fungus-like oomycetes form thalli within the host; zoospores are produced within these thalli and are subsequently released to the outside. Perhaps the most remarkable example of a parasitic infection structure is the 'gun cell' of *Haptoglossa* spp. (oomycete) (Figure 9.10).

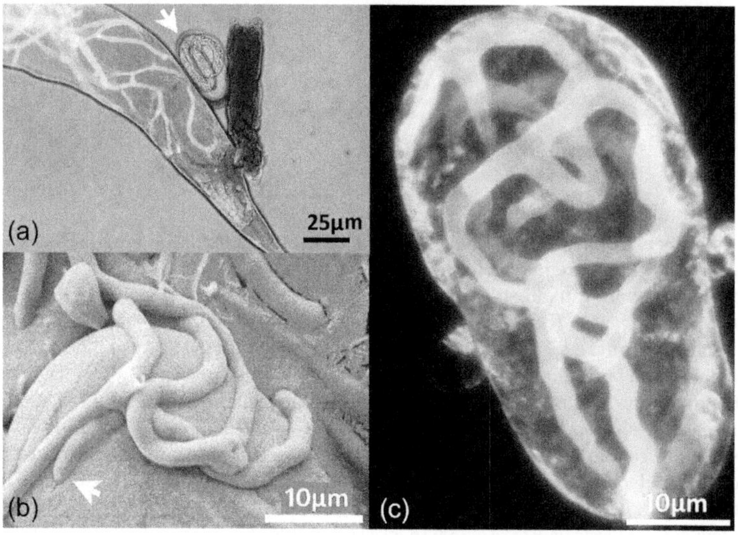

FIGURE 9.9 (a) Hyphae (fluorescence stained) of the wood decay fungus *Amylostereum areolatum* within the body of a nematode *Deladenus siricidicola*, having entered through the vulva. Note the healthy, unparasitised egg (arrowed). (b, c) Parasitism of *Deladenus siricidicola* eggs by *Amylostereum areolatum*. (b) Cryogenic scanning electron micrograph of *Amylostereum areolatum* hyphae growing over the egg surface. The point of penetration is arrowed. (c) Fluorescence stained hyphae within a parasitized egg. *Source: From Morris, E.E., Hajek, A.E., 2014. Eat or be eaten: fungus and nematode switch off as predator and prey. Fungal Ecol. 11, 114–121.*

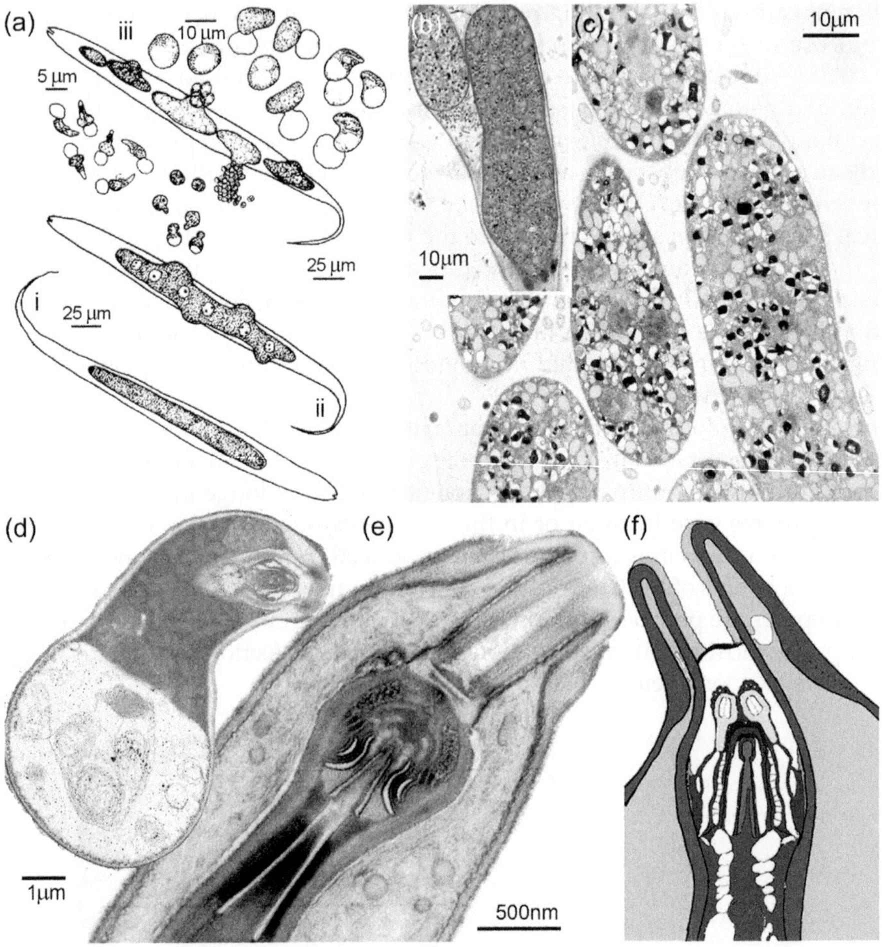

FIGURE 9.10 Fungus-like oomycete gun-cell-forming parasites of nematodes. Fungus-like species in the oomycete genus *Haptoglossa* produce extremely complex infection structures. (a) Life-cycle of the parasite; light (b) and electron (c) micrographs of young thalli showing typical densely packed cytoplasm; Gun cell of (d) an unnamed species and (e) *Haptoglossa erumpens*; (f) Diagrammatic summary of gun cell needle chamber apparatus in *Haptoglossa dickii*. Once infection has occurred a sausage-shape thallus develops within the body of the nematode (a(i), b, c), sporangia start to form (a(ii)) and large and small spores emerge from mature sporangia (a(iii)). The two spore types give rise to the two different infection gun cells. The spores germinate to form an infection cell (d, e, f), which matures to form an intracellular needle-shaped 'missile', within a looped, invaginated, infection tube, with a large vacuole at the base (d). The inverted tube containing the needle apparatus has associated restraining apparatus (f). This tube everts within a fraction of a second, penetrating the host's cuticle, and injecting the parasite's cytoplasm into the nematode. *Source: (a) From: Glockling, S.L., Beakes G.W., 2000. Video microscopy of spore development in Haptoglossa heteromorpha, a new species from cow dung. Mycologia 92, 747–753; (b, c) © Gordon Beakes; (d, f) from Beakes, G.W., Sekimoto, S., 2009. The evolutionary phylogeny of oomycetes – insights gained from studies of holocarpic parasites of algae and invertebrates. In: Lamour, K., Kamoun, S. (Eds.), Oomycete Genetics and Genomics: Diversity, Interactions and Research Tools. John Wiley & Sons, Hoboken, pp. 1–24; (e) from Beakes, G.W., Glockling, S.L., Sekimoto, S., 2012. The evolutionary phylogeny of the oomycete "fungi". Protoplasm 249, 3–19.*

Parasites of nematode eggs and cysts are taxonomically different to those that parasitize adults, but are usually soil saprotrophs found in plant roots, though some wood decay fungi can also parasitise nematode eggs (Figure 9.9). Specific morphological features are not apparent on the mycelium, but appressoria-like swellings are sometimes seen on hyphae. Zoospores have a role in others, and these are attracted to their hosts.

Aspergillosis of Coral

Marine invertebrates also suffer from fungal disease. In the Caribbean and Florida Keys, there is an ongoing epizootic among the sea fan corals (*Gorgonia* species), which was first reported in 1995, though an epidemic in the 1980s may have been the same disease. Lesions, galls, and a purpling of the coral tissue occur, that can lead to death. Over 50% of sea fan tissue has been lost due to mortality arising from infection with *Aspergillus sydowii*, a soil saprotroph, though the marine strain is different. In inoculation experiments, *A. sydowii* from terrestrial sources was not pathogenic to Gorgonian coral. Infection induces a generalised defence response in the host, including production of antifungal compounds in the vicinity of infection. Like opportunistic pathogens of humans (see above), pathogenicity depends on host immune status. Elevated temperature has been hypothesised to drive outbreaks of the disease; production of antifungal compounds is much greater at elevated temperatures, but *A. sydowii* grows optimally at 30 °C. However, it appears that prevalence of the disease has declined steadily since the 2000s.

FUNGI AS FOOD AND HABITAT FOR ANIMALS

Obtaining carbon and energy from photosynthetic organisms often necessitates the breakdown of complex molecules (e.g. cellulose), plant defence compounds (e.g. polyphenols) and lignin, yet few organisms aside from basidiomycete fungi possess the necessary biochemical machinery to breakdown the latter. Further, the nutritional value of plants is often relatively low. The carbon:nutrient ratio of different species of plants and different plant parts vary considerably, but carbon:nitrogen (C:N) and carbon:phosphorus (C:P) ratios of leaves are usually in the range 25:1 to 100:1 and 450:1 to 1850:1, respectively, and for wood range from 350:1 to 500:1 and 1250:1 to >3500:1. Thus, a lot of resources must be consumed to obtain small amounts of nitrogen and phosphorus, which are essential building blocks for making enzymes, proteins, DNA, etc. When fungi decompose plant tissues the nutrient content increases, as the carbon is lost during respiration as CO_2. Also, mycelium itself has seven times lower C:N and C:P ratios (35:1 and 505:1) than undecayed wood. It, therefore, comes as no surprise that many organisms feed directly on fungi, mycophagy being most prevalent amongst members of the phylum Arthropoda, although there are also many examples within Mollusca, Enchytraeidae, Annelida, Collembola, and Nematoda. Many invertebrates also consume fungi indirectly when they eat decaying plant material. The benefits of feeding on fungal-decayed rather than undecayed plant material are seen with the death watch beetle, *Xestobium rufovillosum* (Anobiidae), which typically attacks wood colonised by certain basidiomycetes (e.g. *Laetiporus sulphureus*, *Trametes versicolor*, and *Coniophora puteana*). The length of time it takes for the beetle to complete its lifecycle is related to the state of decay of the wood: in undecayed wood the larvae develop very slowly or not at all, but in decayed wood the life cycle is completed within 10–17 months. This is not only because nutrients are more concentrated in the decayed wood but also because softer decayed wood is easier to consume. Many mutualistic associations have evolved between invertebrates and fungi, based on fungal improvement of nutrition (e.g. ants, higher termites, ambrosia beetles, and wood wasps, pp. 330–334).

Mycelium

Not all mycelia are equally palatable to invertebrates, as some contain toxic chemicals or produce toxic volatiles. When mycelium is eaten, its morphology, foraging patterns, physiology, and biochemistry often change dramatically, though effects differ depending on fungal species, resource status, grazing intensity (density), and invertebrate species (Figure 9.11), probably reflecting differences in patterns of grazing. Some invertebrates preferentially feed on individual hyphal tips at the colony margins, some on swathes of hyphal tips, and others in patches within the colony. Nematodes feed, not by severing hyphae like most, but by inserting stylets and 'sucking out' contents. Changes take place in the immediate vicinity of grazing and also several cm from the site of grazing. Counterintuitively, grazing does not always decrease mycelial growth; very low density grazing sometimes stimulates hyphal coverage, and when heavy grazing ceases there can be catch-up or even over-compensatory growth.

Mycelium is also eaten by man, as part of fermented foods (e.g. tempeh, pp. 410–411). Meat substitutes are also made from fungal mycelium – *Fusarium venenatum* is cultured on an industrial scale, formed into the texture of meat and marketed as Quorn® (p. 416).

Fruit Bodies

Some macroscopic fruit bodies are perennial (e.g. polypore brackets) while others are ephemeral lasting for only a few days (e.g. Agaricales). Both provide food sources and breeding grounds for a diversity of invertebrates, including nematodes, enchytraeids, mites, Collembola (springtails), Coleoptera (beetles), and Diptera (flies). Insects that feed on agarics are mostly polyphagic (i.e. a wide variety of species are eaten) because the fruit bodies are unpredictable and ephemeral resources. On the other hand, half of the insects feeding in polypore brackets are monophagous, which probably results from evolution of physiology to cope with the specific chemical defences, and mouth parts to cope with the physical structure of the host bracket. Some *Drosophila* species are resistant to the toxin amanitin, found in some poisonous fleshy fruit bodies (e.g. the death cap *Amanita phalloides*) and are able to feed on it, whereas other *Drosophila* are not. While some invertebrates use living fruit bodies, others use those that are decaying, with different invertebrate species found at different stages of decay. The invertebrates that are polyphagous on brackets tend to colonise them after they have been decaying for some time, by which time the defence chemicals have presumably decreased. Early colonisers, that tend to be monophagous, have the advantage of less competition and a more nutritious environment, but they have a narrower choice of fungi. Many fruit bodies produce a bouquet of volatile organic compounds (VOCs), the composition of which can change during ageing, which are attractive to the flies and beetles that feed and breed in the fruit bodies, and are at least partly responsible for the partitioning of resource use between species.

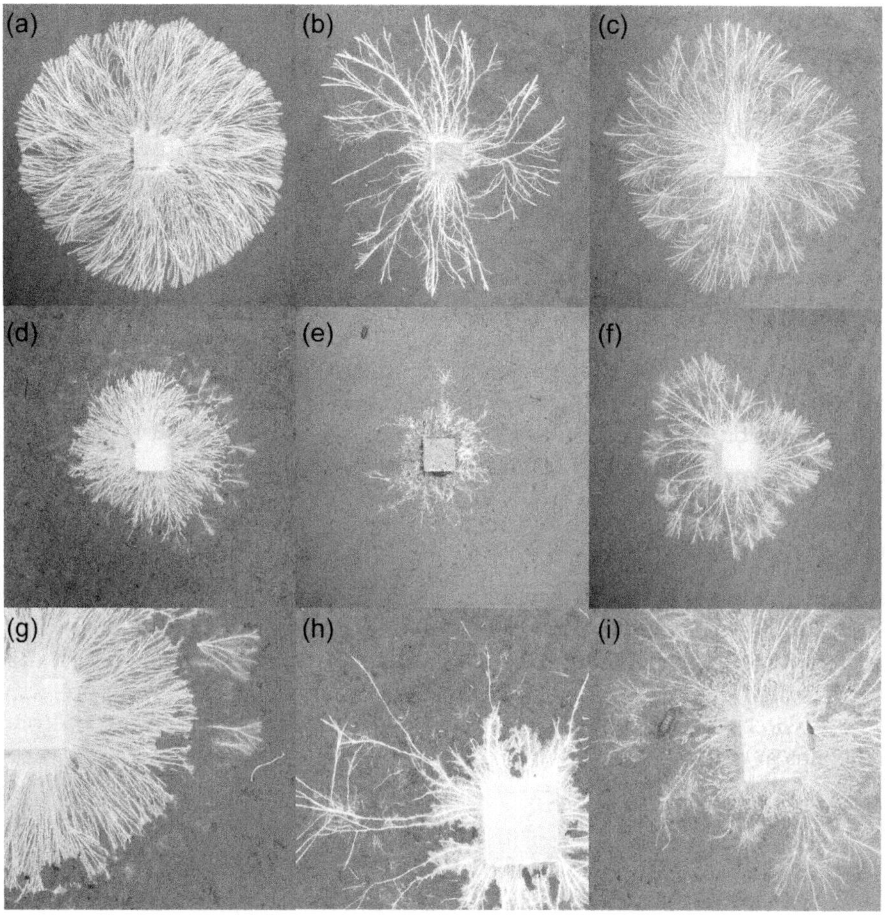

FIGURE 9.11 Many invertebrates graze on mycelium because of its high nutritional quality. Grazing can have dramatic effects on the fungus, but this depends on the extent of grazing, the invertebrate and fungus species, and the size of the mycelium. Digital images showing un-grazed mycelia of (a) *Hypholoma fasciculare*, (b) *Resinicium bicolor* and (c) *Phanerochaete velutina*, and the respective effects of 10 days of grazing by (d) 5 millipedes *Blaniulus guttulatus*, (e) 5 woodlice *Oniscus asellus* and (f) 60 collembola *Folsomia candida* on 24 × 24 cm trays of woodland soil. Invertebrates vary in their styles of grazing; (g) millipedes (Myriapoda), seen here on *Hypholoma fasciculare*, graze in an arc rather like a windshield wiper; (h) collembola, seen here on *Resinicium bicolor*, graze on finer cords and hyphae; and (i) woodlice (Isopoda), seen here on *Phanerochaete velutina*, often graze in straight lines, like a lawn-mower. *Source: All images © Tom Crowther.*

Invertebrate feeding on fruit bodies can decrease reproductive fitness of fungi by decreasing the area of functional hymenium, causing altered morphology due to gall formation, and by damaging/destroying spores during passage through the gut. Fungi have evolved a variety of defence mechanisms, including chemicals, as already mentioned, and stephanocysts that kill nematodes, Collembola and other small invertebrates. In contrast, some fungi are dependent on invertebrates for spore dispersal; for example, species of stinkhorn (Phallales) smell of rotting flesh, which attracts Diptera to feed on the sticky head of spores, dispersing them elsewhere on their legs and bodies (p. 92). Some species can only germinate following passage through the gut of an appropriate invertebrate (e.g. spores of a *Ganoderma* spp. must pass through a fly larva gut).

Fruit bodies also form an important part of the diet of some vertebrates. Some are opportunistic mycophagists, others (e.g. squirrels and several marsupials) concentrate on fruit bodies when they are available, and a few are obligately mycophagous (e.g. the marsupial rat kangaroos – *Potorous longipes* and *Potorous gilbertii*). Some marsupials and rodents are the main dispersers of spores of hypogeous mycorrhizal fungi – truffles. Like other animals, humans also eat fungi because they are nutritious – high in protein and low in fat – and some are tasty or of medicinal value (pp. 403–404). *Agaricus bisporus* is cultivated in 70 countries worldwide, accounting for 38% of mushroom sales. It is the fungus traditionally cultivated in Europe and North America but others, favoured more in Asian cuisines, are gaining in popularity, including the wood-rotting oyster fungi (*Pleurotus* spp., 25% of world sales with China leading in cultivation and consumption), the paddy straw mushroom (*Volvariella volvacea*, 16% of world sales), and shiitake (*Lentinula edodes*, 10% of world sales).

Lichen Thalli

Lichens are grazed by many invertebrates, including gastropods, and are often inhabited by invertebrates. Lichens produce carbon-based secondary compounds (CBSCs) that defend against lichenivory. Gastropods are selective in the regions of thalli upon which they graze, avoiding parts high in CBSCs, such as the sorelia of *Lobaria scrobiculata*, and preferring parts low in CBSCs, such as the cephalodia of *Nephroma arcticum* and the cortex and photobiont layers of lichens in general.

MUTUALISTIC ASSOCIATIONS BETWEEN FUNGI AND ANIMALS

Many invertebrates and ruminants have taken feeding on fungi one step further and over millennia, have coevolved with specific fungi forming innumerable liaisons in which both partners benefit (mutualism). High quality food is made available to the animals, and sometimes they also acquire fungal enzymes (e.g. cellulases) that remain active in the gut. Occasionally animals also benefit from an improved environment. Benefit to the fungus usually accrues from animals either bringing organic resources to the fungus or taking the fungus to suitable organic resources – often in specialised organs, and in some cases includes the animal creating a suitable microclimatic environment and even reducing competitors. Some of the most studied and intriguing associations are described below. Symbioses based largely on nutrition of both partners include fungi in the guts of invertebrates and vertebrates, and the farming of fungi by leaf-cutting ants and higher termites. Symbioses based on vectoring fungi to appropriate habitats and nutrition of the vector include those with ambrosia beetles and siricid woodwasps. Other symbioses are based on fungal nutrition and provision of suitable environment for the invertebrate, including the relationship between *Septobasidium* and scale insects (p. 318), and the jet black ant (*Lasius fuliginosus*) – aphid – fungus association (p. 332).

Fungi as Gut Symbionts

Some fungi live only in the guts of invertebrates and vertebrates. The *Asellariales* and *Harpellales* (in Subphylum **Kickxellomycotina**; formerly the Trichomycetes, though this grouping was phylogenetically heterogeneous, also containing protozoa) are obligate symbionts in the guts of freshwater, marine and terrestrial crustaceans, insects and millipedes, distributed worldwide. They have a small determinate body, which attaches to the gut lining by means of a holdfast. The vast majority do not penetrate the gut peritrophic membrane, though there are exceptions, e.g. *Stachylina minuta*, but it does not reach the cuticle. They appear to feed by absorption of nutrients as gut fluids flow over them, and may aid in food digestion, so the majority are commensal (cause no harm) or mutualistic. However, a few gut-inhabiting fungi are parasites/pathogens; some produce cysts in the ovaries of adult black flies, others cause sterilisation of the host, and *Smittium morbosum* kills mosquito larvae by preventing them from moulting. The invertebrate gut is a harsh and transient environment, because of digestive enzymes, the lining is shed at ecdysis (moulting), and older regions of the peritrophic membrane break up and are expelled with faeces. To cope with this problem, Kickxellomycotina grow rapidly between moults and membrane breakdowns, converting almost all thallus material to spore production. Spores expelled at ecdysis are likely to be reingested when the host eats its shed skin. Kickxellomycotina (pp. 32–33) growing in aquatic hosts often have adaptations, such as long appendages on spores, that keep them in the vicinity of hosts.

Over 650 species of yeasts, mostly (>75%) Saccharomycetes (Ascomycota) and a few Tremellales (Basidiomycota), have only been cultured from the guts of beetles. As more and more insects are examined, more yeast species new to science are being discovered. The location of yeasts in invertebrate guts varies between species, some being found in the crop at the anterior end, others in the midgut and others in the hindgut. In anobiid beetles the fore part of the midgut comprises a group of blind sacs like a bunch of grapes – **mycetomes** – lined with large cells that contain many yeast cells. The yeasts provide enzymes for digestion, detoxify toxic plant metabolites, and provide essential nutrients, as many of them are able to synthesise a wide variety of B-complex vitamins. The symbioses can be intra- or extracellular, and yeasts are not only found in the gut, but sometimes in the fat body, eggs, haemolymph, and blood.

The symbiosis seems to be more important to the insect than to the yeasts; the yeasts obviously have a nutrient rich environment, but a major benefit may be as a means of dispersal to different habitats. In *Drosophila*, no more than three yeast species are usually isolated from individual flies, and this varies seasonally, apparently reflecting what species are present on feeding substrata. In rice planthoppers, yeasts are found in the fat body and are vertically transmitted by movement of the yeast to the primary oocyte, where eggs become infected. In some beetles, vertical transmission occurs when the female smears yeast cells onto the egg shells, which are then consumed when the larvae hatch.

The rumen and the rest of the digestive tract of herbivorous vertebrates provide a rich supply of food to microbes, though conditions are unfavourable to most fungi. In the rumen, temperatures are 39–40.5 °C, raised above mammal body temperature by fermentation; conditions are largely anaerobic (headspace atmosphere of around 65% CO_2, 30% CH_4, 4% N_2, H_2, 0.6% O_2), with any O_2 rapidly used by facultative anaerobes; the pH is continually

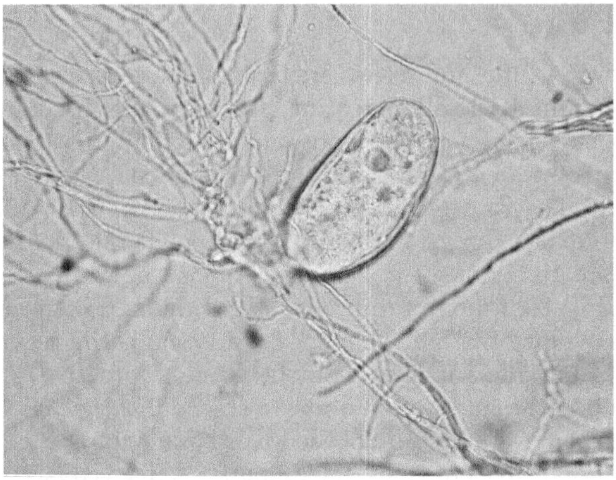

FIGURE 9.12 Neocallimastigomycota inhabit the rumen. The rhizoids grow into and breakdown plant fragments. *Source: © Gareth W. Griffith.*

being modified by host food and microbial metabolism but is buffered around pH 6.4–6.8 by bicarbonate rich saliva; and redox potential is extremely low. In the rumen fluid, anaerobic bacteria (10^4–10^5 ml^{-1}) and ciliate protozoa (10^{10}–10^{11} ml^{-1}) predominate, but in the entire rumen content Neocallimastigomycota (p. 30) can form up to 20% of the microbial biomass. Anaerobic fungi in six genera – *Anaeromyces, Caecomyces, Cyllamyces, Neocallimastix, Piromyces,* and *Orpinomyces* – are invaders of newly ingested plant material. Zoospores are attracted to plant material. They encyst and form a thallus with rhizoids that penetrate into the plant material (Figure 9.12). The rhizoids secrete a range of enzymes that break down cellulose, xylans, starch, and other polymers. The nucleus from the zoospore is retained in the cyst and then in the sporangium, which is separated from the anucleate rhizoids by a septum. Zoospores are subsequently released from the sporangium, to infect more plant material. Newborn ruminants do not have the complex rumen microbial community found in mature animals, but colonisation occurs rapidly before the rumen becomes functional, probably from resistant stages of Neocallimastigomycota in faeces and possibly from saliva during licking by the mother. Also, several species are found as spores in air samples.

Ants and Higher Termites

Obligate mutualisms between fungi and social insects, in which the former are housed in the nest of the latter, have evolved independently twice; once in attine ants on the American continent, 50 million years ago (mya), and once in higher termites (Macrotermitinae) of central African tropical rainforest, 24–34 mya, later spreading to savanna and to Asia. The leaf-cutting genera of the attine ants are associated with a diversity of basidiomycete lineages, including *Attamyces, Leucoagaricus* spp., and *Lepiota* spp., whereas the termites are associated with a single genus *Termitomyces*. No reversions to the ancestral ant and termite life style are known, indicating the huge benefit of these mutualisms to the animals. There have, however, been fungal reversals from the ant/fungus mutualism, though not for termite fungi. The basis of the mutualisms is that the worker ants and termites provide the fungus in the nest with organic matter (Figure 9.13). In the case of the ants, this is portions of leaves, and with the termites it is comminuted (i.e. broken into fine fragments), dead grass and wood. The insects do not have the enzymes necessary to digest lignocellulose, but the fungi do. Fungal decomposition of the organic matter results in nitrogen- and phosphorus-rich fungal biomass that provides the insects with most of their food. For example, the N concentration in the nests of *Macrotermes bellicosus* with *Termitomyces* was 2 times greater in the fungus comb (the main part of the fungus garden constructed from primary faeces that is subsequently completely consumed) and 20 times greater in the mycotêtes (fungal nodules; asexual fruit bodies – coremia) than in the original food. The fungus-growing termites eat the mycotêtes, which mix with consumed organic matter in the gut and are then deposited in the faeces, on the

FIGURE 9.13 Mutualism between fungi and ants and termites. (a) Fungus garden from the nest of *Macrotermes* belicosus. (b) A close up showing the mycotêtes (white spheres of fungal material) that are consumed by the termites. While in some species sexual fruit bodies are only produced when the termites abandon the nests, in others they are regularly produced, as in (c) *Termitomyces reticulates*, seen here growing straight from the underground fungus gardens of *Odontotermes badius*. Note the pseudorhiza connecting the fruit body with the fungus garden. *Source: (a) © Karen Machielsen; (b) Photo taken by Prof. Renoux © Duur Aanen; (c) from Aanen and de Beer (2007).*

top of the fungus comb. In contrast, with the fungus-growing ants, all fruiting is suppressed and the fungus spreads vegetatively from the older, bottom to the newer, upper regions of the fungus garden; small colonised fragments are added, together with faecal droplets. These droplets stimulate growth of the fungus and contain incompatibility compounds that are antagonistic towards genetically different fungal symbionts from other ant colonies brought in by foraging workers, maintaining a monoculture. Both ants and termites acquire cellulases and hemicellulases when they consume the fungus. These acquired enzymes survive gut passage and are concentrated in the faecal droplet that is deposited on fresh plant material, preparing it for fungal colonisation and increasing the initial mycelial growth. Contaminant fungi are kept out by: physical removal of spores by licking; chemical secretions from the ants; and antibiotics secreted by actinomycetes (*Streptomyces*) growing on the ants' bodies, targeted at a virulent Ascomycota mycoparasite (*Escovopsis*) that specialises in attacking fungus gardens. Similar grooming and antibiotic secretions maintain a monoculture in the nest of termites. However, *Leucoagaricus* spp., *Lepiota* spp., and other Basidiomycota associated mutualistically with the ants are more competitive than in the termite-fungus mutualism, being able to survive for approaching 12 days following abandonment of nests, as opposed to almost immediate replacement in termite nests. The fungal symbionts also benefit from a fairly constant microclimatic environment. In African savanna, above-ground temperatures can vary by 35 °C between winter night and summer daytime, but in the nests they remain in the narrow range of 29–31 °C, maintained by the ventilation system within the mounds.

When new colonies are founded it is essential that the fungal symbiont establishes within the new nest. Young ant queens take asexual propagules, in an infra-buccal pocket, from the natal nest on their mating flights, and use this to start the garden in the new colony (vertical transmission). Some termites also transmit the fungus to the new colony via a single parent in a similar way. *Termitomyces* spp. are carried to new colonies by queens in the genus *Microtermes* and by kings of *Macrotermes bellicosus*, but other Macrotermitinae acquire their symbionts while foraging: sexual spores are consumed and survive passage through the gut, and are then deposited in a faecal pellet on the new fungus comb. With such horizontal transmission, it might be expected that it would be easy to exchange fungal symbionts between termite lineages but this rarely happens between genera. Basidiomycota symbionts of ants fruit sexually only rarely. *Termitomyces* spp., however, produce large basidiocarps on the soil surface. These connect, via pseudorhiza, with the nest sometimes 2 m below ground (p. 64, Figure 9.13). *Termitomyces titanicus* produces the largest known edible fruit bodies, weighing up to 2.5 kg with caps greater than 60 cm diameter.

The highly successful mutualistic relationships between fungi and leaf-cutting ants and termites are based on invertebrate nutritional requirements and both provision of food and habitat for the fungus. An intriguing mutualistic relationship between the ascomycete *Cladosporium myrmecophilum* and the jet black ant appears to be based on improvement of the ant's environment rather than on nutrition. This common European ant, which lives in hollow tree trunks or beneath tree stumps, makes its nest from a cardboard-like material called carton. Carton comprises small wood and soil particles, honeydew (secreted by aphids that are farmed by the ant) and mycelium of the fungus. The fungus is inoculated into new walls by workers adding fragments from old nest walls. The ants do not appear to eat the fungus, though mycelium only emerges from nest walls in the absence of ants, implying that the ants prevent mycelial emergence in some way.

Ambrosia Beetles, Bark Beetles, and Wood Wasps

Bark and wood are difficult to break down and are relatively poor in nutrients, thus again it comes as no surprise that many invertebrates that feed on these plant tissues associate with fungi. In doing so, they obtain nutritional benefits of concentrated nitrogen sources and essential nutrients, including vitamins and sterols. The extent of dependence on fungi ranges from opportunistic to facultative to obligate. In bark beetles (that feed on fungi) and ambrosia beetles (that feed in subcortical tissues), mycophagy has evolved many times. In many species, when the adults lay their eggs in wood, fungi are also deposited. These fungi have been transported in specialised structures on the body of the beetle, termed **mycangia**. The most developed mycangia are invaginations of the beetle integument that are lined with secretory cells or glands. Less developed structures include shallow pits, deeper pits, and tubes not associated with glands, and setae. The glandular secretions contain amino acids, fatty acids, phospholipids, and sterols that support growth of fungal propagules, protect them from desiccation and may act against fungi not symbiotic with the beetle. Most mycelial fungal symbionts are *Ophiostoma* species (Ascomycota), with some *Ceratocystis* species (Ascomycota) and a few *Entomocorticium* species (Basidiomycota). These ascomycetes are well-adapted to arthropod dispersal, extruding ascospores from long-necked perithecia at a height where they are likely to contact the invertebrate's body, and having sticky spores shaped to allow multiple contact points. Asexual conidia are also vectored by the beetles, and in those beetles that produce sac mycangia, it is only these spores that are carried; asexual spores are found only in the galleries housing pupae, though ascomata form in old, disused galleries. Chlamydospores and yeasts of the basidiomycetes are found in mycangia, but not conidia or basidiospores. The mycelial fungi often have species-specific associations with beetles. Ascomycete yeasts are also symbionts of bark beetles, though adult beetles often carry several species; *Pichia capsulata* and *Pichia pini* are prevalent yeast associates with most bark beetle species.

The larvae of ambrosia beetles (of which there are around 3400 species) develop in woody xylem. The fungal symbionts penetrate the wood beneath the beetle tunnels, and line the walls with a thin mycelial cover or separate mycelial cushions, in both cases comprising short erect hyphae with swollen tips or chains of cells. Both larvae and adults feed only on fungi. Unlike ambrosia beetles, bark beetles lay their eggs in the inner bark of trees and feed on the phloem, which is rich in nutrients. Nonetheless many bark beetles, especially species of *Dendroctonus* and *Ips*, are associated intimately with fungi, as both larvae and adults feed on mycelium, yeasts, and conidia. Young adults of *Ips avulsus* and *Ips calligraphus* also seek out and ingest entire perithecia of *Ophiostoma ips*. The species of mycangial fungus carried by a bark beetle can have differential effects on the beetle. For example, many more progeny are produced by *Dendroctonus ponderosae*, and emerge sooner, when *Ophiostoma clavigerum*, rather than *Ophiostoma montium*, is the food source.

Woodwasps (Siricidae), like the wood-boring beetles, carry basidiomycete fungal symbionts (*Amylostereum* species) in a pair of pouches – mycangia, at the base of the ovipositor, and inoculate the fungus together with eggs into wood. Three species of *Amylostereum* – *Amylostereum areolatum*, *Amylostereum chailletii* and *Amylostereum laevigatum* – are involved, and the relationship is obligatory and species-specific. These fungi cause white rot, softening the wood and improving the relative nutrient content, so that wood wasp larvae can feed by burrowing through decomposing wood. The fungus benefits by being carried to and inoculated directly into a suitable resource. Asexual oidia are carried by the woodwasps, but the fungi can also spread via basidiospores. In the northern hemisphere clones are common amongst isolates of *Amylostereum areolatum*, but less common in *Amylostereum chailletii*, indicating that woodwasps are more important in spreading the former than the latter.

Further Reading

General

Bourtzis, K., Miller, T.A. (Eds.), 2003. Insect Symbiosis. CRC Press, Boca Raton. 2006, 2008.
Evans, H.C., Boddy, L., 2010. Animal slayers, saviours and socialists. In: Boddy, L., Coleman, M. (Eds.), From Another Kingdom: The Amazing World of Fungi. Royal Botanic Garden, Edinburgh.
Heitman, J., 2011. Microbial pathogens in the fungal kingdom. Fungal Biol. Rev. 25, 48–60.
Heitman, J., Filler, S.G., Edwards Jnr., J.E., Mitchell, A.P., 2006. Molecular Principles of Fungal Pathogenesis. ASM press, Washington.
Kurzai, O. (Ed.), 2014. Human Fungal Pathogens. The Mycota, Vol. XII. Springer, Heidelberg.
Vega, F.E., Blackwell, M. (Eds.), 2005. Insect-Fungal Associations. Oxford University Press, Oxford.

Medical mycology

Allergies

Simon-Nobbe, B., Denk, U., Pöll, V., Rid, R., Breitenbach, M., 2008. The spectrum of fungal allergy. Int. Arch. Allergy Immunol. 145, 58–86.
Pringle, A., 2013. Asthma and the diversity of fungal spores in air. PLoS Pathog. 9. e1003371.

Mycotoxicoses

Bennett, J.W., Klich, M., 2003. Mycotoxins. Clin. Microbiol. Rev. 16, 497–516.

Mycoses

Abu-Elteen, K.H., Hamad, M., 2011. Anti-fungal agents for use in human therapy. In: Kavanagh, K. (Ed.), Fungi: Biology and Applications, second ed. John Williamson, Chichester, pp. 191–217.
Brown, G.D., Denning, D.W., Goew, N.A.R., Levitz, S.M., Netea, M.G., White, T.C., 2012. Hidden killers: human fungal infections. Sci. Transl. Med. 4, 1–9.
Colombo, A.L., Tobon, A., Restrepo, A., Queiroz-Telles, F., Nucci, M., 2011. Epidemiology of endemic systemic fungal infections in Latin America. Med. Mycol. 49, 785–798.
Gow, N.A.R., van de Veerdonk, F.L., Brown, A.J.P., Netea, M.G., 2012. *Candida albicans* morphogenesis and host defence: discriminating invasion from colonization. Nat. Rev. 10, 112–122.
Haynes, K., 2011. Emerging fungal pathogens. Microbiol. Today 38, 26–29.
Hector, R.F., Laniado-Laborin, R., 2005. Coccidiomycosis – a fungal disease of the Americas. PLoS Med. e2, 0015–0018.
Heitman, J., 2011. Microbial pathogens in the fungal kingdom. Fungal Biol. Rev. 25, 48–60.
Johnson, L., Gaab, E.M., Sanchez, J., Bui, P.H., Nobile, C.J., Hoyer, K.K., Petersen, M.W., Ojcius, D.M., 2014. Valley fever: danger lurking in a dust cloud. Microbes Infect. 16, 591–600.
Kwon-Chung, K.J., Sugui, J.A., 2013. *Aspergillus fumigatus* – what makes the species a ubiquitous human fungal pathogen? PLoS Pathog. 9. e1003743.
Lin, X., Heitman, J., 2006. The biology of *Cryptococcus neoformans* species complex. Annu. Rev. Microbiol. 60, 69–105.
Nucci, M., Marr, K.A., 2005. Emerging fungal diseases. Clin. Infect. Dis. 41, 521–526.
Soubani, A.O., Chandrasekar, P.H., 2002. The clinical spectrum of pulmonary aspergillosis. Chest 121, 1988–1999.
Sullivan, D.J., Moran, G.P., Coleman, D.C., 2011. Fungal infections of humans. In: Kavanagh, K. (Ed.), Fungi: Biology and Applications, second ed. John Williamson, Chichester, pp. 171–190.

Pathogens of other vertebrates

Blehert, D.S., Hicks, A.C., Behr, M., Meteyer, C.M., Berlowski-Zier, B.M., Buckles, E.L., Coleman, J.T.H., Darling, S.R., Gargas, A., Niver, R., Okoniewski, J.C., Rudd, R.J., Stone, W.B., 2009. Bat white-nose syndrome: an emerging fungal pathogen? Science 323, 227.

Fisher, M.C., Garner, T.W.J., Walker, S.F., 2009. Global emergence of *Batrachochytrium dendrobatidis* and amphibian chytridiomycosis in space, time and host. Annu. Rev. Microbiol. 63, 291–310.

Fisher, M.C., Henk, D.A., Briggs, C.J., Brownstein, J.S., Madoff, L.C., McCraw, S.L., Gurr, S.J., 2012. Emerging fungal threats to animal, plant and ecosystem health. Nature 484, 186–194.

Gleason, F.H., Chambouvet, A., Sullivan, B.K., Lilje, O., Rowley, J.J.L., 2014. Multiple zoosporic parasites pose a significant threat to amphibian populations. Fungal Ecol. 11, 181–192.

Van den Berg, A.H., McLaggan, D., Diéguez-uribeondo, J., van West, P., 2013. The impact of water moulds *Saprolegnia diclina* and *Saprolegnia parasitica* on natural ecosystems and the aquaculture industry. Fungal Biol. Rev. 27, 33–42.

Pathogens of invertebrates

Hajek, A.E., St. Leger, R.J., 1994. Interactions between fungal pathogens and insect hosts. Annu. Rev. Entomol. 39, 293–322.

Lundgren, J.G., Jurat-Fuentes, J.L., 2012. Physiology and ecology of host defense against microbial invaders. In: Vega, F.E., Kaya, H.K. (Eds.), Insect Pathology, second ed. Academic Press, San Diego, pp. 461–480.

Thorn, R.G., Barron, G.L., 1984. Carnivorous mushrooms. Science 224, 76–78.

Tunlid, A., Ahrén, D., 2011. Molecular mechanisms of the interaction between nematode-trapping fungi and nematodes: lessons from genomics. In: Davies, K., Spiegel, Y. (Eds.), Biological Control of Plant-Parasitic Nematodes: Building Coherence Between Microbial Ecology and Molecular Mechanisms. Springer, Dordrecht, pp. 145–170.

Vega, F.E., Goettel, M.S., Blackwell, M., Chandler, D., Jackson, M.A., Keller, S., Koike, M., Maniania, N.K., Monzón, A., Ownley, B.H., Pell, J.K., Rangel, D.E.N., Roy, H.E., 2009. Fungal entomopathogens: new insights on their ecology. Fungal Ecol. 2, 149–159.

Vega, F.E., Kaya, H.K. (Eds.), 2012. Insect Pathology, second ed. Elsevier, Amsterdam.

Vega, F.E., Meyling, N.V., Luangsa-ard, J.J., Blackwell, M., 2012. Fungal entomopathogens. In: Vega, F.E., Kaya, H.K. (Eds.), Insect Pathology, second ed. Academic Press, San Diego, pp. 171–220.

Weir, A., Blackwell, M., 2005. Fungal biotrophic parasites of insects and other arthropods. In: Vega, F.E., Blackwell, M. (Eds.), Insect-Fungal Associations: Ecology and Evolution. Oxford University Press, Oxford, pp. 119–145.

Mutualistic associations between fungi and animals

Aanen, D.K., Boomsma, J.J., 2006. Social-insect fungus farming. Curr. Biol. 16, R1014–R1016.

Aanen, D.K., de Beer, W., 2007. Farming fungi: termites show the way. Quest 3, 22–25.

Currie, C.R., Scott, J.A., Summerbell, R.C., Malloch, D., 1999. Fungus-growing ants use antibiotic-producing bacteria to control garden parasites. Nature 398, 701–704.

Harrington, T.C., 2005. Ecology and evolution of mycophagous bark beetles and their fungal partners. In: Vega, F.E., Blackwell, M. (Eds.), Insect-Fungal Associations: Ecology and Evolution. Oxford University Press, New York, pp. 257–291.

Mueller, U.G., Schultz, T.R., Currie, C.R., Adams, R.M.M., Malloch, D., 2001. The origin of the attine ant-fungus mutualism. Quart. Rev. Biol. 76, 169–197.

Six, D.L., 2003. Bark beetle-fungal symbioses. In: Bourtzis, K., Miller, T.A. (Eds.), Insect Symbiosis. CRC Press, Boca, Raton, pp. 97–114.

Slippers, B., Coutinho, T.A., Wingfield, B.D., Wingfield, M.J., 2003. A review of the genus *Amylostereum* and its association with woodwasps. S. Afr. J. Sci. 99, 70–74.

Vega, F.E., Dowd, P.F., 2005. The role of yeasts as insect endosymbionts. In: Vega, F.E., Blackwell, M. (Eds.), Insect-Fungal Associations: Ecology and Evolution. Oxford University Press, New York, pp. 211–243.

Animal feeding on fungi

Boddy, L., Jones, T.H., 2008. Interactions between Basidiomycota and invertebrates. In: Boddy, L., Frankland, J.C., van West, P. (Eds.), Ecology of Saprotrophic Basidiomycetes. Elsevier, Amsterdam, pp. 153–177.

Claridge, A.W., Trappe, J.M., 2005. Sporocarp mycophagy: nutritional, behavioural, evolutionary, and physiological aspects. In: Dighton, J., White, J.F. Jr., Oudemans, P. (Eds.), The Fungal Community: Its Organization and Role in the Ecosystem. Taylor & Francis, Boca Raton, pp. 599–611.

Jonsell, M., Nordlander, G., 2004. Host selection patterns in insects breeding in bracket fungi. Ecol. Entomol. 29, 697–705.

CHAPTER

10

Interactions Between Fungi and Other Microbes

Lynne Boddy
Cardiff University, Cardiff, UK

In all but the most sparsely inhabited environments (e.g. cold deserts or hot dry deserts), fungi continuously interact with other organisms. This can dramatically affect the size of the interacting populations or the fitness of interacting individuals. Commonly, several or many populations/individuals will interact with one another simultaneously but, for simplicity, it is usually only interactions between two species/individuals that have been studied. For the most part we will adopt this approach here, but it is essential to remember that the situation in the natural environment is far more complex. The overall effects on each member of the interacting pair of populations/individuals can be positive, negative, or neutral, resulting in six possible eventualities for the pair of interacting organisms (Table 10.1). This has been simplified to three scenarios: (1) the outcome of the interaction is beneficial to either or both, but detrimental to neither, (2) the outcome is detrimental to either or both, and (3) neither organism is affected. The last situation seems unlikely to occur very often if organisms are truly interacting, rather than passively coexisting.

Fungi participate in a wide range of interactions that result in benefit to one or both partners, especially when they are interacting with organisms from other kingdoms, for example, with plants (Chapter 7) as mycorrhizas and endophytes, with algae and cyanobacteria to form lichens (pp. 228–234), with invertebrates and ruminants (Chapter 9), and with some bacteria (pp. 356–357). However, no such intimate interactions have yet been described between different species of fungi in nature, although undoubtedly they exist. Mutualism can occur when physiological activity of a fungus and another organism are complementary, such as when one fungus synthesises or releases a compound which the other fungus requires but cannot otherwise obtain, and vice versa. For example, *Nematospora gossypii* and *Bjerkandera adusta* can grow in mixed culture in the laboratory, but not individually, on a medium lacking biotin, inositol, and thiamine. Whilst the former can synthesise thiamine, but not biotin or inositol, the latter can synthesise biotin and inositol but not thiamine. Another example might be the cogrowth of fungi which produce different cellulases, resulting in more rapid utilisation of cellulose. However, this might not necessarily be mutualism if more rapid utilisation

The Fungi
http://dx.doi.org/10.1016/B978-0-12-382034-1.00010-4

TABLE 10.1 Classification of the Possible Interactions Between Two Species/Individuals

Original scheme			**Simplified scheme**
Effect on interacting species/individuals		**Name of interaction**	
++	Both are positively affected	Mutualism	Benefit to either or both, but detrimental to neither
+0	One is positively affected while the other is unaffected	Commensalism	
−−	Both are negatively affected	Competition/combat	Detrimental to either or both
+−	One is positively affected to the detriment of the other	Parasitism/predation	
−0	One is negatively affected, while the other is unaffected	Amensalism	
00	Activities of neither have any effect on the other	Neutralism	Neither are affected

The effects, either positive (+), negative (−), or neutral (0), on population size, population growth rate or individual fitness of two interacting populations/individuals. EP Odum first used a scheme like that depicted on the left side of the table in the early 1950s. This was simplified (right hand columns) for fungi in the early 1980s by Alan Rayner & Joan Webber.

of the resource is not in keeping with a fungus' particular life strategy. Likewise, the induction of fruit body production and of cord or rhizomorph formation (pp. 58–60) by the presence of another organism have sometimes been considered to be examples of mutualism or neutralism, but these are more likely to occur as a result of deleterious action.

All fungi will interact negatively with other organisms at some time in their lives. For some it is a way of life, and the sole or main way of obtaining nutrition, for example, the parasites and pathogens of plants (Chapter 9), animals (Chapter 11), and other fungi (see below). For others it is also usually directly or indirectly related to obtaining food, a situation often described as competition. The definition of competition used for macroorganisms, and applicable to fungi, is the negative effects of one organism on another resulting from the consumption of a resource of limited availability, or from controlling access to a resource. Macroecologists consider two types of competition: interference competition and exploitation competition. Interference competition refers to the situation where one organism inhibits another by, for example, producing allelopathic chemicals or shading out slower growing or smaller plants. Exploitation competition, on the other hand, is the situation when one organism uses a resource and thereby reduces its availability to another organism. These terms have sometimes been used in relation to fungi; certainly some fungi do massively interfere with other microbes, as outlined later in this chapter. Also, fungal mycelia do sometimes directly compete with each other and with other microbes for dissolved nutrients, for example, while growing through soil. However, for saprotrophic or necrotrophic fungi that grow in solid plant tissues (e.g. wood or leaf litter), the idea of exploitative competition and interference competition are not easy to separate as these fungi gain access to nutrients by competition for space/territory. Thus, for them it is not sensible to use the terms interference and exploitation competition.

Competition between fungi in solid, organic resources can be divided into two types – primary resource capture and secondary resource capture. The former refers to the situation where a resource is uncolonised, and success in capturing this type of resource is determined by factors such as good dispersal, rapid spore germination and growth, as well as the ability to use the substrates present. Such fungi are said to have R-selected or ruderal characteristics. Secondary resource capture refers to the situation where resources are already colonised by other fungi; aggressive antagonistic interactions (often called combat) are necessary to capture occupied territory or to defend it against aggressors.

This chapter begins by detailing negative interactions amongst fungi, brought about by antagonism at a distance and following contact – both mycoparasitism and larger scale mycelial interactions, including describing how morphology, biochemistry and gene expression changes, and how these affect physiology and ecology. We then go on to consider negative, positive, and mutualistic interactions of fungi with bacteria, archaea, viruses, and protists.

FUNGAL–FUNGAL INTERACTIONS

Fungal hyphae, mycelia, and yeasts can interact both with individuals of the same and different species (Figure 10.1). Intraspecific (i.e. between members of the same species) interactions, including those related to sexual reproduction and those that are not (i.e. somatic), are described in Chapter 4. Here we confine ourselves to **aggressive interspecific** interactions. These interactions can occur when fungi (1) are at a distance from each other (i.e. there is no contact between hyphae), (2) make contact at the hyphal level, or (3) when large parts of mycelia meet with other large mycelia. Overall there is a range of possible ultimate outcomes to antagonistic interaction between different fungal species, in terms of the territory that they occupy: (1) deadlock, where neither species gains headway, (2) replacement, where one species wrests territory from the other, (3) partial replacement, where one species captures some but not all of the opponent's territory, and even (4) mutual replacement, where one species obtains some of the territory formerly occupied by the other and vice versa (Figure 10.2). Aggressive interactions are not only major determinants of fungus community development, but are also of considerable interest because of the potential of some fungi to be used as biocontrol agents of plant pathogenic fungi (Chapter 8).

Antagonism at a Distance

Antagonism between fungi that have not made physical contact occurs when one or other, or both of the opponents produces volatile and/or diffusible chemicals, including enzymes, toxins, other anti-fungal metabolites, or alters the pH of the environment. It is easy to see the effects of diffusible chemicals in agar culture, as colony size is reduced when opponents are not touching, and often mycelia are unable to meet. Effects of volatiles can be seen by taping a culture of a fungus above a culture of an antagonist and comparing growth with that of the same fungus above uncolonised medium. Reactions often vary depending on the species against which an antagonist is paired, suggesting a reciprocal exchange of chemical signals and subsequent recognition. Chemical signalling by volatile and/or diffusible organic compounds (VOCs and DOCs) plays a major role in fungal recognition systems. These chemicals can be termed infochemicals or semiochemicals.

Microorganisms produce a range of different VOCs, with some of the main compounds detected from soil listed in Table 10.2. Different microbial species have characteristic VOC profiles, though these can vary in amount and composition depending on growth substrates (including amino acids), climate, and as the antagonistic interactions progress. It is

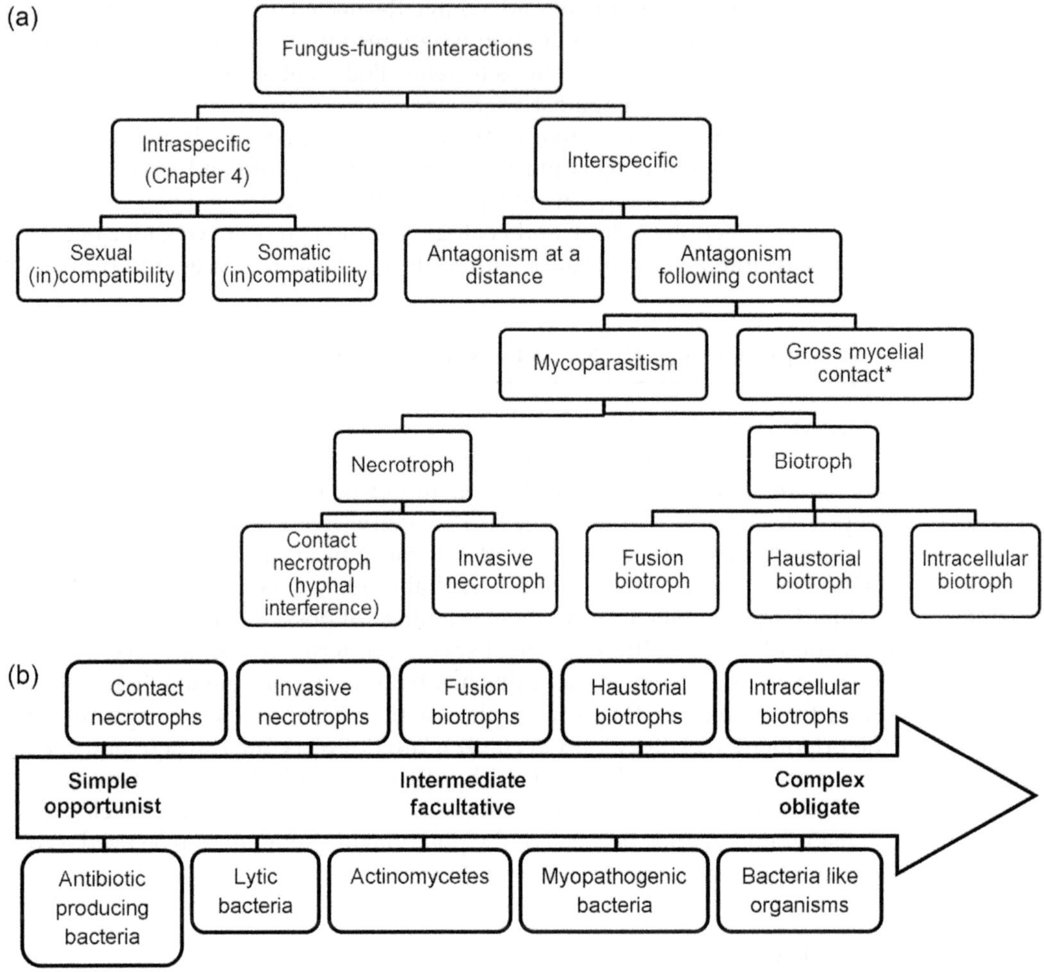

FIGURE 10.1 (a) Spectrum of fungus-fungus interactions. *The lack of further categories within gross mycelia contact reflects lack of knowledge and understanding rather than lack of complexity. (b) Evolution of fungus host–parasite interactions, showing mycoparasites on the top of the arrow, and bacterial antagonists beneath. *Source: Panel (b) is based on Kobayashi and Hillman (2005).*

not a single chemical that is responsible for recognition and antagonistic effects but several or many chemicals in the bouquet. Nonetheless, some classes of VOCs have greater effects than others. For example, aldehydes and ketones, including decanal, heptanal, 2-propanone, 2-methyl-1-butanol, and octanal, produced by *Trichoderma* spp. are particularly inhibitory to mycelial extension of wood decay basidiomycetes. VOCs affect gene expression, and can alter the profile of proteins produced by an affected fungus.

There are a wide variety of types of DOCs, including aromatic compounds. DOCs must operate over shorter distances than VOCs in soil and organic substrata. On artificial media they have been shown to affect spore germination, mycelial foraging, mycelial morphology,

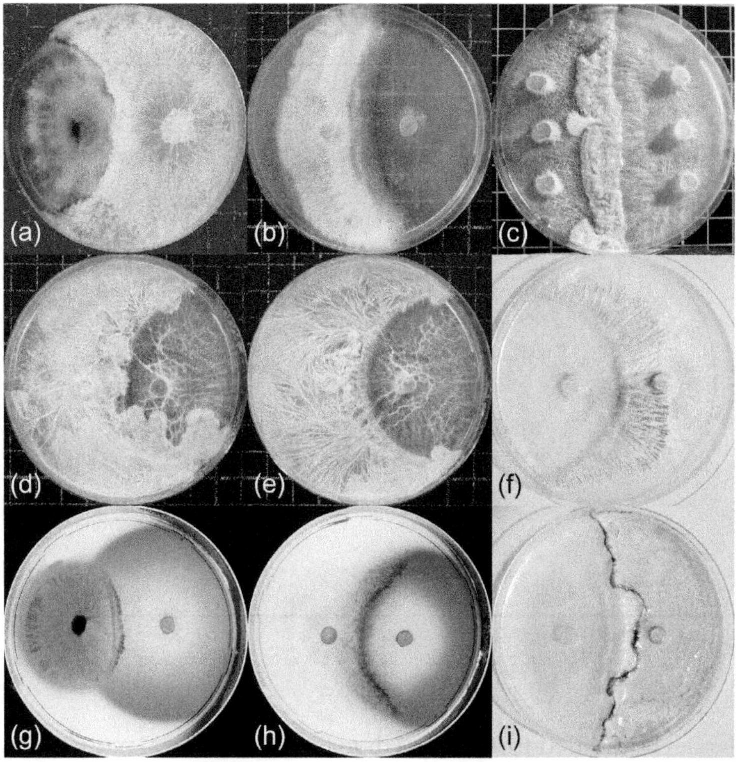

FIGURE 10.2 Examples of interspecific interactions between mycelia of wood rotting basidiomycetes (unless stated differently) in agar culture, showing mycelia barrages, fans and cords. (a) *Bjerkandera adusta* (right) starting to replace *Hypoxylon fragiforme* (Ascomycota); (b) *Trametes versicolor* (right) has almost completely replaced *Stereum gausapatum*; (c) *Hypholoma fasciculare* (right) replacing *Resinicium bicolor*. The reddish brown pigment is produced by *Resinicium bicolor*. (d) and (e) *Trametes versicolor* (left) and *Hypholoma fasciculare* (right). In (e) *Hypholoma fasciculare* has completely over grown *Trametes versicolor* as cords, and eventually completely replaces it, but in (d) while *Hypholoma fasciculare* has overgrown its opponent as cords in some places, *Trametes versicolor* is replacing *Hypholoma fasciculare* at the bottom right. (f) Production of superoxide by *Hypholoma fasciculare* (left) replacing *Trametes versicolor* (right), indicated by purple staining (nitroblue tetrazolium). Intense laccase activity during interactions between (g) *Hypholoma fasciculare* (left) and *Trametes versicolor* (right) and (h) *Bjerkandera adusta* and *Trametes versicolor*, indicated by purple staining (ABTS: 2,20-azino-bis(3-ethylbenzothiazoline-6-sulfonic acid) diammonium salt). *Source: Panel (c) © Timothy Rotheray; all other images © Jennifer Hiscox.*

and to increase production of ligninolytic enzymes in some fungi. DOCs leaching from wood previously colonised by fungi sometimes inhibit and sometimes stimulate extension rates of other fungi. Release of DOCs by fungi in wood is likely to influence which fungi follow during community development (p. 254).

Antagonism Following Contact: Mycoparasitism

There is a spectrum of mycoparasitic relationships with biotrophs and necrotrophs at the two extremes, and many intermediate types between (Table 10.3, Figure 10.1). The biotrophs

TABLE 10.2 Volatile Organic Compounds (VOCs) Evolved from Soil Under Aerobic and Anaerobic Conditions

Alcohols	Aldehydes	Aromatics	Butyl, Ethyl and Methyl esters	Ketones	Sulphides
Butan-1-ol	2-Methyl-butan-1-al	Benzene	Acetic acid	Butan-2-one	Dimethyl sulphide
Butan-2-ol	3-Methyl-butan-1-al	Benzaldehyde	Butanoic acid	3-Hydroxy butan-2-one	Dimethyl disulphide
Ethanol		Dimethyl benzene	2-Methyl-butanoic acid	Pentan-2-one	Dimethyl trisulphide
2-Methyl propan-1-ol		Ethyl benzene	3-Methyl-butanoic acid	Pentan-3-one	2-Methyl propylsulphide
2 Methyl propan-2- ol		Trimethyl benzene	2-Methyl propanoic acid	Propane-2-one	
Propan-1-ol				4-Methyl pentan-2-one	
				5-Methyl heptan-2-one	

The organisms generating the VOCs include both fungi and bacteria.
From Wheatley, R.E., 2002. The consequence of volatile organic compound mediated bacterial and fungal interactions. AAnoton v Leeu. 81, 357–364.

and necrotrophs can be divided further, based on the physiological nature of the interactions. As with necrotrophic pathogens of plants and animals (Chapters 8 and 9), necrotrophic mycoparasites tend to have a broad host range, with relatively unspecialized parasitic mechanisms, and kill their host. In contrast, as with biotrophic relationships between fungi and other organisms (Chapters 7–9), the biotrophic relationships between one fungus and another are complex, controlled and relatively non-destructive, and often, but not always, with narrow host ranges that have co-evolved. Biotrophic mycoparasites are dependent on the host fungus for survival. Mycoparasitic species can often not be defined as causing a specific type of parasitism, as some change their behaviour towards the host as the mycoparasitic interaction progresses. For example, parasitism that is initially biotrophic can later change to necrotrophy. Some fungi grow biotrophically on certain hosts but necrotrophically on others, for example, *Hypomyces chrysospermus* is necrotrophic on fleshy fruit bodies of Basidiomycota, but it can grow biotrophically within *Botrytis cinerea* and *Trichothecium roseum* hyphae when they colonise a mushroom that it has already parasitized. Some mycoparasites form different hyphal structures when they contact different hosts, and can have different effects on different hosts, due to different mechanisms being employed and to differential sensitivity to toxins. *Trichoderma virens*, for example, produces an antibiotic, gliovirin, which is strongly inhibitory to the fungus-like *Pythium ultimum*, but not to the basidiomycete *Rhizoctonia solani* (=*Thanatephorus cucumeris*), nor to the zygomycete *Rhizopus arrhizus*, nor to the ascomycete *Verticillium dahliae*.

It is not known when mycoparasitism first evolved, but it is likely an ancient mode of nutrition. *Glomus*-like spores have been found in the Lower Devonian Rhynie chert, from 400 million years ago (mya). These contained chytrid-like structures within. The oldest known fossil of an agaric was found in amber from the Early Cretaceous (100 mya), and this contained not only a mycoparasite but a hypermycoparasite (i.e. the parasite of a parasite). Mycoparasitism of basidiomycetes likely extends much further back, but they have not fossilised well. It is often assumed that the relationship between obligate biotrophs and their hosts is evolutionarily more advanced than that of necrotrophs, because the former relationships are more complex, but this is not the case. Biotrophic endoparasitism is a way of life for some of the most primitive fungal groups, and may even be the ancestral mode of nutrition in fungi.

There are three main types of biotrophic relationships: (1) intracellular parasites, (2) haustoria-producing parasites, and (3) fusion parasites (Table 10.3, Figure 10.1). With both fusion and haustoria-producing biotrophs, intimate connections with the host are necessary, allowing nutrient transfer from host to parasite. Haustoria-producing biotrophs actually penetrate host hyphae, but fusion biotrophs do not, and the latter relationship may be less complex.

With the **intracellular parasites**, the entire fungal thallus enters the host fungus. For example, many aquatic mycoparasitic Chytridiomycota penetrate the host wall and discharge their entire cytoplasm into that of the host. Nutrients are absorbed directly through the host–parasite interface. The presence of some intracellular mycoparasites, such as *Ampelomyces* on powdery mildew (pp. 271–272), can be seen by the naked eye in the natural environment, and were first described in the early 1800s (Figure 10.3). *Ampelomyces* condia land on leaves and germinate; germination is inhibited at high spore concentrations, but stimulated by the presence of the host, and germ tubes grow towards the host. Germ tubes can form appressoria-like structures, and penetrate the hyphal or spore walls of powdery mildew species (Figure 10.3) using enzymatic and mechanical pressure. The parasite grows intracellularly as a biotroph initially, but kills the host mycelia after 5–8 days, after which time it produces pycnidia, containing conidia, within the hyphae, conidiophores and immature ascomata of the host. In moist environments, spores are released and can spread to other leaves and powdery mildew colonies by rain splash and water run-off. *Ampelomyces* conidia can also be spread for long distances in parasitized hyphal fragments and condia of the host. These powdery mildew conidia can sometimes germinate and give rise to new colonies with *Ampelomyces* already inside. As well as eventually killing its host, *Ampelomyces* suppresses asexual and sexual reproduction of the host by killing conidia and immature ascomata, though conidia can be produced after the mycelium has been invaded. Though host death is not rapid, *Ampelomyces* was among the first fungi to be tried as a biological control agent, with commercial formulations available today (p. 424).

Haustorial biotrophs form an appressorium on the host surface, from which a fine peg develops, that penetrates the host hypha (Figure 10.4). When it has passed through the wall, it branches to form a lobed haustorium that invaginates the host plasmalemma. The haustorium is the presumed site of nutrient transfer to the parasite. *Piptocephalis fimbriata* is a haustorial biotroph. Its conidia germinate in the absence of the host, but its germ tube makes limited growth unless a suitable Mucorales host mycelium is sufficiently close to enable chemotropically directed growth, with subsequent penetration and haustorium formation. **Fusion biotrophs** produce specialised hyphae or buffer cells, which are closely adpressed to those of the host hyphae. Direct contact of parasite and host is made through channels or micropores in the cell walls. The plasmalemma of the host and parasite make contact and fuse, making their cytoplasm contiguous.

TABLE 10.3 Types of Mycoparasitism (Including Fungus-Like Oomycota)

Type of parasitism	How the parasite interfaces with the host	Examples: Parasite species	Parasite (sub)phyllum	Host species	Host (sub) phyllum
Contact necrotroph (hyphal intereference)	Parasite contacts but does not penetrate the host hyphae. Host cytoplasm degenerates and lysis may occur	*Phlebiopsis gigantia*	Basidiomycota	*Heterobasidion annosum*	Basidiomycota
		Coprinellus heptemerus	Basidiomycota	*Ascobolus crenulatus*	Ascomycota
		Panaeolus sphinctrinus	Basidiomycota	*Bolbitius vitellinus*	Basidiomycota
		Cladosporium sp.	Ascomycota	*Exobasidium camelliae* (basidia)	Basidiomycota
Invasive necrotroph	Following contact, the parasite penetrates and enters the host; host cytoplasm rapidly degenerates and hyphal lysis often occurs	*Rozella* species	Cryptomycota	*Allomyces, Chytridium, Rhizophlyctis, Rhyzophydium, Zygorrhizidium*	Chytridiomycota
		Syncephalis californicus	Zoopagomycotina	*Rhizopus oryzae*	Mucoromycotina
		Nectria inventa	Ascomycota	*Alternaria brassicae* (hyphae and conidia)	Ascomycota
		Coniothyrium minitans, Talaromyces flavus	Ascomycota	*Sclerotinia sclerotiorum* (sclerotia)	Ascomycota
		Cladosporium uredinicola	Ascomycota	*Puccinia violae* (uredospores)	Basidiomycota
		Fusarium merismoides	Ascomycota	*Pythium ultimum* (oospores)	Oomycota
		Mycogone perniciosa	Ascomycota	*Rhopalomyces elegans* (conidia)	Mucoromycotina
		Mycogone perniciosa	Ascomycota	*Agaricus* and *Pluteus* fruit bodies	Basidiomycota
		Trichoderma spp.	Ascomycota	Many, e.g. *Rhizoctonia solani* (=*Thanatephorus cucumeris*), *Corticium rolfsii*	Basidiomycota
		Trichoderma harzianum	Ascomycota	*Botrytis cinerea*	Ascomycota
		Rhizoctonia solani (=*Thanatephorus cucumeris*)	Basidiomycota		Mucoromycotina
		Pythium acanthicum	Oomycota	*Phycomyces blakesleeanus*	Mucoromycotina

Intracellular biotroph	Entire thallus of the parasite enters the hypha of the host; host cell remains functional	*Ampelomyces* spp.	Ascomycota	*Arthrocladiella mougeotii*, *Blumeria graminis*, *Sawadaea bicornis* (all powdery mildews)	Ascomycota
Haustorial biotroph	A short haustorial branch from a parasite hypha penetrates the host; host cell remains functional	*Piptocephalis* spp.	Zoopagomycotina	At least 20 genera of Mucorales	Mucoromycotina
		Dimargaris spp.	Kickxellomycotina		Mucoromycotina
		Filobasidiella depauperata	Basidiomycota	*Verticillium lecanii*	Ascomycota
Fusion biotroph	Host and parasite are in intimate contact; micropore(s) form between the adpressed host and parasite hyphae, or from a short penetrative branch from the parasite hypha; host cell remains functional	*Gonatobotrys simplex*	Ascomycota	*Alternaria alternata*	Ascomycota
		Dicyma parasitica	Ascomycota	*Physalospora obtusa*	Ascomycota

Abstracted from Jeffries (1995) and other sources.

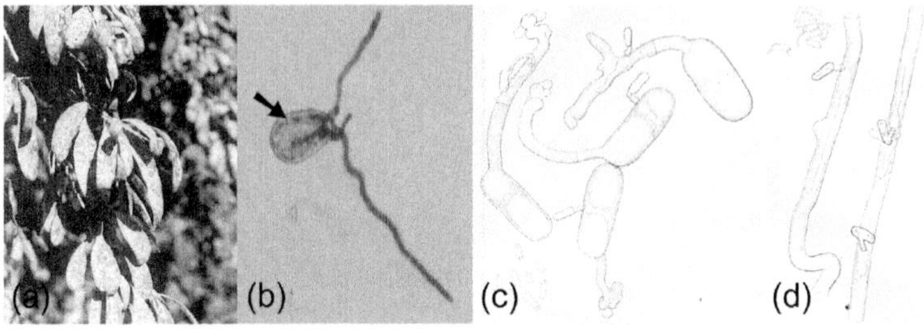

FIGURE 10.3 Intracellular mycoparasites. *Ampelomyces* species are initially biotrophic, though eventually they kill the host cell. *Ampelomyces* spp. are parasitic on powdery mildew fungi. (a) The brown patches are masses of intracellular pycnidia of *Ampelomyces* within white powdery mildew colonies, which are parasitic on *Lycium halimifolium* (Solanaceae). Hyphae of the mycoparasite penetrate into conidia (b, c) and hyphae (d) of the host. (b) Hyphae of *Ampelomyces* (stained with cotton blue) within (arrowed) and emerging from a conidium of *Erysiphe syringae-japonicae*. *Source:* (c, d) Drawings from early work on Ampelomyces by De Bary (1870), showing germ tubes extending from conidia (arrowed) and penetrating into (c) germ tubes of *Erysiphe heraclei* and (d) into hyphae of *Neoerisiphe galeopsidis*. *From Kiss (2008)*. (See the colour plate.)

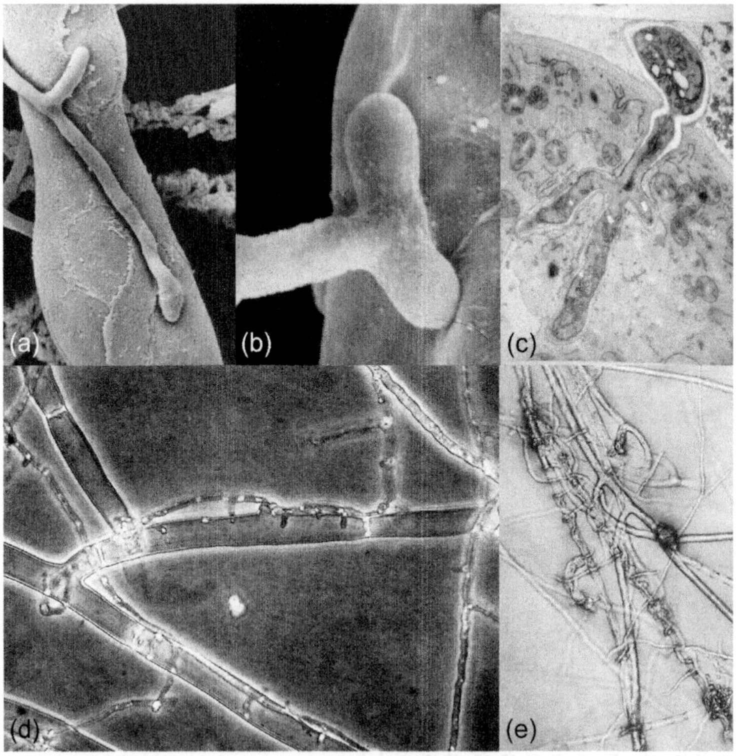

FIGURE 10.4 Mycoparasitism. (a, b) Scanning and (c) Transmission Electron Micrographs of *Piptocephalis unispora* parasitizing *Cokeromyces recurvatus*. (a) Narrow hyphae of *Piptocephalis unispora* closely adpressed to the surface of its host, and bulging at the tip to form an appressorium. (b) An appressorium developing lateral hyphae. (c) An appressorium (top right) on the surface of the host, with haustorial apparatus developing inside the host. (d) *Trichoderma virens* hyphae penetrating into large hyphae of *Rhizoctonia solani*. (e) Hyphae of *Trichoderma virens* coiling around the larger hyphae of *Rhizoctonia solani*. Coiling of hyphae around a host is often a feature of mycoparasitism. *Source: Panels (a, b) © Peter Jeffries; (c) from: Jeffries, P., Young, T.W.K., 1976. Ultrastructure of infection of Cokeromyces recurvatus by Piptocephalis unispora (Mucorales). Arch. Microbiol. 109, 277–288; (d, e) from Howell (2003).*

While for some fungi, mycoparasitism is the major or only way in which nutrition is obtained, for many necrotrophs it is more opportunistic. Parasitism can also be temporary, serving as a means to obtain a different food source (see below). Necrotrophic mycoparasites acquire nutrients from hosts in a much less controlled way than biotrophic mycoparasites, and are hence more destructive, and have broad host ranges, as seen with *Trichoderma* species and *Pythium* species (oomycetes). Necrotrophic parasites can be invasive or non-invasive.

Non-invasive (contact) necrotrophic mycoparasites make contact or almost (within a few micrometre) contact host hyphae. This type of parasitism is sometimes called **hyphal interference** and was first described by John Webster and colleagues in the early 1970s for *Coprinellus heptemerus*. Hyphal interference is most effective at the hyphal tips of the mycoparasite, and the affected fungus is also most sensitive in the tip region. Hyphal extension of the fungus being interfered with slows dramatically, membrane function is impaired, organelles lose contents, the plasmalemma invaginates, and the contacted hyphal compartment dies. Death of the whole mycelium will occur if multiple contacts are made. A single hypha of *Panaeolus sphinctrinus* can prevent a colony of *Bolbitius vitellinus* from advancing in agar culture. Hyphal interference is so effective in causing death of the opponent that *Phlebiopsis gigantea* is used as a biocontrol agent against the conifer tree pathogens *Heterobasidion annosum* and *Heterobasidion parviporum*. cDNA analysis of interference of *Heterobasidion parviporum* by *Phlebiopsis gigantea* showed up-regulation of expression of genes encoding a wide range of proteins, including those important for nutrient acquisition and use, enzymes involved in glycolysis and gluconeogenesis, breakdown of pectic compounds, and in nitrogen metabolism. Hydrophobins were up-regulated, and these are implicated in cell wall assembly and the monomers may act as toxins.

With **invasive necrotrophs**, hyphae of the mycoparasite contact those of the host, sometimes coiling around and often penetrating. *Trichoderma* species are examples of invasive necrotrophs, and their parasitic ability and bio-control ability has been studied since the 1930s. Mycoparasites are often able to sense the presence of a potential host and direct their growth towards it (e.g. *Trichoderma* spp. towards *Rhizoctonia solani*), and the presence of a potential host can stimulate spore germination (e.g. conidia of *Coniothyrium minitans*) are stimulated by the presence of sclerotia of *Sporidesmium sclerotivorum*. After contact, recognition, binding and morphological changes occur. *Trichoderma* spp. hyphae typically coil around those of the host (Figure 10.4) and in some penetrate into the host (e.g. *Trichoderma virens* penetrates the hyphae of *Rhizoctonia solani*) (Figure 10.4). Both *Trichoderma* spp., and some of its hosts (e.g. *Corticium* (*Sclerotium*) *rolfsii*), produce lectins which interact with carbohydrates on the surface of the opposing fungus and are involved in recognition and differentiation of structures involved in the parasitism. Signals from the host are recognised by receptors on the surface of the mycoparasite. This elicits an internal signal transduction cascade causing transcription of genes relevant to parasitism. With *Trichoderma* spp., heterotrimeric G proteins of the parasite transmit signals from G-protein-coupled receptors, to the cAMP and the MAP kinase pathways, which govern production of antifungal metabolites, lytic enzymes, and infection structures. Host cytoplasm is then completely disrupted (Figure 10.4). When *Trichoderma* (= *Gliocladium*) *roseum* attacks *Botrytis* species, for example, vacuolation occurs and hyphal walls and organelles are then lysed. *Trichoderma* spp. are such vigorous necrotrophs because of the toxic substances and enzymes, including chitinases, glucanases, and proteases, that they produce.

Some fungi are **temporarily parasitic** on others. They first obtain nutrition by parasitizing a host and then, more importantly, taking over the resource occupied by the host. When the parasite has killed its host and occupied its resource, it can then defend and expand its territory by non-parasitic, antagonist interactions with adjacent fungi. This occurs with several wood decomposing species. For example, *Lenzites betulina* is temporarily mycoparasitic on *Trametes* species, and *Trametes gibbosa* on *Bjerkandera* species. In both cases the mycoparasite is relatively uncommon, while the host is very common, occupying large volumes of decomposing wood which, following death of the host becomes available to the temporary mycoparasite.

Not only can vegetative hyphae be parasitized or killed, but so too can **hyphae within fruit bodies**. *Hypomyces aurantius* parasitizes *Trametes versicolor* and other wood-decay basidiomycetes, by producing a powerful toxin, which causes destruction of host organelles and coagulation of cytoplasm. Initially lipid bodies accumulate, then mitochondria and cisternae of the endoplasmic reticulum swell, and the plasma membrane invaginates, shrinking away from the cell wall. Other presumed parasites of fruit bodies include the basidiomycetes *Asterophora lycoperdoides* and *Pseudoboletus parasiticus* that fruit on *Russula nigricans* and *Scleroderma citrinum* respectively, and *Spinellus fusiger* (Mucoromycotina) on *Mycena* species (Figure 10.5). The basidiomycete *Squamanita odorata* reduces the fruiting tissues of its basidiomycete host *Hebeloma mesophaeum* to unrecognisable galls. Most, perhaps all, species of Tremellales (Basidiomycota) are mycoparasitic, including on the hymenia of basidiomycete mushrooms, brackets and Dacrymycetales, and the ascomata and stromata of pyrenomycetes (Ascomycota). Parasitic *Tremella* species form haustoria with micropores connecting the cytoplasm of host and parasite.

Fungal spores are also parasitized. **Mycoparasitism of spores** has been mostly studied on those of arbuscular mycorrhizal fungi (AMF, pp. 206–215), which are the largest within kingdom Fungi. Many, often the majority, of AMF spores extracted from soil have walls perforated by fine canals where mycoparasitic fungi, or sometimes amoebae (p. 358), have penetrated. These canals are commonly associated with ingrowths similar to the papillae produced in response to the entry of penetration pegs of *Piptocephalis* spp. on Mucorales. *Spizellomyces* and *Pythium*-like zoosporic oomycetes are common on Glomales spores, sporulating on the surface or within. Many genera of Ascomycota (e.g. *Trichoderma* and *Verticillium*) that parasitize hyphae have species that parasitize spores. Older spores and less melanised spores appear to be less able to resist invasion than younger, darker spores of Glomales.

FIGURE 10.5 Fruit bodies of mycoparasitic fungi on fruit bodies of basidiomycete hosts. (a) *Spinellus fusiger* (Mucoromycotina) on *Mycena* sp. (b) *Asterophora lycoperdoides* on *Russula nigricans*. (c) *Pseudoboletus parasiticus* on *Scleroderma citrinum*. *Source: Panels (a, b) © Penny Cullington and (c) © Alan Hills*. (See the colour plate.)

Mycoparasitism of sclerotia – survival structures – is also common, and of particular interest when the host is a plant pathogen, as again there is potential for biocontrol. For example, hyphae of *Sporodesmium sclerotivorum* and *Teratosperma oligocladium* penetrate the rind of *Sclerotinia sclerotiorum* and *Sclerotinia minor* and then proliferate intercellularly within the medulla, but not within sclerotial cells. The parasites probably use glucose and other monosaccharides released from hyphae within the medulla of the sclerotium. Subsequently, hyphae of the mycoparasite grow to the surface of the sclerotium and sporulate profusely.

Antagonism Following Contact: Gross Mycelial Contact

Mycoparasitism can clearly ultimately affect the whole mycelium of the parasitized fungus and, in agar culture, the mycelium of the mycoparasite can sometimes be seen macroscopically, advancing unhindered through the colony of the host. The rather imprecise term, gross mycelial contact, is used to describe the situation where dramatic changes, obvious to the naked eye, occur when mycelia of different species meet. The growth of one or both mycelia slow, the morphology of the mycelium changes, pigments are often produced and zones of lysis are sometimes seen. These changes are particularly evident in interactions between basidiomycetes and also xylariaceous ascomycetes on nutrient rich agar media and also on the surface of soil, where morphological changes such as 'barrages' of mycelium resistant to invasion, invasive mycelia fans and aggregated mycelia structures – cords and rhizomorphs – can be seen (Figure 10.2). These morphological shifts are often accompanied by redistribution of mycelial biomass from elsewhere, with a reduction in hyphal density distant to the interaction zone, which can result in greater susceptibility to invasion if the opposing mycelium reaches this area.

Interactions, or at least the site of interactions, can be seen in natural organic resources as 'interaction zone lines' (Figure 10.6). In wood, these appear in cross section as narrow, often dark-coloured though sometimes brightly coloured (e.g. orange with *Oudemansiella mucida*), lines in cross section. These lines are pseudosclerotial plates (PSPs; p. 61). They often extend many centimetres longitudinally and completely surround the territory occupied by the decay fungus that produced them. A narrow (one or a few wood cells thick) 'no man's land' is present between PSPs surrounding the territory of adjacent fungal individuals. This region is often occupied by small colonies of dematiaceous ascomycetes (e.g. *Chaetosphaeria myriocarpa* and *Rhinocladiella* spp.), which may partly explain the vast number of different DNA sequences revealed by high throughput sequencing of colonised wood. These barriers can sometimes be breached by decomposer fungi occupying large adjacent territories, and the fungus that produced them can be replaced. This can be seen in wood as 'relic' zone lines (i.e. PSPs) that have been partly decomposed. Mycelial biomass is often unevenly distributed within decay columns in the wood. This can be seen when wood is incubated in moist conditions; prolific mycelial outgrowth occurs at the edges of decay columns close to interaction zone lines, while outgrowth from inner regions is often sparser.

There are likely to be many different mechanisms covered by this single catch-all term, gross mycelial interactions. Indeed, a fungus may operate different combative mechanisms against different opponents. These mechanisms involve the release of enzymes, toxins, and

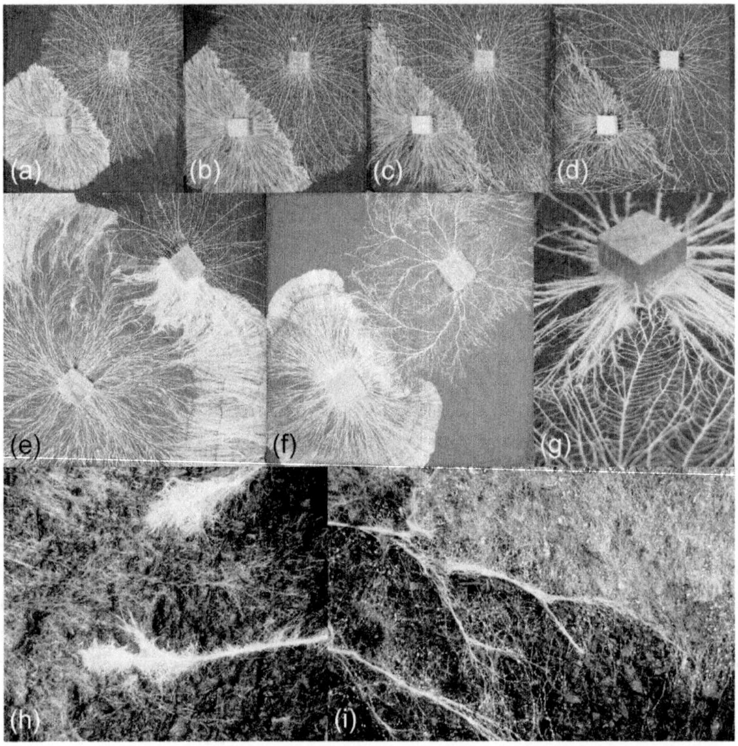

FIGURE 10.6 Interactions between basidiomycete mycelia in soil and wood. (a–d) Mycelium of *Hypholoma fasciculare* (left) extending from a colonised wood block interacting with that of *Phallus impudicus* on the surface of compressed, non-sterile soil in 24×24cm trays, 2 (a), 10 (b), 32 (c), and 84days (d) after contact, respectively. After 32days cords of *Phallus impudicus* have managed to pass the defences of *H. fasciculare* and to grow over it. By 84days *Phallus impudicus* has reach the wood block colonised by *H. fasciculare*, and from then on interaction occurs within the wood block. Damage to the *Phallus impudicus* mycelium, resulting from grazing by invertebrates (Collembola) is evident at 84days in regions away from the opponent. (e) A different strain of *H. fasciculare* (left) replacing *Phallus impudicus*. (f) *Hypholoma fasciculare* (left) being replaced by *Phanerochaete velutina*. Note that in some interactions replacement of mycelium occurs by a fungus as it progresses across soil (f) whereas in others (d) overgrowth occurs first. (g) Close up of mycelium of *Phallus impudicus* (bottom) replacing *Resinicium bicolor*. (h) Mycelial cords of *Phanerochaete velutina* overgrowing (from the right) extraradical mycelium of the ectomycorrhizal *Paxillus involutus*. Tips of *Phanerochaete velutina* cords are truncated (cf. with cords on soil in (f), so are also being affected antagonistically). (i) Cords of *Phanerochaete velutina* (bottom) preventing access to territory by the extraradical mycelium of the ectomycorrhizal *Paxillus involutus* (top). *Source: Panels (a–g) © Timothy Rotheray; (h, i) © Damian Donnelly.*

other antifungal compounds (see below). Which species ultimately dominate in antagonistic interactions is determined by the relative abilities of the opponents to capture and defend territory. Fungi exhibit a hierarchy of combative ability, some species being more aggressive than most, others being replaced by most antagonists, and the combative ability of many others in between (Figure 10.7). Further, some fungi are good at 'defence', others at 'attack', and others at both. This is similar to the concepts of competitive effect and competitive response in plant ecology, the former being able to suppress resource levels for other plants, and the latter

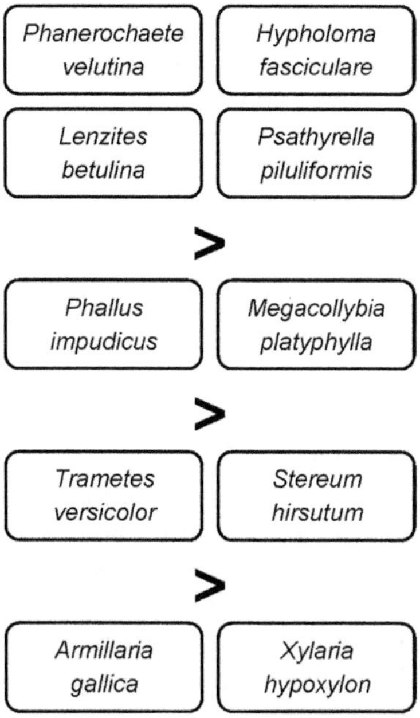

FIGURE 10.7 The hierarchy of combative ability for some of the dominant fungi involved with decomposition of felled beech (*Fagus sylvatica*) wood.

having the ability to tolerate suppression. So within the spectrum of combative abilities there are: (1) transitive hierarchies, where species A is more combative than species B, which in turn is more combative than species C (i.e. A>B>C); and (2) intransitive hierarchies, where species A is more combative than species B, and species B is more combative than species C, but species C is more combative than species A (i.e. A>B, B>C, C>A). Such intransitivity may result from different species employing different combative and defensive mechanisms. This hierarchy of combative ability is a bit like a sports league, in which the team at the top of the league table beats most of the others, but it can, too, occasionally be beaten. Likewise, teams at the bottom of the league sometimes beat teams higher up and are occasionally 'giant killers', beating the very best teams. The situation is further complicated by the fact that outcomes between particular combinations of species vary depending on abiotic regime (e.g. temperature, water potential, nutrient status), fungal strain, size of territory already held, location of the interaction (e.g. wood or soil), presence of other microorganisms and invertebrate grazing, and even sometimes under apparently identical conditions.

Interaction Chemistry and Gene Expression

As well as changes in mycelial morphology during contact between mycelia, up-regulation of genes involved in antagonism results in the production of additional chemicals (pp. 339–341) in hyphae, which are released into the surrounding environment as volatile or diffusible organic compounds (VOCs and DOCs), as already mentioned in the section on 'Antagonism at a distance' (pp. 339–341). The production of antibiotic compounds by fungi (e.g. penicillin) has long been known, but the extent of production and scale over which they exert an effect in the natural environment is still unclear. Some antimicrobial compounds are definitely persistent; large quantities of methylbenzoates are produced *in vivo* in wood decayed by *Sparassis crispa* – a brown rot (pp. 147–152) basidiomycete that is a tree pathogen – and persist for many years. Production of such high concentrations represents a significant biosynthetic commitment by the fungus, implying considerable benefit to it. Other antifungal chemicals have also been readily detected in wood, including lipid-soluble compounds from the majority of fungi isolated from stumps of Sitka spruce (*Picea sitchensis*). The activity of these compounds, however, varied against other fungi. For example, three compounds from culture of *Stereum sanguinolentum* on artificial medium were antifungal to *Cladosporium cucumerinum*, but only one of them inhibited mycelial growth of *Hypholoma fasciculare*, *Heterobasidion annosum*, and *Resinicium bicolor*.

Enyzme activity varies in different regions of interacting mycelia. Some enzymes are hugely up-regulated in the interaction zone, with small increases in activity in other regions of the mycelia. Production of reactive oxygen species, phenoloxidases and sometimes β-glucosidase increase widely in mycelia, whereas laccases and manganese-dependent lignin peroxidases increase in the contact zones between interacting wood decay-causing species. The laccases may well have a defensive role, since they are involved in the production of melanins and similar compounds which are frequently formed during interspecific interactions. Chitinases also play a role since they cause cell lysis and release sequestered nitrogen from the hyphae of fungi whose territory is being taken over.

Antagonistic Effects on Physiology and Ecology

In the natural environment, fungi almost always exist in multispecies communities, so interspecific mycelial interactions and competition for resources occur continually. Not only do interactions occur between mycelia of saprotrophs, but also between saprotrophs and mycorrhizal fungi (Chapter 7), and amongst mycorrhizal fungi. Many key enzyme systems are common to ectomycorrhizal (ECM) and saprotrophic basidiomycetes, and since their mycelia are often found together in wood at very late stages of decay, and in organic soil, they will inevitably compete. In Boreal forests, however, there may be separation of niches which reduces competition, with saprotrophs dominating upper, organic soil horizons and ECM fungi dominating in lower, mineral layers.

During interactions between saprotrophic fungi, CO_2 evolution sometimes increases, reflecting an increase in metabolic cost when defending territory or wresting it from an opponent. Movement and partitioning of carbon within mycelia is also affected; in wood when one species replaces another, the replacing fungus can switch from reliance on carbon available in the originally occupied resource to that available in the captured territory, including use of the mycelium of the fungus that has been killed. When saprotrophic basidiomycetes interact with ECM basidiomycetes, there are marked effects on allocation of photosynthetically derived carbon to the ECM mycelium growing from the roots (Figure 10.8). Carbon exchange between mycelia can also occur; during antagonistic interactions carbon compounds leak from damaged hyphae in the interaction zone, even where there is deadlock, and these can be absorbed opportunistically by the opponent.

Mineral nutrient uptake, movement, partitioning, and release are altered by interspecific mycelial interactions. As with carbon, mineral nutrient exchange between mycelia occurs, probably via leakage in the interaction zone. Nutrients also move within mycelia during

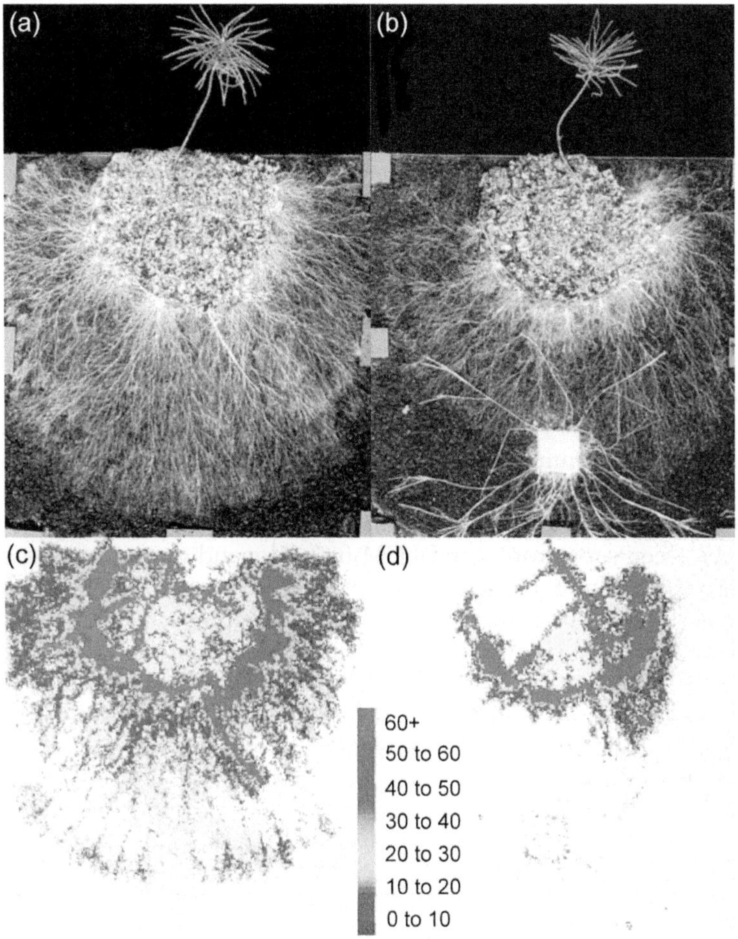

FIGURE 10.8 Saprotrophic basidiomycetes can dramatically affect ectomycorrhizal mycelial spread and allocation of carbon to the extraradical mycelium of the mycorrhizal fungus. Here the wood decay fungus *Phanerochaete velutina*, growing from a piece of wood, is interacting with mycelium of *Suillus bovinus* in association with *Pinus sylvestris* in soil microcosms. (a) and (b) show, respectively, ectomycorrhizal mycelial growth in the absence and presence of the saprotroph. Plants were pulse-labelled with ^{14}C, and this was quantified in a 20 × 24 cm below-ground area by digital autoradiography (c, d). The colour scale represents counts mm^{-2} over 45 min. Very little carbon from the host plant is allocated to ectomycorrhizal mycelium in the area of territorial combat, and its growth is inhibited. There was a 60% reduction in ^{14}C allocated to mycelium of *Suillus bovinus* when interacting with *Phanerochaete velutina*, up to 30 h after pulse labelling. Presence of ^{14}C (0.03 %) was detected in *Phanerochaete velutina* after 5 days. *Source: Modified from: Leake, J.R., Donnelly, D.P., Saunders, E.M., Boddy, L., Read, D.J., 2001. Carbon flux to ectomycorrhizal mycelium following ^{14}C pulse labelling of Pinus sylvestris L. seedlings: effects of litter patches and interaction with a wood-decomposer fungus. Tree Physiol. 21, 71–82. By permission of Oxford University Press.* (See the colour plate.)

interactions. In some cases there is movement of phosphorus towards the interaction zone and in others away from it, the latter probably occurring when a mycelium is losing a confrontation. During interactions, nutrients are not only released in the interaction zone, but the whole mycelium can become 'leaky', and this is probably one of the main ways by which nutrients are released from mycelia into the soil.

Interspecific Interactions and Fungal Community Development

The outcomes of fungal interactions are often the major causes of change in fungal communities. This has been studied most intensively in decaying organic matter, especially wood, but also in respect of establishment of mycorrhizas. Position in the hierarchy of combative ability (pp. 349–351) is broadly, but not completely, correlated with position of fungi in community succession. Initial colonisers of dead organic resources do not have to be successful combatants to gain territory, and hence access to resources, but those which arrive later do have to be, as other fungi are already present. Combative interactions are not, however, always the main drivers of community change. Disturbance or environmental stress can alter communities. Thus, certain fungi can be poor combatants, but have characteristics that allow them to cope with the disturbance or stress, and thus to dominate.

Interactions between saprotrophs and ECM fungi depend on the species, suppression of either sometimes occurring. For example, there are reports of *Suillus variegatus* and *Paxillus involutus* ectomycorrhizal with pine seedlings being antagonistic against the wood decaying cord-forming basidiomycete *Hypholoma fasciculare*, whereas *Hypholoma fasciculare* can inhibit *Pisolithus tinctorius* ectomycorrhizal with chestnut. *Hypholoma fasciculare* can inhibit root colonisation by *Pisolithus tinctorius*, if present at the time when colonisation begins, and even influences the mycorrhizas formed when arriving 30 days after the start of root colonisation by the ECM fungus.

INTERACTIONS BETWEEN FUNGI AND BACTERIA

Interactions between fungi and bacteria are many and varied, and they can affect the growth, survival, and virulence of each other. These effects can be negative, positive or mutualistic to the interacting organisms, and can result from fungi being sources of nutrition for bacteria, and vice versa, as well as other stimulatory and inhibitory effects.

Negative Interactions Between Bacteria and Fungi

Some fungi are bacterivores, for example, the cultivated mushroom, *Agaricus bisporus*, utilises bacteria as a major source of nitrogen from the microbe-rich compost on which it is grown. Those fungi that are able to lyse bacteria appear to be attracted to bacterial colonies. On the other hand, some bacteria are mycophages – parasites/pathogens of fungi. Some filamentous *Streptomyces* strains can coil around and penetrate hyphae (e.g. of *Aspergillus niger*) and also parasitize spores of the arbuscular mycorhizal fungus *Gigaspora gigantea*. Non-filamentous bacteria can also attack hyphae; *Collimonas* strains attack hyphal tips using a combination of chitinases and antibiotics; myxobacteria destroy yeasts and penetrate hyphae of soil fungi, including plant pathogenic *Rhizoctonia* species.

Bacterial parasitism of cultivated mushroom species has been known for almost 100 years. There is a range of mushroom diseases of fruit bodies, but brown blotch disease of *Agaricus bisporus*, caused by *Pseudomonas tolaasii*, has been most studied. The bacterium grows in soil and compost, but can also attack fungal hyphae and fruit bodies. Zones of attack are seen as brownish spots; tolaasin and, to a lesser extent, other secondary metabolites cause membrane disruption which releases nutrients from within fungus cells. A cascade of events follows, resulting ultimately in the production of melanins and quiniones that form chemical barriers to protect inner fruit body tissues from becoming infected. Soft rots are another set of fruit body diseases caused by bacteria, with symptoms such as pitting, sticky blotches, and even complete dissolution. Rapid soft rot of the cap and stipe are caused by *Burkholderia gladioli* pv. *agricicola* and *Janthinobacterium agaricidamnosum*, while a *Pantoea* sp. causes a soft rot accompanied by water-soaked lesions in *Pleurotus eryngii*. Mummy disease has very different symptoms (fruit bodies fail to reach maturity but, rather than decomposing, become mummified) but the cause has not yet been verified by fulfilling Koch's postulates, though several *Pseudomonas* species have been implicated. As well as there being a range of pathogenic bacteria, there is also variation in the diseases caused by the same species. *Pseudomonas agaricus* causes: a mild blotch disease of *Agaricus*; yellow blotch disease of *Pleurotus* species, characterised as yellow droplets on the surface of fruit bodies; and dippy gill disease which manifests itself as exudates from longitudinal splits in the stipe where bacteria colonise inter- and intra-cellularly.

There are many examples of bacterial pathogens of fungi seen in agriculture, not least because of the possibility of using these as biocontrol agents of plant pathogenic fungi. *Lysobacter enzymogenes* is pathogenic on a diverse range of fungi as well as on nematodes, bryophytes, and other organisms. Following attachment, it infects its hosts colonising intracellularly, it replicates and is subsequently released. Enzymes produced during pathogenesis include chitinases, β-1,3-glucanases and proteases, which are involved in cell wall degradation. It also employs antibiotics, including a heat-stable antifungal factor, and it uses a type III secretion system (a protein appendage used to secrete proteins into the host) in pathogenesis of some organisms. As well as lytic enzymes and antibiotics, a well-characterised virulence mechanism during bacterial pathenogenesis of fungi is the T4 pilus (a filamentous projection that aids in attachment to the host).

Communication has a critical role in pathogenesis, as shown by the pathogenic capabilities of *Pseudomonas aeruginosa* on *Candida albicans. Pseudomonas aeruginosa* produces quorum sensing molecules, which prevent *Candida albicans* switching from the yeast to the mycelial phase (p. 167), but it is only pathogenic on the mycelial stage. In the yeast phase, *Candida albicans* can produce its own quorum sensing molecule – farnesol – that self-regulates conversion to the mycelial phase, and prevents *Pseudomonas aeruginosa* from producing its quorum sensing signal quinolone and other virulence factors. The human bacterial pathogen, *Acinetobacter baumannii*, also inhibits *Candida albicans* from forming a mycelial phase, which is an important feature of *Candida albicans* pathogenicity to humans (pp. 306–307).

Clearly, production of lytic enzymes and antibiotics are important pathogenic mechanisms, though they are also employed during transient, non-specific interactions conducted at a distance. Bacteria tend to produce mixtures of antagonistic metabolites rather than single antibiotics, which prevents development of resistance in the target. Some fungi do have defence mechanisms against some antibiotics, and may employ different mechanisms in concert. Such mechanisms include the ability to detoxify and degrade antibiotics. Transport by membrane bound efflux pumps enables release of toxic compounds from the fungus cell (e.g. *Botrytis cinerea* uses the efflux pump BcAtrB to expel the antifungal compound 2,4-diacetylphloroglucinol (DAPG); *B. cinerea* uses tannic acid as a mediator for DAPG degradation). In contrast, *Fusarium oxysporum* converts DAPG to a less toxic derivative. *Fusarium* species also produce fusaric acid, which inhibits the production of DAPG and other antifungal compounds by bacteria, as well as being toxic to bacteria and eukaryotes. However, the situation becomes yet more complex as some biocontrol bacteria (e.g. *Pseudomonas fluorescens* WCS365) are attracted to fusaric acid-producing *Fusarium oxysporum*, and so colonise the fungus.

Competition between bacteria and fungi for nutritional resources that they both need is another negative interaction. Not only do they compete for simple carbon compounds, but also for products of extracellular digestion of lignocellulolysis. Because they lack the appropriate extracellular enzymes, bacteria are unable to breakdown complex lignocellulose-rich resources, but many fungi can, releasing water-soluble sugars and phenolic compounds that form carbon and energy sources for the fungus (pp. 146–152). Intense bacterial competition for breakdown products could deprive the fungus of these soluble resources. Bactericidal effects have not surprisingly, therefore, evolved widely in fungi, some of which have already been mentioned. Fungi produce many antibacterial compounds – not least penicillin, cephalosporins, and griseofulvin – that are widely used in medicine (pp. 418–421). Other anti-bacterial strategies include extra production of hydroxyl radicals and acidification of the environment. The extremely hydrophobic nature of the mycelium of some fungi (e.g. *Pleurotus* species) can also prevent bacteria from penetrating colonised substrata.

Another widespread negative effect of microbes on fungi is the ubiquitous phenomenon of mycostasis (also called fungistasis), whereby the majority of fungal spores that land on soil fail to germinate. Mycostasis is correlated with microbial activity, and can be alleviated if soil is sterilised and/or easily available energy sources are added. The effect is, thus, probably mediated by soil microbes rapidly sequestering any soluble nutrients that become available in soil and also by the production of inhibitory metabolites. Susceptibility of spores to mycostasis may actually be advantageous to fungi, preventing germination until local microbial activity is reduced and substrates are available.

Positive and Mutualistic Effects

Despite the anti-bacterial effects mentioned above, bacteria are commonly present on the surface of hyphae and spores, including species in the genera *Pseudomonas*, *Burkholderia*, and *Bacillus*, and non-culturable Archaea. Bacterial attachment is regulated by species specificity and fungal vitality, but the molecular and biochemical controlling mechanisms are largely unknown. Organic acids, oxalic acid, and antibiotics may all have a selective role, as do lectins, which have been shown to be produced by truffle (*Tuber*) species and bind *Rhizobium* sp. to them. As well as the negative effects alluded to in the previous section, bacteria can have positive effects on fungi, the most well-known being the so-called mycorrhiza helper bacteria (MHB). Some pseudomonads, bacilli, paenibacilli, and streptomycetes stimulate ectomycorrhiza formation (pp. 215–220). The mechanisms involved include effects on the fungus and on the host plant. The bacteria detoxify antagonistic substances in soil, inhibit competitors/antagonists, and produce growth factors that stimulate spore germination and mycelial growth.

Bacteria associated with fungi can also influence formation of fruit bodies and production of spores. With *Agaricus bisporus*, fruit body initiation depends on the presence of *Pseudomonas putida*, and the latter also promotes mycelial growth and fruit body formation. However, though as humans we may view these responses as beneficial to the fungus, fruiting and increased mycelial extension rate can occur as a result of stress. Though bacteria commonly inhibit spore germination (p. 356), some spore-associated bacteria can stimulate germination, perhaps as a result of volatiles, breakdown of germination-inhibiting compounds and/or by enzymatic weakening of the spore wall.

As in plants, animals, and protists, bacteria are found living within fungal cells (i.e. as endosymbionts). These have been most studied in arbuscular mycorrhizal (AM) fungi. Sequence analysis of the 16S rDNA has shown that the unculturable bacteria-like organisms (BLOs) within *Gigasporaceae* are related to the genus *Burkholderia.* They can occur in groups or singularly, often within vacuoles, in both hyphae and spores. Loss of these bacteria severely affects hyphal elongation and branching. Endobacteria of other fungi also have major effects on their host: *Rhizopus microsporus*, a fungal pathogen of rice, owes its pathogenicity to an endosymbiotic *Burkholderia* strain. In contrast, a biocontrol strain of *Fusarium* becomes pathogenic when its endosymbionts are removed.

The fungus cell wall is a physical barrier that largely prevents entry of bacteria. The wall is, however, not rigid at the hyphal tip (p. 37) and bacterial acquisition could take place there. *Geosiphon pyriforme* (Glomeromycota) incorporates primordia of free-living *Nostoc* (cyanobacteria) at the hyphal tips, which then swell to form bladders containing *Nostoc* cells that photosynthesise. *Burkholderia* sp. can invade germinating AM spores via weakened fungal germ tubes (e.g. by bacterial lytic enzymes). Acquisition of the endosymbionts by *Gigasporaceae*, however, appears to have been a unique event that occurred in an ancestral fungus; the bacteria are now spread vertically through generations, and are probably obligate endosymbionts.

Fungal-bacterial interactions can have positive effects on bacteria, in addition to the nutritional effects previously mentioned. As single cells that can only move in water films, bacterial movement across air gaps in soil is extremely difficult. This is not a problem for fungal hyphae, and some bacteria use hyphae to 'hitchhike' through soil (e.g. species of *Burkholderia, Dyella*, and *Ralstonia* on a *Lyophyllum* sp.). It is not that bacteria simply attach to a hyphal wall and move through soil as a hypha grows; because hyphae grow from the tips, the bacteria must move actively along the hyphae as they grow, and probably do so in a biofilm on the surface of hyphae.

VIRUSES OF FUNGI

Viruses are widespread in fungi in all phyla, though the host range of a virus type is very narrow, and frequency of infection within a species is variable, but sometimes over 80%. They were first discovered in the early 1960s, as the cause of La France disease, which causes malformed fruit bodies of the cultivated mushroom *Agaricus bisporus.* Soon after, they were found to be responsible for the production of interferon in the culture filtrates of some *Penicillium* species. Though mycoviruses have been relatively little studied, they occur in fungi that feed in all ways, including saprotrophs, plant and animal pathogens, endophytes, mycorrhizal fungi, and lichens. Most mycoviruses are spherical, have double-stranded RNA (dsRNA) genomes with isometric particles, 25–50 nm diameter, and are found in three families, based on the number of genome segments – *Chrysoviridae*, *Partitiviridae*, and *Totiviridae*. Mycoreoviruses (family *Reoviridae*), however, have double-shelled particles, 80 nm diameter. Fungi do not usually contain large molecules of dsRNA, so presence of the latter is a sign of viral infection.

Mycoviruses are intracellular within the fungus, are transmitted between hosts by cell-to-cell contact/fusion, and can be disseminated within spores. No natural vectors are known. Transmission within a species is often restricted by vegetative incompatibility (pp. 102–104). Transmission between species is even less common, though phylogenetic analyses have revealed that it does occur. Also, infection experiments show that transmission between species can occur within genera (e.g. *Aspergillus*, *Cryphonectria*, and *Heterobasidion*). There has probably also been horizontal transfer between plants and plant pathogenic fungi, since Partitiviruses that infect fungi and plants are closely related taxonomically. Also, there are some homologs of partitivirus and totivirus genes in some plants and fungi. To be a true mycovirus, infection between fungi must occur; so, as many dsRNA elements are not transmissible, they are referred to as virus-like particles.

Most viruses cause few or no obvious symptoms, but sometimes effects are severe, and both beneficial and adverse effects have been reported. In plant pathogenic fungi, some mycoviruses lead to increased fungal virulence (hypervirulence) and some to decreased virulence (hypovirulence), including in *Cryphonectria parasitica* (cause of chestnut blight), *Botrytis cinerea*, *Helminthosporium victoriae* and *Sclerotinia sclerotiorum*. The hypovirus CHV1 is now used successfully to control chestnut blight. Viruses have effects other than on virulence. In a three-way symbiosis, the presence of the virus CThTV in *Curvularia protuberata* endophytic within a tropical grass (*Dichanthelium lanuginosum*) allows the plant and fungus within to grow at the high soil temperatures found in Yellowstone National Park. Single virus strains can sometimes have different effects on different species of fungi, and different effects on the same species in different situations. For example, virus HetRV3-ec1 increases, decreases or has no effect on the competitive ability of various species of *Heterbasidion* against other fungi.

INTERACTIONS BETWEEN FUNGI AND PROTISTS

Protists and fungi are likely to encounter each other often in soil, aquatic ecosystems and in the guts of ruminants, but interactions between them have received little attention. There is certainly evidence of protists feeding on fungi. Some testate amoebae (e.g. *Geococcus vulgaris*), fasten to the walls of fungal spores and hyphae and suck out the contents. There is also evidence that protists may reduce ectomycorrhizal colonisation of roots and reduce the amount of mycorrhizal mycelium in the mycorrhizosphere. Considerable interactions between fungi and protists occurs in the rumen (pp. 329–330) – ciliate protozoa ingest Neocallimastigomycota zoospores, and their predatory activity can reduce overall cellulolytic activity and alter the fermentation products formed.

Plasmodia of myxomycetes have been shown to consume fungi in lab culture, the susceptibility of the latter varying between species of myxomycetes and fungi. Plasmodia will feed on mycelia mats that emerge from wood under moist conditions, and on resupinate fungal fruit bodies. The fungi in decay columns extending for many centimetres through wood can be completely devoured by plasmodia. For example, *Badhamia utricularis* has been seen to consume the ascomycete *Xylaria hypoxylon*, and *Comatricha nigra* consumes the basidiomycete *Stereum hirsutum*. Myxomycetes can be prevalent in wood; in one study myxoflagellates emerged from about 50% of fallen dead angiosperm branches sampled, having been active vegetatively or having emerged rapidly from microcysts; *Stemonitis fusca* was particularly common.

The tables are turned, however, by *Dactylella passalopaga*, amongst others, which produces bulbous outgrowths which trap testate amoeba. Several *Zoopagales* (zygomycete) adhere to amoeba and feed on them. Also, some isolates of *Heteroconium chaetospira* (a dark septate root endophyte; pp. 238–239) can control clubroot disease of Brassicacae, caused by the soil-borne protozoan *Plasmodiophora brassicae*. The fungus infects root epidermal cells, but the mechanism of disease control is unclear.

Some fungi colonise the fruit bodies of myxomycetes. Hyphae penetrate the spore masses and kill the spores. Other fungi (e.g. the ascomycete *Gliocladium album*), parasitize the calcium-rich fruit bodies of the Physarales; *Nectriopsis violacea* is even more specific, only colonising species of *Fuligo*. Others are specific to non-calcareous myxomycetes, yet others colonise a wide host range, *Nectria exigua* being recorded on all of the major myxomycete groups. In lab culture on agar, the hemiascomycete yeast, *Dipodascus utricularis*, is able to live in the slime trail of *Badhamia utricularis*. If the yeast is ingested it is not digested and lives parasitically, multiplying within the myxomycete plasmodium.

From a medical mycology viewpoint, interactions between amoebae and fungi which are saprotrophic in the natural environment, but which can cause systemic infection in humans (pp. 303–309) and other mammals, are of particular interest. The yeast form of the pathogens *Blastomyces dermatidis*, *Cryptococcus neoformans*, *Histoplasma capsulatum*, and *Sporothrix schenckii* are ingested by amoebae, but inhibit amoebal growth or kill them. Outcomes of interactions, however, vary depending on the combination of species interacting. For example, growth of *Candida albicans* is enhanced by the amoeba *Hartmannella vermiformis* but killed by *Acanthamoeba castellanii*. Similarly, spores of *Aspergillus fumigatus* (p. 306) are ingested by amoebae; the spores can germinate within the cytoplasm and the fungus is released to the environment. These interactions between amoebae and yeasts and spores are similar to those with human macrophages, and the interaction of these saprotrophic fungi with amoebae in the natural environment over evolutionary time has been described as a 'training ground' for overcoming the macrophage defences of vertebrates.

Further Reading

General Text

Kobayashi, D.Y., Hillman, B.I., 2005. Fungi, bacteria, and viruses as pathogens of the fungal community. In: Dighton, J., White, J.F., Oudemans, P. (Eds.), The Fungal Community: Its Organization and Role in the Ecosystem. third ed.. Taylor & Francis, Boca Raton, pp. 399–421.

Interspecific Fungal Interactions

Boddy, L., 2000. Interspecific combative interactions between wood-decaying basidiomycetes – a review. FEMS Microbiol. Ecol. 31, 185–194.

Gams, W., Diederich, P., Poldmaa, K., 2004. Fungicolous fungi. In: Mueller, G.M., Bills, G.F., Foster, M.S. (Eds.), Biodiversity of Fungi: Inventory and Monitoring Methods. Elsevier, Amsterdam, pp. 343–392.

Howell, C.R., 2003. Mechanisms employed by *Trichoderma* species in the biological control of plant diseases: the history and evolution of current concepts. Plant Dis. 87, 4–10.

Jeffries, P., 1995. Biology and ecology of mycoparasitism. Can. J. Bot. 73, S1284–S1300.

Kennedy, P., 2010. Ectomycorrhizal fungi and interspecific competition: species interactions, community structure, coexistence mechanisms, and future research directions. New Phytol. 187, 895–910.

Kiss, L., 2008. Intracellular mycoparasites in action: interactions between powdery mildew fungi and *Ampelomyces*. In: Avery, S.V., Stratford, M., van West, P. (Eds.), Stress in Yeasts and Filamentous Fungi. Elsevier, Amsterdam, pp. 37–52.

Omann, M., Zeilinger, S., 2010. How a mycoparasite employs G-protein signalling: using the example of *Trichoderma*. J. Signal Transduct. 2010, 123126. http://dx.doi.org/10.1155/2010/123126

Whipps, J.M., 2001. Microbial interactions and biocontrol in the rhizosphere. J. Exp. Bot. 52, 487–511.

Woodward, S., Boddy, L., 2008. Interactions between saprotrophic fungi. In: Boddy, L., Frankland, J.C., van West, P. (Eds.), Ecology of Saprotrophic Basidiomycetes. Elsevier, Amsterdam, pp. 123–139.

Interactions Between Fungi and Bacteria

Bonfante, P., Anc, I.-A., 2009. Plants, mycorrhizal fungi, and bacteria: a network of interactions. Ann. Rev. Microbiol. 63, 363–383.

de Boer, W., Folman, L., Summerbell, R.C., Boddy, L., 2005. Living in a fungal world: impact of fungi on bacterial niche development. FEMS Microbiol. Rev. 29, 795–811.

Hoffman, M.T., Arnold, A.E., 2010. Diverse bacteria inhabit living hyphae of phylogenetically diverse fungal endophytes. Appl. Env. Microbiol. 76, 4063–4075.

Kobayashi, D.Y., Crouch, J.A., 2009. Bacterial/fungal interactions: from pathogens to mutualistic endosymbionts. Annu. Rev. Plant Physiol. Plant Mol. Biol. 47, 63–82.

Lackner, G., Partida Martínez, L.P., Hertweck, C., 2009. Endofungal bacteria as producers of mycotoxins. Trends Microbiol. 17, 570–576.

Leveau, J.H., Preston, G.M., 2008. Bacterial mycophagy: definition and diagnosis of a unique bacterial-fungal interaction. New Phytol. 177, 859–876.

Tarkka, M.T., Sarniguet, A., Frey-Klett, P., 2009. Inter-kingdom encounters: recent advances in molecular bacterium-fungus interactions. Curr. Genet. 55, 233–243.

Wargo, M.J., Hogan, D.A., 2006. Fungal-bacterial interactions: a mixed bag of mingling microbes. Curr. Opin. Microbiol. 40, 309–348.

Interactions Between Fungi and Viruses

Ghabrial, S.A., Suzuki, N., 2009. Viruses of plant pathogenic fungi. Annu. Rev. Plant Physiol. Plant Mol. Biol. 47, 353–384.

Herrero, N., Sánchez Márquez, S., Zabalgogeazcoa, I., 2009. Mycoviruses are common among different species of endophytic fungi of grasses. Arch. Virol. 154, 327–330.

Hyder, R., Pennanen, T., Hamberg, L., Vainio, E.J., Piri, T., Hantula, J., 2013. Two viruses of *Heterobasidion* confer beneficial, cryptic or detrimental effects to their hosts in different situations. Fungal Ecol. 6, 387–396.

Marquez, L.M., Redman, R.S., Rodriguez, R.J., Roossinck, M.J., 2007. A virus in a fungus in a plant: three-way symbiosis required for thermal tolerance. Science 315, 513–515.

Interactions Between Fungi and Protists

Vohnik, M., Burdíková, Z., Veyhnal, A., Koukol, O., 2011. Interactions between testate amoebae and saprotrophphic microfungi in a Scots pine litter microcosm. Soil Microbiol. 61, 660–668.

CHAPTER

11

Fungi, Ecosystems, and Global Change

Lynne Boddy
Cardiff University, Cardiff, UK

INTRODUCTION

Autotrophic plants and heterotrophic fungi colonised the land together between 600 and 460 million years ago (mya), and Ascomycota and Basidiomycota diverged around 550 mya, according to molecular clock data. Evolution of trees and forests from the Devonian to the Jurassic period provided a wealth of nutritional and spatial niches which fungi diversified to fill. Today, hundreds of thousands of fungi are associated with plants as essential mutualistic symbionts (Chapter 7), as pathogens (Chapter 8), and as decomposers that break down dead wood, leaves, and other plant tissues, and conserve and recycle their nutrients. Moreover, there are many mutualistic interactions between fungi and animals (Chapter 9), and fungi are pathogens and decomposers of these, too. So while fungi are widely associated with rot, mould, blight, and disease, they are less appreciated than they deserve for their essential part in the evolution, productivity and sustainability of life on land. Without the contributions of fungi to ecosystem processes, life as we know it would never have begun and would soon cease.

Uses of fungi in pharmaceutical, chemical, food, drink, decontamination, biorefinery, and biofuel applications (Chapter 12) stem from unique processes and products evolved by fungi in the course of adapting to their niches in the environment. Understanding how fungi operate in ecosystems helps to predict promising locations and fungal groups to explore among the estimated 5 million undiscovered species.

In this chapter, we firstly outline the part played by fungal processes in element cycles, bioconversions and energy flows in the Earth's biosphere, and then we briefly consider the global distribution of fungi – their biogeography. We then concentrate on how fungi and fungal communities are being affected by global change, how in turn this will affect ecosystems, and finally the need for, and difficulties faced in, fungal conservation.

http://dx.doi.org/10.1016/B978-0-12-382034-1.00011-6

THE ROLES OF FUNGI IN ELEMENT CYCLES, BIOCONVERSIONS, AND ENERGY FLOWS

From what we have already said in other chapters it should be evident that fungi are essential to the functioning of the planet's ecosystems. Here we pull together knowledge of the many and varied ways in which fungi live and feed, so that we can understand the major roles that fungi play in the flow of energy and carbon through ecosystems, and in mineral nutrient cycling and soil. Ecosystems can be considered to operate as three interconnected subsystems dominated by the activities of autotrophs, herbivores, or decomposers (Figure 11.1). The autotrophic subsystem in most terrestrial ecosystems is dominated mainly by photosynthetic plants or in the Arctic and Antarctic by lichens (Chapter 7). Even in this subsystem where plants dominate, fungi are major players, as all plant species examined so far contain fungal endophytes (Chapter 7) and over 85% have mycorrhizal associations (Chapter 7) with their roots; the mycorrhizal fungi absorbing water and mineral nutrients from soil, as well as playing protective roles. Further, pathogenic fungi can reduce plant productivity but increase plant species diversity (Chapter 8). In marine ecosystems, algae are major contributors to photosynthetic energy input, and again these have fungal endosymbionts.

The herbivore subsystem largely comprises animals that feed directly on photosynthetic organisms, or that feed on organisms that have already fed on autotrophs, ranging from microscopic nematodes to large ruminant mammals, and the carnivores that feed on them. In this sense, some fungi can be considered to be part of the herbivore subsystem, feeding directly on photosynthetic organisms, and on herbivorous and carnivorous animals. These fungi are the biotrophic and necrotrophic pathogens (Chapters 8 and 9) that abstract nutrients from their host at a cost to the host. Like herbivores and carnivores, the bodies of these fungi will eventually enter the decomposer subsystem where they will be decomposed by other fungi and bacteria.

In the decomposer subsystem, energy flows from plant and animal remains through a trophic web of heterotrophic organisms. Fungi dominate the decomposer subsystem in habitats where the main inputs are lignocellulosic plant remains. These are combined in the upper layers of soil with other detritus – the dead bodies, tissues, cells, and exudates of other organisms. The fungi of this habitat include not only those that specialise in saprotrophy (feeding from dead organic matter), but also many mycorrhizal fungi that have some saprotrophic ability (Chapter 7), and necrotrophic pathogens that continue feeding saprotrophically in the tissues that they have killed (Chapters 8 and 9). The mycelial nature of fungi (Chapter 2) allows them to ramify not only over the surface, but also to penetrate within bulky substrata, which sets them apart from unicellular organisms. Further, the ability of many species, especially basidiomycetes, to break down recalcitrant compounds (e.g. cellulose and lignin, described in detail in Chapter 5) means that fungi are the major agents of decomposition of dead organic matter, and hence of recycling of the mineral nutrients bound within dead organic matter in terrestrial ecosystems.

Without decomposer fungi, life on earth would probably cease after a few decades because carbon and mineral nutrients would be locked up in dead tissues and unavailable to autotrophs for continued primary production. While some plant remains are decomposed by bacteria or combusted in fires, most are decomposed by fungi which can depolymerise lignin and cellulose synthesised by plants, thereby replenishing the carbon dioxide of the

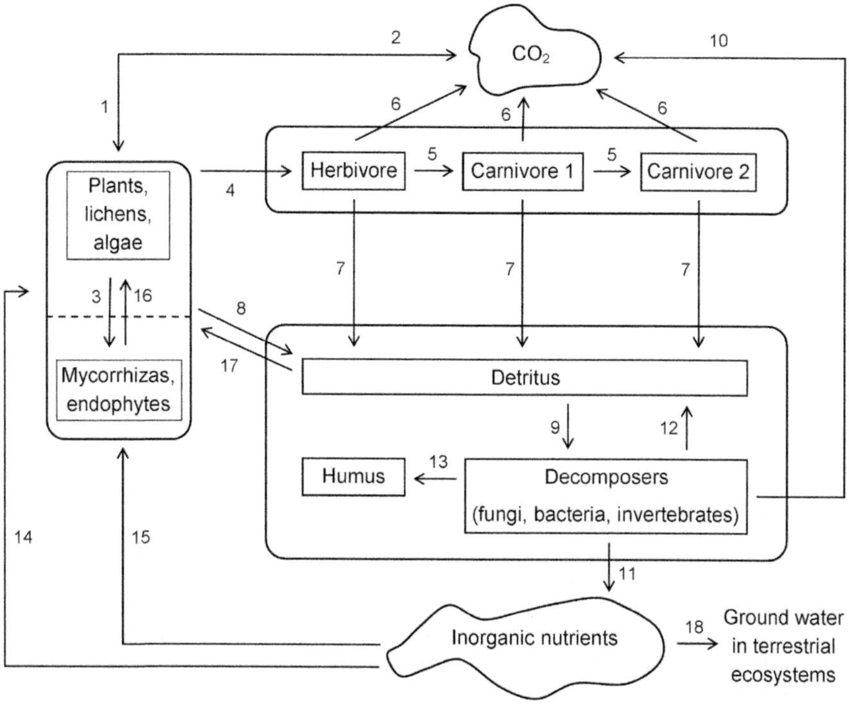

FIGURE 11.1 A general model of ecosystem structure comprising the autotroph, herbivore, and decomposer subsystems. Autotrophs fix carbon from CO_2 during photosynthesis (1). Some photosynthate is respired and returns to the atmosphere as CO_2 (2), up to 20% supports mycorrhizal growth and activity (3), and the remainder forms plant biomass – net primary production (NPP). Living material from autotrophs enters the herbivore subsystem, by vertebrate and invertebrate grazing, and plant pathogens (4). In forests, usually <5% above ground NPP is grazed, though occasionally it is >10%, and complete defoliation can sometimes occur during pathogen outbreaks or population explosions of grazing invertebrates. In grasslands, up to 50% NPP can be grazed, though usually it is <25%. Within the herbivore subsystem, organisms are consumed by others (5), C is lost to the atmospheric pool by respiration (6), and dead animals and pathogens, excreta, sloughed cells, skin, hair, etc. enter the decomposer subsystem (7). Dead organic matter from the autotroph subsystem will go directly to the decomposer subsystem (8), where it and that from the herbivore subsystem (7) is broken down by decomposers (9) ultimately to CO_2 (10) and H_2O with the release of mineral nutrients (11). Within the decomposer subsystem, the bodies of decomposers themselves die (12) and are decomposed. Decomposer bacteria and fungi also produce humus (13). Mineral nutrients released during decomposition are taken up by the autotrophs, occasionally directly (14), but usually via mycorrhizal mycelium (15) and then passed to the plant (16), and mycorrhizal fungi can also decompose organic materials to some extent, effectively short- circuiting the system (17). Some nutrients released into the soil nutrient pool may be lost via ground water (18), but the presence of mycorrhizal fungi keeps this to a minimum. *Source: Adapted from: Swift, M.J., Heal, O.W., Anderson, J.M., 1979. Decomposition in Terrestrial Ecosystems. Blackwell, Oxford.*

atmosphere (Figure 11.2). Bacteria and nematodes are important in decomposing animal tissues, including bodies of vertebrates, and plant remains in aquatic and in less aerobic environments. Invertebrates do not, however, have the enzymatic ability to catabolise as many complex molecules, though some have significant roles in the decomposition of wood and leaf litter. For example, some invertebrates have mutualistic partnerships with fungi (e.g. the attine ants and higher termites, Chapter 9), while hundreds of other arthropods are associated

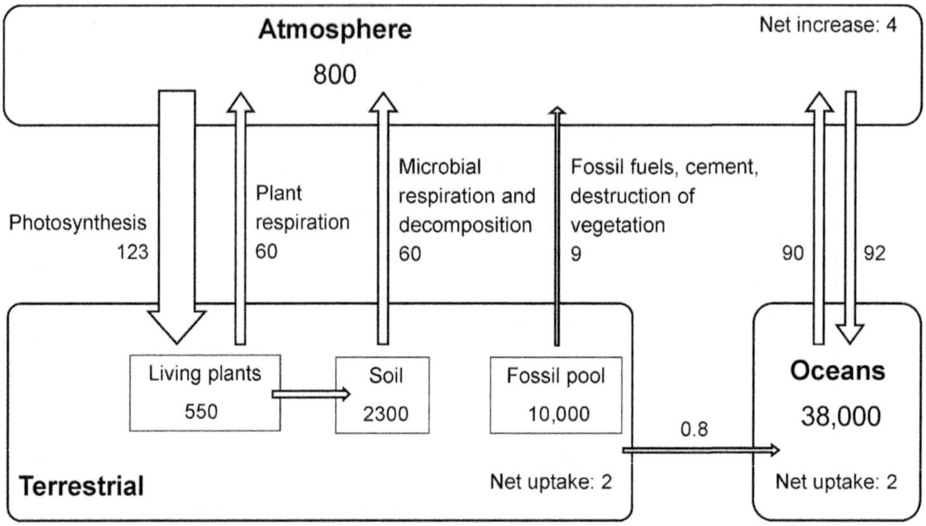

FIGURE 11.2 The global carbon cycle. The numbers in boxes are values for pools expressed as 10^{15} g C and the fluxes (indicated by arrows) and net uptake are expressed as 10^{15} g C/year. Atmospheric carbon is in the form of CO_2, and that in plants is largely lignified cellulose. The fungi make major direct contributions in transfer from living plants to soil as pathogens and mycorrhizas, and from soil to the atmosphere via decomposition. In boreal and arctic regions, lichens make major contributions to photosynthesis. *Source: DOE, 2010. Grand Challenges for Biological and Environmental Research: A Long-Term Vision; A Report from the Biological and Environmental Research Advisory Committee March 2010 Workshop. US Department of Energy Genomic Science: Biological and Environmental Research Information System (BERIS).*

with wood decay fungi as commensals. The main role of invertebrates is in comminution – the physical (not chemical) breakdown of large litter components into small particles that can be decomposed more quickly and alter the community of decomposer microorganisms and fungi.

Within the decomposer subsystem fungi, together with bacteria, produce humus, which at 1600×10^{15} g globally, represents the main carbon reservoir in terrestrial ecosystems. Humus is not just the slowly decomposing remains of complex organic materials; its accumulation also involves synthesis of complex molecules based on substances such as lignin, though its biochemistry is still poorly understood. The importance of humus lies in its water-holding capacity, its capacity for cation exchange and nitrogen binding and hence ability to store nutrients, and its improvement of soil structure. Fungi, particularly those that form arbuscular mycorrhizas (Chapter 7), also play a direct role in promoting soil aggregation and improving soil structure. Not only do their extensive hyphal networks entwine soil particles, they also produce the carbon-, nitrogen-, and iron-rich glycoprotein glomalin, which joins small soil particles promoting aggregation and stability (see also p. 215).

The huge importance of fungi to carbon cycling in terrestrial ecosystems is evident when we consider the quantities involved. Photosynthetic carbon, entering the soil via mycorrhizal fungi, fuels half of the belowground microbial activity, the other half coming from dead organic matter. Between 27% and 68% of plant net primary productivity (NPP) is allocated belowground to roots and mycorrhizal fungi, 1–21% of NPP being allocated directly to mycorrhizal fungi. The soil carbon pool is three times greater than the atmospheric carbon pool and four times greater than the plant carbon pool. Fungal processes thus represent a potential control point in the global carbon cycle, and the implications of environmental change for the contributions of fungi to ecosystems are considered later in this chapter. The huge role of fungi in mineral nutrient cycling is evident from the forgoing, since it is the fungi that release nutrients from lignocellulosic substrata, and the nutrient requirements of most plants are met via mycorrhizal mycelium that sequesters nutrients from soil solution, and sometimes directly by decomposition of dead plant material and from inorganic rock phosphate (Chapters 5 and 7). Some fungi also perform transformation of inorganic nitrogen, phosphorus, and sulphur oxides (pp. 157–159), though in ecosystem terms this is usually negligible compared with prokaryote transformations. Geotransformation processes effected by fungi (pp. 181–184), on the other hand, including transformations of rocks and minerals, bioweathering, mycogenic mineral formation, and the interactions between fungi and clay particles and between fungi and metals, can be of major significance.

BIODIVERSITY AND BIOGEOGRAPHY OF FUNGI

The biogeography of plants and animals has been studied in detail for over a century, but patterns of fungal biogeography have only started to emerge in the last few years. One major impediment has been our lack of knowledge of fungal phylogeny, and our inability to discriminate between morphologically similar taxa (pp. 107–109). For example, the wood decay fungus *Hyphoderma setigerum* was considered to have a worldwide distribution, but it is now clear from molecular phylogenetic studies that, although fruit bodies are morphologically indistinguishable, this morphological 'species' actually comprises at least nine taxa. These have geographically distinct ranges: four restricted to northern Europe, two to North America, two to East Asia, one to Greenland, and one found in the Caucasus, East Asia, Greenland, and North America. Another hurdle to realising that fungi do exhibit biogeographic differences was the incorrect perception that fungi have few barriers for dispersal because they have airborne spores. The dictum of microbial ecology 'everything is everywhere, the environment selects' is now seen not to apply to fungi.

Species of fungi can have different geographical distributions, and to some extent these distributions do depend on the environment. Thus, fungal distributional patterns at geographical scales from landscape to continents depend on suitable conditions for establishment and growth, and these are determined by climate and vegetation (Figure 11.3). Not surprisingly, geographical distributions of mycorrhizal fungi (Chapter 7), biotrophic pathogens (Chapter 8), and saprotrophs with substrate preferences (Chapter 5) match that of their hosts and/or preferred resources. Superimposed on this are direct climatic effects on fungal growth and reproduction. Finally, biogeographic distribution also reflects the ability of fungi to disperse.

Our knowledge of how these patterns have changed historically is even more scant. The poor fungus fossil record has not helped; however, historical accumulations of fungal spores are an important component of palynological records that are dominated

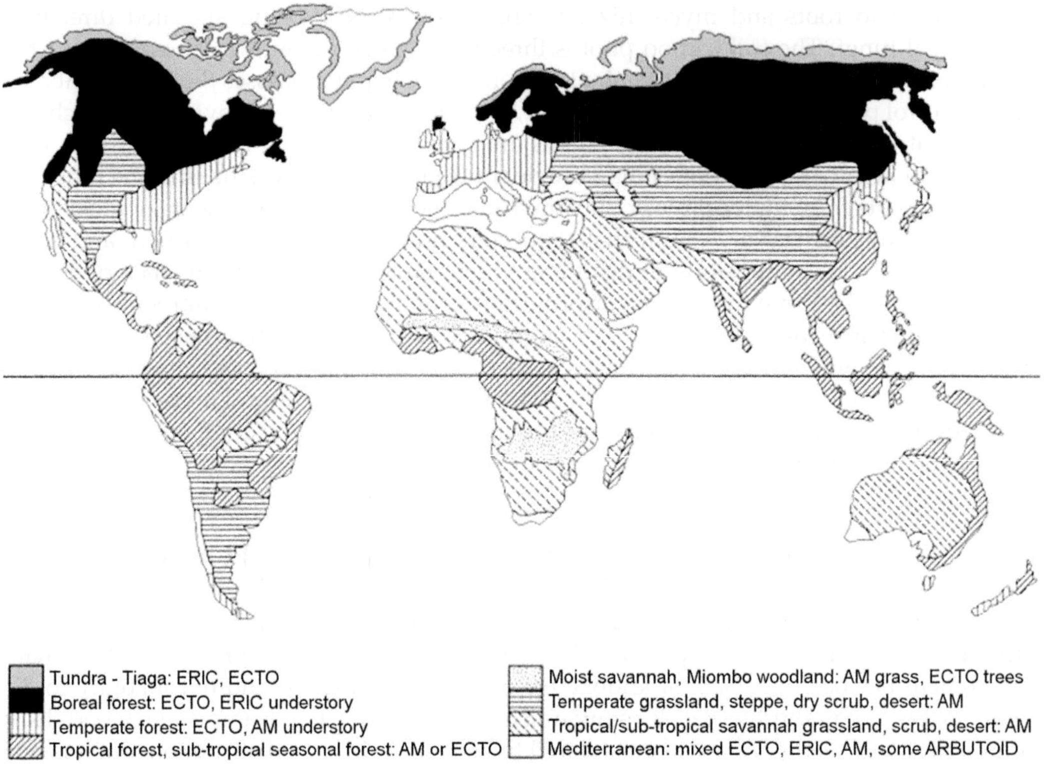

FIGURE 11.3 Biomes are regions where climate and vegetation are relatively constant over an area of land surface, largely reflecting climate. Fungal diversity and function are believed to be characteristic of different biomes. The figure gives types of mycorrhizas found in different biomes as an example. In very broad terms, arbuscular mycorrhizal symbioses (pp. 206–215), for example, predominate where decomposition is rapid and nitrogen supply is good relative to that of phosphorus, notably tropical regions and temperate grasslands. Major exceptions are the dipterocarp forests of Southeast Asia that are dominated by ectomycorrhizal symbioses (pp. 215–225). Ectomycorrhizal symbioses dominate the cool temperate and boreal forests of deciduous angiosperms and evergreen conifers, where decomposition is slower and nitrogen is usually limiting. Ericaceous mycorrhizas dominate in tundra and heathlands where the vegetation is largely Ericales, and decomposition and nitrogen cycling are extremely slow. The distribution of major forest types has a strong influence on distribution patterns of ectomycorrhizal and wood decay species, differences occurring amongst, for example, Betulaceae, Fagaceae, and Pinaceae. The types of decay performed by the dominant fungi vary too, with white rot (pp. 146–150) predominating in angiosperm woodlands and brown rot (pp. 150–152) in conifers. In the boreal zone, many species have a circumpolar distribution occurring in North America, Europe, and Siberia, reflecting the long time that the taiga zone has been coherent. Many wood decay species of temperate forests in Europe appear also to be found in similar forest types in East Asia and North America. Likewise, many species of the Mediterranean are also listed in subtropical and tropical regions. On the other hand, many species have been recorded only in geographically small areas, though this may reflect insufficient searching elsewhere. *Source: Read, D.J., Leake, J.R., Perez-Moreno, J., 2004. Mycorrhizal fungi as drivers of ecosystem processes in heathland and boreal forest biomes. Can. J. Bot. 82, 1243–1263.*

otherwise by spores produced by seedless plants and pollen from seed plants. Changes in the proportions of different kinds of fungal spores can provide crucial information on the biological communities and climatic conditions that supported these fungi. The abundance of ascospores of the dung-fungus *Sporormiella* serves as a proxy for the mass extinction of mega-herbivores toward the end of the Pleistocene (12,000–13,000 years ago). In North America, sediment samples from lakes and wetlands show a precipitous decline in the abundance of these ascospores, which is thought to have been caused by the decimation of the woolly mammoth (*Mammuthus primigenius*), mastodon (*Elephas americanus*), and rhinoceros by human hunters and consequent disappearance of their dung. A rich, 130,000 year old palynological record from Queensland, Australia, shows an abrupt loss of *Sporormiella* spores 41,000 years ago, following human immigration. Human hunters are identified as the cause of the collapse of animal populations. Changes in the concentrations of *Sporormiella* ascospores have also been linked to the extinction of Madagascan mega-herbivores 1700 years ago and the seventeenth century extirpation of species of moa (*Dinornis*) – the giant flightless birds of New Zealand.

The distribution of fungi is changing as a result of man's activities. The current geological epoch is sometimes described as the anthropocene, reflecting the huge changes that man has made to the planet. The start of the anthropocene is perhaps best defined as about the time when James Watt invented the steam engine. Though man has changed the environment on a local scale for thousands of years (e.g. by gradual forest clearance for agriculture), major features of the anthropocene are changes on a global scale, with land use change, climate change, nitrogen deposition, movement of biota and CO_2 increase threatening biodiversity in decreasing order of impact. The relative importance of these threats, however, varies between ecosystems (Figure 11.4). Land use change is by far the strongest driver of biodiversity loss in tropical forests and the temperate forests of South America, the weakest driver in Arctic and alpine ecosystems, and intermediate in grasslands and Mediterranean ecosystems. Climatic warming is expected to be most dramatic at high latitudes and least in the tropics. Nitrogen deposition is greatest in the northern temperate zone in the vicinity of cities, and, perhaps surprisingly, is predicted to be the biggest driver of biodiversity change in this region.

While large scale loss of habitat (e.g. for creation of agricultural land, p. 381), and large scale pollution (e.g. from burning fossil fuels, (p. 383), are obvious causes of species loss, small scale habitat loss and pollution can also endanger species. For example, *Poronia punctata* (whose sexual fruit bodies look like the heads of carpenter's nails; Figure 11.5a) is one of the few ascomycetes worldwide to be recognised as needing conservation (though there are likely to be very many not yet recognised). Its demise follows the rise in automobiles, not because of pollution but because of loss of habitat – it is found in old dung of horses, donkey, ponies, and exceptionally, elephants, none of which are much used now for transport! Further, it seems only to thrive in dung from animals feeding on 'unimproved' pasture (p. 382) (i.e. meadows to which no fertilisers, pesticides or high yield grass varieties have been introduced). The effects of global change on fungi are considered below under three headings: (1) climate change responses of fungi; (2) land use change; and (3) pollution, pesticides, fertilisers, nutrient distribution, and recycling. However, none of these factors act in isolation.

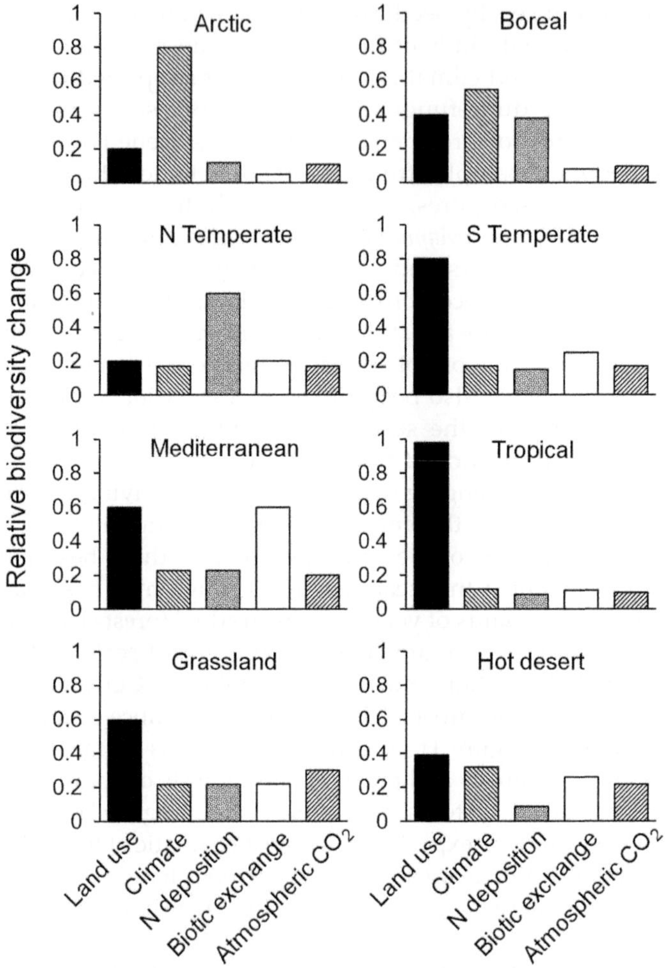

FIGURE 11.4 Many factors affect biodiversity loss. The predicted relative effect of changes in land use, climate, nitrogen deposition, biotic exchange, and atmospheric CO_2 on total biodiversity change, in major biome types is shown. Estimates were obtained by multiplying the expected change caused by each driver by the predicted impact of a large change in each driver. The scale is relative to the maximum possible value. *Source: Modified from: Sala, O.E., Chapin, F.S. III, Armesto, J.J., Berlow, E., Bloomfield, J., Dirzo, R., Huber-Sanwald, E., Huenneke, L.F., Jackson, R.B., Kinzig, A. et al., 2000. Global biodiversity scenarios for the year 2100. Science 287, 1770–1774.*

CLIMATE CHANGE RESPONSES OF FUNGI

There is no doubt that Earth's climate is changing, and that this has largely been brought about by man. By 2100, atmospheric CO_2 concentration is predicted to reach 540–970 ppm. This rise in CO_2 and some other gases (e.g. CH_4) is the cause of the so-called greenhouse effect, which results in a rise in surface temperature. Temperature increases ranging between 1.1 and 6.4 °C are predicted, varying across the globe. A 1 °C increase in mean annual temperature in the temperate zone will result in a northward shift in isotherms of approximately 140 km, and

FIGURE 11.5 Rare and endangered species. (a) The nail fungus *Poronia punctata* on the Red List of many European countries © Martyn Ainsworth; (b) The wax cap *Hygrocybe collucera* endangered in New South Wales Australia © Ray and Elma Kearney; (c) *Ramariopsis pulchella* near threatened in the British Red List © Martyn Ainsworth; (d) *Hericium coralloides* © Martyn Ainsworth; (e) *Hericium erinaceus* protected by law in the UK © Martyn Ainsworth; (f) *Hygrocybe lanecovensis* endangered in New South Wales Australia © Ray and Elma Kearney; (g) *Phellodon melaleucus* © Martyn Ainsworth; (h) young *Piptoporus quercinus* © Martyn Ainsworth; (i) *Zeus olympius* (Ascomycota) found only on Bosnian Pine (*Pinus heldreichii*) in northern Greece © Stephanos Diamandis; (j) *Pleurotus nebrodensis* found only in Sicily © David Minter; (k) *Ophiocordyceps sinensis* (Ascomycota) on a caterpillar (arrowed) © Paul Cannon (CABI/RBG Kew); (l) critically endangered *Erioderma pedicellatum* (boreal felt lichen) © Christoph Scheidegger. (See the colour plate.)

an upward altitude shift of 170 m. The incidence of extremely hot days is likely to increase and extremely cold days to decrease. Patterns and amount of precipitation will alter dramatically, and UV-B penetration and the incidence of extreme events will increase, though the magnitude of these changes is uncertain. 30–50% of glacier mass is predicted to disappear by 2100, and permafrost will shrink by 16%. Sea level will rise by 15–95 cm by 2100.

Effects on fungal distribution, community composition, ecophysiology, activity, and times and extent of reproduction are all likely to alter in response to climate change, both directly through effects on physiology and indirectly through changes in habitat and interactions with other organisms. Since fungi are major agents of decomposition and nutrient cycling (pp. 362–365), are mycorrhizal with about 90% of plants (Chapter 7) and can cause diseases of plants (Chapter 8) and animals (pp. 310–325), changes in their activity will have major knock-on effects to ecosystem functioning.

On a global scale the distribution of plant communities is likely to change dramatically as a result of climate change, and associated saprotrophic, mycorrhizal and pathogenic fungi are likely to change with them. For example, over the next century in the northern hemisphere, boreal and temperate forests are likely to move northwards with huge reduction in taiga and Arctic tundra. Concomitantly, grassland and shrubland will expand northwards. For a 4 °C increase in temperature over the next century plants will probably have to move 500 km northwards or 500 m upwards to remain in the same climatic envelope as now. This far outpaces historical tree migration rates of 100–200 m per annum at low altitude.

Effects on Saprotrophs

The general effects of temperature, moisture, and CO_2 on the metabolic activity and timing of different events within the fungal lifecycle (e.g. growth and spore production) are well known (Chapters 3 and 5). For example, metabolic activity increases with rise in temperature up to an optimum, above which it decreases, as a result of effects on enzyme catalysed reactions. Moisture inhibits activity when there is both too little and too much: low water potential inhibits water uptake and retention, and enzyme function; high water is inhibitory because it decreases rate of diffusion of O_2 to hyphae and of CO_2 away from hyphae. However, despite understanding the ecophysiology of effects of individual abiotic factors on individual fungal species, it is extremely hard to extrapolate to effects of climate change on fungi in the field, in mixed communities, and in fluctuating environments. That dramatic changes are occurring is clear from long-term datasets on fruiting of macrofungi. For example, the first of this type of study analysed records of 200 saprotrophic basidiomycetes within a 30-mile radius of Salisbury, in the south of the UK between 1950 and 2005. For autumn fruiting, this revealed that the mean date when fruiting of a species was first recorded (averaged across all species) is now significantly earlier than it was prior to the late 1970s, while the mean last fruiting date is significantly later, resulting in a doubling of the length of the fruiting season (Figure 11.6). Response varies between species even within genera, 47% showing an advancement and 55% continuing fruiting later. The response also differs depending on habitat type, with only 13% of grassland species fruiting earlier and 48% having later last fruiting, but with 53% of wood decay fungi fruiting earlier and 20% having later last fruiting. These changes in autumn fruiting times are correlated with late summer temperature and rainfall. Changes in fruiting time have also been reported in other European countries, China and Japan, and are likely occurring worldwide, though there is often considerable variation between countries.

As well as changes to autumn fruiting patterns, many species that previously only fruited in autumn now also fruit in spring, and spring fruiting is getting earlier in northern Europe. Times and extent of basidiomycete fruiting depend on environmental triggers that initiate the various processes, plus sufficient water, nutrients, and energy sources. Earlier autumn

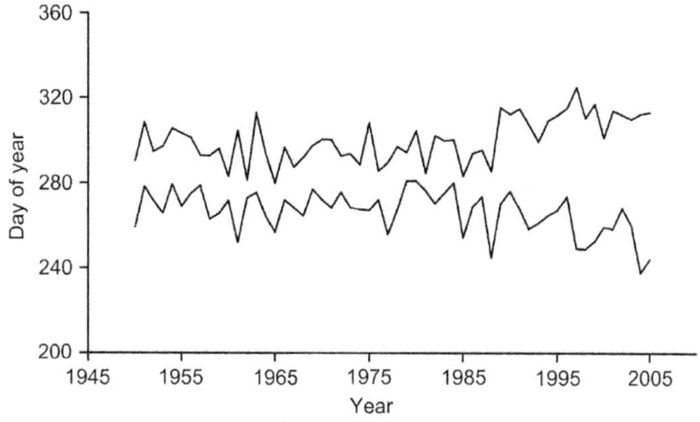

FIGURE 11.6 Fungal fruiting phenology has changed dramatically in the last 30 years or so as a result of climate change. This is shown for the mean first fruiting date (lower line) and mean last fruiting date (upper line) for 200 saprotrophic basidiomycetes around Salisbury in the south of the UK, over 56 years. The fruiting season is now significantly longer than it was pre-1970s. This change is correlated with changes in temperature and rainfall. *Source: Moore, D., Gange, A.C., Gange, E.G., Boddy, L., 2008. Fruit bodies: their production and development in relation to environment. In: Boddy, L., Frankland, J.C., van West, P. (Eds.), Ecology of Saprotrophic Basidiomycetes. Elsevier, Amsterdam, pp. 79–102.*

fruiting is likely to relate to changes in the timing of triggers to fruiting, but may also reflect the fact that mycelia have already acquired sufficient resources to fruit. Extended fruiting implies that environmental conditions remain conducive to fruiting for longer – hard frosts come later, and fungi can obtain water, nutrients, and energy to maintain fruit body production for longer. Spring fruiting suggests that some fungi are now more active in winter and spring than they were in the past. So these phenological changes that are related to climate change imply that decomposer activity of fungi is increasing. It follows that decay rate of organic matter is increasing, as is evolution of CO_2. Provided that plant primary productivity increases at the same rate then this will not be a problem, but if decay rate increases more than primary productivity then the greenhouse effect will be exacerbated. The latter may well occur in soils which are very high in organic matter, such as boreal forests, tundra soils and peat, where in the past decay has been limited by low temperature and/or lack of aeration due to high water content; because of elevated temperatures, these ecosystems may become warmer and drier. There is now evidence that Arctic soils are changing from being net carbon sinks to net carbon sources, with the worrying potential to amplify climate change.

Overall decay rate of dead organic matter in ecosystems depends on the sum of the decay rates brought about by all of the individual fungi. However, climate change will create different niches; plants grown in CO_2 enriched atmospheres tend to release more labile compounds into soil and tissues are more lignified, have greater phenolic content and higher C:N ratios, making them more recalcitrant. Also, the relative competitiveness of species will change and will affect the outcome of combative interactions (pp. 339–354), and hence fungal community composition will change. If the new communities predominantly comprise fungi more active in decomposition, then organic matter will decay more rapidly, and vice versa.

Fungi in aquatic environments are also affected by climate. Temperature preferences of species are broadly correlated with their geographical distribution. On a more local scale, distinct winter and summer communities are often evident; for example, *Flagellospora penicillioides* and *Lunulospora curvula* do not grow at less than 5 °C, and disappear from streams in colder seasons. In temperate streams, where temperatures are below the optimum for fungal growth, an experiment in which the temperature was artificially raised by 4.3 °C increased the rate of litter decomposition, decreased the number of leaf fragments (the main substrata of aquatic hyphomycetes), and decreased the number of fungal species on leaf fragments. As in terrestrial environments, climate-induced changes to riparian vegetation communities will affect the composition and diversity of the fungal decomposers of this litter in water.

Mycorrhizal Responses

Climate change affects both plants and their root symbionts. Ectomycorrhizal (ECM) associations are formed with forest trees, and the distribution of tree species within tropical, temperate, and boreal biomes will alter as climate changes, as paleoecological studies have shown for the past. In the Quaternary, trees migrated or adapted in the face of ecological change. Those that did not became locally extirpated or extinct globally. Trees move towards the poles and higher elevations when climate warms and in the opposite direction when it cools. In the southern Santa Rosa Mountains, California, United States, white fir (*Abies concolor*) and Jeffrey pine (*Pinus jeffreyi*) have moved about 96 and 28 m upwards, respectively, in the last 30 years. The current dominance of ECM associations in temperate and boreal forests is clearly a result of their dispersal patterns in response to changing climate over millennia, and changes in their distribution will continue. In tundra, warming experiments show that shrub biomass and cover will increase, and moss and lichen cover will decrease. Successful plant migration and survival of mycorrhizal fungi will depend on successful co-migration of the latter. This may not pose a problem for fungi that produce copious wind-dispersed spores in above ground fruit bodies. However, for resupinate forms that produce spores in the organic soil horizon dispersed by soil invertebrates, and also for hypogeous fungi, such as truffles, that are dispersed by small mammals, migration may depend on the ability of their animal dispersal agents to move too.

Since altitudinal gradients are paralleled by climatic and vegetation gradients, they can be used as models of communities that might be expected following climate change. Studies in the Canadian Rocky Mountains have shown that ECM species richness and diversity decreases with increasing altitude; host-specific ECM are present in subalpine forests, but in alpine regions ECM tend to be generalists. This reflects not only climate but also the plant communities present. However, there are also clear differences in mycorrhizal communities on a single plant species, depending on elevation. The ericaceous shrub *Vaccinium membranaceum* grows from valley bottoms to alpine habitats; at lower elevations the alpha diversity is lower than at higher elevations and the most frequently isolated species is *Phialocephala fortinii*, whereas at higher altitudes *Rhizoscyphus ericae* dominates.

The activity of mycorrhizal fungi is also changing, and immediately evident from changes in timing of fruit body production. Like saprotrophs, ECM basidiomycete fruiting phenology has changed since the late 1970s, but in different ways. Also there are differences between those in association with conifers and deciduous trees; in the UK, 59% of deciduous mycorrhizal species now fruit later than in the 1970s, but coniferous mycorrhizal species have changed little. This is probably because mycorrhizal fruiting responds to cues from the host trees, rather than environmental factors alone, and cues from evergreens are different to those from deciduous trees. Deciduous trees now frequently remain in leaf, and hence supply fixed carbon to their fungal partners, for much longer; thus any cue for fruiting when photosynthate declines is delayed, and carbohydrates to fuel mycelia and fruiting are available for longer. This implies that, as with saprotrophic fungi, fungi ECM with deciduous trees are active for much longer periods.

Combinations of increasing temperature and changing water regimes will impact directly on mycorrhizal fungal activity and interactions with each other and their hosts. Climatic changes and increasing CO_2 will also affect photosynthesis, elevated CO_2 causing increases; this will in turn affect the mutualistic symbioses. As a result of increased photosynthesis, mycorrhizal abundance is increasing by an average of 47% with a much larger increase for arbuscular mycorrhizal (AM) fungi (84%) than for ECM fungi (19%). However, abundance is not increasing for all species (e.g. ECM in black spruce, *Picea mariana*). There appear to be functional shifts in mycorrhizal community composition. An increase in ECM species that are long-range explorers (i.e. produce copious mycelial cords (p. 217), was found in *Betula nana* root tips in warmed Arctic tundra soils, and Douglas fir (*Pseudotsuga menziesii*) in warmer sites. Experiments show that elevated CO_2 also results in a shift in ECM community composition, again favouring morphotypes that produce large amounts of extraradical mycelium, cords and/or thicker mantles; later successional species dominate fruit body production. With elevated CO_2, mycorrhizal biomass increases, as does plant and fungal exudates and carbon inputs to soil. Soil respiration rate increases, as a result of increased mycorrhizal activity and activity of other microbes capitalising on increased carbon inputs.

Effects on Lichens

Climate largely exerts effects on lichens by affecting the amount of time that thalli are fully hydrated, and by temperature effects on metabolic processes including photosynthesis. Though lichens are autotrophic, their responses to climate will differ from those of higher plants, since lichens are poikilohydric. Climate change effects on lichens are likely to be particularly important in the Arctic and Antarctic where they are dominant primary producers. Though lichen physiological activity in the Antarctic is limited by low temperature, long-term monitoring of the most prominent fruticose lichen, *Usnea aurantiaco-atra*, has revealed that counterintuitively, an increase in thallus temperature of 0.5–1 °C may lead to a 90% reduction of annual biomass increase, due to increased losses through respiration. This occurs particularly in winter where solar radiation, and hence photosynthesis, is almost zero but respiration is higher because of elevated temperature. Also, when there is light for photosynthesis, elevated temperature results in lower atmospheric humidity, and hence longer periods when thalli are not hydrated sufficiently for photosynthesis to occur. In the long-term, *Usnea-Himantormia*-lichen heath may be replaced by competitive grass species that do benefit from elevated temperature.

The distribution of lichens that live on inorganic substrata will be influenced by direct effects of climate on a particular species and indirectly by effects on interacting species. The distribution of lichens that live on particular living substrata will additionally be affected by changes in host distribution and tree species composition in forests. Northward shifts in the boundary between taiga and tundra, will lead to northward shifts in lichens, and upward shifts in snowline and glacial fronts will provide new habitats for lichens.

The strong correlation between lichen communities and climate has been clearly shown in large- and small-scale monitoring along environmental gradients. With harsher conditions lichen cover increases, and in alpine areas lichen species richness is usually higher than plant species richness. In the Northern Ural Mountains, lichen cover and diversity are stable at 1000–1400 m above sea level, but at higher altitudes dominant lichen species have changed, and lichen cover has increased.

Lichen community composition in most regions of the globe, including the temperate zone, is changing in response to changing climate. In The Netherlands, long-term monitoring indicates that species typical of cold environments are decreasing or have recently disappeared, while warm temperate species have increased. In Western Europe, patterns of species change depending on habitat: many terricolous species are declining, whereas many epiphytic species are increasing. Interestingly, species increasing most rapidly in forests are from different genera, but all contain *Trentepohlia* algae as photobiont. Potential changes in lichen distribution have been elegantly demonstrated for Britain, where the bioclimatic envelope in which different species can grow was modelled using different climate change scenarios. Even using climatic scenarios based on low greenhouse gas emissions, considerable changes in distributions of species are predicted (Figure 11.7). This type of predictive approach can be used as a valuable tool in developing conservation strategy.

Changing Patterns of Plant Disease

Climate change, particularly elevated temperature and CO_2, and amount and patterns of precipitation, with consequent changes in air and soil humidity, are having or will have major effects on plant disease. Almost all research concerns diseases of plants of agricultural importance, because of the impact on food security. As with mutualistic symbioses, one consequence of climate change is the poleward shift of agro-climatic zones, and the concomitant expansion of the geographic range of pathogens to higher latitudes. Both host and pathogen biology, including life cycles, are affected by changing climate, and it is the interactions between the two that will determine how patterns of disease will alter. Not surprisingly, therefore, effects vary between different host–pathogen combinations (Table 11.1), and between different host and pathogen strains. Also, even for the same host–pathogen combination, effects of changing one or a few climate variables are not always the same if another variable changes slightly, hence effects vary amongst and between growing regions. Climate change can also have indirect effects on plant disease; for example, by prompting man to change cropping strategies, such as planting new varieties better adapted to heat stress during flowering. Planting 'Mediterranean-type' wheat varieties has been suggested for the UK, as they typically flower 2 weeks earlier than the cultivars currently used. This may have unexpected consequences for diseases to which the plants are susceptible during flowering, such as fusarium head blight (FHB). Changes in cultivation practice, such as including new crops in crop rotations, may provide additional hosts for pathogens.

Temperature is the single most significant environmental variable affecting spread and development of fungal pathogens. Elevated precipitation and humidity are important for spore production, release, germination and infection for most fungal pathogens, apart from those adapted to hot dry conditions (e.g. *Ustilago* species, pp. 16–18). Though average temperature and moisture values are often used in studies and predictive modelling, single extreme maximum and minimum values can be much more important.

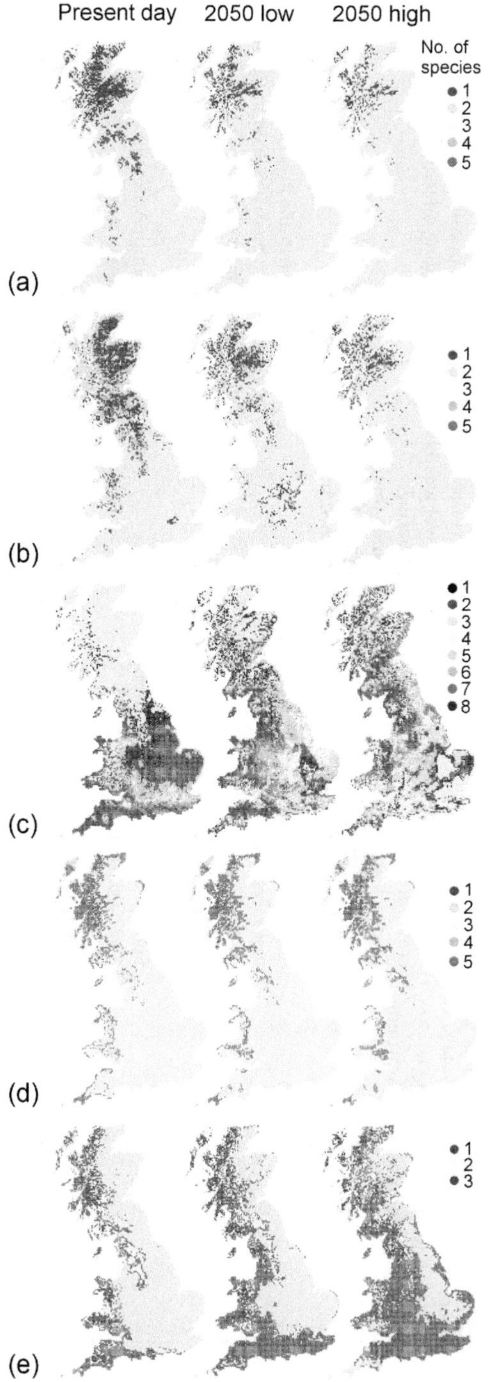

FIGURE 11.7 Predictive modelling is an extremely useful tool for revealing possible impacts of climate change on species distributions. A modelling approach used information on the bioclimatic growth envelope of 26 species of lichens whose distributions are well defined within Britain. Even using climatic data for a low greenhouse gas scenario resulted in modelled bioclimatic growth envelopes for 2050 that are very different from those now. The lichen species were considered separately under five biogeographic groups: (a) northern-montane; (b) northern-Boreal; (c) southern-widespread; (d) oceanic-northern; and (e) oceanic-widespread. *Source: Modified from: Ellis et al. (2007).*

TABLE 11.1 Climate Change Effects on Disease of Major Energy Food Crops

Crop		Examples of major diseases	Pathogen mode of nutrition	Main weather factors affecting epidemics	Climate change effects (actual or predicted)
Wheat	The most important source of carbohydrate, providing 20% of the calorie intake of the global population. Cultivated worldwide, 10 main producing regions: China, India, USA, Russia, France, Canada, Australia, Germany, Pakistan, Turkey. In the United States, a projected decrease in production of 25–44% under slowest warming, and 60–79% decrease under most rapid warming	Rust: *Puccinia striiformis* f. sp. *tritici* (Pst)	Biotrophic	Moisture, temperature, wind	Elevated temperature may decrease development and survival in some wheat-growing regions, but most reports of increase with rising temperature. New Pst strains, better adapted to higher temperatures, now dominate in south central United States
		Fusarium head blight (FHB) and crown rot (CR): *Fusarium* and *Microdochium* spp.	Necrotrophic	FHB favoured by warm wet weather at flowering. CR when there is drought post-flowering	The species responsible for disease change depending on temperature
		Crown rot: *Fusarium pseudograminearum*	Necrotrophic	–	Prevalent in Australia in southern regions , where there has been drought since 2000. Under elevated CO_2 (825 ppm) fungal biomass increased in wheat stems
		Spot blotch: *Cochliobolus sativus*	Necrotrophic	–	Increased in South Asia with higher night time temperatures
Rice	Staple food crop of the world, mainly produced in Asia and Brazil, with demand increasing. Rising sea levels, changing patterns of rainfall and elevated temperatures will lead to major alterations in land and water resources for rice production	Rice blast: *Magnaporthe grisea/oryzae* complex	Biotrophic	Temperature	Severity predicted to increase. In cool subtropics of Japan and northern China epidemic severity is predicted to increase, but in warm/cool subtropics fewer epidemics are predicted following warming. Elevated temperature can increase effectiveness of some genes, e.g. *Xa7* (most effective against *Xanthomonas oryzae* pv. *oryzae*) With elevated CO_2 leaf blast severity increased
		Sheath blight: *Rhizoctonia solani*	Necrotrophic	–	With elevated CO_2 there were more diseased plants, but lesion size was unaltered

Barley	Less produced than for other cereals; mainly Europe	Powdery mildew: *Blumeria graminis*	Biotrophic	–	Decrease at elevated CO_2
		Smut: *Ustilago hordei*	Biotrophic	–	Decrease at elevated CO_2
Maize	Particularly produced in the Americas	Smut: *Ustilago maydis*	Biotrophic	–	Inhibited at elevated CO_2
Soybean	A major crop and now rapidly expanding, as an important source of protein as well as carbohydrate. Four main producers: Argentina, Brazil, China and USA	Downy mildew: *Peronospora manshurica*	Biotrophic	–	Elevated CO_2 alone or with O_3 reduced severity by 39-66%
		Brown spot: *Septoria glycines*	Necrotrophic	–	Elevated CO_2 alone or with O_3 caused a small increase, occasionally
		Sudden death syndrome (SDS): *Fusarium virguliforme*	Necrotrophic	–	Elevated CO_2 and O_3 had no effect
		Asian soybean rust (ASR): *Phakopsora pachyrhizi*		–	This extremely aggressive pathogen has rapidly spread, since 2000, from Asia to Africa to South America, and was introduced to 8 US states by hurricane Ivan
Potato	Grown worldwide, but more now grown in developing than industrialised countries. China is the main grower, with over 20% global production. 18–32% predicted decrease with climate change. Elevated temperature will lead to change in planting time and location of growing regions, and longer growing seasons	Late blight: *Phytophthora infestans* (Oomycota)	Necrotrophic	–	In Canada: predicted increase in duration of epidemics and levels of inoculum, but decrease in disease progression In Finland: 17 fold more outbreaks in 1998–2002 than in 1933–1962 and 1983–1997 due to more conducive conditions; 2–4 week earlier start to outbreaks; 1OC rise predicted to extend the infection period.
		Verticillium spp.	Necrotrophic	–	In Canada: predicted to increase due to elevated temperature

Data from Luck et al. (2011).

Temperature and water regime affect differently the various stages of the fungal pathogen life cycle, including spore production, survival, germination, mycelial growth, and disease progress. Hence, the timing of these weather events in relation to different stages of the fungus lifecycle is important. Mild, damp winters, for example, increase survival of pathogens that overwinter on infected seed or crop debris. In contrast, warmer, drier summers tend to reduce incidence of pathogenesis.

Most pathogens are active over a wide range of temperatures, but the extent of the time under optimum conditions may extend or narrow, and the infection window may alter. The relative importance of different spore types in infection and survival between crops may change. For example, *Leptosphaeria maculans* (=*Phoma lingam*) – the cause of blackleg and stem canker of oilseed rape (*Brassica napus*) – produces ascospores and conidia; conidia require higher temperatures for germination, so their importance in the disease cycle is predicted to increase with climate warming in Northern Germany. Temperature may also influence which species are actually dominant causes of particular disease types. FHB of small grain cereals is caused by a complex of many different species of *Fusarium* and *Microdochium* (Figure 11.8). *Fusarium graminearum* has temperature optima for growth at 24–28 °C, and predominates in regions having relatively hot summers, such as Australia, North America, and parts of continental Europe, whereas *Fusarium culmorum*, with lower optima for growth (20–25 °C), is found in cooler maritime areas such as northwest Europe. In northern Europe, there has recently been a shift from *Fusarium culmorum* to *Fusarium graminearum*. In maize, *Fusarium graminearum* was the predominant species but with warmer temperatures this has shifted to *Fusarium proliferatum*, *Fusarium subglutinans*, and *Fusarium verticilliodes*.

Development of FHB is related to temperature during the 6 weeks prior to flowering, and also to rainfall during flowering, as high humidity promotes sporulation. Not only is there greater infection from spores during warm (15–30 °C), wet weather but also mycelia spread to more florets. Modelling predicts that under climate change scenarios, rainfall will be considerably reduced at the current flowering time suggesting that FHB would be decreased; however, UK flowering is predicted to occur 2 weeks earlier when there will still be high rainfall, hence no predicted reduction, and even a slight increase, in FHB. This earlier and more rapid development will occur in many crop plants in existing production regions. Clearly, interactions between host, pathogen, and climate are extremely complex, and different effects will be expected in different locations. Modelling suggests that FHB will increase in the UK (Figure 11.8), though studies in other areas such as Ontario suggest no increase (Table 11.1).

Elevated atmospheric CO_2 has been reported to increase growth and fecundity of some fungal pathogens, but to cause decrease in pathogenesis of others (Table 11.1). It also indirectly affects fungal pathogens by affecting plant growth. Plant yield increases by over 15% as a result of the 'fertiliser' effect. Increase in the thickness of the epidermis and of leaf waxes under elevated CO_2 results in increased resistance to some pathogens. As with elevated temperature, increased rate of photosynthesis as a result of increased CO_2 may lead to earlier growth flushes, which may affect pathogen colonisation. More luxuriant canopy will develop with associated different microenvironment for pathogen development. The larger aerial biomass and increased biomass of crop debris could lead to larger reservoirs of pathogens.

Currently, there are major gaps in our understanding of how plant diseases will be influenced by climate change, with a lack of information in the field and at cellular and genomic level. Since several interacting factors influence disease outcome, modelling is an important

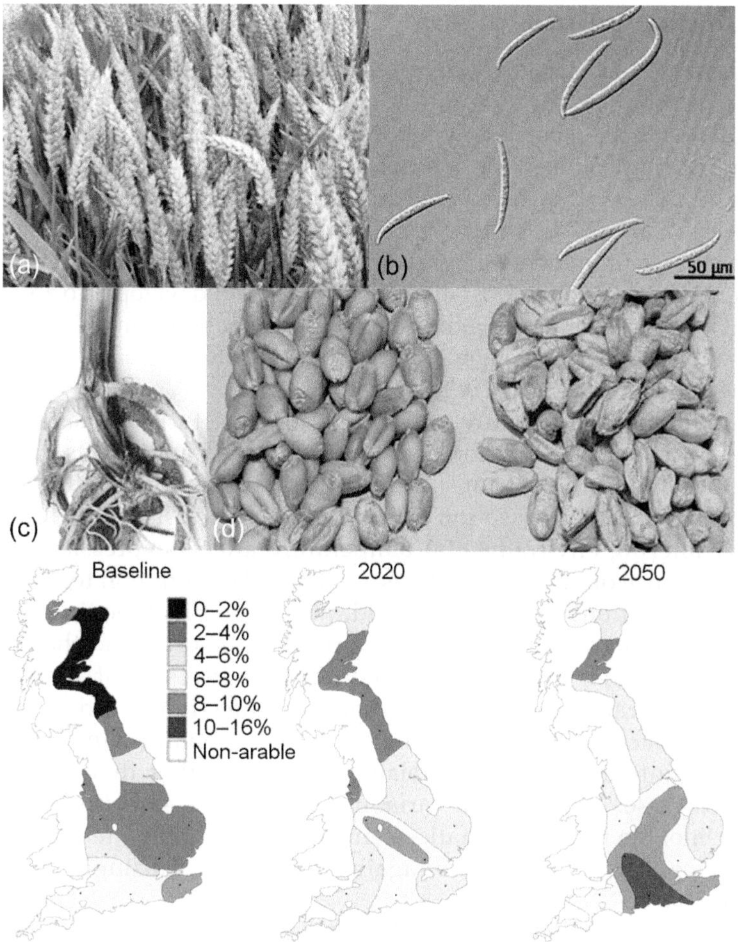

FIGURE 11.8 Fusarium head blight, also known as fusarium ear blight or scab, is a serious disease of cereals, caused by many different species of *Fusarium* and *Microdochium*. (a) Fusarium ear blight in winter wheat crop; (b) Macroconidia of *Fusarium graminearum*; (c) fusarium foot rot; (d) healthy harvested grain (left) and *Fusarium* infected grain (right). Models can be used to predict disease outbreaks in the short term, and changing patterns of incidence in the longer term. The maps show the predicted current incidence (% plants affected) of fusarium head blight disease in Great Britain, by a model that combined crop growth with a weather-based disease model using simulated future climate data. The baseline scenario prediction used weather from 1960 to 1990, and the maps for 2020 and 2050 both used a high emission scenario. *Source: West, J.S., Holdgate, S., Townsend, J.A., Edwards, S.G., Jennings, P., Fitt, B.D.L., 2011. Impacts of changing climate and agronomic factors on fusarium ear blight of wheat in the UK. Fungal Ecol. 5, 53–61. Maps adapted from: Madgwick, J.W., West, J.S., White, R.P., Semenov, M.A., Townsend, J.A., Turner, J.A., Fitt, B.D.L., 2011. Impacts of climate change on wheat anthesis and fusarium ear blight in the UK. Eur. J. Plant Pathol. 130, 117–131.*

predictive approach (Figure 11.8). Little consideration has yet been given to strategies for managing disease under a changing climate, but this is extremely important for long-term planning. It takes, for example, at least 10 years to develop annual crop varieties resistant to any particular disease. Clearly, much remains to be done to safeguard the food supply of the planet's human population.

Changing Patterns of Animal Disease

As with diseases of plants, climate change affects the biology of both the pathogen and the host, including distribution, survival, rate of growth/development, and extent and timing of reproduction. Perhaps the most widely reported fungal disease of animals in relation to climate change is the chytridiomycosis of amphibians caused by *Batrachochytrium dendrobatidis* (Bd) (pp. 311–312). The biology of Bd is clearly affected by temperature; growth in the lab is optimal between 17 and 25 °C, so rising environmental temperatures in montane environments are likely to result in increased growth of the pathogen. Interestingly, however, while at these temperatures infectious zoospores of Bd encyst and develop into zoosporangia more rapidly than at lower temperatures, more zoospores per sporangium are produced at 7–10 °C, and these remain infectious for longer. Life-history tradeoffs, therefore, mean that Bd maintains high fitness over a wide range of temperatures.

Changes in climatic temperature are widely thought to be a major driver in host/pathogen dynamics, but the situation is more complicated and effects of temperature are often confounded with other factors. Increasing mean and minimum temperatures are associated with increased disease incidence. With some hosts, though not with others, experimentally altering environmental temperature affects expression of host immune response, maintenance of infection and host survival. The common toad, *Bufo bufo*, which inhabits temperate regions of the Old World, has shifted its breeding phenology in response to warmer winters; the overwintering period is shorter and the post-hibernation body condition of females is poorer, as is reproductive investment and survival. Higher overwintering temperature increases the chance of infection by Bd. However, once established, the proliferation of Bd in toadlets is better in colder winters, and overall, the survival of overwintering toadlets is related to factors other than winter temperature and infectious disease. Temperature variability also seems sometimes to be a driver of amphibian decline; the El Nino climatic events result in regional temperature variability which may reduce amphibian defences against pathogens, causing widespread losses in the genus *Atelopus*.

The distribution and incidence of fungal diseases of humans (pp. 293–310) is also likely to change along with changing climate. Valley fever, caused by *Coccidioides* spp. (Ascomycota; pp. 303–305), which occurs in very dry regions of western North America, is more prevalent 1 or 2 years after an abnormal increase in rainfall, following prolonged drought, associated with El Nino. Regional warming, changes in the rain season and increasing frequency of extreme weather events, as a result of climate change, are likely to alter the distribution of the disease and to increase the risk of infection. Less obviously, fungi may affect hosts differently when they have grown outside the host under different abiotic conditions. *Aspergillus fumigatus*, which grows saprotrophically but is also an opportunistic mammal pathogen and an airborne allergen (Chapter 9), expresses genes encoding for major allergens more highly at 17 °C than at 32 °C, resulting in a 12-fold difference in allergenicity per spore. Atmospheric CO_2 concentrations also affect allergenicity of spores: spores of *Aspergillus fumigatus* produced under ambient CO_2 (392 ppm) have over 8 times the allergenicity of those produced under preindustrial (280 ppm) conditions, as reflected by the allergic protein Asp f1. Extremely elevated CO_2 (560 ppm), however, reduces spore allergenicity.

Extinctions Due to Climate Change

Some habitats and the fungi within them are clearly threatened by global warming, including arctic and alpine, mangrove swamps, coastal areas, and coral reefs. As climate warms, fungi inhabiting the coldest areas, for example, will find their niches moving progressively poleward and higher up mountains until their habitats disappear and the fungi are replaced by fungi more competitive in the slightly warmer environments. The ascomycete cup fungus, *Lachnellula pini*, causes a canker disease of Scots pine in Scandinavia. It is restricted to the far north and mountainous areas with low January temperatures, and is apparently dependent on snow damage of trees for infection. This species is likely to be threatened as snow damage decreases. It also raises an interesting moral and political dilemma that fungi that are viewed as harmful to our crops will enter our lists of endangered species. Indeed, there is a Red List for rust fungi in Wales, UK.

LAND USE CHANGE

Since the human lifestyle changed from hunter-gatherer to farming, the vegetation landscape of much of the planet has changed from continuous forest cover at ca. 4000 BP to vast swathes of arable land and grassland as grazing pasture. Though forests have been lost, a need for wood remains, and so man plants and manages forests to provide the types of timber required. In grassland, arable land and plantation forests, productive and pathogen-resistant species and varieties are commonly selected and often grown as a monoculture. Further, mineral nutrient fertilisers, pesticides, and herbicides are applied to enhance yield. Habitat has been lost, and all of these management practices dramatically alter the microbial communities and activities within these ecosystems. As seen in some of the examples that follow, intensification of agricultural practices can sometimes result in less diverse fungal communities. These may be less able to adapt to other global change factors, particularly climate. Maintenance of biodiversity provides an 'insurance effect' with respect to ecosystem functioning, as a more diverse community is more likely to contain some species able to adapt to change, which is a good argument for conservation (pp. 391–398).

Grasslands

Grasslands cover about 20% of Earth's land surface, usually where there is low or seasonal rainfall (250–1500 mm/year). Grasslands are found at a range of latitudes and altitudes, e.g. tropical east African savannah to temperate steppes, prairies and pampas, montane alpine meadows, and Andean Pármano, often merging into tundra and heathland. The grassland habitat type reached its maximum extent in Europe between the fifteenth and twentieth centuries. Since then, dry mesic grassland has been converted to arable crops, conifer plantations and scrubland, and in many countries grassland areas have decreased by more than 90%, though even in Europe there are still some countries with significant areas of semi-natural grassland (e.g. The Faeroe Isles, Norway, Romania, Scotland and Wales). Most grasslands are semi-natural, grazing preventing succession to scrubland and woodland. Most dry grasslands are nutrient poor, because grazers remove nitrogen and phosphorus, and have often been 'improved' by addition of fertiliser and sowing with a few fast growing grasses.

Grasslands are very different ecosystems to forest and woodland. In grasslands, litter components input to soil are smaller, and contain less lignin and fungitoxic compounds. Mammalian herbivores eat 40–70% of above ground NPP in grasslands, hence a large amount of material enters the soil in a comminuted, partially digested form as dung. A high proportion of plant biomass is underground, especially when grazing pressure is high and regular grazing results in a high root turnover and large input of dead organic matter to soil. It comes as no surprise then that grasslands and woodlands have very different communities of fungi. In 'unimproved' dry grassland with long continuity in Europe, characteristic fungi include the waxcaps (*Hygrocybe* and *Camarophyllus* species), and pink gills (*Entoloma* spp.), some club fungi (*Clavaria, Clavulinopsis, Ramariopsis*), and earth tongues (*Geoglossum, Microglossum, Trichoglossum*). In several countries, the number of waxcap species has been suggested as a good indicator of valuable 'unimproved' grassland sites: 1–3 of limited importance; 4–8 of local importance; 9–16 of regional importance; 17–21 of national importance; and ≥ 22 of international importance. The exact habitat and nutritional requirements of most grassland fungi is unknown, but it is generally accepted that some species seem to occur in grasslands with long continuity (e.g. *Entoloma anatinum, Entoloma longistriatum, Entoloma mougeotii, Hygrocybe intermedia, Hygrocybe citrinovirens, Hygrocybe ovina, Hygrocybe aurantiosplendens*, and *Hygrocybe ingrate*), whereas some rare species (e.g. *Entoloma formosum, Entoloma hispidulum, Entoloma xanthochroum, Hygrocybe subpapillata, Hygrocybe glutinipes, Hygrocybe vitellina*, and *Hygrocybe spadicea*) are found in grasslands with long and shorter continuity. A more broadly applicable indicator system than that based only on *Hygrocybe* species, assigns a different point score to each grassland species, so that the value of a locality is calculated by summing the value assigned to each of the species found there. Many other groups are also found in dry grasslands, both 'improved' and 'unimproved' and other habitats too, including species of *Agaricus, Galerina*, and *Macrolepiota*, and also puff balls.

Not only are there major differences in the communities of fungi that produce macroscopic fruit bodies but also in microfungi, depending on vegetation type and management intensity. Considering yeasts as an example, while forests harbour predominantly basidiomycete yeasts (e.g. *Cryptococcus terricola* and *Trichosporon porosum*), those of grasslands are largely ascomycetous (e.g. *Schwanniomyces castelli*). The proportion of ascomycetes in grasslands increases with management intensity, but community composition also changes, *Schwanniomyces castelli*, for example, decreasing and *Barnettozyma californica* increasing. These changes reflect the fact that land management, grazing, and application of fertilisers alter the soil nutrient status, providing readily assimilable carbon and mineral nutrient sources.

Arable and Plantation Forests

In the developed world, though land use is always changing, agricultural land has been established for a long time. In the developing world, however, there is a very high rate of conversion from forest to cultivated land, often with monoculture of crops. As well as loss of biodiversity of macrobiota and loss of soil quality, there are changes in the microbial communities. As an example, in Central Mexico, forest is being converted to avocado (*Persea americana*) plantations and to fields of maize (*Zea mays*). In a study comparing the arbuscular mycorrhizal fungal (AMF) communities in all three vegetation types, it was found that AMF species diversity was only slightly less in avocado plantations than in forest, despite the use of copper-based fungicides and fertilisers in the former, perhaps reflecting a similar microclimate and the presence of many similar herbaceous plants which provide additional hosts for AMF. In maize fields, however, there were 50% less AMF species, probably due to lack of other herbaceous plants, no irrigation or canopy, low soil organic matter, and low available phosphorus. There were more AMF species unique to forest than to avocado plantations, and more unique to avocado than to maize, highlighting the importance of preserving forests. The geographical region in which the vegetation types were found had a major impact on the AMF communities, suggesting that climate change may have a bigger effect on AMF communities than changes in land use.

POLLUTION, PESTICIDES, FERTILISERS, NUTRIENT DISTRIBUTION, AND RECYCLING

Pollutants can arrive by air, water, or directly deposited by man. There are three main classes of air pollutants: (1) primary pollutants that remain in the same form as they were emitted (e.g. SO_2, NO_2, and fluoride compounds); (2) secondary pollutants formed by the primary pollutants in the atmosphere (e.g. ozone (O_3), peroxyacetyl nitrate, nitric (HNO_3), and sulphuric (H_2SO_4) acids); and (3) other compounds – pesticides, industrial organic compounds, metals, metalloids, and radionuclides.

Fungal hyphae feed by absorption (Chapters 2 and 5), and can accumulate both micronutrient and non-nutrient elements. Lichen thalli, in contrast, can also accumulate absorbed nutrients, non-nutrients and other particles that land on their surface both as dry deposition and deposited in precipitation. Very high element concentrations tend to result from entrapment of particles, which can accumulate in large intercellular medullary spaces. These particles can range in size from microscopic aerosols (<1 μm) to macroscopic dust (>1 mm). Lichens and other fungi vary in their sensitivity to pollutants, pesticides, and fertilisers, resulting in changes in communities, including species loss. Differences in sensitivity, ability to accumulate, and the perennial nature of lichens and some macrofungi make them useful monitors of pollutants and other elemental deposits.

Lichens and Air Pollution

Large scale pollution, especially from SO_2 and NO_x, resulting from the burning of fossil fuels started in the industrial revolution during the late eighteenth century. Since then the lichen biota of almost all of Europe has undergone considerable change. These effects are well illustrated by the classic study of Hawksworth and Rose on the distribution of lichen communities in Britain in relation to atmospheric SO_2 (Figure 11.9 and Table 11.2). The lichen communities around industrial conurbations became depauperate, with only very resistant species present, while less polluted areas had more species rich communities. These atmospheric pollutants declined following targeted emission controls, and pollution-sensitive lichen species have begun to recolonise areas from which they had become extinct. Habitats seem to become suitable for lichens relatively soon (10–100 year) after removal of SO_2 deposition and

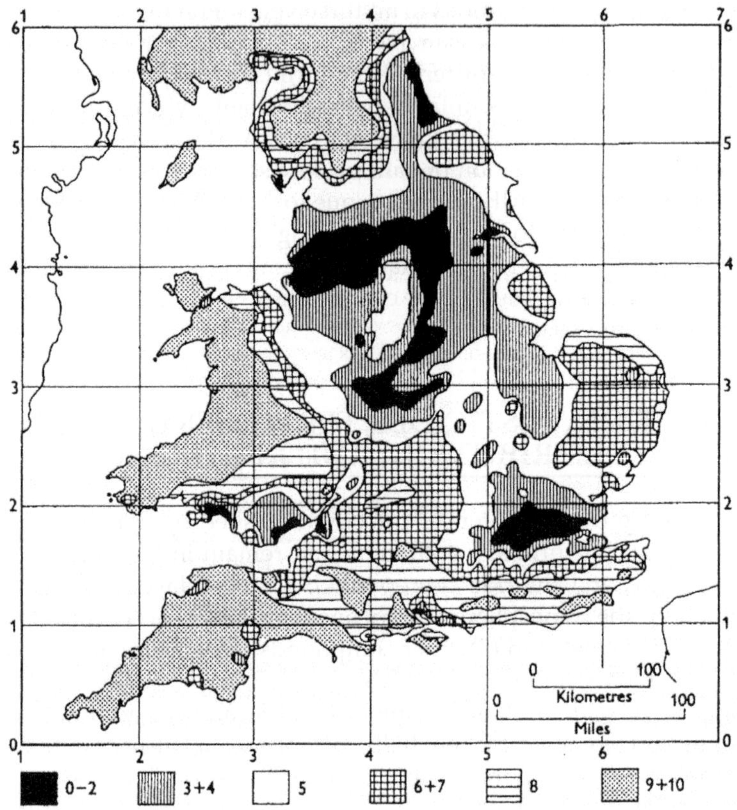

FIGURE 11.9 Lichen communities are dramatically affected by airborne pollutants. Shown here are zones of lichens in England and Wales in the late 1960s. The communities are highly correlated with atmospheric SO_2 concentration. Typical lichens found in each zone are indicated in Table 11.2. The key beneath the map indicates the zone number, and corresponds to zone numbers in the table. Notice the restricted lichen biota around the industrial conurbations of London, the midlands and south Wales, and the diverse biota including sensitive lichens in the nonindustrial south west, west Wales and northern UK. *Source: Hawksworth, D.L., Hill, D.J., 1984. The Lichen-Forming Fungi. Blackie, Glasgow.*

accompanying acidity. However, unfortunately, the decline in atmospheric SO_2 has been accompanied by an increase in N-pollution from intensive agriculture, and not just in Europe but also in biodiversity hotspots, including Southeast Asia, Central America, and India. Lichen community composition changes in response to even small increases in nitrogen deposition. In the Pacific Northwest of the USA, in polluted areas, sensitive species (e.g. *Alectoria sarmentosa, Bryoria fuscescens, Sphaerophorus globosus, Usnea filipendula*) and N-fixing cyanolichens (e.g. *Lobaria oregana*) have declined or disappeared, while nitrophilous species (e.g. *Xanthoria polycarpa, Physcia adscendens,* and *Candelaria concolor*) are common.

Since lichens are long-lived and able to bioaccumulate atmospheric deposits, their thalli are valuable archives, providing useful tools for monitoring changes in atmospheric pollutants. The residence times of pollutants and nutrients within thalli vary among elements. Macronutients (e.g. N, P, K, Ca, Mg, S) are relatively mobile and easily leached, so that

TABLE 11.2 Relationship between lichens and SO_2 pollution in England and Wales in the late 1960s, on deciduous trees, particularly *Fraxinus* and *Quercus*, with moderately acidic, rough bark.

Zone	Not eutrophicated: nitrogen-loving communities have not developed	Eutrophicated: nitrogen-loving communities have developed	Mean winter SO_2 ($mg\,m^{-3}$)
0	Lichens absent	Lichens absent	?
1	*Desmococcus viridis* at trunk base	*Desmococcus viridis* extends up trunk	about 170
2	*Desmococcus viridis* extends up trunk; *Lecanora conizaeoides* at trunk base	*Lecanora conizaeoides* abundant; *Lecanora expallens* occacionally on base of trunk	about 150
3	*Lecanora conizaeoides* extends up trunk; *Lepraria incana* frequent at trunk base	*Lecanora expallens and Buellia punctata* abundant; *Buellia canescens* sometimes present	about 125
4	*Hypogymnia physodes* and/or *Parmelia saxatilis* and/or *Parmelia sulcata* at trunk base; *Chaenotheca ferruginea, Hypocenomyce scalaris and Lecanora expallens often present*	*Buellia canescens* common; *adscendens, Physcia tribacia* (south facing) and *Xanthoria parietina* sometimes present.	about 70
5	*Hypogymnia physodes* or *Parmelia saxatilis* extends up trunk; *Lecanora chlarotera, Parmelia glabratula, Parmelia subrudecta* or *Parmeliopsis ambigua* present; *Calicium viride, Chrysothrix candelaris* and *Pertusaria amara* may occur; *Ramalina farinacea* and *Evernia prunastri* may be present but usually only at trunk base; *Platismatia glauca* may be present on horizontal branches	*Buellia canescens* and *Xanthoria parietina* common; *Buellia alboatra, Haematomma coccineum, Opegrapha varia, Opegrapha vulgata, Parmelia acetabulum* (in East), *Physconia grisea, Physconia farrea, Physica orbicularis, Physcia tenella, Ramalina farinacea, Schismatommma decolorans, Xanthoria candelaria* sometimes present	about 60
6	*Parmelia caperata* present at least on the base; rich in species of *Pertusaria* (e.g. *Pertusaria albescens, Pertusaria hymenea*) and *Parmelia Graphis elegans* appearing; *Pseudevernia furfuracea* and *Bryoria fuscescens* present in upland areas	*Physcia orbicularis, Physconia grisea, Opegrapha varia* and *Opegrapha volgata* abundant; *Arthopyrenia alba, Caloplaca luteoalba, Lecania cyrtella, Pertusaria albescens, Physconia pulverulenta, Physciopsis adglutinata, Xanthoria polycarpa* sometimes present.	about 50
7	*Parmelia caperata, Parmelia revoluta* (except in North east), *Parmelia tiliacea, Parmelia exasperatula* (in North) extend up the trunk; *Pertusaria hemisphaerica, Rinodina roboris, Usnea subfloridana* (in South) and *Arthonia impolita* (in East) appear	*Anaptychia ciliaris, Arthopyrenia biformis, Bacidia rubella, Candelaria concolor, Physcia aipolia* sometimes present	about 40
8	*Usnea ceratina, Parmelia perlata* or *Parmelia reticulata* (South and West) appear; *Rinodina roboris* extends up the trunk (in South); *Normandina pulchella* and *U. rubicunda* (in South) usually present	*Physcia aipolia abundant; Anaptychia ciliaris* occurs with apothecia; *Desmaziera evernioides, Gyalecta flotowii, Parmelia perlata, Parmelia reticulata (in south and west) Ramalina obtusata and Ramalina pollinaria sometimes present.*	about 35
9	*Dimerella lutea, Lobaria pulmonaria, Lobaria amplissima, Pachyphiale cornea,* or *Usnea florida* present; if these are absent crustose flora well developed with often more than 25 species on larger well lit trees	*Caloplaca aurantiaca, Caloplaca cerina, Physcia leptalea, Ramalina calicaris, Ramalina fraxinea, Ramalina subfarinacea* sometimes present	<30
10	*Lobaria amplissima, Lobaria scrobiculata, Pannaria* spp., *Sticta limbata, Teloschistes flavicans, Usnea articulate or Usnea filipendula* present and locally abundant.	*Caloplaca aurantiaca, C. cerina, Physcia leptalea, Ramalina calicaris, Ramalina fraxinea, Ramalina subfarinacea* sometimes present	'unpolluted'

Modified from Hawksworth, D.L., Rose, F., 1970. Qualitative scale for estimating sulphur dioxide air pollution in England and Wales using epiphytic lichens. **Nature** *227, 145–148. Hawksworth, D.L., Hill, D.J., 1984. The Lichen-Forming Fungi. Blackie, Glasgow.*

measurable changes can be detected over weeks or months. Total nitrogen in thalli of *Cladonia portentosa* ranged from 7 to 13 mg/g in low to high nitrogen deposition sites across the UK. Some environments are naturally rich in nitrogen (e.g. bird perching stones). Up to 3% of thallus dry weight was N in *Xanthomendoza borealis* (a *Xanthoria*-like species) close to a penguin roosting site in Antarctica. In lichens near industrialised areas in the 1960s and 1970s, 1000 µg/g or more of sulphur was commonly found in thalli near pollution sources, which compares with 300 µg/g or less in green algal lichens in clean sites.

Toxic metals (e.g. cadmium, lead, and zinc) are more tightly bound within lichen thalli and hence more slowly released. The use of lichens as monitors of heavy metals is nicely shown by a study in which freshly collected samples and herbarium specimens, from a site within 15 km of the centre of Washington, DC, from 1907 to 1992 were analysed for lead. Lead content peaked in 1970 and then declined with the advent of catalytic converters in cars, despite the increase in road traffic.

Lichens also accumulate radionuclides, and have been used to monitor fallout from nuclear bomb testing and following the Chernobyl power station accident. This accumulation of radionuclides was a particular problem in arctic tundra, where lichens dominate the vegetation, and are the main food source of caribou/reindeer (*Rangifer tarandus*) (See also p. 233). The radionucludies accumulated in the *Cladonia* lichens and, therefore, accumulated in wild and domestic herds and ultimately in the North American Inuit and Scandinavian Sàmi people that use them as a major food source.

From the forgoing it is clear that lichens can be used for monitoring environmental pollution in several different ways: (1) community surveys as done by Hawksworth and Rose (Figure 11.9); (2) analysing pollutant content of resident lichens, including those collected historically; and (3) transplanting lichens, attached to twigs, rocks, etc., especially into areas where there are no suitable indigenous lichens. Transplanting also has the advantage that procedures can be standardised.

Metal and Radionuclide Pollution Effects on Saprotrophic and Mycorrhizal Fungi

Non-lichenised fungi accumulate pollutants, pesticides, etc., by absorption through hyphae, and elevated levels in mycelium are mirrored in the fruit bodies of macrofungi. It is the fruit bodies that have usually been analysed. The ability to accumulate metals varies widely between taxa, but can be considerable in species growing on polluted sites (Table 11.3). Generally, concentrations of cadmium, lead, mercury, and zinc, for example, are higher in fruit bodies from urban and industrialised regions than from rural areas. So, fruit bodies could be used as bioindicators of soils polluted by these and other metals. There are two sorts of bioindicators: (1) reaction indicators – individual species or communities that are sensitive or tolerant and respectively decline/disappear or increase on polluted sites; and (2) accumulation indicators that are analysed for the pollutant. Some mycorrhizal basidiomycetes are particularly tolerant of heavy metals, including *Amanita muscaria*, *Laccaria laccata*, and several *Boletus* species, while some *Russula* species are sensitive. There is no simple positive relationship between fruit body content and soil content of cadmium, lead and mercury but some species have been proposed as bioindicators based on metal analyses. For example: species of *Agaricus* and *Mycena pura* for mercury and cadmium; *Amanita rubescens*, *Amanita*

TABLE 11.3 Examples of Heavy Metal Accumulation in Saprotrophic and Ectomycorrhizal Basidiomycetes Growing on Metal Polluted Sites

Metal	Species	Nutritional mode	Fruit body or rhizomorph	Concentrations (mg/kg dry wt)	Comments
Aluminium Al	*Armillaria* spp.	Saprotroph	Rhizomorph	3440	
Antimony	*Chalciporus piperatus*	Mycorrhizal	Fruit body	Maximum 1423	
Arsenic As			Fruit body	100-200	
Cadmium Cd	*Laccaria amethystina*	Mycorrhizal	Fruit body	Average 5, maximum 40	
Copper Cu	*Amanita muscaria*	Mycorrhizal	Fruit body	93	
Copper Cu	*Armillaria* spp.	Saprotroph	Rhizomorph	15	
Gold Au	*Boletus edulis*	Mycorrhizal	Fruit body	Maximum 235 ng/g	in nongold sites 10 ng/g
Gold Au	*Langermannia gigantea*	Saprotroph	Fruit body	160 ng/g	in nongold sites 10 ng/g
Lead Pb	*Armillaria* spp.	Saprotroph	Rhizomorph	680	
Lead Pb	*Hypholoma fasciculare*	Saprotroph	Fruit body	7	
Silver ^{110}Ag	*Agaricus bisporus*	Saprotroph	Fruit body	Maximum 167	Concentration factor up to 40
Manganese Mn	*Polyporus squamosus*	Saprotroph	Fruit body	Mean 138	
Mercury ^{203}Hg	*Agaricus bisporus*	Saprotroph	Fruit body	Maximum 75	Concentration factor up to 3.7
Mercury Hg	*Hydnum repandum*	Mycorrhizal	Fruit body	1	
Zinc Zn	*Armillaria* spp.	Saprotroph	Rhizomorph	1930	
Zinc Zn	*Polyporus squamosus*	Saprotroph	Fruit body	Mean 200	

Examples of highest accumulating species given unless otherwise indicated.
Source: Gadd (2007).

strobiliformis, *Coprinus comatus*, *Lycoperdon perlatum*, and *Marasmius oreades* for mercury; and *Agaricus* sp. and *Lycoperdon* sp. for lead, zinc, and copper. Non-lichenised fungi also accumulate radionuclides; the mycorrhizal *Hebeloma cylindrosporum* has particularly high transfer factors for uranium and thorium, which might make it a good bioindicator of radioactive content of soil. *Hebeloma cylindrosporum* and the saprotrophic *Lycoperdon perlatum* have been suggested as bioindicators for ^{239}Pu, ^{240}Pu, and ^{241}Am.

Aquatic fungi can also face elevated levels of metals, if streams run through rock, are fed by soils rich in a particular element, and most particularly as a result of mining activity. Effluent from coal, copper, gold, zinc, and uranium mining has negative effects on fungal species richness, biomass, leaf decomposition rates, and spore production. At relatively low pollution levels, though diversity is strongly affected, biomass, growth and decomposition activity are affected relatively little, implying some compensation by tolerant species or strains.

Nitrogen Enrichment Effects on Saprotrophic and Mycorrhizal Fungi, and Nutrient Cycling

Though dinitrogen (N_2) makes up over 70% of Earth's atmosphere, it cannot be used directly by plants. Plants can only obtain nitrogen as nitrate (NO_3^-) or ammonium (NH_4^+) ions, directly from soil solution, from nitrogen-fixing bacteria in the case of legumes, and via mycorrhizal associations for over 90% of plants (Chapter 7). Nitrogen is often limiting for plant growth, so man adds fertilisers (N, P, K) to agricultural ecosystems, from natural sources and fixed industrially. Burning fossil fuels releases vast quantities of nitrogenous oxides and other compounds; nitrous oxide (N_2O) in the atmosphere has rapidly increased since the 1940s and is now about 310 ppm compared with 285 ppm from preindustrial times, as revealed by analysis of air bubbles in glacier ice. As a result of all of these inputs, nitrogen deposition has increased by 3–5 times. This nitrogen pollution affects fungal species diversity, abundance and activity, and hence also ecosystem functioning.

N-pollutant effects on mycorrhizal fungi are evident. Though mycorrhizal abundance is increasing as a result of climate change (see previous section), N-pollution has opposite effects with an average 15% decline in mycorrhizal biomass (AM 25%, ECM 5%). The diversity and richness of ECM fungi also declines in coniferous forests and to a lesser extent in deciduous forests, as shown by studies along nitrogen deposition gradients, natural fertility gradients and experiments adding N-fertiliser, based on root tips, soil hyphae and fruit body studies. Sensitivity varies between species and genera. Species of *Cortinarius*, *Piloderma*, *Suillus*, and *Tricholoma* respond negatively to elevated N, while *Paxillus involutus* and species of *Laccaria*, *Lactarius*, and *Thelephora*, which tend to be early successional species, increase. Sporocarp declines in the genera *Cantharellus*, *Cortinarius*, *Suillus* and *Tricholoma*, first noticed in Europe in the 1980s, have now largely been backed up by soil and root tip studies there and elsewhere. Further, some species, for example stipitate hydnoids (*Bankera*, *Hydnellum*, *Phellodon*, and *Sarcodon*), are now endangered as a result of nitrogen increase (see p. 367). Mycorrghizal fungi do not all respond to elevated nitrogen in the same way, depending on functional traits. Species with medium distance fringe category exploration types (p. 217) (e.g. species of *Cortinarius*, *Piloderma*, and *Tricholoma*) have declined dramatically in response to elevated nitrogen. This exploration type is characterised by dense proliferation of hyphae into loose undifferentiated aggregations (cords) that ramify around and within patches of organic matter. In contrast, short distance exploration types are favoured under elevated N.

Because of effects on ECM fungi, nutrient cycling in polluted forest ecosystems is altered (Figure 11.10). In non-polluted ecosystems, ECM not only sequester inorganic nitrogen from soil solution but also short-circuit the N cycle by utilising organic N, as a result of their abilities as decomposers. This restricts the supply of N in soil available to soil microorganisms and to non-ECM plants, suppressing arbuscular mycorrhizal herbs and some tree saplings whose AM partners are less effective at obtaining organic N. Since little N is available in soil solution, little N is lost through leaching of ammonium and nitrate, and through denitrification. In N-polluted ecosystems, however, the situation is very different as there is a decline in ECM fungi: organic and inorganic (ammonium and nitrate) N pools are much larger resulting in losses through leaching; utilisation of organic N by ECM is reduced; and use of inorganic N increases.

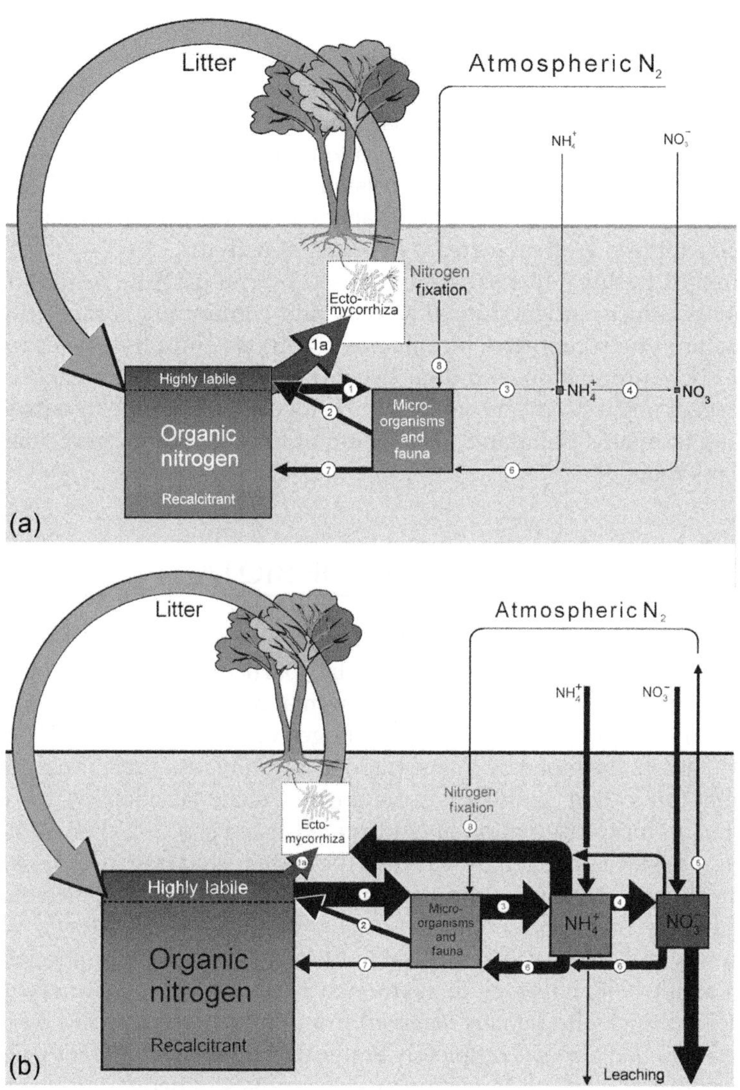

FIGURE 11.10 The nitrogen cycle in unpolluted, N-limited, forest ecosystems with ectomycorrhizal (ECM) trees (a) is very different from in N-polluted forests. In N-limited forests, trees obtain organic nitrogen directly from dead organic matter through the activities of their fungal partners (pathway 1a; largest arrow), short-circuiting the microbial mineralization pathways (pathways 3 and 4). The main microbially driven pathways are: (1) fungal depolymerisation and assimilation of organic N, (2) release of nutrients from dead microbes, (3) mineralisation (ammonification), (4) nitrification, (6) microbial immobilisation, (7) humus formation, and (8) nitrogen fixation. (b) In N-polluted forests dominated by ectomycorrhizal trees, organic and inorganic (ammonium and nitrate) N pools are much larger, utilisation of organic N by ECM is reduced, and use of inorganic N increases. Rates of mineralization, nitrification and ensuing N losses through nitrate leaching and denitrification, all increase. *Source: Leake, J.R., Johnson, D., Donnelly, D.P., Muckle, G., Boddy, L., Read, D.J., 2004., Networks of power and influence: the role of mycorrhizal mycelium in controlling plant communities and agro-ecosystem functioning. Can. J. Bot. 82, 1016-1045.*

N deposition affects saprotrophic fungi too. This is important because changes in decomposition rates will affect net carbon sequestration in ecosystems. Because organic matter with a low C:N ratio decomposes more rapidly than that with a high C:N ratio, decomposition is thought to be N-limited. Effects of increased nitrogen seem to be quite variable depending on sites, though leaf litter decomposition often increases along increasing N-deposition gradients in both terrestrial and aquatic ecosystems. However, additional N suppresses basidiomycete decomposition of lignin-rich organic matter, due to inhibition of the lignolytic enzymes, which contrasts with an increase in cellulose activity.

Eutrophication of freshwater ecosystems, caused by run-off from fertilised soil and from domestic effluents, affects aquatic fungal activity and community composition. Variable effects have sometimes been reported, but increased nitrates and phosphates usually increase fungal growth, spore production, and litter decomposition, though fungal diversity tends to decrease. Reports of negative effects are probably not due to increase in nutrients per se, but to accompanying inorganic pollutants, deposition of fine sediments, and changes in activity of competitive microbes.

MOVEMENT OF BIOTA

Fungi can invade new areas by wind, water, and animal dispersal of spores. Sometimes movement can be rapid and widespread, even global, particularly as a result of man's activities. There are numerous examples of both plant and animal pathogens spread by man (Chapters 8 and 9). To give two examples: Bd, the chytrid pathogen of amphibians (pp. 311–312), has probably been spread across the globe by mans' trade in amphibians. There is clear evidence, from whole genome-amplification quantitative polymerase chain reaction (PCR), that the IUCN (International Union for Conservation of Nature) red-listed (pp. 392–394) Mallorcan midwife toad (*Alytes muletensis*) was infected by Bd when the endangered frog *Xenopus gilli* was brought into breeding facilities in Mallorca from Western Cape, South Africa. Dutch elm disease, (pp. 274–275) caused by several species of *Ophiostoma*, though spread locally via carriage by bark beetles in the genus *Scolytus*, was first noticed in Europe in 1910, but the species had relatively mild effects. In about 1967, however, the extremely virulent *Ophiostoma novo-ulmi* was transported to the UK in Rock elm (*Ulmus thomasii*) from North America, and by the early 1980s *Ophiostoma novo-ulmi* had almost completely eradicated English elm (*Ulmus glabra*) as mature trees in Southern Britain, though English elm still survives in hedgerows from suckers (root sprouts). Introduction of pathogens into new biogeographic zones is increasingly occurring, and as a tree pathologist once described it, every introduced pathogen is an open-ended experiment in evolution. Host/environment clones evolve where there are isolating mechanisms, and this leads to emergence of new species (pp. 135–137).

It is not only pathogenic fungi that invade new areas – though they are most obvious because of their dramatic effects on hosts, but also mutualistic fungi including mycorrhizas. Ectomycorrhizas have been frequently introduced to new locations along with their hosts. Though they may not always spread, being restricted to a few locations and their original hosts, there are examples of mycorrhizal fungi that become invasive, the best documented example probably being the death cap *Amanita phalloides*. It is native to Europe and North Africa, has been reported in Asia and definitely introduced to southern Africa, Australia,

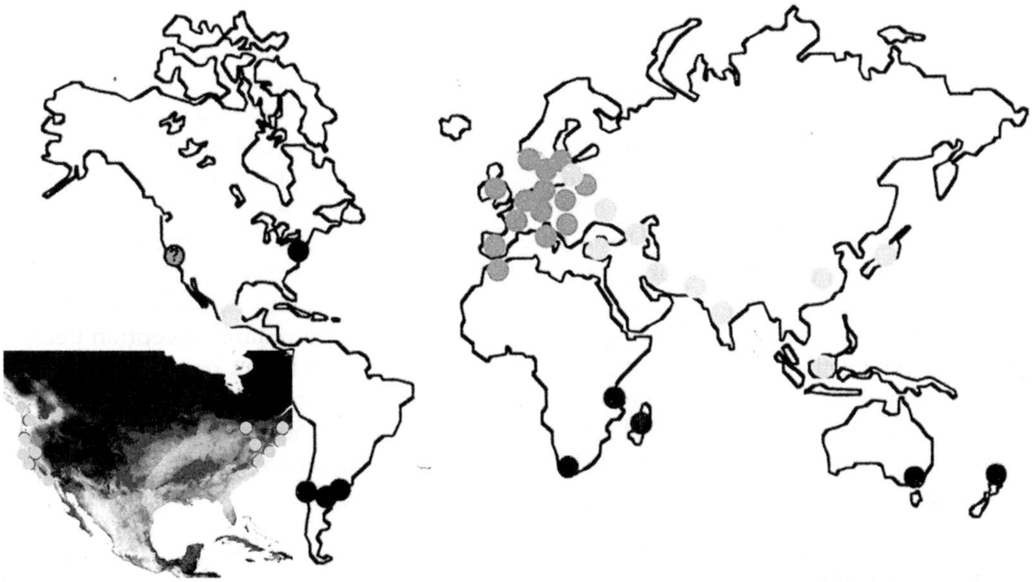

FIGURE 11.11 Current biogeography of the ectomycorrhizal death cap *Amanita phalloides*. Green (dark grey in the print version) dots indicate locations where the fungus is known to be native. Pale green (light grey in the print version) dots are places where the fungus has been reported but the description does not match the current concept of the species. Red (black in the print version) dots indicate where the species has been introduced. Dots with a question mark are places where the fungus is definitely found and has probably been introduced, but this is uncertain. From: Pringle, A., Vellinga, E.C., 2006. Last chance to know? Using literature to explore the biogeography and invasion biology of the death cap mushroom *Amanita phalloides*. Biol. Invas. 8, 1131–1144. The inset shows the known occurrences of *A. phalloides* (red (grey in the print version) dots), and the shading indicates the probability of occurrence: white 50–100% to dark shading <1%. *Source: Wolfe B.E., Richard, F., Cross, H.B., Pringle, A., 2010. Distribution and abundance of the introduced ectomycorrhizal fungus, Amanita phalloides, in North America. New Phytologist 185, pp. 803–816.*

New Zealand, South America, and the United States (Figure 11.11). From its native Europe, it was introduced to North America's east and west coasts, but the distribution and abundance on these coasts are different. It has a large range on the west coast (Figure 11.11 inset), and in California it is common in native forests where it is sometimes the dominant ECM species, but on the east coast it is most common in plantations of native and nonnative trees, and has a smaller range; temperature is probably the major factor affecting its distribution.

FUNGAL CONSERVATION

The loss of biodiversity is a pressing crisis across the globe. As with plants and animals, many fungal populations most likely have and are experiencing large declines due to destruction, fragmentation, and loss of habitat, pollution, invasive species, climate change and other human impacts. Again, as with plants and animals, mass extinctions of fungi are likely to occur, but when they do many will be unrecorded, as so many species are still unknown (Chapter 1). Six species of ascomycetes new to science were found associated with the living leaves of *Coussapoa floccosa*, a red listed Brazilian tree, until recently thought to be extinct in the wild. When plants become extinct, fungi that are specific to them also become extinct. Large population declines of some of the larger fungi were first shown beyond doubt in The Netherlands in the 1970s, by comparing annual records from the same sites. The decline can also be seen in records of the amounts of locally collected wild mushrooms on sale in the markets of some European towns. For example, in Saarbrucken (Germany) in the mid-1950s, annual sales of fruit bodies of the chanterelle (*Cantharellus cibarius*), a mycorrhizal fungus, were 6–8 tonnes, but had decreased to about 2% of this by the mid-1970s. Since then evidence has accumulated for the decline of many species to critical levels in many countries.

Though loss of diversity is a problem at least as large for fungi as for plants and animals, fungi are not a high profile group in most of the world, a notable exception being in Fennoscandia. Further, fungi are not explicitly mentioned in the Bern Convention and most people are unaware of their significance and decline. A first step towards halting this loss is to determine which species are threatened, and then to determine and implement appropriate conservation measures. The most widely used scheme for assessing and documenting the current status of biodiversity is the IUCN red-listing system.

Fungus Red Lists

It is important to present the biodiversity status of all organisms with one voice, so that the public and decision-makers can easily understand the situation. The IUCN red-listing system allows comparisons to be made between species from different taxonomic groups and in different regions, and it is increasingly being used to set national, regional, and global priorities for conservation action. Around 15,000 fungal species have been evaluated at the national level, though not all Red Lists fully meet the IUCN criteria. As with several groups of animals and plants, it is challenging to apply the IUCN criteria to fungi. With fungi this is because, for most of the time, they are hidden from sight, revealing themselves only occasionally as fruit bodies. Many of the macrofungi fruit annually, but irregularly, depending on weather and many other factors, and there is low correlation between the presence of fruit bodies and the presence and extent of mycelium. Difficulties in applying the IUCN criteria stem from difficulties in defining mature individuals, generation length, location, and uncertainty of how absence data should be handled (because absence of fruit bodies does not necessarily imply absence of a fungus).

There are nine IUCN Red List categories (Table 11.4). To assess the conservation status of a species (i.e. to determine to which Red List category a species belongs) the following data are necessary: (1) its geographic distribution; (2) an estimate of its population size; and (3) how geographic distribution and population size are changing over time. However, species with very small population sizes can be red-listed (Table 11.5 criterion D) even if there is no evidence that a population is in decline or will decline. The assessment aims to give a likely estimate of the total population and, if possible, also the population trend. The yet unrecorded number of localities or individuals always needs to be considered, and factors such as whether a species is conspicuous or inconspicuous, and how much it has been looked for, etc., must be taken into account.

For red listing, the size of a population is the number of mature individuals, defined by the IUCN Standards and Petitions Subcommittee (2010) as 'individuals known, estimated or

TABLE 11.4 The IUCN Red List Categories Applied for Fungi

A species is	When
Extinct (EX) or Regionally extinct (RE)	The species has previously been resident and there is no reasonable doubt that the last individual has died. Due to the cryptic nature of fungi. RE should be used only if exhaustive surveys during an adequate time period have failed to record an individual. Extinct (EX) refers to the global scale and RE to any lower geographical scale
Critically endangered (CR)	The best available evidence indicates that it meets any of the criteria A[a] to E for CR, and it therefore faces an extremely high risk of regional extinction
Endangered (EN)	The best available evidence indicates that it meets any of the criteria A[a] to E for EN, and it therefore faces a very high risk of regional extinction
Vulnerable (VU)	The best available evidence indicates that it meets any of the criteria A[a] to E for VU, and it therefore faces a high risk of regional extinction
Near threatened (NT)	It has been evaluated against the criteria and does not qualify for CR, EN, or VU, but is likely to qualify for a threatened category in the near future
Least concern (LC)	The species has been evaluated against the criteria and does not qualify for CR, EN, VU, or NT. Species in this category are normally widespread and abundant
Data deficient (DD)	There is inadequate information to make assessment of the species risk of extinction based on its distribution and population status. DD is not a threat category and species designated DD are rarely targets for conservation action
Not applicable (NA)	The taxon is not native to the region or of lower taxonomic rank than considered eligible for red-listing within the region. This category is not used at the global level
Not evaluated (NE)	The species has not yet been evaluated against the criteria

[a]*See Table 11.5.*
Modified from: Dahlberg and Mueller (2011).

inferred to be capable of reproduction'. The concept of a fungal individual is complex (pp. 100–101), and the IUCN unit of red-listing is the ramet (p. 100) rather than the genet (p. 100). Unfortunately, there is little information on the number and dynamics of mycelial ramets in nature, and it would be extremely hard and costly to collect data on the number of mycelial ramets in large geographical areas over many years. For practical purposes, the concept of a functional individual is used, which is based on numbers of fruit bodies. For purposes of red-listing, a functional individual of a wood-inhabiting fungus is defined as all conspecific fruit bodies on an individual tree, log, twig, etc., and for fungi found on the ground as all conspecific fruit bodies within a diameter of 10 m. Since functional individuals on the ground are likely to be fragmented into several ramets, for IUCN purposes each functional individual can be considered to correspond with 10 mature individuals (ramets) if fruit bodies on the ground occur as scattered gregarious patches, and two individuals if fruit bodies occur solitary. On a log, a few aggregated fruit bodies are counted as two mature individuals, but when widely scattered they can be counted as 5–10. This approach will usually underestimate the number of mature individuals, so it will provide a conservative, but realistic, indication of species of conservation concern.

To determine if populations are declining, the IUCN evaluates changes in population size not over years but over numbers of generations, for equivalence between different organisms.

Length of a generation is defined as the average age of the parents of the current cohort. This has not been estimated for any species of fungi (nor for most plants and animals), but must be greater than the age at first fruiting and less than the age at final fruiting. It will vary considerably between fungi with different ecological strategies. For example: mycelia growing in small, discrete, rapidly decomposing substrata (e.g. dung) have shorter generation times than do wood decay fungi; mycelia growing through continuous substrata such as soil and leaf litter layers are able to persist longer than mycelia growing in discrete, resource-limited patches such as twigs and logs; ruderal fungi colonising freshly felled sapwood have shorter generation times than heartrot fungi. The time frame for evaluations of population size vary between different IUCN criteria (Table 11.5): for criterion A, it is 3 generations or 10 years, whichever is longer; for criterion C, it is 1–3 generations. Different operational times for assessment of population change have been proposed for fungi in different habitats and having different life strategies (Table 11.6).

Conservation Measures, Future Needs, and Prospects

Since the causes of fungal species decline are destruction, fragmentation, and loss of habitat, pollution, invasive species, climate change and other human impacts, remedying these should halt decline. Indeed, strict controls on SO_2 emissions have seen the return of sensitive lichen species to industrialised areas, as mentioned earlier (p. 383). A decrease in nitrogen pollution in The Netherlands is heralding a slight increase in the rare stipitate hydnoids, though there has been no similar increase in the UK. In hay meadows, management that

TABLE 11.5 The Main Criteria (A–D) for IUCN Red Listing

IUCN criteria code	Criteria	Summary criteria
A	Severe population reduction	Reduction
B	Small geographic range (and reduction and fragmentation)	Reduction; few individuals
C	Small population (and reduction)	Reduction; few individuals
D	Very small or restricted population	Few individuals
E	Quantitative analysis	

TABLE 11.6 Operational Estimates of the Time Over Which Population Changes Should be Assessed for Fungi in Different Habitats

Habitat/ecological role	Assessment period
Ephemeral, short-lived substrata, e.g. dung	10 year, the minimum assessment period for IUCN
Wood	20–50 year (=3 generations) depending on decay rate. *Quercus* and *Pinus*, 50 year; *Picea* and *Fagus*, 30 year; *Betula*, *Alnus*, and *Populus*, 20 year
Ecomycorrhizal fungi	50 year (=3 generations)
Soil and litter inhabiting fungi	20–50 year (=3 generations)

involves cutting followed by maintenance of a short sward increases fruiting of grassland fungi, including rare waxcaps, though whether this reflects increase in presence of mycelia is not known. There was rapid recovery of lichens in overexploited reindeer winter grazing pastures in northern Norway when protective fencing was erected. Clearly, most causes of decline will be halted very slowly, so measures need to be taken to protect declining or rare and endangered species, including: red-listing; protection by law; responsible harvesting; formation of reserves; habitat management; action programmes; and research.

Rare and endangered fungi are protected by law in some countries, including 14 European countries, but the amount of species protected and the extent of that protection is variable, ranging from four species in the UK to over 300 species in Croatia. Usually, picking or destroying protected species is prohibited. While there is little or no evidence that removing fruit bodies results in a reduced yield of fruit bodies, at least for the more common fungi, there are certainly negative effects of habitat destruction, trampling and raking to reveal fruit bodies. Currently, there is only one basidiomycete on the International IUCN Red List – *Pleurotus nebrodensis* categorised as Critically Endangered (Table 11.4), and three lichens – *Cladonia perforata* (endangered), *Erioderma pedicellatum* (critically endangered), and *Gymnopholus lichenifer* (vulnerable). Known only from an area of less than 100 km^2 in limestone mountain meadows in Sicily, *Pleurotus nebrodensis* populations are becoming fragmented by road building and other developments, and it is overpicked as it is a desirable edible species. It has no legal protection, but is easily cultivated and is now grown commercially which will hopefully relieve pressure from wild populations. Another example of overpicking as a threat to a fungus is the ascomycete *Ophiocordyceps sinensis*, which parasitizes and kills Lepidoptera (Figure 11.15k). It is highly prized as an aphrodisiac and Chinese medicinal treatment for various ailments, and fetched a price of US $18,000/kg with annual sales of around 100 tonnes in Tibet in 2008, and an estimated US $20 million in the black economy of Bhutan in the early 2000s. This over-collecting, which is largely beyond the law and is the fungal equivalent of poaching, is not sustainable. However, again the fungus can be grown in culture, and commercial cultivation may save its wild populations.

An overarching principle should be to protect and manage biodiversity in general, and to focus on certain plants, animals, or fungi when appropriate. Fungi, however, in common with some plant and animal species, often suffer from not being considered by the public as charismatic. Useful general principles are laid out in Table 11.7. Conservation of habitats likely to be important for fungi is usually likely to be implemented by non-mycologists whose foremost interest is often other organisms; a forerunner has been the integration of mycological considerations with forest management in Fennoscandia since the mid-1980s. In The Netherlands, common activities in urban planning, land management, etc., are classified as favourable or unfavourable to fungi. These approaches have a general benefit, though some fungal species will require specific and detailed management guidelines. Six European countries have species action programmes (SAP) for threatened fungal species. There are SAPs for 19 nonlichenised species in Estonia, 10–15 in Finland, 27 in Sweden, 150 in Switzerland and 77 in the UK. Conservation science has been described above, but when legalities and money are involved then conservation becomes political. Conservation has a greater political impact if species are red-listed at a continental or global scale. Thirty three threatened macrofungi were proposed for inclusion in the Bern Convention in 2001, but political difficulties prevented their adoption. There is not even an official European fungal Red List, though all

TABLE 11.7 Principles and considerations for conservation and management of fungi

Principle	Justification	Management Considerations
Maintain habitat diversity at landscape scales	Fungal species have evolved within a shifting mosaic of plant communities, and periodic disturbances across broad landscapes over millennia	• Protect and restore disappearing habitat (e.g. old-growth forests, 'unimproved' grassland) • Maintain a diversity of forest successional stages • Ensure that habitat patches are located sufficiently closely to allow for fungal dispersal and population establishment
Maintain habitat diversity	Fungal species often reside in unique niches and respond to myriad microhabitat conditions	• Maintain or develop habitat diversity within ecosystems (e.g., diverse plant composition, retain coarse woody debris in forests, continue grazing and refrain from fertilisation of old grasslands)
Maintain host diversity	Many ectomycorrhizal and saprotrophicic fungi associate with specific host plants	• Plant mixtures of plants that resemble natural assemblages • In forests, diversify the understory vegetation which acts as hosts or create unique microhabitat
Maintain soil health	Most soil fungi are aerobic and utilise diverse organic and mineral resources.	• Avoid soil compaction and hot surface fires that destroy soil structure • Avoid removal of the forest floor litter layer and minimise disturbance • Maintain natural levels of soil organic matter • Reduce nitrogen inputs to soil from emissions and fertiliser application
In forests, maintain legacy trees and limit size of timber harvest units	When all tree hosts are removed (e.g. clearcuts), fungal populations are reduced and slow to recover compared to forest thinnings or partial cuts	• Retain some living trees in felled areas to maintain variety of age structure and live mycorrhizal populations on roots • Avoid large clearcuts; adopt thinning approaches or aggregate uncut trees to create reservoirs of fungal diversity and allow fungal dispersal into disturbed areas • Maintain refuge understory plants that may act as mycorrhizal hosts or create microhabitat within the future forest • Plant seedlings soon after harvest while there is residual fungal inoculum in the soil
Protect known locations of rare fungi and fungal diversity hot spots	Repeated inventorying of fungi has revealed locations of rare species and areas where fungal richness remains high (e.g. parks or reserves)	• Protect known sites of rare species by minimising disturbance and maintaining critical habitat elements • Identify fungal diversity hotspots and work with responsible managers to protect these areas or designate them as fungal reserves
Other activities essential for fungal conservation:		
Monitor fungal populations	Data collection spanning several years is needed to detect trends in fungal populations	• Establish permanent monitoring locations for targeted species or fungal communities • Include fungal monitoring within long-term biodiversity monitoring programmes • Include citizen scientists in the design and implementation of fungal monitoring programmes

TABLE 11.7 Principles and considerations for conservation and management of fungi—cont'd

Principle	Justification	Management Considerations
Develop partnerships with the public, other scientists and resource managers	Species conservation is a complex and expensive undertaking. Mycologists should take advantage of other ongoing biodiversity monitoring and conservation programmes	• Educate the public and resource managers as to the importance of fungi, and principles for their conservation • Integrate fungal conservation goals within ongoing multitaxa conservation programmes • Work directly with resource managers to include fungi in management programmes

Conservation efforts can concentrate on maintaining the species currently present at particular sites, focus on what species we want to have, or a combination of both.
Modified from: Molina et al. (2011).

Red Lists from European countries have been compiled, and 1644 species suggested for a European Red List, though these still need to be evaluated. The major challenges for fungal conservation are to: (1) raise both public and political awareness; (2) integrate fungi into national, continental and global conservation strategies for plants and animals; and (3) perform the science that will provide understanding of why each species is declining or rare, and how to counteract declines. Protection is not the only answer, rather we need to manage habitats and substrata to create appropriate conditions in time and space for the diversity of fungi we want to have in the long term.

Rare and Endangered Species – Case Studies

To determine if a species is rare and endangered, or common but declining, and to determine its conservation needs, it is essential to be able to distinguish one species from another, and to understand its ecology, particularly the biotic and abiotic factors involved in its reproductive success and decline. This is illustrated by case studies of two groups of endangered basidiomycetes in Europe: (1) saprotrophic *Hericium* species and (2) ECM stipitate hydnoids.

Hericium cirrhatum, *Hericium coralloides*, and *Hericium erinaceus* (Figure 11.5) are on the Red List of several European countries, and *Hericium erinaceus* was one of the species proposed for inclusion in the Bern Convention, though it is common in Japan and North America, and easily cultivated. All three are saprotrophs on wood of angiosperm trees, especially *Fagaceae*. They fruit prolifically in culture in the lab, perhaps indicating that the few fruit bodies seen in the field is a true indication of lack of mycelial individuals, though the fungi may be trapped in central regions of wood with no access to the surface for production of fruit bodies. When fruit bodies are formed, they are large and sporulation is prolific, though germination on agar is poor. Following germination in nature, it is suspected that the fungus may exist in the homokaryotic state (p. 113) for a prolonged time, due to low frequency of contact between mating compatible partners. This is a major difficulty faced by all rare fungi. Both homokaryotic and heterokaryotic mycelia grow rapidly and are combative against other wood decay fungi, so this is unlikely to be the cause of their rarity in Europe, but poor germination and infrequent establishment may be. They have been detected latently present within functional sapwood, but gaining entry to trees and encountering appropriate conditions to develop overtly may be limiting factors.

The decline of stipitate hydnoids, an informal grouping of basidiomycetes whose fruit bodies have a stalk and cap with spore-producing spines (*Bankera*, *Hydnellum*, *Phellodon*, and *Sarcodon*), has been alluded to on several occasions earlier in this chapter. They are ECM partners with a range of woody angiosperms and gymnosperms, especially within *Fagaceae* and *Pinaceae*, though the suites of fungal species are associated with each differ. Typically they fruit on sparsely vegetated soil with low nitrogen and organic matter content, and sloping ground (e.g. animal burrows, hillsides, glacial moraines, ditches, and the margins of tracks and roads). They are well represented in Red Lists of Europe, and *Sarcodon fuligineoviolaceus* was proposed for inclusion in the Bern Convention. There are two main difficulties when studying them: (1) it is not currently possible to culture them on artificial media and (2) distinguishing one species from another. The fruit bodies of some species are very similar to others, yet it is essential to be able discriminate one from another, as even closely related taxa can have different nutritional modes and lifestyles. DNA sequencing, together with morphological analysis, has now made this possible. Also, use of PCR primers, specific to different species, on DNA and RNA extracted directly from soil, have revealed that mycelium is present in soil where the fungus has fruited previously, even though there has been no evidence of fruit bodies for 5 or more years. It is obviously much harder to unravel the ecophysiology of fungi, such as the stipitate hydnoids, that grow in intimate association with a living host, making culturing difficult or impossible, but molecular sequencing will allow researchers to ascribe host to fungal species by analysis of root tips, as well as detection in soil.

Further Reading

General

Crutzen, P.J., Stoermer, E.F., 2000. The "anthropocene". Int. Geosph. Biosph. Program. Newsl. 41, 17–18.

Krauss, G.-J., Solé, M., Krauss, G., Schlosser, D., Wesenberg, D., Bärlocher, F., 2011. Fungi in freshwaters: ecology, physiology and biochemical potential. FEMS Microbiol. Rev. 35, 620–651.

Roles of Fungi in Ecosystems

Bardgett, R.D., Wardle, D.A., 2010. Aboveground-Belowground Linkages: Biotic Interactions, Ecosystem Processes and Global Change. Oxford University Press, Oxford.

Dighton, J., 2003. Fungi in Ecosystem Processes. Marcel Dekke, New York.

Gadd, G.S. (Ed.), 2006. Fungi in Biogeochemical Cycles. Cambridge University Press, Cambridge.

Biogeography

Gill, J.L., Williams, J.W., Jackson, S.T., Lininger, K.B., Robinson, G.S., 2009. Pleistocene megafaunal collapse, novel plant communities, and enhanced fire regimes in North America. Science 326, 1100–1103.

Heilmann-Clausen, J.H., Boddy, L., 2008. Distribution patterns of wood-decay basidiomycetes at the landscape to global scale. In: Boddy, L., Frankland, J.C., van West, P. (Eds.), Ecology of Saprotrophic Basidiomycetes. Elsevier, Amsterdam, pp. 263–275.

Hibbett, D.S., 2001. Shiitake mushrooms and molecular clocks: historical biogeography of *Lentinula*. J. Biogeogr. 28, 231–234.

Peay, K.G., Bidartondo, M.I., Arnold, A.E., 2010. Not every fungus is everywhere: scaling to the biogeography of fungal-plant interactions across roots, shoots and ecosystems. New Phytol. 185, 878–882.

Rule, S., Brook, B.W., Haberle, S.G., Turney, C.S.M., Kershaw, A.P., Johnson, C.N., 2012. The aftermath of megafaunal extinction: ecosystem transformation in Pleistocene Australia. Science 335, 1483–1486.

Taylor, J.W., Turner, E., Townsend, J.P., Dettman, J.R., Jacobson, D., 2006. Eukaryotic microbes, species recognition and the geographic limits of species: examples from the kingdom Fungi. Philos. Trans. R. Soc. B 361, 1947–1963.

Climate Change

Bebber, D.P., 2015. Range-expanding pests and pathogens in a warming world. Annu. Rev. Phytopathol. 53, 335–356.

Boddy, L., Büntgen, U., Egli, S., Gange, A.C., Heegaard, E., Kirk, P.M., Mohammad, A., Kauserud, H., 2014. Climate variation effects on fungal fruiting. Fungal Ecol. 10, 20–33.

Chakraboty, S., Newton, A.C., 2011. Climate change, plant disease and food security: an overview. Plant Pathol. 60, 2–14.

Ellis, C.J., Coppins, B.J., Dawson, T.P., Seaward, M.R.D., 2007. Response of British lichens to climate change scenarios: trends and uncertainties in the projected impact for contrasting biogeographic groups. Biol. Conserv. 140, 217–235.

Fransson, P., 2012. Elevated CO_2 impacts ectomycorrhiza-mediated forest soil carbon flow: fungal biomass production, respiration and exudation. Fungal Ecol. 5, 85–98.

Garrett, K.A., Forbes, G.A., Savary, S., Skelsey, P., Sparks, A.H., Valdivia, C., van Bruggen, A.H.C., Willocquet, L., Djurle, A., Duveiller, E., Eckersten, H., Pande, S., Vera Cruz, C., Yuen, J., 2011. Complexity in climate-change impacts: an analytical framework for effects mediated by plant disease. Plant Pathol. 60, 15–30.

Helfer, S., 2013. Rust fungi and global change. New Phytol. 201, 770–780.

Luck, J., Spackman, M., Freeman, A., Trebicki, P., Griffiths, W., Finlay, K., Chakraborty, S., 2011. Climate change and diseases of food crops. Plant Pathol. 60, 113–121.

Mohan, J.E., Cowden, C.C., Baas, P., Dawadi, A., Frankson, P.T., Helmick, K., Hughes, E., Khan, S., Lang, A., Machmuller, M., Taylor, M., Witt, C.A., 2014. Mycorrhizal fungi mediation of terrestrial ecosystem responses to global change: mini review. Fungal Ecol. 10, 3–19.

Pickles, B.J., Egger, K.N., Massicotte, H.B., Green, D.S., 2012. Ectomycorrhizas and climate change. Fungal Ecol. 5, 73–84.

Simard, S.W., Austin, M.E., 2010. The role of mycorrhizas in forest stability with climate change. In: Simard, S.W., Austin, M.E. (Eds.), Climate Change and Variability. Sciyo, Rijeka, pp. 275–302.

Treseder, K.K., 2004. A meta-analysis of mycorrhizal responses to nitrogen, phosphorus, and atmospheric CO_2 in field studies. New Phytol. 164, 347–355.

Land Use Change

Griffith, G.W., Roderick, K., 2008. Saprotrophic basidiomycetes in grasslands: distribution and function. In: Boddy, L., Frankland, J.C., van West, P. (Eds.), Ecology of Saprotrophic Basidiomycetes. Elsevier, Amsterdam, pp. 277–299.

Pollution, Pesticides, Fertilisers, Nutrient Distribution, and Recycling

Gadd, G.M., 2007. Geomycology: biogeochemical transformations of rocks, minerals, metals and radionuclides by fungi, bioweathering and bioremediation. Mycol. Res. 111, 3–49.

Lecerf, A., Chauvet, E., 2008. Diversity and functions of leaf decaying fungi in human-altered streams. Freshw. Biol. 53, 1658–1672.

Lilleskov, E.A., Hobbie, E.A., Horton, T.R., 2011. Conservation of ectomycorrhizal fungi: exploring the linkages between functional and taxonomic responses to anthropogenic N deposition. Fungal Ecol. 4, 174–183.

Nimis, P.L., Scheidegger, C., Wolseley, P.A. (Eds.), 2002. Monitoring with Lichens – Monitoring Lichens. Kluwer Academic Publishers, Dordrecht.

Movement of Biota

Wolfe, B.E., Richard, F., Cross, H.B., Pringle, A., 2010. Distribution and abundance of the introduced ectomycorrhizal fungus, *Amanita phalloides*, in North America. New Phytol. 185, 803–816.

Conservation

Dahlberg, A., Genney, D.R., Heilmann-Clauesen, J., 2010. Developing a comprehensive strategy for fungal conservation in Europe: current status and future needs. Fungal Ecol. 3, 50–64.

Dahlberg, A., Mueller, G.M., 2011. Applying IUCN red-listing criteria for assessing and reporting on the conservation status of fungal species. Fungal Ecol. 4, 147–162.

European Council for Conservation of Fungi, 2003. 33 Red-listed fungi from Europe. Available at: http://www.artdata.slu.se/Bern_Fungi/ECCF%2033_T-PVS%20(2001)%2034%20rev_low%20resolution_p%201-14.pdf

Heilmann-Clausen, J.H., Vesterholt, J., 2008. Conservation: selection criteria and approaches. In: Boddy, L., Frankland, J.C., van West, P. (Eds.), Ecology of Saprotrophic Basidiomycetes. Elsevier, Amsterdam, pp. 325–347.

IUCN Standards and Petitions Subcommittee, 2010. Guidelines for Using the IUCN Red List Categories and Criteria, Version 8.0. Prepared by the Standards and Petitions Subcommittee in March 2010. Available at: http://cms-data.iucn.org/downloads/redlistguidelines.pdf (accessed 16.04.10).

Minter, D., 2010. Safeguarding the future. In: Boddy, L., Coleman, M. (Eds.), From Another Kingdom. Royal Botanic Garden, Edinburgh, pp. 144–152.

Molina, R., Horton, T.R., Trappe, J.M., Marcot, B.G., 2011. Addressing uncertainty: how to conserve and manage rare or little-known fungi. Fungal Ecol. 4, 134–146.

CHAPTER

12

Fungi and Biotechnology

Nicholas P. Money
Miami University, Oxford, OH, USA

THE INDUSTRIAL SIGNIFICANCE OF FUNGI

The term **fungal biotechnology** is often reserved for modern industrial processes involving genetically modified organisms, but a broader reading includes the baking and brewing practices that originated in the ancient world. Humans have employed fungi for thousands of years. For most of our history this was unconscious: people followed preparative methods that produced the desired result and knew nothing about yeasts that made dough rise and fermented cereals. Mushroom cultivation on logs and horse dung is another example of early biotechnology that has flavoured the omnivorous diet of our species for generations. Methods for controlling fungal fermentations have developed in parallel with advances in the field of microbiology since the nineteenth century. The use of fungi to produce antibiotics and other pharmaceutical products is a more recent part of this endeavour and the introduction of molecular genetic manipulation of fungal strains has revolutionised the business of biotechnology. It is important to recognise that fungal fermentations are not limited to alcohol production by yeasts and other anaerobic processes. **Fermentation** refers to any of the biochemical transformations catalysed by fungi that are commercially significant and most of these processes are powered by aerobic metabolism.

THE CULTIVATION OF MUSHROOMS FOR FOOD AND PHARMACEUTICALS

Cultivation Methods

The global market for edible mushrooms has increased in recent decades and the marketplace has embraced an ever broader range of cultivated species. Most cultivated mushrooms are derived from natural varieties of wood-decay species of Basidiomycota. **Shiitake,** *Lentinula edodes*, is a white rot fungus that has been raised on logs for more than 1000 years. More than 1 million tonnes of shiitake are produced every year and China dominates the market. The **white button mushroom,** *Agaricus bisporus*, remains the most popular cultivated

The Fungi
http://dx.doi.org/10.1016/B978-0-12-382034-1.00012-8

FIGURE 12.1 Cultivated mushrooms. (a) Shiitake, *Lentinula edodes*, on logs. (b) White button mushroom, *Agaricus bisporus*, on beds of compost. *Source: panel (a) http://www.sharondalefarm.com/cultivation/ and (b) http://modernfarmer.com/2014/05/welcome-mushroom-country-population-nearly-half-u-s-mushrooms/.* (See the colour plate.)

mushroom, with a global crop exceeding 2 million tonnes. Production of this mushroom in the United States approaches 400,000 tonnes and is valued at $1 billion, but this is eclipsed by Chinese growers who process buttons for export. The methods of production for these market leaders offer a study in contrasts (Figure 12.1).

Mushroom cultivation begins with the production of a rich culture or spawn of the vegetative fungal mycelium, which is used as the inoculum. Shiitake spawn is raised on plugs of wood or on mixtures of sawdust, cereal bran, and other ingredients. Once the wood plugs or sawdust are colonised, the spawn is pressed into holes drilled into hardwood logs. The logs are kept in piles and covered with straw or with bags to maintain the high moisture levels that promote mycelial growth and wood decomposition. After 8–12 months, the logs are soaked with cold water to stimulate fruiting. Logs can produce multiple crops at intervals of a month or more for up to 6 years. Shiitake is also cultivated in plastic bags containing sawdust and a variety of waste agricultural products.

Agaricus bisporus spawn produced on cereal grains is used to inoculate beds of wheat or rice straw composted with animal dung and other ingredients. Before inoculation with the fungus, piles of these mixtures are fermented by thermophilic bacteria that change the nutrient content to favour the subsequent growth of the *Agaricus* mycelium. After up to 2 weeks of this cooking process, the compost is spread into wooden pallets and pasteurised by steaming to kill nematodes, insects, and other pests. The aim of steaming is to kill pests and competitors while minimising the loss of beneficial microorganisms. Fungi that can attack the mushroom mycelium and those that compete for nutrients are killed along with fungi that are potential pathogens of humans. Nitrifying bacteria that remove ammonia are unharmed by the pasteurisation process and residual ammonia is driven off by steaming. After pasteurisation, the compost is inoculated with spawn. Colonisation of the compost is stimulated by careful control of temperature, relative humidity, and carbon dioxide levels in the growing room. After 2–3 weeks, the compost is cased with a layer of peat and limestone in preparation for fruiting. The role of the casing is not known, but it is essential for fruiting. Peat is low in nutrients and the fungus produces short cords in the casing before mushroom primordia develop.

Fruiting is stimulated by reducing the temperature in the growing room from 25 to 18 °C and increasing air circulation. In earlier stages of cultivation, the CO_2 concentration in the growing room may be set as high as 3000 ppm. Airing reduces the levels of CO_2 and volatile organic compounds. Growers used to think that CO_2 served a major regulatory role in the initiation of fruiting, but recent studies show that a decrease in the concentration of volatile compounds (particularly 1-octen-3-ol, known as mushroom alcohol) above the mushrooms is the primary switch. Fruiting begins in 3 weeks and up to three flushes of mushrooms can be picked from a single bed. Although many of the details of production have evolved, the fundamental practice of cultivating mushrooms on a compost of animal dung has not changed since the technique was developed 300 years ago in France.

Mushrooms as Food and Medicine

Mushrooms have about the same calorific value as lettuce: more than 90% of the fresh weight of a mushroom is water and the remaining 10% is split between protein and carbohydrate in the form of insoluble fibre. The fat content of mushrooms is very low and the mixture of vitamins and minerals is unremarkable. Much of the culinary value of cultivated mushrooms derives from their flavour and the aroma of wild species is a determinant of their pricing. The most important volatile flavourings of mushrooms are C_8-derivatives, along with a variety of terpenoids and sulphur-containing compounds. The flavour profiles of mushrooms are exceedingly complex. More than 150 different volatiles have been identified in *Agaricus bisporus* using GC/MS and other analytical methods. The 'mushroomy' smell of the common edible species is produced by a mixture of compounds dominated by 1-octen-3-ol, which is also synthesised by fruits and potatoes. Other aromatic compounds in *Agaricus bisporus* include benzyl alcohol, benzaldehyde and cyclo-octenol. In some mushrooms, the level of benzaldehyde is higher than the C_8 compounds. Polyunsaturated fatty acids are the precursors for the synthesis of C_8 compounds. The lipid composition of mycelia and fruit bodies is very similar, but volatile production varies considerably between the caps and stalks of mushrooms. The natural functions of the volatile flavourings of mushrooms are unclear. In light of the importance of volatile levels in mushroom cultivation, it is possible that mushrooms, or clusters of mushrooms, use these compounds to suppress the fruiting of neighbours (conserving water and nutrients for their own mycelium). Another possibility is that the volatiles attract animals that serve as vectors for spore dispersal. This mechanism is crucial for the dispersal of truffles and many other fungi (Chapter 3). For mushrooms whose spores are spread by wind this seems paradoxical, but insects may play an accessory role in dispersal in some instances.

There are many claims about the medicinal properties of mushrooms. Mushrooms have been used in traditional medicines in China for centuries and there is a growing market for medicinal mushrooms outside Asia. Shiitake is the best-known medicinal mushroom and its dried basidiomata are used to treat a variety of ailments. Among its purported health benefits are its anti-tumour properties, anti-viral efficacy, and its utility in lowering serum cholesterol levels. Stimulation of the immune system is another of the therapeutic properties ascribed to shiitake. The majority of the claims about the health benefits of fungi are anecdotal, but shiitake has been the subject of a vast body of research and the mushroom's pharmacological properties have been linked to specific compounds in the fruit body. **Lentinan** is a water-soluble **beta-glucan** in the cell wall that appears to enhance the activity of dendritic cells that are involved in the recognition of cancer cells. Other studies have suggested that by stimulating an inflammatory response, lentinan acts as a powerful anti-viral agent. Studies on a cell wall **proteoglycan** called polysaccharide K, from *Trametes versicolor*, show that this operates as an antioxidant and may inhibit tumour development in mice. Other mushrooms, including *Polyporus umbellatus* and *Hericium erinaceus*, have a broad range of medicinal uses and are recommended for the treatment of cirrhosis, hepatitis B, gastric ulcers, oesophageal cancer, and, like every other medicinal mushroom, are supposed to stimulate an impaired immune system. The majority of experiments that support these claims have been performed on tissue-cultured cells and mice. There are very few placebo-controlled double-blind clinical trials on human patients that demonstrate any benefit from treatment with medicinal mushrooms. Nevertheless, the business of medicinal mushrooms is booming.

Uses of Lichens

Lichens have been used as a source of dyes for centuries. They provided the reddish-brown, purple, and orange colours in Harris Tweed until synthetic dyes were adopted by Scottish manufacturers of this superb wool fabric. Lichens have also been used as food and in traditional medicine, and served as embalming agents in ancient Egypt. *Evernia prunastri*, or oakmoss, is a lichen that provides musky and woody scents in iconic fragrances formulated by French perfume houses. Compounds in extracts from this lichen can cause skin allergies and regulations have been introduced to reduce their concentration in perfumes.

PRODUCTION OF FOOD AND DRINK USING YEASTS AND FILAMENTOUS FUNGI

Wine and Beer

When oxygen is available, yeasts metabolise sugars to form carbon dioxide and water. If oxygen is scarce or absent, or if the sugar concentration is very high, metabolism follows a different fermentative pathway with the production of carbon dioxide and ethanol. This is the basis of the great variety of alcoholic drinks consumed by humans. Species of the ascomycete yeast, *Saccharomyces*, are responsible for the majority of alcoholic fermentations: *Saccharomyces cerevisiae* is used for fermenting beer and wine (Figure 12.2); lager production involves *Saccharomyces pastorianus*, and *Saccharomyces bayanus* is used for cidermaking and for producing Champagne and other sparkling wines. The association of these fungi with different beverages is complicated by the genetic diversity within single species of *Saccharomyces* and frequent errors in identification. *Saccharomyces pastorianus* (also known as *Saccharomyces carlsbergensis*) and *S. bayanus* are the products of complex hybridization events between other *Saccharomyces* species.

Wine and cider are made by fermenting plant juices rich in sugar. Beer is produced from starchy plant materials and the starch must be converted to sugars before fermentation. Drinks containing 10–12% alcohol are produced readily by natural fermentation. Higher concentrations of alcohol inhibit the metabolic activity of yeasts and the production of these more potent beverages involves distillation processes to boost the alcohol content after the initial fermentation. Spirits from sugary plant products include brandy (from grape juice), rum (from sugarcane), tequila (from agave), a variety of fruit brandies, and distilled versions of palm wines.

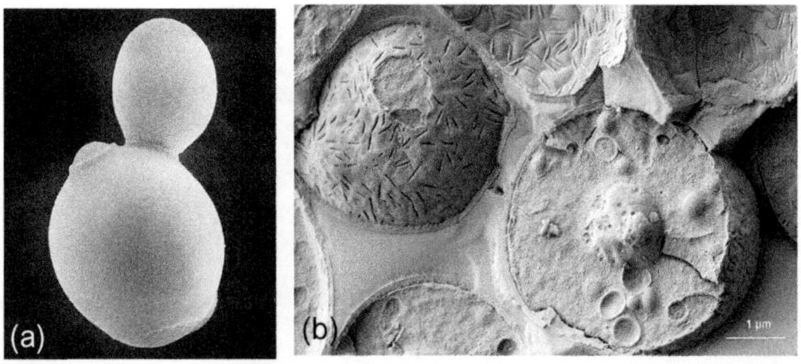

FIGURE 12.2 Electron micrographs of baker's yeast, *Saccharomyces cerevisiae*. (a) Scanning electron micrograph of mother cell with bud scar on upper left cell surface and less prominent birth scar at the bottom of the cell. Daughter cell at top right is approaching time of separation. (b) Transmission electron micrograph of freeze fractured yeasts showing outer surface of the cell wall (left) and its impression (top), and interior of whole cell containing prominent nucleus and nucleolus, and vesicles in the granular cytoplasm. *Source: Creative Commons.*

Spirits made from starchy materials include Scotch malt whiskey (from barley malt), gin (from barley, maize [corn], or rye), and vodka from various starchy sources including potatoes.

Wine and Other Beverages Produced from Sugary Juices

Red wines are fermented from grapes with red or purple/black skins whose phenolic pigments colour the wine (Figure 12.3). Grape skins and seeds are included in the first phase of fermentation of red wines. White wines are made from juice expressed from grapes without using the skins and seeds. Most white wines are produced from white grapes, but dark grapes are used for a few white wines because the pigments separate with the skin. After harvesting, grapes are crushed to break their skins and release the juice. True rosé wines are produced by including the grape skins at the beginning of the fermentation and discarding them after they have imparted the desired colour. This is called the skin contact method. The flavour of some white wines, including Sauternes, is enhanced by the colonisation of ripened grapes by *Botrytis cinerea* before picking. The fungus can destroy grapes under wet conditions, but its growth on drying grapes is called noble rot and produces a concentrated sweet white wine.

Wine fermentation is carried out in barrels, open vats, or industrial fermenters. Traditional wine production is carried out by the proliferation of yeasts that occur naturally on grapes. *Saccharomyces cerevisiae* is one of many grape yeasts and very little is known about its ecology and dispersal. It becomes the dominant species once the fermentation proceeds because it has quite a high tolerance to alcohol. Modern wine production usually involves the inoculation of the grape juice with a specific yeast strain. The juice is acidic (pH 2.9–3.9) which discourages the growth of bacteria and fungi that might otherwise spoil the fermentation. The addition of sulphur dioxide is an effective method for inhibiting the development of these undesirable microorganisms during the fermentation. *Saccharomyces cerevisiae* tolerates sulphur treatment and continues to grow until the alcohol concentration reaches 10–12%. Once this threshold alcohol level is reached, yeast metabolism slows and cell numbers decline. The remaining concentration of unfermented sugars determines the level of sweetness of the wine. Yeast metabolism during the fermentation generates hundreds of compounds that affect wine flavour. Glycerol adds 'body' to the wine and volatile ethyl esters are important in determining aroma.

FIGURE 12.3 Modern winery with high capacity stainless steel fermenters. *Source: Fedor Kondratenko©123RF.com* (See the colour plate.)

Following the first phase of the fermentation, the wine is transferred to wooden casks and other containers for maturation and storage. Lactic acid bacteria carry out malolactic fermentation, converting strongly acidic malic acid to blander lactic acid and raising the pH. The wine is filtered before bottling to eliminate bacteria that might spoil the wine by converting the ethanol to acetic acid and carrying out other reactions.

Cider apples are classified according to sweetness, ranging from bittersharp (high in acidity and tannin) to sweet (low acidity and low tannin). The apples are crushed and the pulp is pressed to release the juice. Traditional cidermaking relies upon yeasts carried on the apple skin, but, like winemaking, modern cidermaking involves inoculation with specific yeast strains and addition of sulphur dioxide to exclude other microorganisms. Apple juice has lower sugar content than grape juice, which results in a beverage with a lower alcohol content than wine. Following alcoholic fermentation, lactic acid bacteria may reduce the acidity of the fermentation. Bottled cider is pasteurised or filtered to prevent spoilage. Perry is produced from pears using similar methods.

Wines are made from many kinds of fruit using the same methods used to ferment grape juice, and palm wine made from the sugary sap of palm trees is popular in the tropics. Palm trees are tapped and their sap is fed into gourds or other containers. Rapid fermentation is catalysed by natural yeasts and the wine is consumed swiftly before bacterial spoilage occurs. Pulque is a Mexican wine produced from agave sap.

Beer Production from Starchy Materials

Beers made from a variety of grains and root crops rich in starch have a huge global market (Figure 12.4). European style beer is an alcoholic beverage made from malted barley and flavoured with hops. The German beer purity law, or Reinheitsgebot, dating to the sixteenth century, limits beer ingredients to malted barley, water, hops, and yeast. Brewers with a less stringent attitude to beer-making add other kinds of cereal, fruit juices, and diverse

FIGURE 12.4 Open vat fermentation to produce wheat beer. Rolling fermentation in centre of the tank is caused by carbon dioxide bubbling up from submerged yeasts. *Source: http://brewercameron.wordpress.com/2012/06/28/bavaria-excursion-day-3/.*

chemicals to create different flavours, control foaming, and control other characteristics. Ales and wheat beers (German weissbiers) are produced with **top yeasts** and served at room temperature or slightly cooled; lager is produced by a **bottom yeast** and is consumed after refrigeration. Top yeasts mix with the gas that accumulates as foam at the surface of the traditional open vat. Bottom yeasts tend to sediment to the bottom of the vat. These fundamental differences in behaviour are unimportant when a closed fermenter is used for brewing.

Few yeasts can utilise starch, so the conversion of starch into sugars, or **saccharification**, must be carried out before fermentation can commence. In beer brewing, saccharification is known as **malting**. Barley grains are steeped in water for up to 24h and allowed to germinate under moist aerobic conditions. During germination, proteinases and amylases break down proteins and starch into amino acids and sugars. After 1 week, the germinating grains are dried with a stream of air heated in a kiln. The dried grains are known as **malt**. If the **kilning** process is carried out at 60–79 °C, the embryos in the grain are killed but the enzymes retain their activity. This active malt is used to produce light-coloured beers. Kilning at higher temperatures destroys the enzymes and this kind of malt is added to the more active type to produce darker beers.

After kilning, the malt is milled into a coarse flour or grist. This is added to warm water to produce a mash in which enzyme activity continues to modify the proteins and starch. In beers that ignore the German purity standard, corn or rice grits are added to increase the starch content of the mash. After a few hours, the solids are separated from the mash to leave a sugar-rich liquid wort. Syrups are added to increase the sugar content of some beer worts. In the next step, dried female flowers of the hop plant, *Humulus lupulus*, are added to the wort and the mixture is boiled. Boiling denatures the enzymes, coagulates some of the proteins, and sterilises the wort. Cyclic organic compounds in the hops, called humulones, isomerise to form isohumulones that make the beer bitter, and other molecules impart different flavours and aromas. After the wort is cooled, hop residues and other solids are removed, and the wort is inoculated with a specific yeast strain. The wort is aerated to promote yeast growth and fermentation proceeds in open tanks or in closed stainless steel fermenters. Cylindro-conical fermenters are the most popular designs used in today's breweries, in which circulation of the contents is driven by carbon dioxide bubbles produced by the fermentation, or forced into the tank via a sparger. The carbon dioxide bubbles rise to the surface through the centre of the tank, and cooler liquid descends on the outside toward the conical base.

Temperatures for ale fermentation (15–25 °C) are higher than those used for lager (8–15 °C). Oxygen levels decline very quickly during the fermentation and most of the process is anaerobic with efficient conversion of sugar to alcohol rather than biomass. The yeasts produce a variety of compounds that flavour the beer. Yeasts which are particularly good for brewing flocculate and clump toward the end of the fermentation, which facilitates their removal. Maturation in casks follows and a secondary fermentation of residual sugars is completed by low concentrations of yeast remaining in the beer. The final step of clarification of the beer, or fining, involves chemical treatment to cause sedimentation of particles (**isinglass** prepared from the dried swimbladders of fish is the traditional additive) followed by filtration. Pasteurisation reduces opportunities for spoilage.

Beer with up to 0.5% alcohol, marketed as alcohol-free beer, is made by adapting the mashing process to limit **saccharification** and by reducing the wort gravity (a measure of its sugar content) at the beginning of the fermentation. Yeast strains with limited fermentative performance are also utilised by some brewers and alcohol content can be reduced by vacuum distillation or reverse osmosis once the brewing process is complete.

Baking

Saccharomyces cerevisiae has been used to ferment dough made from wheat and rye flour for more than a millennium. Top yeast harvested as a waste from beer brewing was used in bread making by Romans and this practice was widespread in the nineteenth century. Baker's yeast is manufactured today using fed-batch fermentation with molasses supplemented with a variety of nutrients (p. 413). Fed-batch fermentation allows the operator to add nutrients to the reaction continuously or intermittently to control the metabolic activity of the cells and generate high cell densities. The yeast cultures are aerated to maximise respiration and biomass accumulation and conditions are regulated to stimulate trehalose synthesis. *Saccharomyces* produces the sugar alcohol trehalose as a storage carbohydrate and it protects the cells when they are freeze-dried. The yeast is harvested by vacuum filtration and prepared for baking as freeze-dried granules, compressed cakes, or in a concentrated liquid form called cream yeast. Cream yeast has a limited shelf life, but is preferred by large bakeries. Dough is prepared by mixing flour with water, yeast (2%) and salt (1.5–2%). Amylases in the dough break down the starch and form glucose, maltose (disaccharide of glucose units), and maltotriose (trisaccharide of glucose), which are fermented by the yeast. Maltose is the dominant sugar. Milk, sugar, eggs, and other ingredients are added to the dough to produce different kinds of bread and mixing (kneading) of the dough ensures uniform distribution of the ingredients. As the dough is kneaded, proteins in the flour, called gliadin and glutenin, form strands of gluten that give the dough its springy and elastic texture and are responsible for the chewiness of the baked bread. After kneading, the dough is left for a few hours and the yeast ferments the sugars and releases carbon dioxide that causes the dough to rise. Ethanol produced during this fermentation kills the yeast and is expelled during baking.

Cheeses and Meat Products

Lactic acid bacteria ferment lactose in milk to lactic acid and produce many of the metabolites that impart characteristic flavours to cheeses. The enzyme **chymosin** (also known as rennin) is added to coagulate the milk and form curd. Traditionally, this has been obtained from the stomach lining of unweaned calves. Most of commercial cheese production today relies on **fermentation-produced chymosin** (FPC) from recombinant strains of *E. coli* and the ascomycetes *Aspergillus niger* var. *awarmori* and *Kluyveromyces lactis* (a yeast). Natural proteases obtained from species of *Rhizomucor* offer a 'vegetarian' alternative to the use of these recombinant microorganisms. Yeasts and filamentous fungi also play subsidiary roles in flavouring and ripening cheeses. Brie and Camembert are well known surface-ripened cheeses whose rinds are formed from a dense white mycelium of *Penicillium camemberti* (Figure 12.5). These cheeses are salted during preparation and the halotolerance of the fungus allows it to grow on the surface. Enzymes secreted from the rind penetrate the outermost millimetres of the cheese and these proteinases and lipases add fruitiness and fragrance. Like these soft cheeses, Italian salami is preserved and flavoured by a thinner coating of *Penicillium nalgiovense* and other *Penicillium* species. *Penicillium roqueforti* is the filamentous fungus that proliferates within the blue-vein cheeses including Roquefort, Gorgonzola, Stilton, and Danish Blue. The bacteria that ferment the milk for these varieties produce carbon dioxide that creates irregular cavities in the cheese. *Penicillium* conidia are included in the starter culture or added to the fresh curd. After the curd is compressed, salt is dusted on the surface and diffuses into the cheese. The cheese is spiked after salting and the resulting aeration allows the conidia of the fungus to germinate and colonise the cavities. The blue coloration of the cavities is caused by sporulation, and enzymes secreted by the fungus impart glorious flavours to the mature cheese.

Yeasts are involved in fermentation reactions and maturation processes in cheese production, but many of their effects on flavour and texture are unclear. *Debaryomyces hansenii* grows in semi-soft cheeses and is important in surface ripening of Munster, Limburger, and Port Salut; *Geotrichum candidum* is added to soft cheeses; *Yarrowia lipolytica*, *Saccharomyces cerevisiae*, and *Kluyveromyces* species are also common ascomycete yeasts in cheeses.

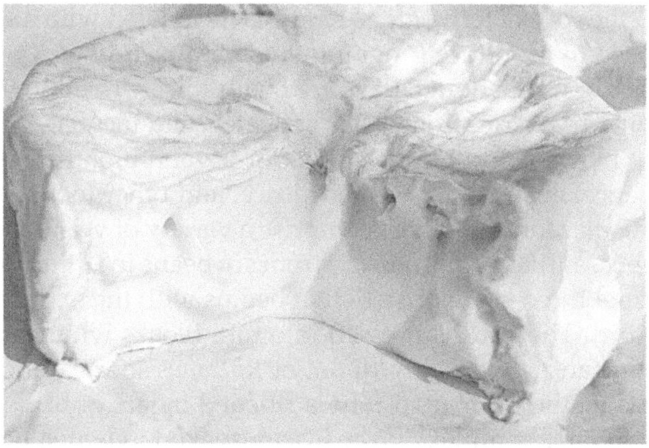

FIGURE 12.5 Camembert cheese with characteristic white rind of *Penicillium camemberti* mycelium. *Source: http://www.pnwcheese.com/2011/07/kurtwood-farms-dinahs-cheese-coming-soon-to-portland.html*

Chocolate

Cacao seeds, or 'beans', are embedded in white pulp inside the colourful pods of *Theobroma cacao* trees. To manufacture chocolate, the pods are cut open, the pulp-bean mass is arranged in heaps, boxes, or baskets and allowed to ferment in the open air for 4–7 days. Fermentation of sugars and citric acid in the mucilaginous pulp kills the cacao embryos and reduces the astringency of the seeds. Cacao fermentation involves a complex community of yeasts and bacteria. *Hanseniaspora guilliermondii* and *Hanseniaspora opuntiae* dominate the initial stages of the fermentation and are followed by multiple strains of *Saccharomyces cerevisiae*, *Pichia kudriavzevii*, and other yeasts. The succession of microorganisms reflects changes in the physical and chemical environment within the pulp-bean mass. Bacteria that ferment the citric acid in the pulp create the conditions that favour the growth of *Saccharomyces cerevisiae* and this species outcompetes yeasts that are less tolerant of the increase in temperature and ethanol concentration in later stages of the fermentation. At the end of the fermentation, the seeds are dried, their thin husks are removed, and the remaining tissue is ground to produce nibs. The nibs are combined with other ingredients to produce chocolate.

Asian Fermented Foods

The greatest variety of foods and beverages produced by fungal fermentation are found in Asia. **Tempe**, originating in Java, is produced by fermenting beans or cereals and is a popular meat substitute. The traditional production method begins with the de-hulling of soybeans, followed by soaking and boiling in water before the beans are spread on trays to allow rapid cooling and evaporation of water. After cooling, the beans are inoculated with spores of *Rhizopus* or *Mucor* species (Mucorales), mixed, spread into layered beds, and incubated for 1–2 days allowing the development of a dense mycelium. The mycelium transforms the raw soybeans into a firm, sliceable cake and its enzymes modify the polysaccharide content of the beans and improve the digestibility of tempe. Other biochemical transformations by the fungus include the generation of antioxidants with purported health benefits. Different varieties of *Rhizopus microsporus* (another mucoralean) are the most common fungi used for manufacturing tempe. These grow in soil and on leaves and are prevalent in the air spora. Rather than hoping for natural colonisation with the correct fermenters, tempe is inoculated with a **starter** enriched in the fungus. Starters are used to introduce fungi into the manufacture of a variety of Asian foods. They are made from rice or wheat dough flavoured with spices. The dough is shaped into small biscuits, inoculated with dry starter from an earlier batch, and incubated for up to 5 days. After drying in the sun, the new starters can be stored for several months before they are used to produce tempe, soy sauce, miso, saké, sweetened rice, and other foods. Some starters contain pure cultures of a particular fungus, others comprise a variety of yeasts and filamentous fungi.

Furu or **Sufu** is a cheese-like food produced from soybeans in China. It is manufactured according to a complicated three-stage process that begins with the extraction of soy milk from the beans to form a curd. The curd is pressed into tofu blocks which are heated, inoculated with a starter containing an *Actinomucor*, *Mucor*, or *Rhizopus* species, and fermented for up to 1 week. In the third stage, the fermented tofu is ripened in jars of brine for 2–4 months. **Soy sauce** production is similarly complex. Soybeans are soaked in water, boiled, drained, mixed with roasted wheat and spread on trays, inoculated with *Aspergillus oryzae* or *Aspergillus sojae* and fermented for 5 days to form koji. The koji is mixed with brine to produce moromi which is fermented by yeast and lactic acid bacteria for 1 year. The filamentous fungi are destroyed during this second stage fermentation, but their enzymes remain active in the brine. Other Asian fermented foods include red koji rice, fermented by *Monoascus*, saké and other rice wines, and Chinese liquor or jiu made from sorghum.

FERMENTATION TECHNOLOGY

Fermentation technology is concerned with the large-scale culture of microorganisms in fermenters, and the recovery of useful products from the metabolic activity of the microbial cells. Industrial mycologists study the fungi used to perform chemical transformations of commercial significance, the food sources (feedstocks) on which the fungi are cultured, optimization of fermenter design and operation, and 'downstream' processing to harvest the desired products.

Feedstocks

For commercial success, a fungus must behave in a consistent, predictable fashion, showing similar growth rates and metabolite yields in successive fermentations. The provision of specific nutrients is crucial in controlling the growth and development of the fungus. In the laboratory, this can be achieved by producing a defined medium by mixing pure chemicals in defined concentrations. This is not feasible for a large-scale industrial process and the selection of raw materials for a particular fermentation is based on nutrient content, cost, and availability. Downstream processing costs also vary according to the feedstock. Purification of the product from the partly digested raw materials can be expensive, and waste disposal represents an additional cost and can pose environmental hazards. Glucose and sucrose are excellent carbon sources for most fungi so most feedstocks include sugarcane juice, unrefined sugar, molasses, or hydrolysed starch. Vegetable oils are useful alternative feedstocks for some fermentations. Some feedstocks provide the fungus with a combined carbon and nitrogen source, but a separate nitrogen source is supplied to many processes. In some cases, vitamins are added as yeast extract, as well as additional minerals and trace elements. The cost of the raw materials for the fermentation ranges widely, representing more than 50% of the production budget for industrial ethanol.

Fermenter Design and Operation

Most industrial fermentations are carried out in **stirred tank fermenters** in which sterile liquid medium can be inoculated, aerated, stirred, monitored by sensing instruments, heated or cooled, and sampled or fed with additional materials without introducing contaminants (Figure 12.6). Fermentation tanks or **bioreactors** are manufactured from stainless steel to avoid corrosion and leaching of toxic metals into the medium. Industrial fermentations employ **batch culture** and **continuous culture** methods. In batch culture, the level of nutrients declines as the density of cells increases and the fermentation is stopped to harvest the product. Continuous culture allows the operator to maintain optimal conditions for fermentation for many weeks by programming cycles of nutrient injection and other changes in the fermentation conditions. Fermentation products can be harvested repeatedly or continuously using this method. Because many fermentation products are secondary metabolites, they are formed when the growth rate of the fungus begins to stall. Batch cultures are preferred for these kinds of fermentations because the classic growth kinetics characterised by lag,

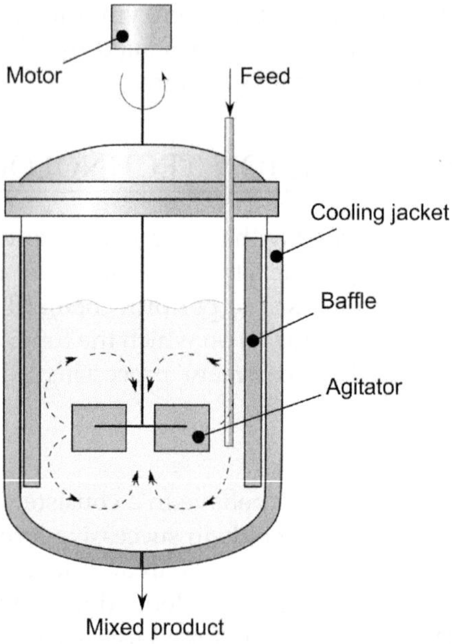

FIGURE 12.6 Stirred tank fermenter or bioreactor. *Source: Creative Commons.*

exponential, and stationary phases occur without manipulating the growth conditions. The same results can be obtained via continuous culture by programming successive cycles of nutrient injection, followed by starvation and secondary metabolite synthesis. This is, however, unnecessarily complicated and expensive when batch culture works so well. Continuous culture is recommended for other processes in which large quantities of a product associated with growth are needed. Examples of continuous culture include the production of ethanol and single cell protein (mycoprotein). **Fed-batch culture** is a hybrid of these methods in which nutrients are added to the culture during the fermentation. This avoids the problems associated with catabolite repression (Chapter 5) in which high levels of sugars added at the start of conventional batch culture inhibit secondary metabolism. Fed-batch culture is not the same as continuous culture, because the fermentation is terminated when the product is harvested. Fed batch culture is used to produce antibiotics (pp. 418–419).

Heat sterilization of the fermenter and growth medium before inoculation is critical. This can be done inside the fermentation tank by heating the surrounding jacket and injecting steam into the reactor. The gas exit valve must be closed to allow pressurization and heating of the system above 100 °C. Alternatively, medium can be sterilised outside the fermenter and supplied via sterile lines. Air for aerobic fermentations is pumped into the medium through a sparger to generate air bubbles that facilitate oxygen diffusion throughout the reactor volume. Agitation of the medium with a stirrer enhances this process.

Downstream Processing

One litre of broth may contain 1 g of an enzyme product or a few grams of antibiotic, which means that a small quantity of product must be separated from a large volume of waste material. Fungal mycelium is separated from broth using a rotary vacuum filter. Yeast cells are removed by flotation and sedimentation in brewing practices and centrifugation has been introduced to standardise many modern biotechnological processes. In fermentations where intact cells are harvested to collect intracellular enzymes, water is removed from the culture and enzymes are recovered by breaking the cells using a variety of methods. Processing of antibiotics and many other products dissolved in the fluid phase of the fermentation involves harvesting the broth and treating the fungal biomass as solid waste.

Solid-State Fermentations

Traditional production methods for Asian foods (see above) are solid state fermentations (also known as **solid-substrate fermentations**) in which fungi are grown on soybeans and cereal grains. The raw materials are treated prior to inoculation to facilitate subsequent transformation by the fungus. These pre-treatments include grinding or milling to increase surface area, soaking to hydrate and soften the food base, and steaming to kill the seeds and eliminate other microorganisms. Because the fungi used in these processes are aerobes, moisture content must be kept quite low during some fermentations to maintain air spaces in the nutrient source. Asian foods are produced on stacked wooden trays that provide aeration and large surface areas for fungal growth. Adaptation of these sorts of methods for industrial fermenters is complicated, but the use of a forced air supply combined with rotating drums can maintain adequate aeration. Mushroom cultivation and composting of agricultural wastes are other examples of solid-state fermentations.

GENETIC MANIPULATION OF FUNGI FOR BIOTECHNOLOGY

The most important filamentous fungi used for protein production are *Aspergillus niger*, *Aspergillus oryzae*, and *Trichoderma reesei*. Beyond the selection of naturally occurring strains of these species that show high levels of secretion of the protein of interest, random mutagenesis can be used to isolate mutants with enhanced synthetic and secretory performance. Genetic engineering of wild type or mutant strains to boost protein production has also proven very effective and dominates modern biotechnological research. Proteins that are natural products of a particular fungus are called **homologous proteins**. A fungus can also be engineered to produce foreign or **heterologous proteins**. Production of heterologous proteins is dependent upon the genetic alteration of a fungus through the incorporation of exogenous genes. This is called **transformation**. Several methods are used to transform *Aspergillus* and other filamentous fungi with genes carried on **plasmids**. Plasmids are small DNA molecules that replicate independently from chromosomal DNA. The plasmids used for transforming fungi originated in bacteria. Natural plasmids are also found in fungal mitochondria, consistent with the bacterial origin of these organelles, and multiple copies of the 'yeast 2-micron plasmid' are found in the nucleus of *Saccharomyces cerevisiae*.

The fungal cell wall is a macromolecular sieve that presents a barrier to plasmid uptake. This is circumvented by digesting the cell wall of germinating conidia with a cocktail of enzymes. The resulting protoplasts absorb plasmids when they are incubated in a solution containing calcium chloride and polyethylene glycol. Other methods for plasmid transfer include **electroporation** and use of the bacterium *Agrobacterium tumefaciens* as a live vector.

Successful transformation requires the integration of the foreign DNA into homologous or non-homologous regions of the host genome. Integration of the heterologous gene can be targeted to a specific locus in the host genome and genetic engineers favour regions of the genome that encode secreted proteins that are already expressed at high levels. If this is done by replacing the native gene, expression of the heterologous gene may be further enhanced by the lack of competition for secretory machinery from the native gene product. Increased production of homologous and heterologous proteins has been achieved by introducing **multiple copies** of a gene and driving their transcription with the use of a **strong promoter**. The gpdA promoter from *Aspergillus nidulans* is part of a gene that encodes the glycolytic enzyme glyceraldehyde-3-phosphate dehydrogenase. This constitutive promoter is valuable in basic research and biotechnological applications because it functions in other fungi (Figure 12.7).

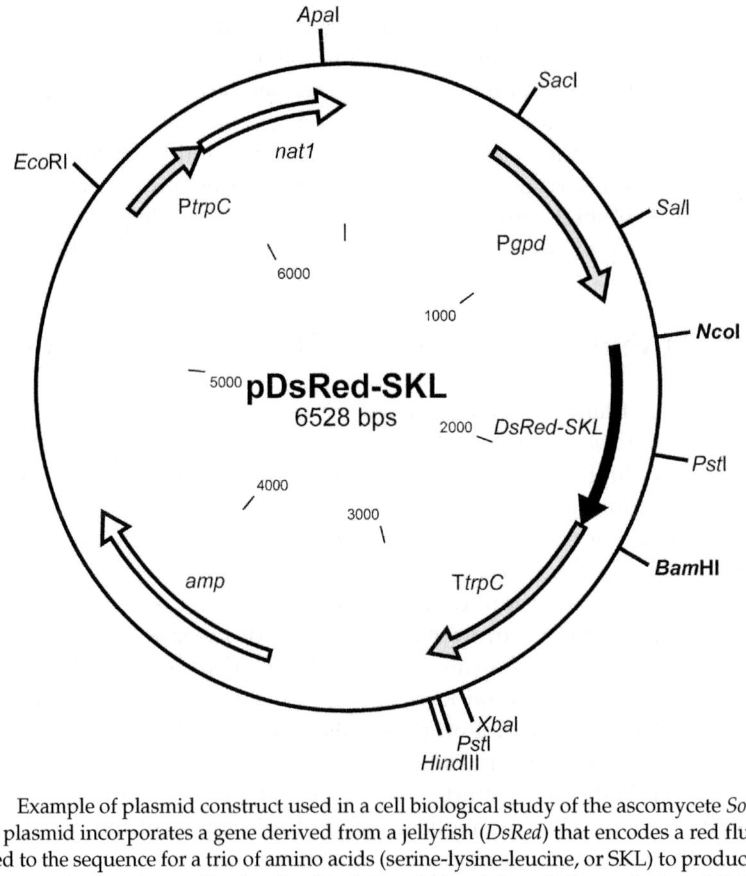

FIGURE 12.7 Example of plasmid construct used in a cell biological study of the ascomycete *Sordaria macrospora*. The DsRed-SKL plasmid incorporates a gene derived from a jellyfish (*DsRed*) that encodes a red fluorescent protein. This gene is fused to the sequence for a trio of amino acids (serine-lysine-leucine, or SKL) to produce a protein that is incorporated into peroxisomes (organelles involved in fatty acid breakdown). Transcription of *DsRed-SKL* is driven by a constitutive promoter (P*gpd*) derived from *Aspergillus nidulans*. T*trpC* is a terminator sequence, derived from *Escherichia coli*, which stops transcription and releases the mRNA that is translated into the DsRed-SKL protein. The positions of restriction sites are indicated with the labels on the outside of the plasmid. Expression of this plasmid in *Sordaria* allowed investigators to study the cellular distribution of peroxisomes using fluorescence microscopy. *Source: Elleuche, M., Pöggeler, S., 2008. Fungal Genet. Rep. 55, 9–12.*

Inducible promoters are used in many applications and these include promoters of genes that encode numerous secreted enzymes. Homologous promoters tend to work better than heterologous promoters.

Several heterologous proteins have been produced in *Trichoderma reesei* at levels of commercial significance including a phytase from *Aspergillus* (2 g/L), glucoamylase from the creosote fungus, *Hormoconis resinae* (0.7 g/L), and xylanase from the thermophilic soil fungus, *Humicola grisea* (0.5 g/L). Comparable yields of human immunoglobulins and interleukin 6 have been produced by recombinant strains of *Aspergillus niger*. An engineered strain of *Aspergillus oryzae* is a source of a heat-stable lipase encoded by a gene from a thermophilic ascomycete, *Thermomyces lanuginosa*. This enzyme is useful as an additive to laundry detergents and has been modified via a single amino acid substitution to be effective at low wash temperatures.

MODERN BIOTECHNOLOGICAL APPLICATIONS OF FILAMENTOUS FUNGI AND YEASTS

Single Cell Protein

The term single cell protein was introduced in the 1960s to describe protein-rich foods manufactured from yeasts that served as dietary supplements for livestock and humans. Single cell protein was viewed as a product category that might address food shortages at a time when it seemed unlikely that agricultural production could keep pace with the skyrocketing human population. Interest in food yeast declined as improvements in plant breeding and agricultural practices led to the contemporary boom in global food production. *Saccharomyces cerevisiae* produced in stirred fermenters on molasses is an example of single cell protein that is manufactured today. The yeast produced in this fashion is not consumed directly, but is used for baking. Marmite is a savoury spread made from yeast extract that has been popular in the United Kingdom for more than a century. Spent yeast from beer brewing is used to produce this sticky dark brown paste. Vegemite is a similar product made in Australia.

Another fungal protein product, called **Quorn**, is a very successful meat substitute manufactured by a single strain of a filamentous saprotrophic ascomycete, *Fusarium venenatum*. Because Quorn is produced from a multi-cellular, filamentous fungus, the term single cell protein is inaccurate and **mycoprotein** is the preferred name. The use of a filamentous fungus makes it possible to produce a meat-like consistency that cannot be replicated with a single-cell protein. The fungus is grown in pairs of 50 m tall air-lift fermenter vessels that contain 230 tonnes of broth (Figure 12.8). The vessels are connected at the top and the bottom to form a continuous loop. Compressed air and ammonia are pumped into the bottom of the first vessel, called the 'riser', oxygenating the culture and circulating the liquid containing the fungus toward the top. Carbon dioxide from the respiring cells is released through a vent at the top of the system, and the liquid falls through the second 'downcomer' vessel and is infused with fresh nutrient solution (glucose plus vitamins and minerals). A heat exchanger at the bottom of the system maintains the temperature at 30 °C and the culture is harvested at a rate of 30 tonnes per hour. The dense mycelium of branched hyphae harvested from the fermenter has a very high RNA content. This is problematic for a food product because its consumption could raise uric acid levels in the blood and lead to gout and other illnesses. This is addressed by heating the mycelium

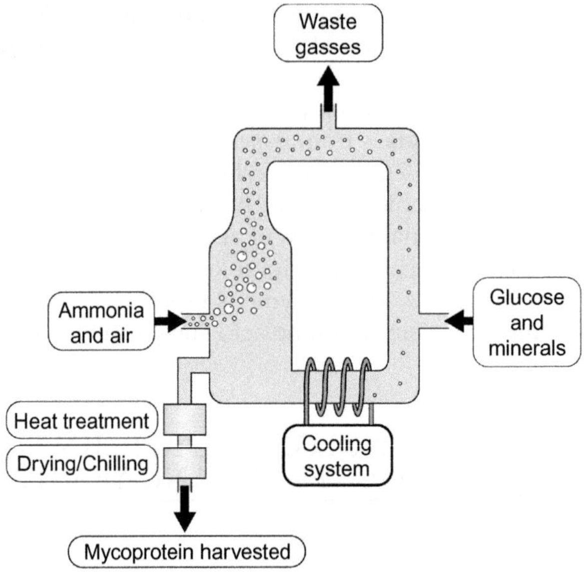

FIGURE 12.8 Air-lift fermentation system used to produce Quorn. *Mark Fischer, Mount St. Joseph University, Cincinnati*

at 68 °C for 20 min, which allows endogenous enzymes to destroy much of the RNA without reducing its protein content. The heated mycoprotein is then dried and bound with egg white. Further processing creates the meaty texture and adds flavourings and colourings.

Fungal Enzymes

The secretion of enzymes that service the absorptive feeding mechanism of the fungi is a huge asset for biotechnology because fungi release many protein products into their culture fluid simplifying purification. A particular advantage of using fungi over bacteria for protein production is that they engage in eukaryotic post-translational processing of proteins which is essential for the biological activity of many heterologous proteins of pharmaceutical relevance. At the same time, fungi are easily cultured like bacteria. The global market for enzymes generated by fermentation exceeds US $5 billion, and fermentation by filamentous fungi accounts for half of this production. Biotechnological applications of specific fungal enzymes impact a vast array of products. More than one hundred enzymes have been commercialized from 25 genera including the ascomycetes *Aspergillus*, *Trichoderma* and *Penicillium*, the zygomycete *Rhizopus*, and the basidiomycete *Humicola*. **Fungal hydrolases** are the most important class of enzymes with applications in the food and beverage industries, personal care products, laundry detergent manufacture, textiles, leather, forest products, animal feed, and biofuels. The use of live fungal cultures in baking, brewing, and the manufacture of other foods was discussed earlier in this chapter, but purified enzymes have other uses in food products. These include mixtures of enzymes that hydrolyse proteins in meat products and vegetables to produce soups, stock cubes, a variety of sauces, and the ubiquitous food flavouring monosodium glutamate. As mentioned earlier (p. 409), fungal enzymes that act as coagulants are used in cheese production and a recombinant version of bovine chymosin produced by *Aspergillus niger* is a substitute for the enzyme obtained from calves. **Lactases** from *Aspergillus* and *Kluyveromyces* increase the digestibility of dairy products for consumers with lactose intolerance. Lactases are also effective at digesting antigens in milk to produce milk formulas for infants.

Fungal enzymes have been used in toothpastes, to combat bacteria active in tooth decay and enhance bleaching, and the addition of a **laccase** from the wood-decaying basidiomycete *Trametes* to breath fresheners destroys the compounds responsible for halitosis. Fungal proteases and amylases in laundry detergents degrade dirt particles, and lipases are active against grease stains. **Cellulases** in detergent formulations increase the softness and brightness of clothing. Enzymes from fungi and from bacteria allow washing at lower temperature, which reduces energy consumption, and water use is reduced by their effectiveness at removing dirt. Cellulases from *Trichoderma reesei* create the popular stonewashed appearance of denim. Processing of raw cotton involves bacterial and fungal enzymes; catalases from *Aspergillus* and other fungi are used to bleach cotton before it is dyed, and cellulases are used in the finishing process to increase the smoothness of the fabric and reduce 'pilling'.

In papermaking, fungal **xylanases** are used in chlorine-free bleaching to hydrolyse the hemicelluloses in wood pulp. This has the effect of opening the molecular structure of the pulp which releases the dark-coloured complexes of insoluble lignin. Lipases, amylases and cellulases are also used in paper manufacture and recycling. Fungal xylanases are vital in many other applications including the degumming of flax, jute and hemp, silage and grain processing, and as additives to wheat flour to improve dough handling. Fermentation of cereals with beta-glucanases from *Trichoderma* and *Aspergillus* increases their nutritional value for monogastric animals like pigs and poultry. Fungal **phosphatases** (**phytases**) that release phosphorus from polymeric stores are also used for processing animal feed.

Finally, fungal enzymes may play an increasingly significant role in the biofuel industry. After milling, whole grains can be cold-cooked with **glucoamylases** and **alpha-amylases** to produce fermentable sugars including glucose and maltose. Dry plant matter, or lignocellulosic biomass, is a huge untapped feedstuff for **bioethanol** whose utilisation will require new approaches for using fungal enzymes to release fermentable sugars from these complex polymers (pp. 421–422).

Organic Acid Synthesis

Aspergillus species are regarded as the most important industrial microorganisms. *Aspergillus niger* is a global source for the production of more than 1 million tonnes of **citric acid** per year, which is used as a preservative and chelating agent, and as an acidic flavouring for foods and soft drinks. *Aspergillus niger* and *Penicillium* species produce **gluconic acid** which is used for cleaning and finishing metal surfaces, and as an additive in cement. Lower concentrations of gluconic acid are also used as a food additive. Annual production is estimated at 50,000–100,000 tonnes. Gluconic acid is generated by a two-step oxidation of glucose by a pair of enzymes secreted by *Aspergillus niger*. **Itaconic acid**, used in polymer synthesis, is produced by *Aspergillus terreus* and *Aspergillus itaconicus* in a submerged fermentation of glucose syrup or molasses. Kojic acid is another *Aspergillus* product that is sold as a cosmetic for whitening skin. It is a byproduct of soy sauce fermentation.

Vitamin B_2 Synthesis

Industrial production of vitamin B_2, or riboflavin, involves the fermentation of soybean oil and soybean meal by the ascomycete *Ashbya gossypii*. Triglycerides in the broth are hydrolysed by a lipase secreted by the fungus. Free fatty acids are then absorbed by the fungus and used as precursors for riboflavin synthesis. Industrial strains of *Ashbya gossypii* have been produced by a combination of classical strain selection and molecular engineering of the riboflavin pathway to stimulate vitamin synthesis.

Antibiotics and Other Pharmacological Agents

β-*Lactam Antibiotics*

Penicillins, secreted by species of the ascomycete *Penicillium*, are powerful anti-bacterial agents that target Gram-positive bacteria. They are members of a class of antibiotics that share a core structure called the β-lactam ring (Figure 12.9). Other **β-lactam antibiotics** include **cephalosporins**, produced by the fungus *Acremonium*, and **carbapenems** produced by bacteria. The production of **penicillin G** (Gold Standard penicillin) by *Penicillium rubens* (identified originally as *Penicillium chrysogenum*) was discovered by Alexander Fleming in 1928 and developed for clinical use by a research team of Oxford University scientists led by Howard Florey. Research to increase production of the antibiotic began in 1938 and the drug was tested on the first patient in 1941. A strain of *Penicillium chrysogenum* isolated from a rotting cantaloupe in 1943 gave much higher yields of the antibiotic and this fungus became the mainstay of industrial production. Penicillins and other β-lactam antibiotics destroy bacterial cells by interfering with the formation of cross-links within their peptidoglycan cell wall. Peptidoglycan is a polymer formed by chains of alternating residues of a pair of amino sugars, *N*-acetylglucosamine and *N*-acetylmuramic acid, with short peptides attached to the *N*-acetylmuramic acid residues. Cross linkages between amino acids in different peptide chains creates a strong three-dimensional crystal lattice. The antibiotic binds to proteins called penicillin-binding proteins or transpeptidases that form the bonds between the cross-linking

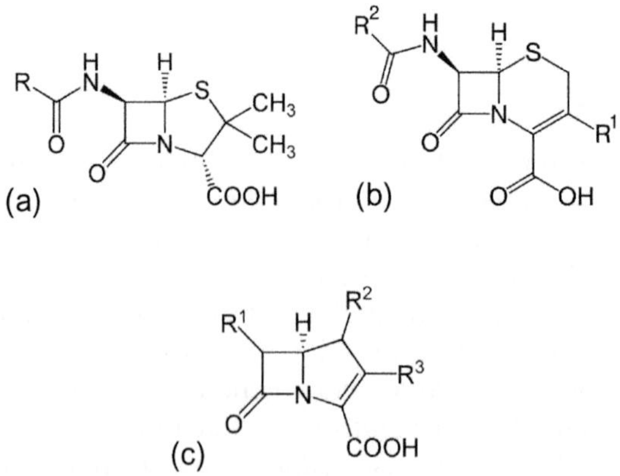

FIGURE 12.9 Structures of β-lactam antibiotic molecules. Core structures of (a) penicillins, (b) cephalosporins, (c) carbapenems. *Source: Creative Commons*

peptides in the wall. Inhibition of these enzymes weakens the cell wall and causes the bacteria to burst. Bacterial resistance to penicillin derives from the evolution of strains that produce enzymes called **β-lactamases** that inactivate the antibiotic by opening the β-lactam ring. Bacterial resistance became a problem within the first decade of penicillin's use. An important advance was made with the discovery that *Penicillium chrysogenum* produces large quantities of 6-aminopenicillanic acid (6-APA), which is the core structure of the antibiotic. This structure lacks antibiotic activity, but can be used to produce semi-synthetic antibiotics by adding a variety of side chains. **Meticillin** (formerly methicillin) and **ampicillin** are examples of these compounds. Meticillin is resistant to β-lactamase due to the steric hindrance offered by its bulky side chain that blocks access to the β-lactam ring. Bacterial resistance to meticillin has led to its replacement with other antibiotics. Ampicillin is a broad-spectrum antibiotic that is active against Gram-negative as well as Gram-positive bacteria.

Cephalosporins are β-lactam antibiotics produced by *Acremonium chrysogenum*. The first cephalosporins, called first generation cephalosporins, were active against Gram-positive bacteria. Later generations of these antibiotics, produced by cleaving the parent molecules and adding various side chains, have shown greater efficacy against Gram-negative bacteria, sometimes at the expense of reduced potency against Gram-positive bacteria. These are exceedingly valuable antibiotics, with the fifth-generation cephalosporins representing the last line of defence against infections caused by meticillin-resistant *Staphylococcus aureus* (MRSA).

Lovastatin

Lovastatin is a fungal polyketide employed as a cholesterol-lowering drug that inhibits (3S)-hydroxy-3-methylglutaryl-coenzyme A (HMG-CoA) reductase. This enzyme catalyses the synthesis of mevalonate, which is the immediate precursor of cholesterol and is the target of all of the drugs classed as statins. Lovastatin was discovered in *Aspergillus terreus* and *Monascus ruber* in the 1970s and is a natural product in oyster mushrooms (*Pleurotus ostreatus*) and red yeast rice (rice fermented by *Monascus*). It is produced by a multi-domain enzyme complex known as a **polyketide megasynthase** encoded by a cluster of genes including *lovA*, *lovB*, *lovC*, *lovD*, and *lovF*. The biosynthetic pathway resembles the pathway of fatty acid synthesis where the polymer is extended with the repetitive addition of acetyl units. Subunits of the enzyme complex are organised in modules that initiate the synthesis (starter module), add acetyl units from malonyl CoA to the growing polyketide chain (elongation module), and terminate the reaction (termination module). Lovastatin is prescribed under the name Mevacor, and a synthetic derivative, called simvastatin, is trademarked as Zocor. These drugs, together with fully-synthetic statins including Lipitor, produce multibillion-dollar sales for Pfizer, AstraZeneca, Merck, and Novartis.

Immunosuppressants

Cyclosporin A is an immunosuppressive, fat-soluble, cyclic peptide that blocks the activation and proliferation of CD4+ and CD8+ T lymphocytes (T helper cells and CD8 killer cells, respectively) by inhibiting the production of the cytokine, interleukin-2 (IL-2). In addition to its effectiveness in supporting patients after bone marrow and organ transplantation, cyclosporin A is used to treat a wide range of illnesses including psoriasis, severe atopic dermatitis, and rheumatoid arthritis. Cyclosporin A was purified in 1969 from a culture of *Tolypocladium inflatum* obtained from a Norwegian soil sample. The fungus is a pathogen of beetle larvae. Though other fungi produce cyclosporin A, industrial production relies on productive strains of this single species. Cyclosporin A is an example of a non-ribosomal peptide that is synthesised by a huge multifunctional enzyme called **cyclosporin synthetase** (simA) that catalyses at least 39 different reaction steps. The enzyme is built from 11 protein modules and each of these functions in the recognition, activation, and modification of one intermediate. A small 12th module is thought to carry out the final cyclization step in cyclosporin A synthesis. The resulting structure of 11 amino acids is too complex to be produced by chemical synthesis, which means that improvements in production are reliant upon manipulation of the fungus. Methods that have been tested include changes in the carbon and nitrogen sources for the fungus, sequential addition of different carbon sources to its growth medium, addition of L-valine to the fermentation, immobilization of the fungus on beads, and cultivation on solid materials (solid state fermentation) rather than in liquid. Efforts to boost cyclosporin production have also involved the study of mutant strains of *Tolypocladium inflatum*.

Ergot Alkaloids

Ergot alkaloids, produced by the ergot fungus *Claviceps purpurea*, have poisoned human populations that consumed bread baked from contaminated rye flour and caused an incalculable number of deaths (Chapter 9). Used in a deliberate fashion as medicines, the same compounds have alleviated the suffering caused by migraine headaches, treated symptoms of Parkinson's disease, and stimulated uterine contractions and stemmed bleeding during childbirth. The toxicity and the therapeutic value of these compounds are due to their affinity for neurotransmitter receptors. The vasoconstriction leading to burning sensations associated with classic ergot poisoning, called St. Anthony's fire in the Middle Ages, is caused by the neuronal stimulation of smooth muscle contraction. The effectiveness of the ergot alkaloids in treating migraine and reducing bleeding lies in the same pharmacological mechanism. Other species of *Claviceps* produce ergot alkaloids and the same pathway for toxin synthesis is found in different members of the family Clavicipitaceae, including endophytes in the genus *Epichloë* (= *Neotyphodium*) that cause toxicosis in animals browsing on contaminated grasses (Chapter 9). The human pathogen, *Aspergillus fumigatus*, and related saprotrophs are also sources of these compounds. The relationship, if any, between the synthesis of the alkaloids and the development of various forms of aspergillosis is not known.

Ergot alkaloids are a diverse category of secondary metabolites that have been classified into three groups as clavines, amides of lysergic acid, and ergopeptines. Lysergic acid diethylamide or LSD is a synthetic derivative of lysergic acid, which is a key intermediate in the biosynthesis of ergot alkaloids. Synthesis of most of the ergot alkaloids involves a common set of initial reactions beginning with the addition of a pyrophosphate group to tryptophan followed by a series of steps to produce a tetracyclic ergoline ring. Genes encoding the enzymes for these pathways are organized in clusters. Ergotamine and ergometrine, produced by *Claviceps purpurea*, are the most important of the alkaloids used in medicine and a number of derivatives of these compounds have been adopted for treating specific conditions. Commercial fermentation is the main source of ergot alkaloids today.

Biotransformations

In addition to the value of natural and genetically modified strains of fungi as sources of pharmaceutical agents, fungi are used to modify the chemical structure of natural products including enzymes and small molecules produced by other organisms. These enzyme-catalysed steps include oxidation and reduction reactions, transfer of functional groups, and hydrolysis. Fungal transformations are used to produce anti-inflammatory cortisones from plant sterols, the hormonal contraceptive gestodene from other sterols, and ketaprofen, which is an important non-steroidal anti-inflammatory drug. Research on the synthesis of lethal amatoxins in species of *Amanita, Galerina,* and other mushrooms shows promise for future drug designs. The activity of these cyclic peptides is unaffected by cooking, stomach acidity, and digestive enzymes. Structural resilience, swift absorption, and precise targeting are among the preferred properties of new generations of chemotherapeutic agents for treating cancer and other illnesses. The design of these drugs may be aided by understanding the mode of action of amatoxins and the process of cyclic peptide synthesis in these mushrooms.

Biofuels

Fungi show tremendous potential for the production of biofuels from a variety of crop plants, including the vast quantities of rice straw and other kinds of **lignocellulosic agricultural waste**. This is a subject of considerable interest given our seemingly limitless need for combustible fuels, limitations to the availability of fossil fuels, the environmental damage caused by oil and gas extraction, and climatic consequences of burning these hydrocarbons. Once sugars are hydrolysed from the parent materials, or feedstock, their fermentation to alcohol by yeasts is a straightforward process. This is the source of bioethanol generated from sugarcane in Brazil and from corn in the United States. Brazilian sugarcane is milled to separate fibre residue from a juice containing 10–15% sucrose. The juice is filtered and concentrated to yield crystallised sugar and molasses; the sugar is refined for the food industry and yeast is used to ferment the molasses to produce ethanol. The fibre residue, called bagasse, is burned to provide the heat and electrical energy to power the carbon neutral – or, at least, energy efficient – biofuel plant. Corn ethanol production is more complicated from a biological point of view, because the corn seeds are rich in starch that must be hydrolysed into sugars prior to fermentation by yeast. This is achieved using purified glucoamylases and alpha-amylases. Some of these enzymes are derived from fungi, including *Aspergillus niger*.

The more significant industrial challenge comes from the treatment of agricultural waste to generate fermentable sugars from the complex polymers that form the bulk of these materials. This is referred to as **second generation biofuel production**, and the raw materials are called cellulosics and lignocellulosics. Rice straw contains a mixture of cellulose, hemicellulose, and lignin. Pre-treatment of the feedstock is designed to optimise the release of sugars and a variety of approaches have been tested. Grinding and milling of the rice straw increases the surface area for subsequent chemical reactions and mixing the straw with alkali and acids increases its subsequent digestibility. The effectiveness of white rot fungi, including *Phanerochaete chrysosporium, Pleurotus ostreatus,* and *Trametes versicolor,* in breaking down the various polymers has been tested. Rice straw treated with *Pleurotus ostreatus* for 60 days showed a 40% reduction in lignin content as well as the partial digestion of cellulose and hemicellulose. Pretreatment of rice straw with *Aspergillus* and other fungi produced the highest ethanol yields, but none of the biological methods compared favourably with the effectiveness of chemical modification of the feedstuff. The discovery of new enzymes effective in lignocellulose degradation through genomic research is one area of research that may aid second generation biofuel production.

Bioremediation

Some of the basidiomycetes considered for second generation biofuel production have been selected as agents for the bioremediation of soil contaminated with **persistent organic pollutants** (POP). The reason that there is so much interest in this application of the white rot fungi is that the peroxidases and laccases that they secrete for lignin decomposition are also effective at degrading resistant aromatic contaminants that threaten human health. The target compounds include polycyclic aromatic hydrocarbons from the oil and gas industries; chlorinated compounds from wood preservatives, as well as discarded transformers and capacitors; halogenated compounds used as flame retardants, and nitroaromatic explosives used for mining and as weapons (Figure 12.10). Pesticides and other agrochemicals are another troubling source of soil contamination. White rot fungi are capable of breaking down compounds in all of these categories and impressive results have been obtained by inoculating sawdust and woodchips impregnated with petroleum hydrocarbons, TNT, chlorophenols, and polychlorinated bipenyls (PCBs). The efficacy of fungi in degrading these compounds in a natural setting is a subject of considerable commercial interest, but few large-scale studies have been published. Some of the largest trials have involved tubes of plastic mesh filled with pine bark and inoculated with fungi. These 'fungal tubes' are buried in the contaminated soil and aerated to promote growth. In addition to adding non-resident wood-decay and litter-decomposing fungi to contaminated sites, it may be possible to stimulate the activity of resident fungi by amending the soil with assorted nutrients. The addition of surfactants can

FIGURE 12.10 Oyster mushrooms (*Pleurotus ostreatus*) grown on soil contaminated with oil. The objective of this experimental form of bioremediation is to use the mycelium of the fungus to break down a variety of hydrocarbons in the oil and export the waste materials to the fruit bodies. If successful, harvesting and destruction of the fruit bodies would leave a cleaner soil. *Source: http://www.fungi.com/blog/items/the-petroleum-problem.html.*

increase the solubility of the contaminants and improve microbial access to the target chemicals. Another promising approach is to plant trees on contaminated sites and allow them to support mycorrhizal partners that are effective at cleansing the soil. The trees will recruit fungi from the existing soil microbiome and can be inoculated with specific mycorrhizal fungi before planting on the contaminated site.

Mycopesticides, Mycoherbicides, and Mycofungicides

Live fungi are used to combat insect pests of crop plants and to control the growth of weeds. *Beauveria bassiana* is an entomopathogen that infects a huge variety of insects and is used to control crop infestations by aphids, thrips, and whitefly. The fungus is cultured in solid state fermentation and formulations of its conidia are sprayed on plants as an emulsion or a wettable powder. *Metarhizium anisopliae* is a related ascomycete used as a mycopesticide against a variety of pests. One of the limitations of mycopesticides is that they do not work on contact, but require a few days to infect and kill the insect. Experiments to increase the virulence of *Metarhizium* include its transformation with a scorpion gene encoding a neurotoxin. A nice feature of this genetically modified fungus is that the neurotoxin gene is controlled by a promoter that is activated within the insect. This limits problems associated with human exposure to the transgenic fungus. Experiments on tobacco hornworm (*Manduca sexta*), and the mosquito, *Aedes aegypti*, which carries dengue and yellow fever, showed that the transgenic strain of the fungus was more effective at killing insects than the unmodified pathogen. *Beauveria* and *Metarhizium* belong to the Clavicipitaceae. *Lecanicillium lecanii* is a third fungus marketed as a biopesticide. It is a member of the family Cordycipitaceae.

Herbicides containing live fungi are called mycoherbicides. The fungi in these products are natural pathogens of the target weed plants. Fungi approved for use in mycoherbicides include *Alternaria destruens* for controlling dodder (*Cuscuta* species), *Colletotrichum gloeosporioides* for controlling low mallow and Virginia jointvetch (*Malva pusilla* and *Aeschynomene virginica*), and rusts (*Puccinia* species) that attack nut grass and woad (*Cyperus esculentus* and *Isatis tinctoria*). Chemical herbicides are very effective against weeds, but can persist in the environment and pollute groundwater. The development of herbicide resistance is another challenge associated with the use of conventional herbicides. Mycoherbicides are favoured by organic farmers who do not use synthetic herbicides. Potential uses of fungi for destroying opium poppy (*Papaver somniferum*), coca (*Erythroxylum coca* and *Erythroxylum novogranatense*), and marijuana (*Cannabis sativa*) have been investigated, but there is no evidence that drug control agencies have employed mycoherbicides. *Pleospora papaveracea* has been proposed as a mycoherbicide against Afghanistan's opium poppy crop. This fungus is a natural pathogen of poppies in Central Asia, which means that opium crops are already vulnerable to epidemic infection. This limits the value of *Pleospora papaveracea* as a mycoherbicide. Prospects for incorporating species of fungi that produce mycotoxins in herbicides are limited by concerns about the toxicity of these compounds toward humans and other animals.

Mycofungicides are a third category of agricultural product that contains live fungi. *Ampelomyces quisqualis* is a hyperparasite (parasite of a parasite) of powdery mildews whose potential as a biocontrol agent has been investigated for many years. A product containing spores of this species has been marketed as a biofungicide for treating powdery mildew of grape, strawberry, and roses.

Further Reading

Alba-Lois, L., Segal-Kischinevzky, C., 2010. Beer and wine makers. Nat. Educ. 3 (9), 17.

Barnett, J.A., Barnett, L., 2011. Yeast Research: A Historical Overview. ASM Press, Washington, DC.

Dickinson, J.R., Schweizer, M., 2004. Metabolism and Molecular Physiology of *Saccharomyces cerevisiae*, second ed. CRC Press, Boca Raton, FL.

Feldman, H., 2012. Yeast: Molecular and Cell Biology, second ed. Wiley, Weinheim, Germany.

Hofrichter, M. (Ed.), 2011. The Mycota, Volume 10, Industrial Applications, second ed. Springer-Verlag, Berlin, Heidelberg, New York.

New articles on fungal biotechology are published in the journal Fungal Biology and Biotechnology (http://www.fungalbiolbiotech.com). The journals FEMS Yeast Research and Yeast are dedicated to research on yeasts.

Appendix

Glossary of Common Mycological (and Related) Terms

The categories used in classifying fungi and the names of major groups are given in Chapter 1. Terms that are widely used in biology are not included. In addition, terms that are used only once in the text and explained are not included, nor are terms that can be located within the index. For terms from molecular biology and genetics, the reader is referred to dictionaries such as: Lackie, J.M. (ed.). 2013. The Dictionary of Cell and Molecular Biology, fifth edition. Academic Press, London.

Aeciospore binucleate infective spore produced by a rust fungus
Air spora the mixture of spores found in the air
Anamorph (Greek: *morphe*, shape) the form of a fungus produced in its asexual phase
Anastomosis fusion between hyphae
Antheridium (-a) male gametangium
Apoplast the space outside the plasma membrane
Apothecium (-a) cup-shaped fruit body of some ascomycetes
Appressorium adhesive pad formed by a pathogenic fungus on the surface of its host to aid penetration
Arbuscular mycorrhiza (AM) a mycorrhiza produced by species of Glomeromycota, in which highly branched haustorial structures are formed within host cells
Arthrospore (Greek: *arthron*, joint) spore formed by breakage of a length of mycelium into segments
Ascocarp a structure bearing asci, general term for the fruit body of an ascomycete
Ascogenous hypha the dikaryotic hypha emerging from an ascogonium after fertilisation, which gives rise to the asci in ascomycetes
Ascogonium cell of ascomycete protoperithecium that takes part in fertilisation
Ascoma (-ata) synonym for ascocarp
Ascospore sexual spore of ascomycetes
Ascus (Greek: *askos*, leather bag) the microscopic sac containing ascospores in ascomycetes
Axenic (Greek: *a*, not; *xenos*, stranger) in the absence of contamination, used to describe a pure culture
Ballistospore the actively discharged spore of basidiomycetes
Basidiocarp the fruit body of basidiomycetes
Basidioma (-ata) synonym for basidiocarp
Basidiospore sexual spore of basidiomycetes
Basidium (-a) the terminal cell of a hypha that bears basidiospores
Basidium initial cell that will become a basidium

Bipolar incompatibility requirement for different alleles at each of two genetic loci for sexual compatibility
Biomass the mass of living material
Biotroph (Greek: *bios*, life; *trophe*, food) fungus deriving its nutrients from living cells of a host
Breeding system see mating system
Cephalodia pockets of cyanobacteria in lichens
Chemostat continuous culture system in which the population of cultured cells is held constant by controlling the rate of nutrient supply
Chemotaxis taxis towards or away from the source of a specific chemical
Chemotropism tropism to or from a specific chemical
Chitin (Greek: *chiton*, coat of mail/garment) a structural component of the fungal cell wall, a polymer of *N*-acetylglucosamine
Chlamydospore (Greek: *chlamys*, cloak) thick-walled, usually asexual, resting spore
Clamp connection short, backwardly directed side branch formed at the time of septum formation in basidiomycetes
Cleistothecium (-a) (Greek: *cleistos*, closed) fruit body of ascomycetes in which the asci are entirely surrounded by a wall of hyphae
Colony an assemblage of hyphae that often develops from a single source and grows in a coordinated way. Synonymous with **mycelium**
Conidiophore (Greek: *phoreo*, I bear) a hypha that gives rise to conidia
Conidium (-a) an asexual spore produced on the surface of a mycelium, not within a sporangium
Contamination growth of unwanted microbes in cultures that should contain a single species
Coprophilous (Greek: *copros*, dung; *phileo*, I love) dung inhabiting
Cryptic species closely related species that are genetically distinct but cannot easily be separated morphologically
Cyst a spherical cell, derived from the swimming spores of zoosporic fungi by cell wall formation (encystment)
Dermatophyte (Greek: *derma*, skin; *phyton*, a plant) a fungus infecting the skin
Dikaryon (Greek: *dis*, two; *karyon*, nut) mycelium containing two genetically different types of nuclei. Usually refers to basidiomycetes with two nuclei of different mating type in each hyphal compartment
Dikaryotization formation of a dikaryon by fusion and nuclear migration between monokaryons
Dimorphism the same species exists in two forms that differ in appearance
Dolipore septum a septum with elaborate pore structure found in basidiomycetes
Duplicative transposition duplication of a DNA sequence followed by insertion of one copy at a different site in the genome
Effectors proteins produced by pathogenic and mycorrhizal fungi that modulate plant immunity and enable colonisation of the host plant
Elicitor a substance derived from a plant pathogenic fungus that induces a plant to resist infection
Encystment formation of a tough wall around a zoospore to form a cyst
Endophyte (Greek: *endon*, within; *phyton*, a plant) fungus that inhabits plant tissues without damaging host
ESTs (expressed sequence tags) partial sequence reads from the 5′- or 3′-end of cDNA clones
Facultative possible, but not obligatory
Fermentation form of catabolism not requiring oxygen or other external electron acceptor. Also used more loosely to describe the chemical transformation of any substrate by the growth of a microorganism
Fermenter vessel used for producing a microbial product by fermentation
Filamentous fungus a fungus with hyphae, not unicellular like yeast
Fruit(ing) body the large spore-bearing structure in ascomycetes and basidiomycetes (e.g. mushrooms, truffles)
Gametangium part of a hypha specialised for fusion in sexual reproduction
Gametes haploid reproductive cells that fuse to form a zygote during sexual reproduction
Germ tube the hypha that emerges from a spore
Hartig net network of ectomycorrhizal hyphae between root cortex cells
Haustorium the part of a symbiotic fungus that feeds inside a host cell
Hemibiotroph (literally half biotroph) a pathogen that establishes itself as a biotrophic parasite within host tissue and later switches to a necrotrophic lifestyle
Heterokaryon hyphae or mycelium containing nuclei of two or more genotypes
Heterothallism (Greek: *heteros*, different) requirement for two compatible mating types for the sexual process. Synonymous with self-sterility
Heterozygous having two different alleles at one or more corresponding loci

Homogenic incompatibility incompatible for mating within a species due to identical mating type alleles
Heterogenic compatibility requirement for different alleles at mating type loci for mating to occur
Homokaryon a mycelium or hypha with nuclei of only one genotype
Homothallism (Greek: *homo*, the same) no requirement for a second mating type for the sexual process. Synonymous with self-fertility
Homozygous a diploid having the same alleles at a locus
Horizontal resistance a form of disease resistance in plants that gives some protection from attack by all strains of a fungal pathogen. Contrast with vertical resistance
Horizontal transmission spread from one organism to another, but not from parent to offspring
Host organism in a parasitic symbiosis that supports the growth of the parasite. Term can also apply to partners in commensal and mutualistic symbioses.
Hydrophobin (Greek: *hydro*, water; *phobos*, fear) fungal protein that can render the hyphal surface unwettable
Hymenium (Greek: *hymenaeos*, wedding) tissue layer of a fruit body on which sexually produced spores are borne
Hypha (Greek: *hypha*, thread) the tubular cell growing at one end which is the developmental unit of the mycelium
Isolate (noun, from verb, to isolate) a strain of a fungus isolated from nature and, often, grown in pure culture
Karyogamy (Greek: *karyon*, nut; *gamos*, wedding) fusion of nuclei preceding the production of sexually-produced spores
Macrofungi fungi that produce large fruit bodies, mostly basidiomycetes and some ascomycetes
Mating system a genetic system that determines whether or not individuals of the same species can mate
Mating type the factor determining whether a strain will or will not be able to mate with another strain
Medium a preparation used for culture of fungi or other microbes. Contains nutrients dissolved in water, and used either in liquid form or gelled with agar
Meiospore spore produced following meiosis
Metapopulation a group of spatially separate populations of the same species
Mildew a plant disease with prominent surface growth of the fungus. Powdery mildews are produced by Erysiphales (Ascomycota), and downy mildews are caused by Peronosporales (oomycetes).
Mitospore a spore produced by mitosis
Monokaryon (Greek: *karyon*, a nut; *monos*, alone) hypha or mycelium with nuclei of a single genotype
Mutualism an interaction that confers a selective advantage on both participants
Mycelial strand or cord linear aggregate of hyphae formed behind an advancing margin in which the hyphae are separated as a fan
Mycelium (Greek: *mykes*, fungus) the mass of hyphae, not in the form of large structures such as mushrooms, of which the fungi are mainly composed. Synonymous with **colony**
Mycorrhiza (pl., strictly mycorrhizae, now usually mycorrhizas) symbiosis between plant root and fungal mycelium
Mycoparasitism parasitism of one fungus by another
Necrotroph (Greek: *necros*, death; *trophe*, food) fungus that kills the cells of a living host and subsequently utilises their remains for food
Obligate the opposite of facultative, a condition in which the fungus has no alternative state
Oidium (-a) a type of asexual spore. The term is used most often spores produced on monokaryons of basidiomycetes that bring about dikaryotization of other monokaryons of the same species
Oospore sexual spore produced by oomycetes
Parasexuality sequence of nuclear fusion and irregular division accompanied by genetic recombination found to occur in some otherwise asexual fungi
Pathotype classification of a pathogen distinguished from other members of the same species by its ability to cause disease in particular host species
Pellets multihyphal structures formed when some fungi are grown in fermenters
Perithecium (-a) small bottle-shaped fruit body of some ascomycetes, from the neck of which one ascus discharges at a time
Petri dish shallow transparent dish with lid for the culture and observation of fungi and other microbes. Interchangeable with culture plate
Phylloplane (Greek: *phyllon*, leaf) microhabitat close to the surface of a leaf occupied by a distinctive population of fungi and other microorganisms
Phylotype group of organisms, described at any level of classification, characterised by a particular level of genetic similarity (typically 97% homology)
Phytoalexin (Greek: *phyton*, plant; *alexo*, defence) substance produced by damaged plants that inhibits fungal growth

Phytoanticipin low-molecular-weight antimicrobial compound produced constitutively by plants
Pileus (Latin: *pileus*, felt hat) the cap of a mushroom
Plasmodium (-a) mass of protoplasm formed by slime moulds
Plasmogamy (Greek: *plasma*, a thing moulded/formed; *gamos*, a wedding) the fusion of cytoplasm from two different hyphae that precedes nuclear fusion during the sexual cycle. In basidiomycetes, plasmogamy and nuclear fusion can be separated by a long time interval.
Ploidy the number of sets of chromosomes in the nucleus of a cell
Primary homothallism homothallism where there is no evidence of a heterothallic ancestor
Primary production the synthesis of organic compounds from carbon dioxide
Primordium the earliest visible stage in the development of a structure
Protoperithecium (-a) the structure produced by ascomycetes that is the site of fertilisation and subsequent fruit body development
Protoplast spherical blob of protoplasm produced by the removal of the fungal cell wall
Pycniospore small spore (gamete) produced by rust fungi whose function is to dikaryotize the mycelium by fusing with receptive hyphae
Quorum sensing a type of decision-making process used by groups of cells to coordinate gene expression and behaviour
Race an informal taxonomic rank, below the level of a species but higher than strain.
Radial growth growth from a centre. The radial growth rate of a colony is the rate at which the hyphal margin advances
Resource unit restricted fungi fungi that are only able to spread to new sources of food via spores
Resting spore a spore with prolonged survival as its main role, or a spore that is in a state of dormancy
Rhizoid a branched hypha that functions like a root in anchoring mycelium growing on a surface
Rhizomorph multihyphal fungal structure (organ) with a root-like apex
Rhizosphere microhabitat close to the surface of a root occupied by a distinctive population of fungi and other microorganisms
Rust fungus basidiomycete in the sub-phylum Pucciniomycotina that causes plant disease and produces reddish urediniospores
Saprotrophic (Greek: *sapros*, rotten; *trophe*, food) using remains of dead organisms as food
Secondary homothallism homothallism, or self-fertility, which has developed from an earlier heterothallic condition
Sclerotium (-a) mass of hyphae with protective rind and containing food reserves
Smut fungus basidiomycete in the sub-phylum Ustilaginomycotina that cause plant disease. Many species produce masses of black spores in infected plant tissues
Somatic (in)compatibility the (in)ability of a fungal thallus to fuse with another of the same species and to then operate as an individual
Somatogamy the sexual fusion of structures which are morphologically no different from other vegetative structures
Soredium (-a) powdery propagule of lichens composed of hyphae wrapped around photobiont cells
Sp., spp. Abbreviations, sing. and pl., for species, used with a generic name. For example, *Agaricus* sp. means a species of *Agaricus*, and *Agaricus* spp. means various species of *Agaricus*.
Speciation the evolutionary process by which new species arise
Spermatium (-a) non-motile male gamete characteristic of rust fungi (syn. Pycniospore)
Spermatization fusion of spermatia with a receptive hypha in pustule or spermagonium of a rust fungus
Spitzenkörper the organelle at the hyphal tip that plays a central role in hyphal growth
Sporangiophore a stalk that bears a sporangium
Sporangiospore asexual spore produced in a sporangium
Sporangium (-a) sac containing sporangiospores
Sporophore a structure that bears spores. This term is used for mushrooms and other sexual fruit bodies, and also for small structures that bear asexual spores
Sporulation the process of forming spores
Sterigma (-ata) microscopic projection from the end of a basidium that bears a basidiospore
Stipe the stem of a mushroom or toadstool
Strain a genetic variety of a fungus, either an isolate from nature or arising by mutation or recombination in the laboratory
Stroma (-ata) mass of hyphae on which spores or fruit bodies are borne

Substratum (-a) the physical surface on or within which mycelium grows and feeds
Symbiosis an intimate relationship between two organisms. It can be mutualistic, parasitic or commensalistic
Teleomorph (Greek: *teleos*, finished; *morphe*, form) the form of a fungus when it produces sexual spores
Teliospore spore formed by rusts and smuts
Tetrapolar incompatibility requirement for different alleles at each of two genetic loci for sexual compatibility. See Bipolar incompatibility
Thallus the body of the fungus, usually applied to a mycelium or lichen
Translocation transport of nutrients within mycelium by processes other than those of growth
Trichogyne receptive hypha involved in ascomycete fertilisation
Trophic of or having to do with nutrition
Tropism (Greek: *tropos*, turn) the bending of a hypha or fruit body towards or away from a stimulus
Urediniospore (Latin: *uredo*, blight upon plants) dikaryotic spore of rust fungi, synonymous with uredospore
Vegetative mycelium mycelium involved in feeding rather than reproduction
Vegetative incompatibility inability of different mycelia of the same species to fuse successfully and function as a single colony
Vertical resistance a form of resistance by plants to fungal attack that gives resistance against some strains of a pathogen but not against others. Contrast with horizontal resistance
Vertical transmission transmission directly from parent to offspring
Virulence factor a feature of a pathogen enabling colonisation of the host and evasion of the host's defense response
Water activity, Water potential measures of water availability
Yeast depending on context, can mean baker's and brewer's yeast (*Saccharomyces cerevisiae*) or any unicellular fungus multiplying by budding, or, in a few instances, fission
Zoosporangium (-a) sac in which zoospores develop, and from which they are released
Zoospore (Greek: zoos, living) spore which swims in water using one or two flagella
Zygophore a specialised hyphal branch that gives rise to a gametangium in zygomycetes
Zygospore (Greek: *zygon*, yoke) spore formed by fusion of two gametangia in zygomycetes

Index

Note: Page numbers followed by *f* indicate figures, *t* indicate tables and *ge* indicate glossary.

D

Q

R

T

U

V

W

X

Y

Z

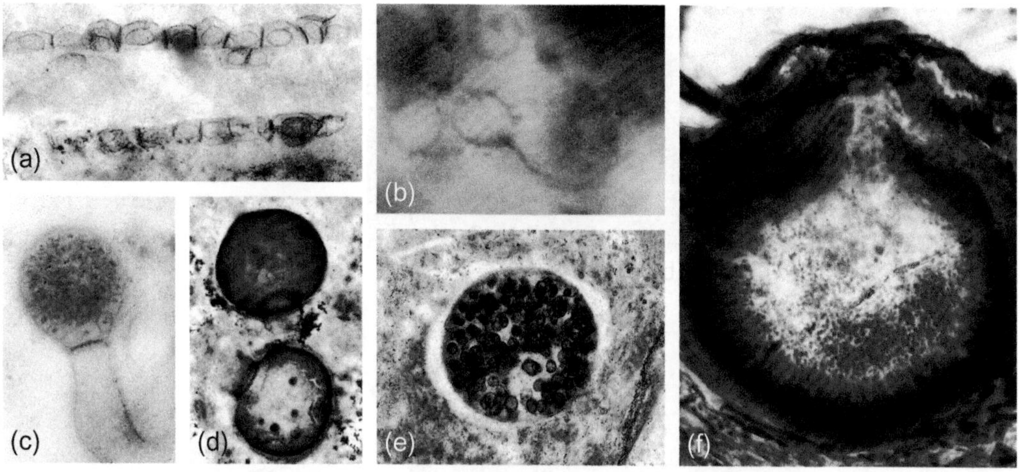

FIGURE 1.2 Fossil fungi in Lower Devonian Rhynie chert. (a) Zoosporangia of a chytrid inside host cells. Note exit tubes in two of the sporangia, through which zoospores were expelled. (b) Chytrid zoospore with (putative) single flagellum. (c) Zoosporangium of *Palaeoblastocladia milleri* (Blastocladiomycota). (d) Sporangia of a zygomycete fungus. (e) Sporocarp of species of arbuscular mycorrhizal fungus containing numerous spores. (f) Perithecium of *Palaeopyrenomycites devonicus* (Ascomycota). *Source: Thomas Taylor, University of Kansas.*

FIGURE 1.4 Surprising relatives. (a, b) Field mushroom, *Agaricus arvensis* and puffball, *Lycoperdon perlatum*. (c, d) Bolete, *Boletus pinophilus* and earth-ball, *Scleroderma michiganense*. *Source: (a, c, d) Michael Kuo and (b) Pamela Kaminskyj.*

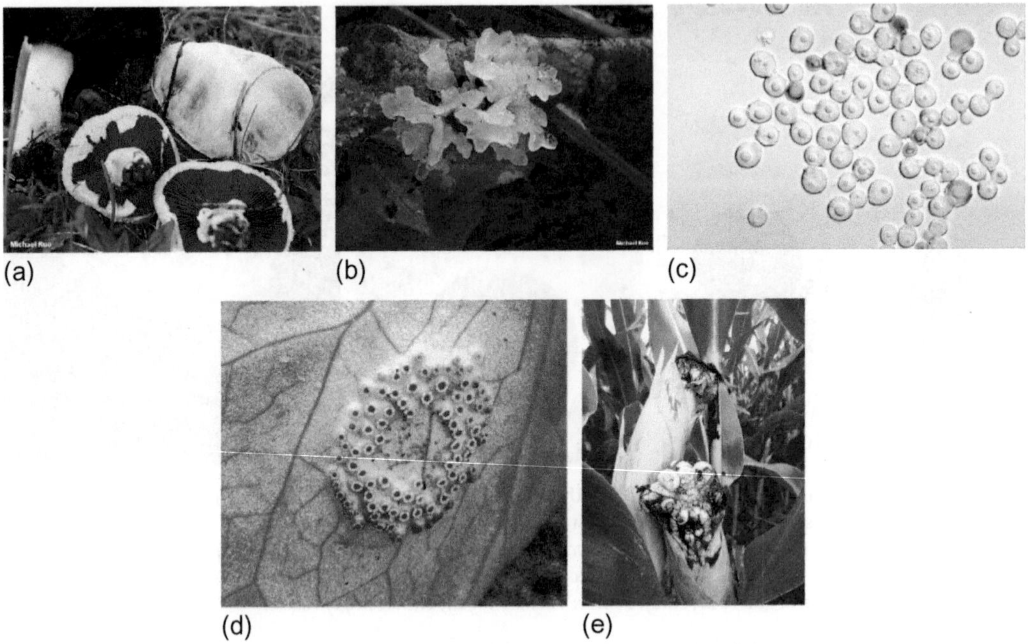

FIGURE 1.6 Diversity of the Basidiomycota. (a) *Agaricus campestris*, the field mushroom. (b) *Tremella fuciformis*, a jelly fungus. (c) *Malassezia globosa*, yeast associated with dandruff. (d) *Puccinia sessilis*, a rust, on *Arum maculatum*. (e) *Ustilago maydis*, corn smut. *Source: (a, b) Michael Kuo, (c) http://www.pfdb.net/photo/nishiyama_y/box20010917/wide/024.jpg, (d) http://upload.wikimedia.org/wikipedia/commons/f/f8/Puccinia_sessilis_0521.jpg, and (e) http://aktuell.ruhr-uni-bochum.de/mam/images/pi2012/begerow_maisbeulenbrand.jpg*

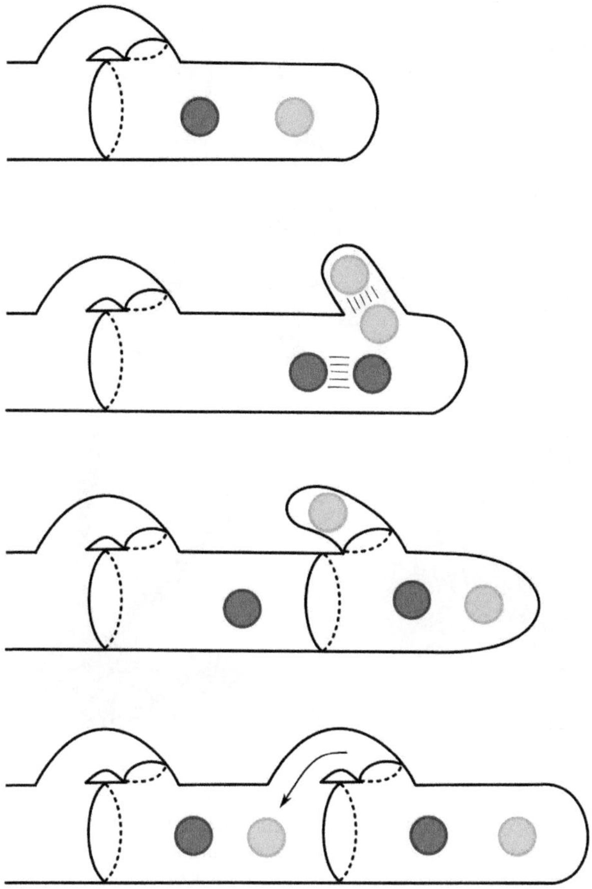

FIGURE 1.8 Formation of clamp connections on heterokaryotic basidiomycete hypha. *Source: Creative Commons.*

(a) (b) (c) (d)

FIGURE 1.13 Examples of fruit bodies, or ascomata, produced by Ascomycota in Subphylum Pezizomycotina. (a) Goblet shaped apothecia of *Urnula craterium*, the Devil's urn. (b) Flattened discoid apothecia on the thallus of a species of the lichen *Xanthoparmelia*. (c) Highly modified apothecium of the summer truffle, *Tuber aestivum*. (d) Perithecial stroma of *Cordyceps militaris* fruiting from parasitized caterpillar. *Source: (a) Michael Kuo, (b) http://en.wikipedia.org/wiki/Lichen#/media/File:Lichen_reproduction1.jpg, (c) http://upload.wikimedia.org/wikipedia/commons/8/89/Tuber_aestivum_Valnerina_018.jpg, (d) http://upload.wikimedia.org/wikipedia/commons/4/44/2008-12-14_Cordyceps_militaris_3107128906.jpg*

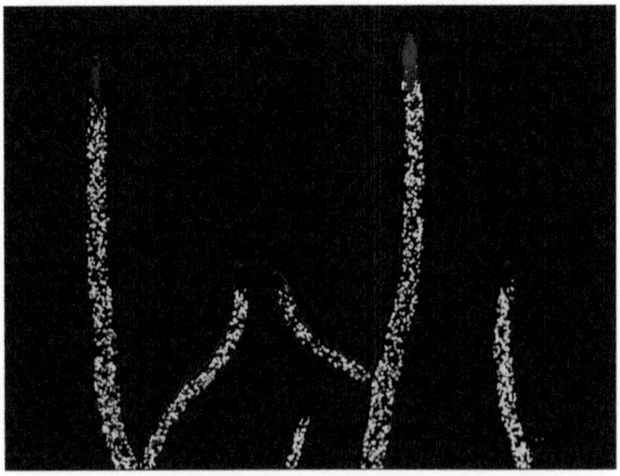

FIGURE 2.9 Confocal image showing tip-growing and branched, multinucleate hyphae of *Neurospora crassa*. The nuclei are shown in green (nuclear-targeted GFP) and membranes, especially the plasma membrane and secretory vesicles, are shown in red (stained with FM4-64). *Source: www.fungalcell.org*

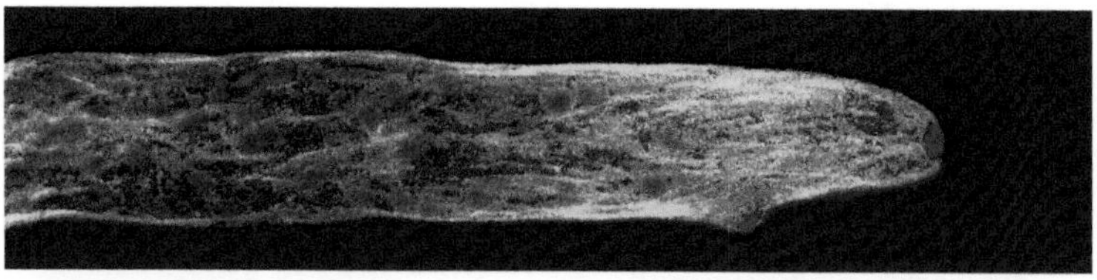

FIGURE 2.11 Confocal image of a growing hypha expressing ß-tubulin-GFP, localised in microtubules (green), and co-labelled with FM4-64 to show distribution of membranes (red). Microtubules extend into the negatively stained Spitzenkörper at the tip. A subapical swelling that will become a hyphal branch is highlighted with a concentration of vesicles that will become a separate Spitzenkörper. *Source: Patrick Hickey.*

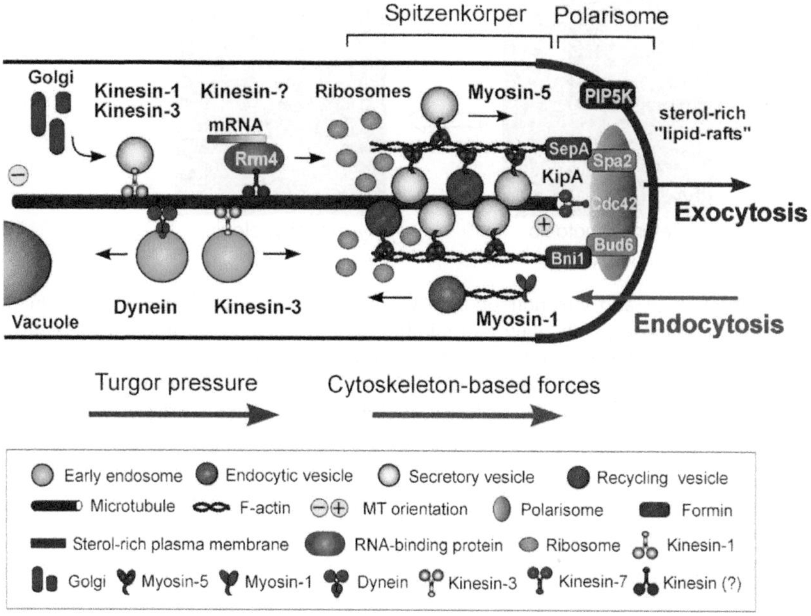

FIGURE 2.14 Model of tip growth showing some of the molecular components in the extending hypha. *Source: Steinberg, G., 2007. Hyphal growth: a tale of motors, lipids, and the Spitzenkörper. Euk. Cell 6, 351–360.*

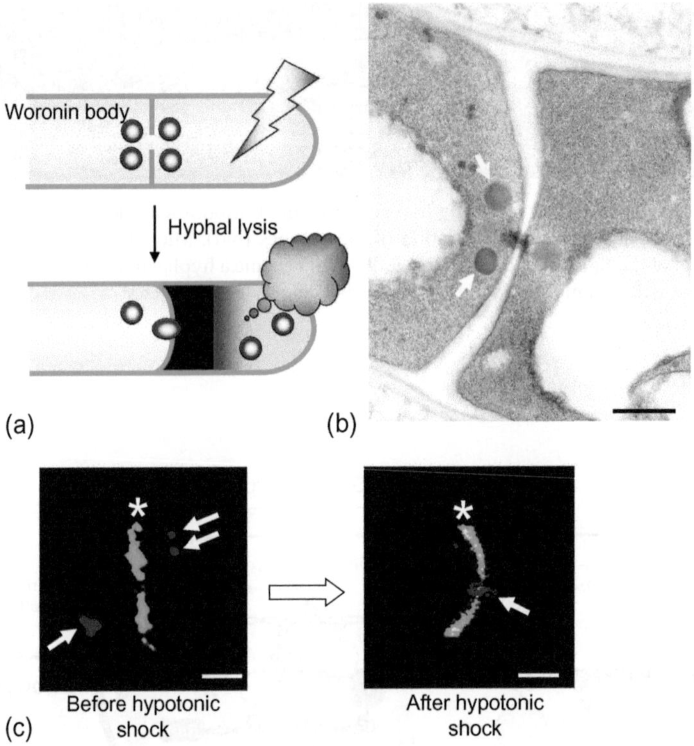

FIGURE 2.15 Morphology and function of the Woronin body. (a) Schematic of Woronin body function. (b) Transmission electron micrograph showing Woronin bodies (arrows) in *Aspergillus oryzae*. Scale bar: 500 nm. (c) Confocal images of Woronin bodies (red arrows) and septa (green asterisks) stained with fluorescent labels before (left) and after (right) hyphal tip bursting. Scale bar = 2 μm. *Source: Maruyama, J., Kitamoto, K., 2013. Expanding functional repertoires of fungal peroxisomes: contribution to growth and survival processes. Front. Physiol. 4, 177.*

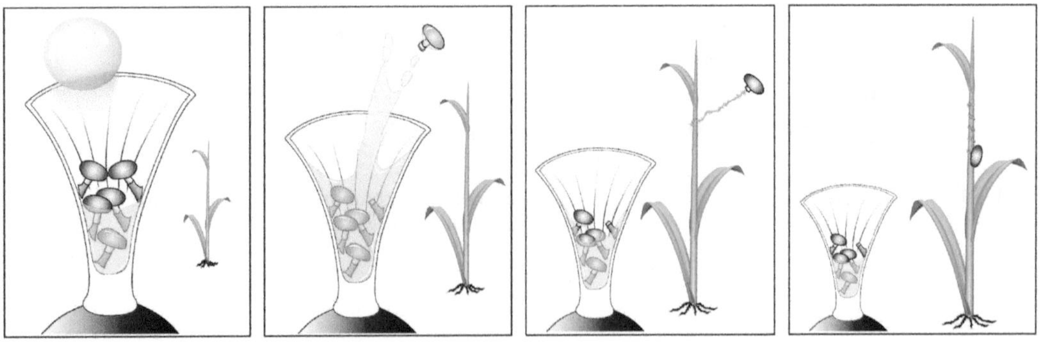

FIGURE 3.8 Diagram showing splash discharge of peridioles of the bird's nest fungus and their mechanism of attachment to vegetation. The funicular cord is packed within a purse before discharge. The force of the raindrop fractures the purse leaving the sticky end of the cord exposed during the flight of the peridiole. Deployment of the funicular cord occurs when the hapteron contacts an obstacle. The process is completed in less than 200 ms. *Source: Hassett, M.O., et al., 2013. Splash and grab: biomechanics of peridiole ejection and function of the funicular cord in bird's nest fungi. Fungal Biol. 117, 708–714.*

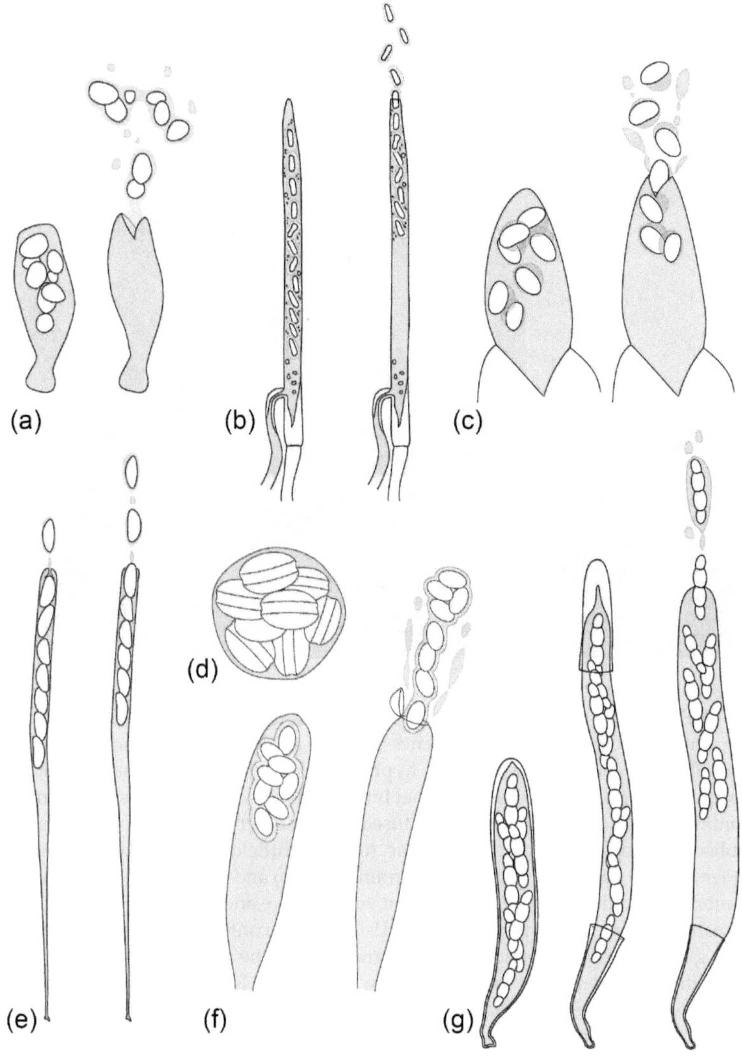

FIGURE 3.9 Diversity of ascus morphology and ascospore discharge mechanisms. (a) *Taphrina deformans* (Taphrinomycotina). Asci of this plant pathogen are exposed on the surface of peach leaves. They split open at the tip and discharge multiple spores in a single shot. (b) *Dipodascus macrosporus* (Saccharomycotina) produces ascospores with a mucilage coating. The spores are extruded slowly through the torn apex of the ascus. This fungus forms exposed asci. (c) Single asci of the powdery mildew *Podosphaera pannosa* (Pezizomycotina, Erysiphales) form in chasmothecia (one ascus per ascoma) and open via a slit at the apex. (d) Multiple non-explosive asci of *Emericella nidulans* (Pezizomycotina, Eurotiales) form inside cleistothecia. The ascospores of this fungus are ornamented with a double flange. (e) Asci of *Xylaria hypoxylon* (Pezizomycotina, Xylariales) discharge spores through constricting ring (apical apparatus). Ascomata of this fungus are perithecia. (f) Operculate asci of *Ascobolus immersus* (Pezizomycotina, Pezizales) expel spores when lid or operculum flips open. Ascomata of this species are apothecia. (g) Ascospore discharge from fissitunicate asci of *Pleospora herbarum* (Pezizomycotina, Pleosporales) occurs after outer wall of ascus ruptures to allow expansion of the inner wall. The fruit body of this species is a pseudothecium. *Source: Mark Fischer, Mount St. Joseph University, Cincinnati.*

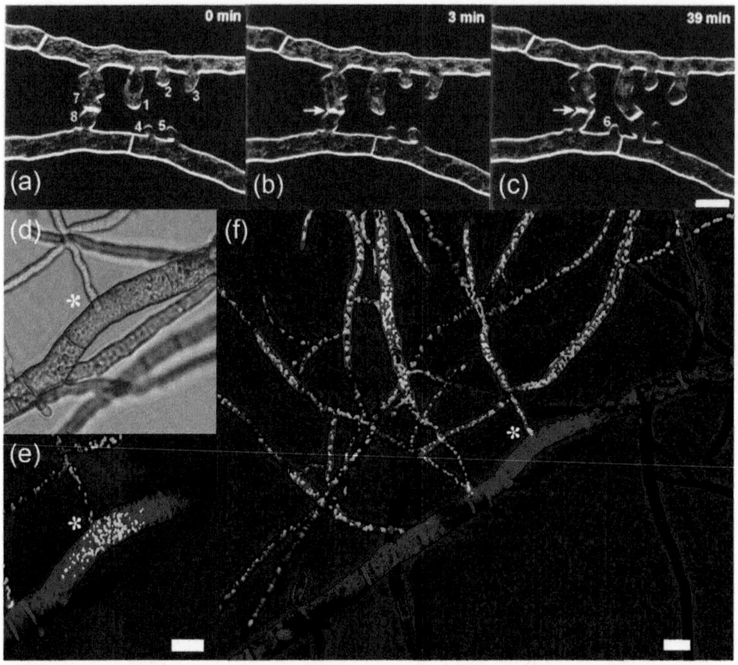

FIGURE 4.3 Vegetative compatibility and incompatibility. Stages of the hyphal fusion process (a–c) and non-self rejection responses (d–f) in *Neurospora crassa* viewed with confocal microscopy. (a–c) Show induction, homing, and fusion of hyphae in a mature colony. (a) Three branches (labelled 1, 2, and 3) from one hypha have started growing towards two short branches (4 and 5) on the opposite hypha. Two branches (7 and 8) have already fused. (b) A fusion pore has started to open between the two fused hyphal branches (arrowed), and in (c) it is completely open, allowing cytoplasmic continuity. (c) Branches 1 and 4 have fused, and another side branch (6) is developing. Bar = 10 μm. (d–f) Fusion and subsequent rejection of non-self hyphae. (d) A brightfield image showing a thin hypha of one strain fusing with the underside of a wide hypha of another strain (*). In (e) and (f) the same interacting hyphae have been labelled with a membrane selective red (grey in the print version) dye and one of the strains has been labelled with a nuclei selective green (light grey in the print version) H1-GFP. (e) A confocal image of (d) showing that nuclei (fluorescing green (light grey in the print version)) have migrated from the narrow hypha into the wider one. (f) One hour later incompatibility is evident. The compartment where fusion has occurred is stained intensely red (grey in the print version) due to increased permeability of the plasma membrane, and the nuclei have broken down, as evidenced by lack of green (light grey in the print version) fluorescence. *Source: (a–c) Hickey, P.C., Jacobson, D.J., Read, N.D., Glass, N.L., 2002. Live-cell imaging of vegetative hyphal fusion in Neurospora crassa. Fungal Genet. Biol. 37, 109–119. (d–f) Read and Roca, 2006.*

FIGURE 5.6 The appearance of wood decayed by: (a) a white rot fungus. This branch has been selectively delignified, leaving a pale, fibrous, cellulose-rich residue. (b) The remains of a fallen tree decomposed by a brown rot fungus. The decomposed wood breaks apart in a cubical pattern and then crumbles to a brown, lignin-rich residue. Inset figure: close-up of brown rotted wood. *Source: (c), Stuart Skeates.*

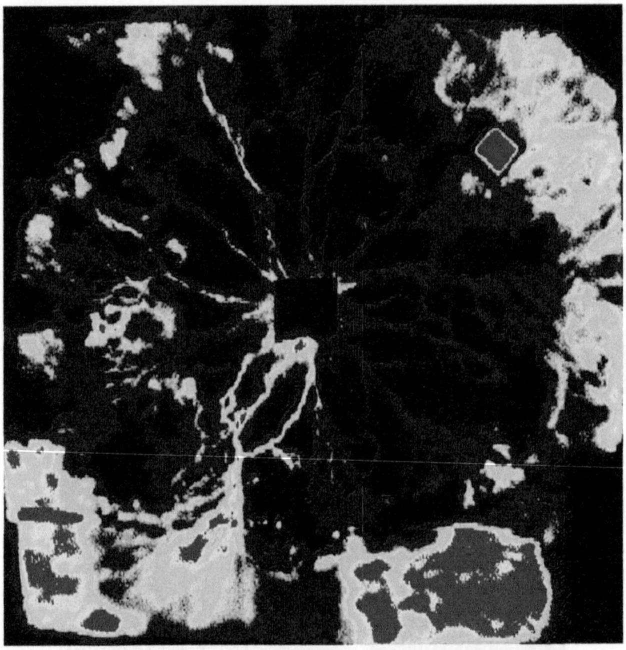

FIGURE 5.7 Photon counting scintillation imaging of ^{14}C-AIB in the mycelium shown in Figure 5.18, during new wood resource capture (Further Reading in Tlalka et al., 2008, Video S2) Video link, see Further Reading. The mycelium shown in Figure 5.18 was supplied with the non-metabolised tracer amino acid AIB (α-aminoisobutyric acid) labelled with ^{14}C. After incubation to allow the AIB to become evenly distributed within the mycelium, a fresh wood block was placed near the colony margin, top right, and the system was imaged by PCSI for the ensuing 10 days. The video shows AIB redistribution from the whole mycelium into the site of colonisation of fresh wood. This is consistent with an ability of the mycelium to sense and respond to a localised fresh carbon supply by importing nitrogen, to achieve a balanced internal C/N ratio for biosynthesis. *Source: Tlalka et al. (2008). http://dx.doi.org/10.5072/bodleian:d217qq90r*

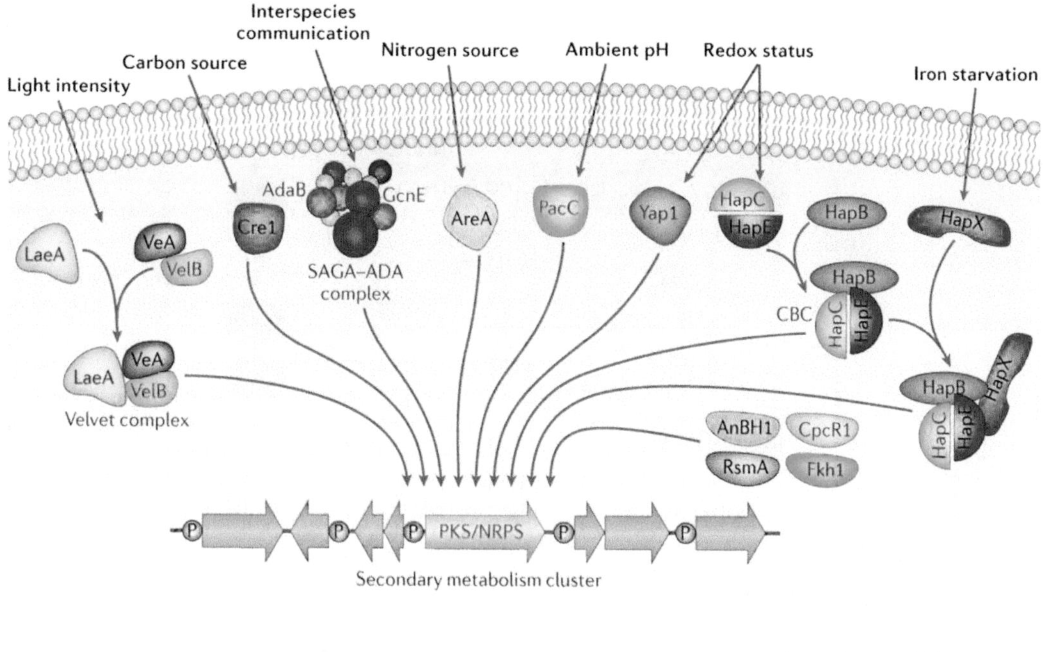

FIGURE 5.13 Regulation of secondary metabolism by environmental factors. Environmental signals can influence the regulation of various secondary metabolism gene clusters through regulatory proteins that respond to these environmental stimuli and, in turn, modulate the expression of the clusters. Shown is a model secondary metabolism gene cluster containing a gene encoding a central non-ribosomal peptide synthetase (NRPS), a polyketide synthase (PKS), or a hybrid PKS–NRPS enzyme. CBC, CCAAT-binding complex; CpcR1, cephalosporin C regulator 1; LaeA, loss of aflR expression A; RsmA, restorer of secondary metabolism A; SAGA–ADA, Spt–Ada–Gcn5–acetyltransferase–ADA. *Source: Brakhage (2013).*

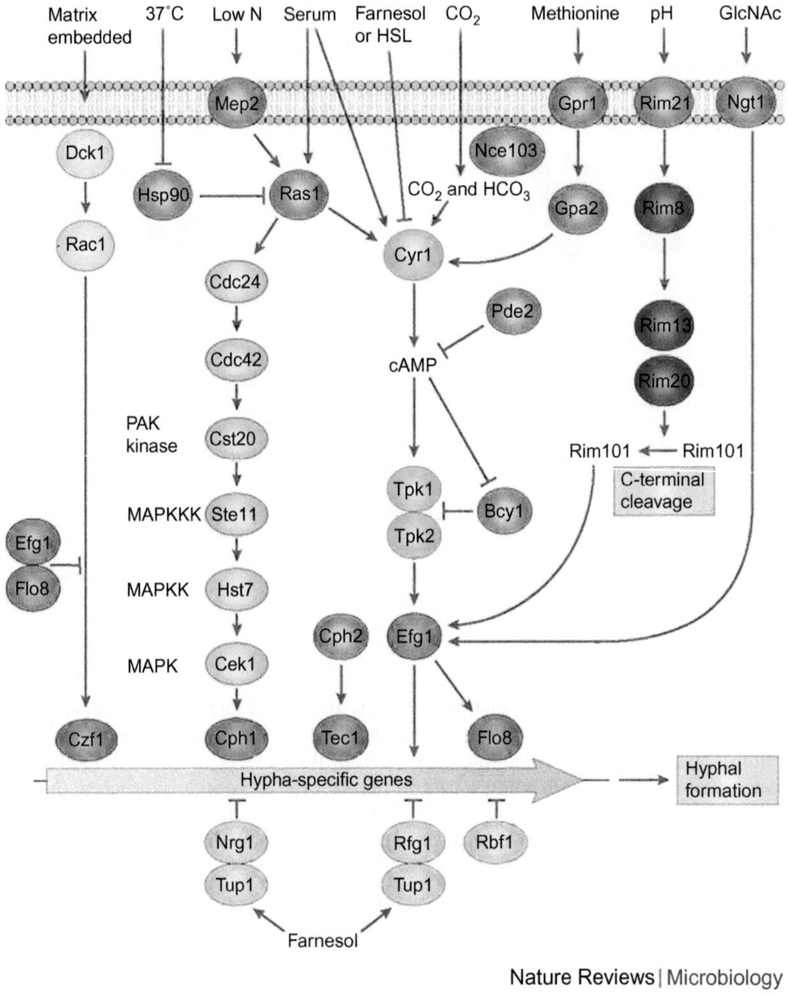

FIGURE 5.15 Signal transduction pathways leading to expression of hypha-specific genes in *Candida albicans*. Multiple sensing and signalling mechanisms that regulate the developmental response of the pathogenic yeast *Candia albicans* to its environment. Environmental cues feed through multiple upstream pathways to activate a panel of transcription factors. The cyclic AMP-dependent pathway that targets the transcription factor enhanced filamentous growth protein 1 (Efg1) is thought to have a major role. In this pathway, adenylyl cyclase integrates multiple signals in both Ras-dependent and Ras-independent ways. Negative regulation is exerted through the general transcriptional corepressor Tup1, which is targeted to the promoters of hypha-specific genes by DNA-binding proteins such as Nrg1 and Rox1p-like regulator of filamentous growth (Rfg1). Protein factors are colour-coded as follows: mitogen-activated protein kinase (MAPK) pathway (green, dark grey in the print version), cAMP pathway (turquoise, dark grey in the print version), transcription factors (orange, grey in the print version), negative regulators (yellow, light grey in the print version), matrix-embedded sensing pathway (light blue, grey in the print version), pH sensing pathway (brown, dark grey in the print version), other factors involved in signal transduction (mauve, dark grey in the print version), C-terminal, carboxy-terminal; Cdc, cell division control; GlcNAc, *N*-acetyl-D-glucosamine; Gpa2, guanine nucleotide-binding protein α-2 subunit; Gpr1, G-protein-coupled receptor 1; HSL, 3-oxo-homoserine lactone; Hsp90, heat shock protein 90; MAPKK, MAPK kinase; MAPKKK; MAPKK kinase; PAK, p21-activated kinase; Rbf1, repressor–activator protein 1. *Source: Sudbery (2011).*

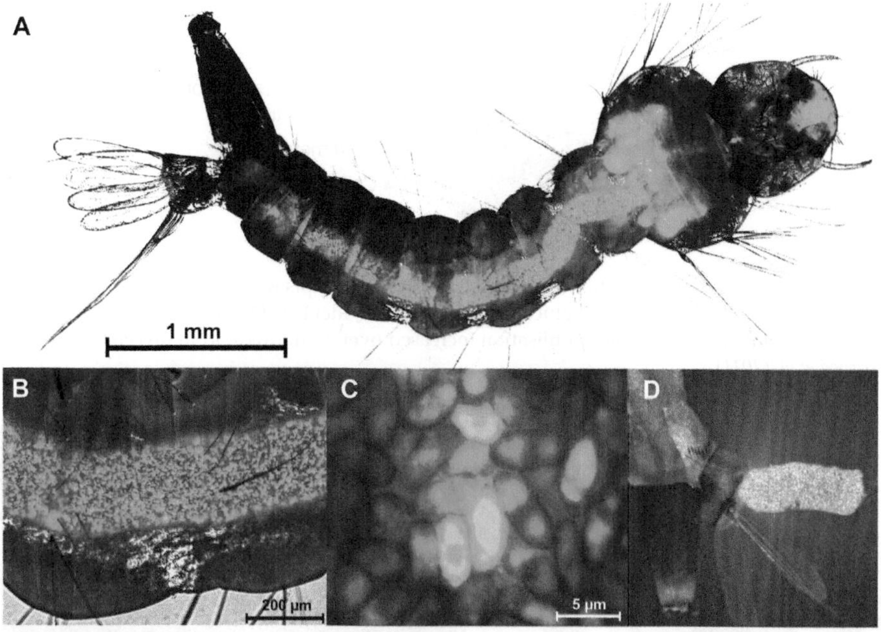

FIGURE 6.4 The insect pathogen *Metarhizium brunnei*, transformed with green fluorescence protein (GFP) and visualised by fluorescence microscopy within the body of its host, a larva of the mosquito *Aedes aegypti*. Species of *Metarhizium* are being investigated for biological control of mosquitoes that are vectors of human disease, including malaria, dengue, and yellow fever. (a) conidia in the gut lumen of the larva, (b, c) conidia expressing GFP in the gut lumen, and (d) active conidia in faecal pellets. *Source: Butt et al. (2013).*

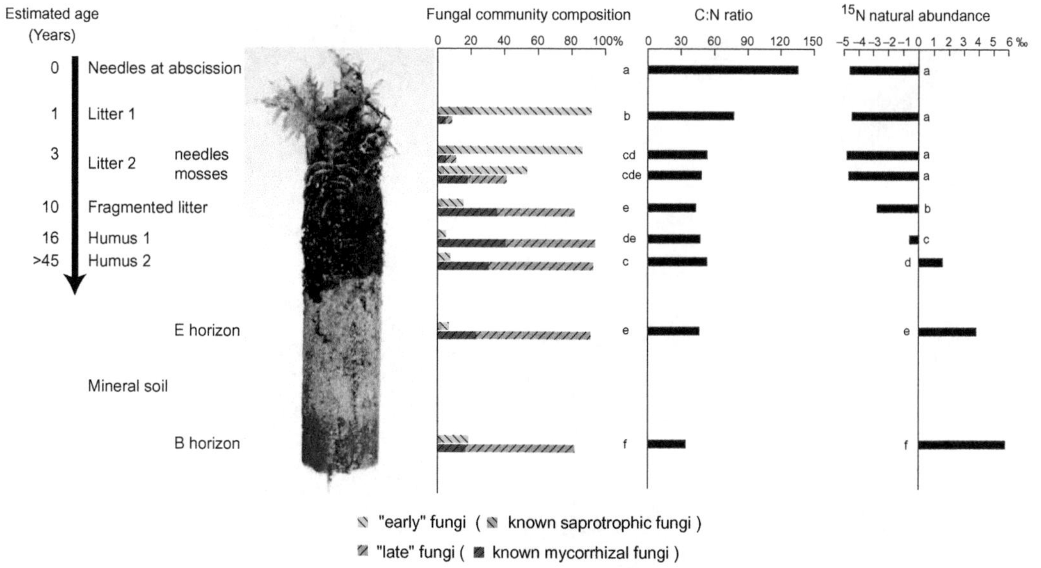

FIGURE 6.5 Fungal community composition, carbon:nitrogen (C:N) ratio and ^{15}N natural abundance throughout the upper soil profile in a Scandinavian *Pinus sylvestris* forest. Molecular methods for identification of fungi, combined with chemical analyses of carbon and nitrogen, indicated different roles of saprotrophic and ectomycorrhizal fungi in the carbon and nitrogen dynamics of separate soil horizons. *Source: Lindahl et al. (2007).*

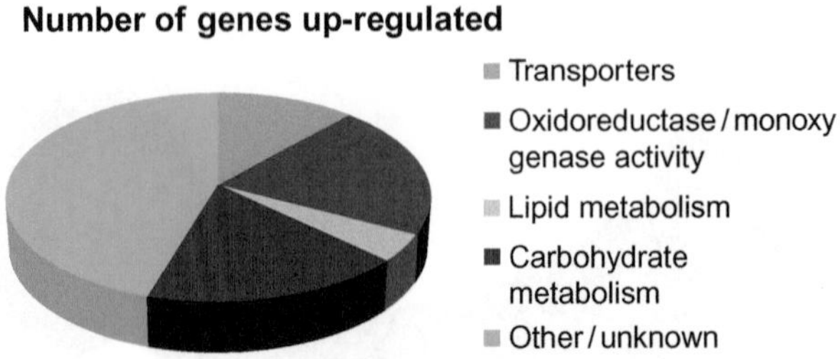

FIGURE 6.7 Functional characterisation of *Serpula lacrymans* transcripts with significantly increased gene expression when grown on wood compared with glucose-based medium, identified by microarray analysis. Expression of glycoside hydrolases involved in cellulose utilisation increased over a hundredfold in mycelium feeding on wood. *Source: Eastwood et al. (2011).*

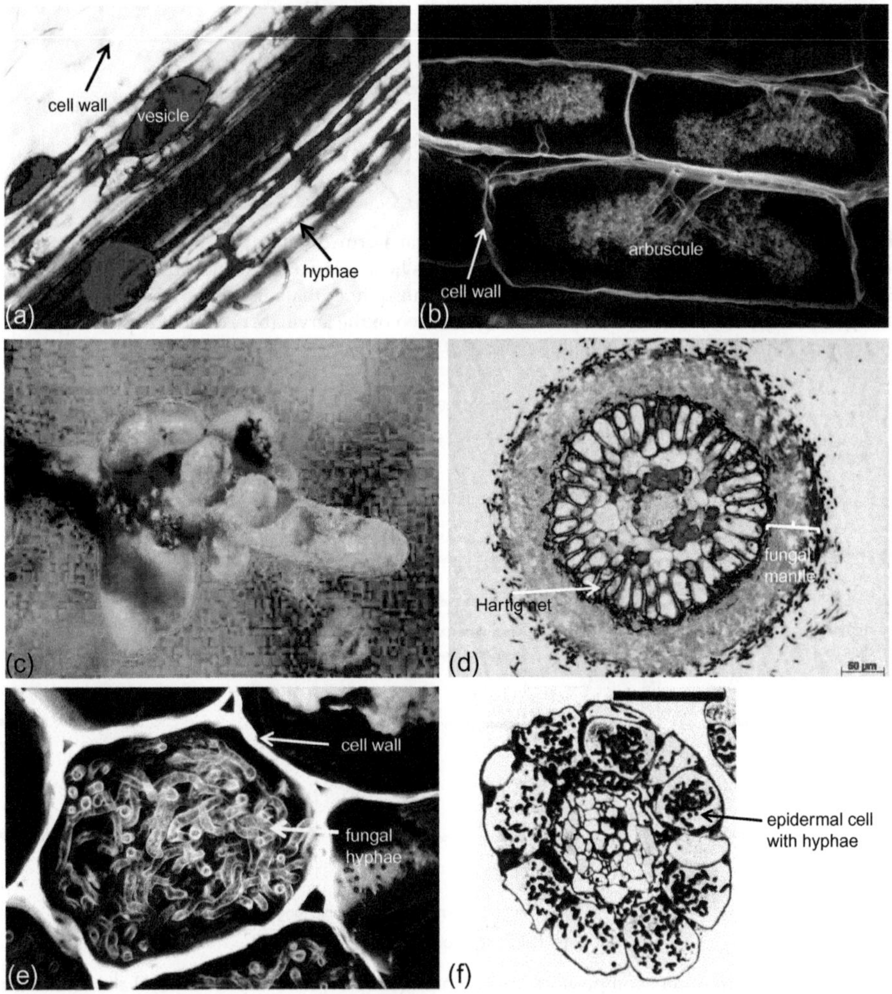

FIGURE 7.2 Typical structures of arbuscular mycorrhizas (a, b), ectomycorrhizas (c, d), and ericoid mycorrhizas (e, f). *Source: van der Heijden et al. (2015).*

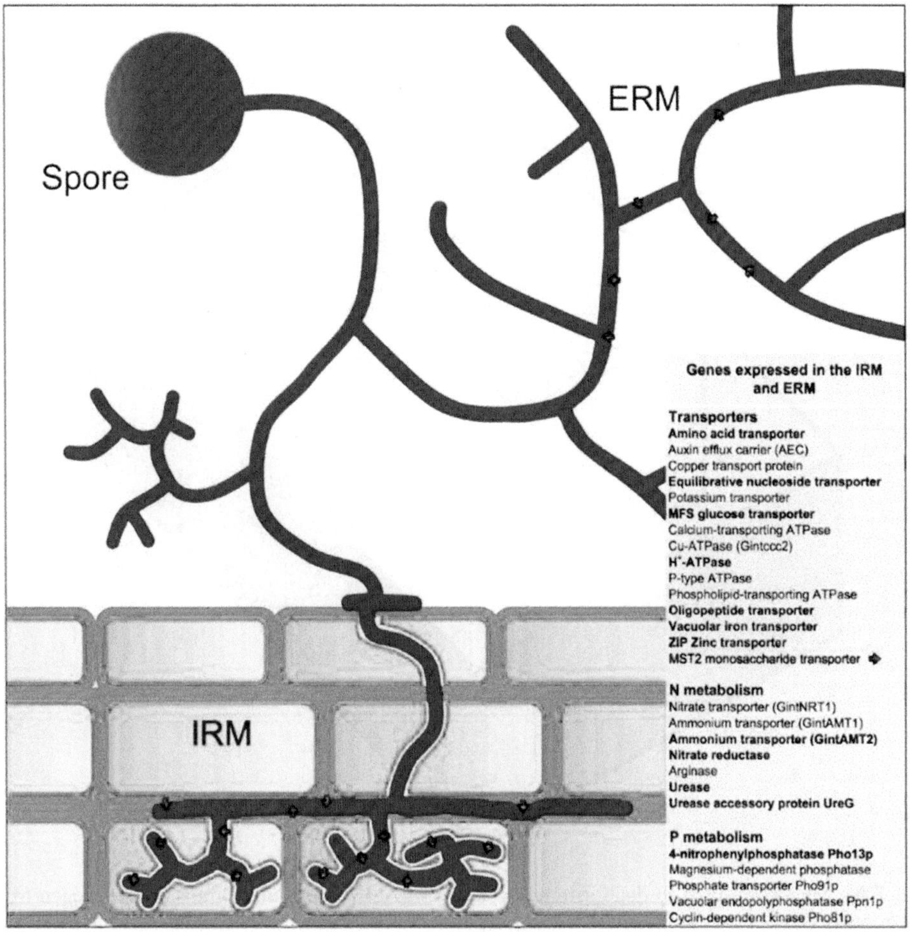

FIGURE 7.4 Gene expression during different life stages of AMF. Comparison of genes expressed in extraradical (ERM) and intraradical IRM mycelium, shows a switch from the expression of genes encoding proteins with transport functions to those involved in metabolism as the host–fungus nutrient exchange interfaces become established. *Source: Lanfranco and Young (2012).*

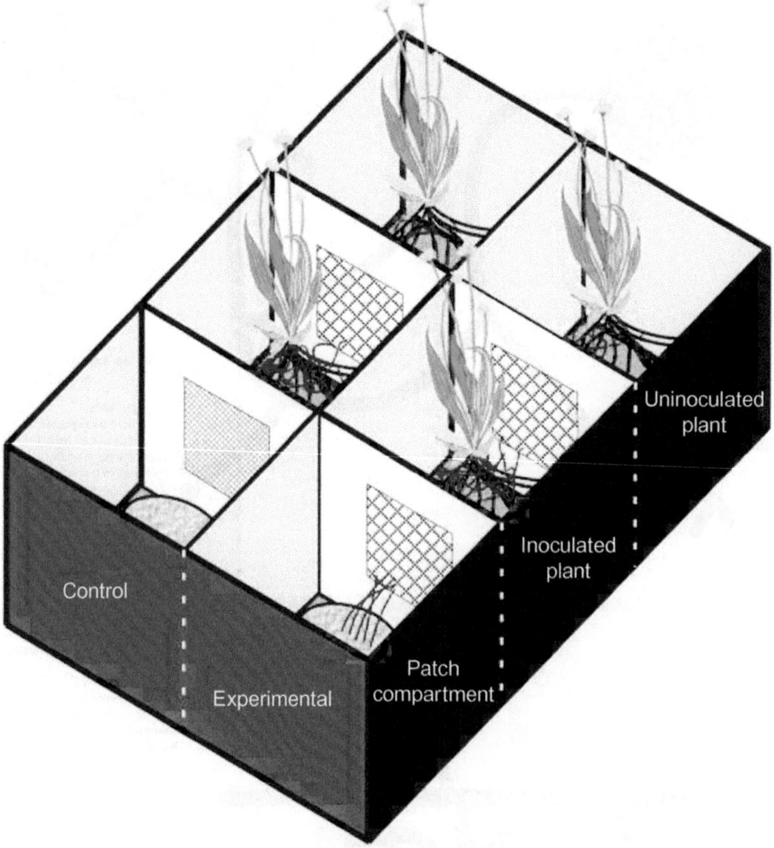

FIGURE 7.6 Microcosm used to demonstrate the ability of AMF mycelium to translocate nitrogen from plant litter into roots of living host plants. Hyphae from experimentally AMF-inoculated plants in the centre compartments of the experimental row grew into the patch of ^{15}N isotope-labelled grass litter in the 'Patch' compartment. In the control row, intervening mesh prevented hyphae growing between compartments but allowed solute diffusion. It was found that AMF-colonised plants allowed to access the grass litter acquired three times as much nitrogen as those where AMF hyphae were prevented from reaching the litter. *Source: Hodge et al. (2001).*

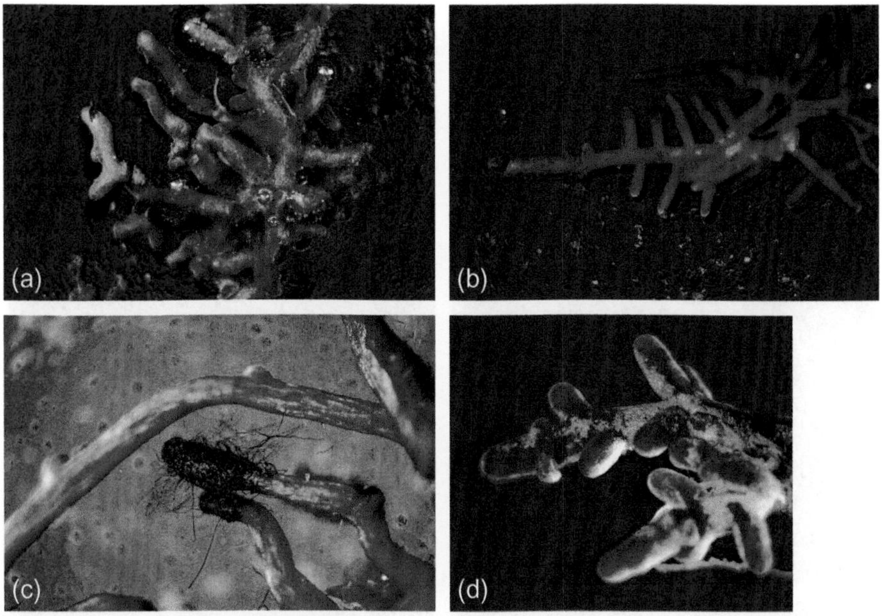

FIGURE 7.7 Ectomycorrhizal morphotypes from beech (*Fagus sylvatica*) woodland soil; (a) Laccaria sp., (b) Lactarius sp., (c) Coenococcum sp., and (d) Russula sp. *Source: John Baker.*

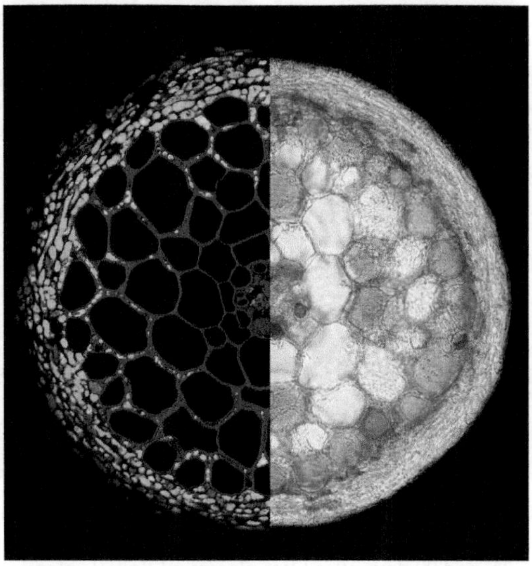

FIGURE 7.8 Immuno-localisation of a highly expressed fungal effector-like protein in a *Populus trichocarpa–Laccaria bicolor* ectomycorrhizal root tip. A transverse cross section of a poplar root colonised by the symbiotic ectomycorrhizal fungus *L. bicolor*. The green signal is an immuno-localisation of the fungal effector protein MiSSP7 highly expressed in the hyphae of *L. bicolor* while staining with propidium iodide highlights the cell walls of the root cells. *Source: Jonathan Plett and Francis Martin.*

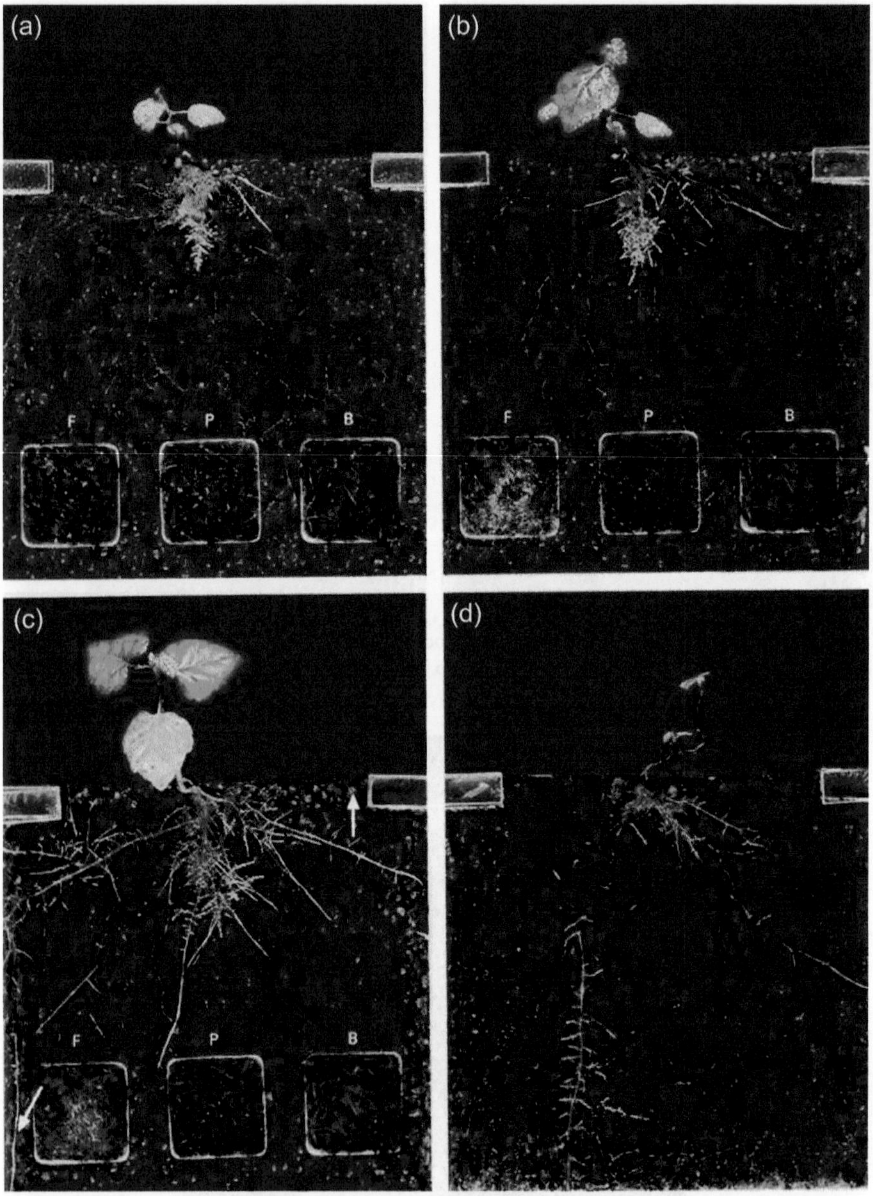

FIGURE 7.9 Photographs (a)–(c) show sequential development of ectomycorrhizal seedlings of birch, *Betula pendula*, associated with *Paxillus involutus* ectomycorrhiza in observation chambers containing trays of litter of beech (F), pine (P), and birch (B) at 8, 35, and 90 days after litter placement. Initial colonisation of litter is followed by increased plant growth, compared with d, a mycorrhizal control plant without litter addition. *Source: Perez-Moreno and Read (2000).*

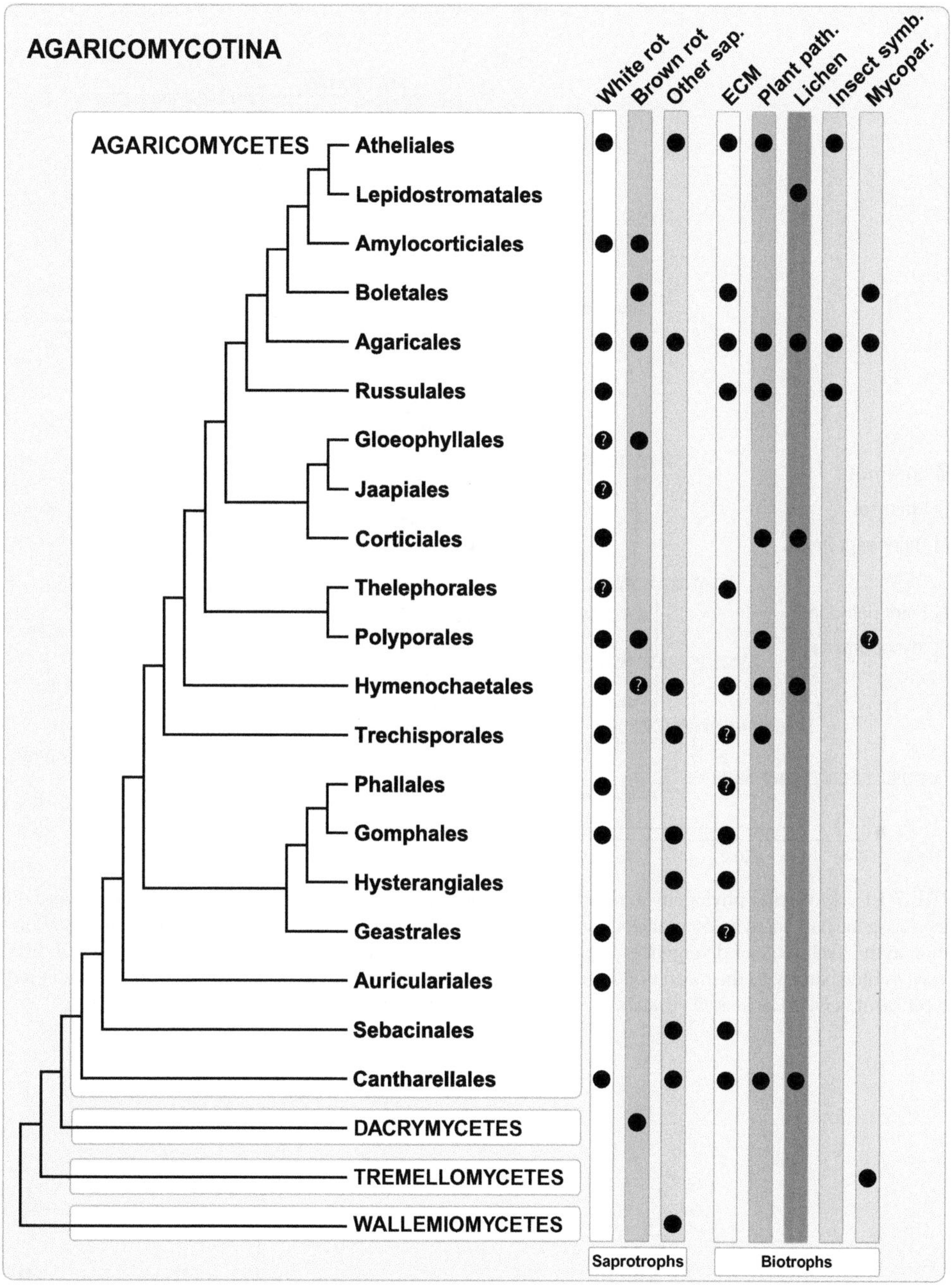

FIGURE 7.10 Phylogenetic distribution of major nutritional modes across the Agaricomycotina. The tree summarises recent phylogenomic and multi-gene phylogenetic studies. Saprotrophs include white rot and brown rot wood decay fungi, and a broad category of 'other' saprotrophs, such as litter, dung, and keratin-degrading fungi. White rot is very widespread and is probably the ancestral condition of the Agaricomycetes, but not the Agaricomycotina as a whole (note that it is absent from Dacrymycetes, Tremellomycetes, and Wallemiomycetes). Brown rot has evolved independently in at least five orders of Agaricomycetes, as well as Dacrymycetes. Biotrophs include ectomycorrhizal symbionts (ECM), plant pathogens, lichen-forming basidiomycetes, insect symbionts and mycoparasites, all of which are ultimately derived from saprotrophic ancestors. Agaricomycotina also include other biotrophs that are not shown, including endophytes, nematode-trapping fungi, bacteriovores, parasites of algae and bryophytes, and animal pathogens. Question marks indicate uncertainty. *Source: See James et al. (2006). Figure David Hibbett.*

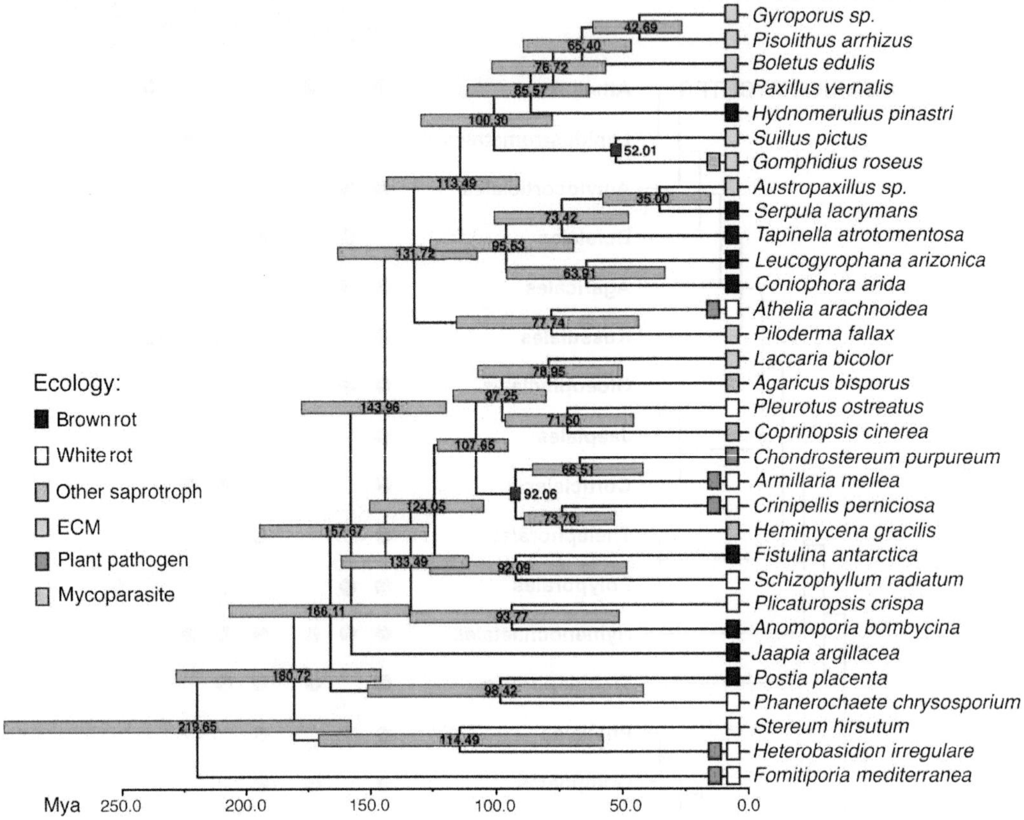

FIGURE 7.11 Molecular phylogeny and the evolution of Agaricomycete ecology. The chronogram, inferred from a combined six-gene data set by molecular clock analysis, illustrates the divergences of nutritional mode in Agaricomycetes in relation to the likely time of divergences in angiosperm and gymnosperm plants. The estimated times of divergence are shown as blue bars, with the mean node ages in the bars. Calibration points with fossil ages are shown in red (dark gray in the print version). *Source: D. Floudas, in Eastwood et al. (2011) Chapter 5, Further Reading.*

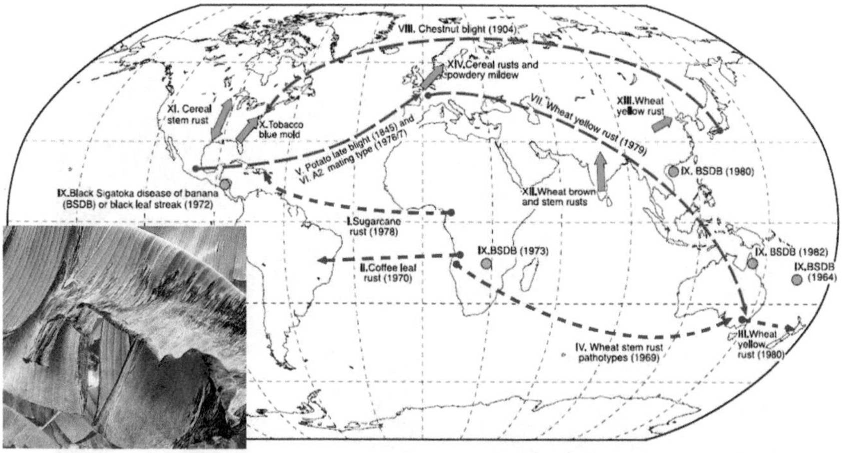

FIGURE 8.4 Plant pathogenic fungi can be transported extremely long distances by wind dispersal of spores. Though travel between continents is a rare event, unusual atmospheric conditions do sometimes occur, leading to dispersal and establishment of disease far from its long established origins. Red (black in the print version) arrows (on lines with short dashes) indicate disease spread by direct movement of airborne spores while blue (dark grey in the print version) arrows (on lines with long dashes) show where pathogens were first spread to new regions in infected plant material or by man and then onwards as airborne spores. Orange circles (grey in the print version) show the global spread of black Sigatoka disease of banana (inset) caused by *Mycospaerella fijiensis*, the first recorded outbreak on each continent is indicated by IX. Green arrows show five examples of diseases which periodically migrate via airborne spores from one region to another in extinction (due to absence of crops over one or more seasons)-recolonisation cycles (X, XI, XII, XIII, and XIV). The inset shows the symptoms of black Sigatoka disease of banana caused by *Mycosphaerella fijiensis* © John Lucas. *Source: Brown and Hovmøller (2002).*

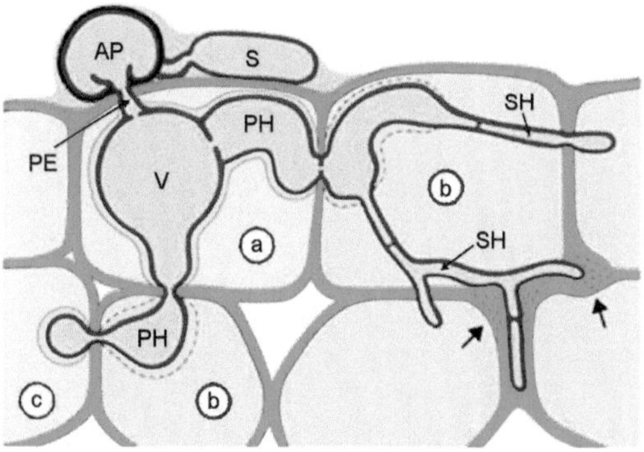

FIGURE 8.7 *Colletotrichum* anthracnose. Most species are hemibiotrophic as seen in this diagram of infection by *Colletotrichum lindemuthianum*. A spore (S) attaches to the surface of the host. When it germinates, it produces a short germ tube, which differentiates into an appressorium (A), from the underside of which develops a penetration peg (PE) which pierces the cuticle and wall of the epidermal cell. The hypha swells to form a vesicle (V) from which develop broad primary hyphae (PH) surrounded by plant plasma membrane. This is the biotrophic stage (a); the plant cell remains alive, and the host and fungal protoplasts remain separated by an interfacial matrix (indicated by yellow (light grey in the print version) colouring). After 1 or 2 days the plant plasma membrane begins to disintegrate and the host cell dies (b). A sequence of colonisation of plant cells by new primary hyphae occurs (c) with subsequent death after a few days. The biotrophic phase ends when narrow secondary hyphae (SH) develop from the primary hyphae. The secondary hyphae are not surrounded by host membrane/interfacial matrix, and secrete plant cell wall-degrading enzymes (indicated by arrows) in this necrotrophic phase. *Source: Mendgen and Hahn (2002).*

FIGURE 8.11 Annosum root and butt rot is caused by several species of *Heterobasidion* (Table 8.7). (b) Jeffrey pine at the edge of a large and still expanding disease centre caused by *Heterobasidion irregulare* in the Sierra Nevada. (c) Fruit body of *Heterobasidion annosum.* (d) Severe white pocket rot decay in sitka spruce (*Picea sitchensis*) caused by *Heterobasidion annosum.* (e) Felled Norway spruce (*Picea abies*) showing heart rot due to *Heterobasidion annosum.* *Source: (b) Modified from Asiegbu et al. (2005); (c) © Alan Outen; (b, d, e) © Stephen Woodward.*

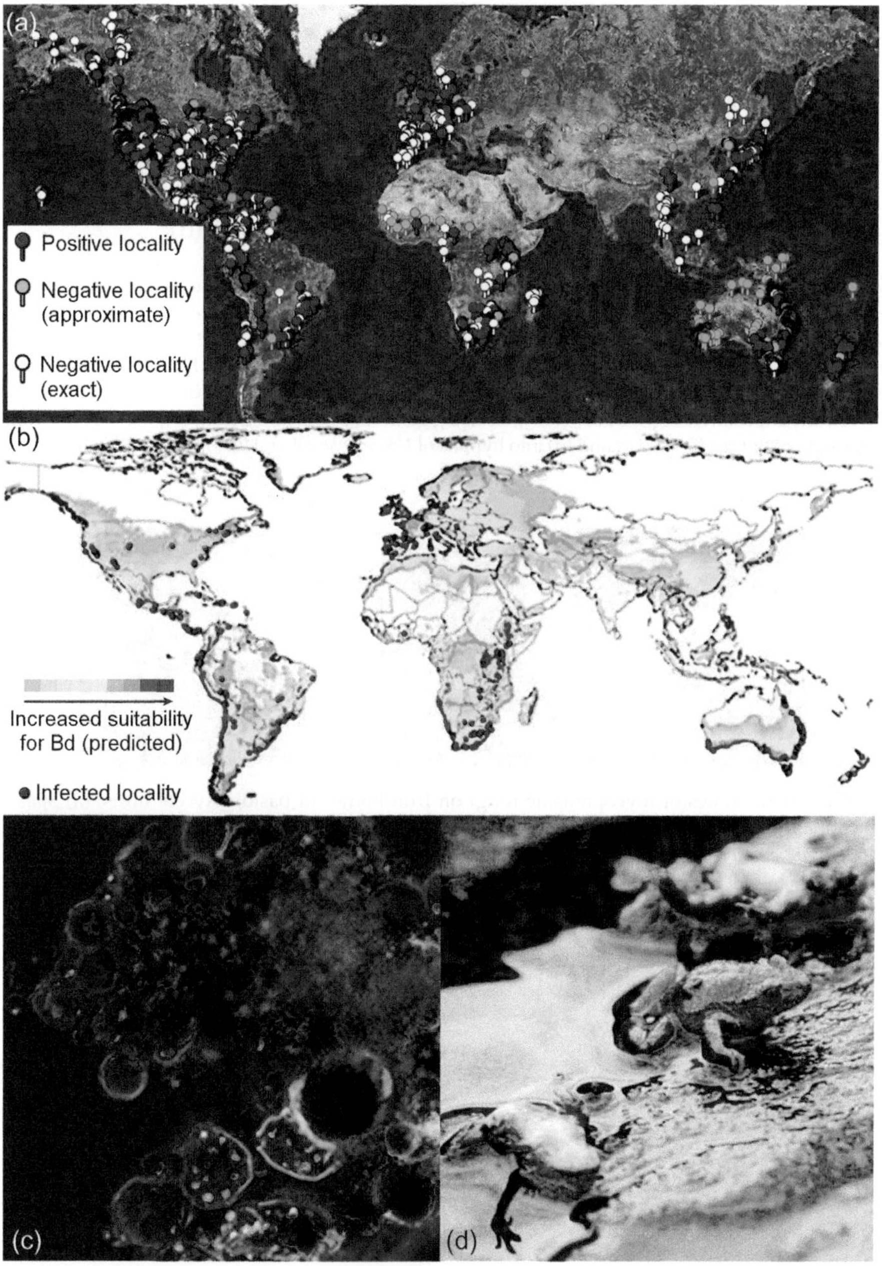

FIGURE 9.3 (a) Chytridiomycosis of amphibians caused by *Batrachochytrium dendrobatidis* (Bd) is now widespread across much of the globe, though Southeast Asia is still largely free of the disease, as seen in this screen shot from the Global Mapping Project (http://bd-maps.net). (b) The disease is predicted to spread even more widely by the Climate Envelope Model developed by D. Rödder, J. Kielgast, J.B. Schmidtlein, et al. (unpublished data). (c) Laser-scanning confocal micrograph of Bd in culture. Blue stained structures are metabolically active sporangia. (d) Pyreneen Midwife toads *Alytes obstetricians* sufferening from lethal chytridiomycosis. *Source: Panel (b) from Fisher et al. (2009), (c) from Fisher et al. (2009), (d) © Mat Fisher.*

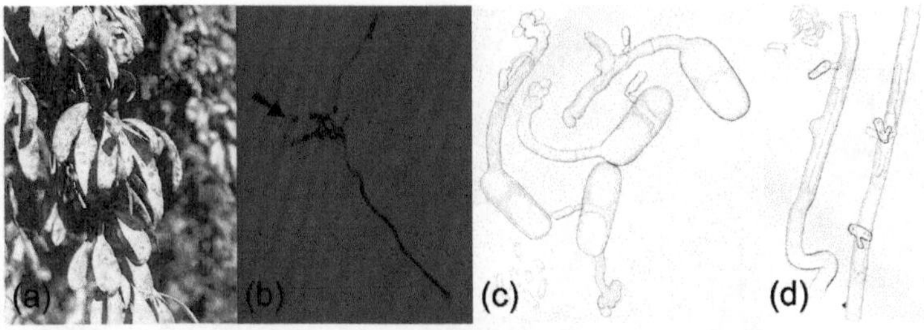

FIGURE 10.3 Intracellular mycoparasites. *Ampelomyces* species are initially biotrophic, though eventually they kill the host cell. *Ampelomyces* spp. are parasitic on powdery mildew fungi. (a) The brown patches are masses of intracellular pycnidia of *Ampelomyces* within white powdery mildew colonies, which are parasitic on *Lycium halimifolium* (Solanaceae). Hyphae of the mycoparasite penetrate into conidia (b, c) and hyphae (d) of the host. (b) Hyphae of *Ampelomyces* (stained with cotton blue) within (arrowed) and emerging from a conidium of *Erysiphe syringae-japonicae*. *Source:* (c, d) Drawings from early work on Ampelomyces by De Bary (1870), showing germ tubes extending from conidia (arrowed) and penetrating into (c) germ tubes of *Erysiphe heraclei* and (d) into hyphae of *Neoerisiphe galeopsidis*. *From Kiss (2008).*

FIGURE 10.5 Fruit bodies of mycoparasitic fungi on fruit bodies of basidiomycete hosts. (a) *Spinellus fusiger* (Mucoromycotina) on *Mycena* sp. (b) *Asterophora lycoperdoides* on *Russula nigricans*. (c) *Pseudoboletus parasiticus* on *Scleroderma citrinum*. *Source: Panels (a, b) © Penny Cullington and (c) © Alan Hills.*

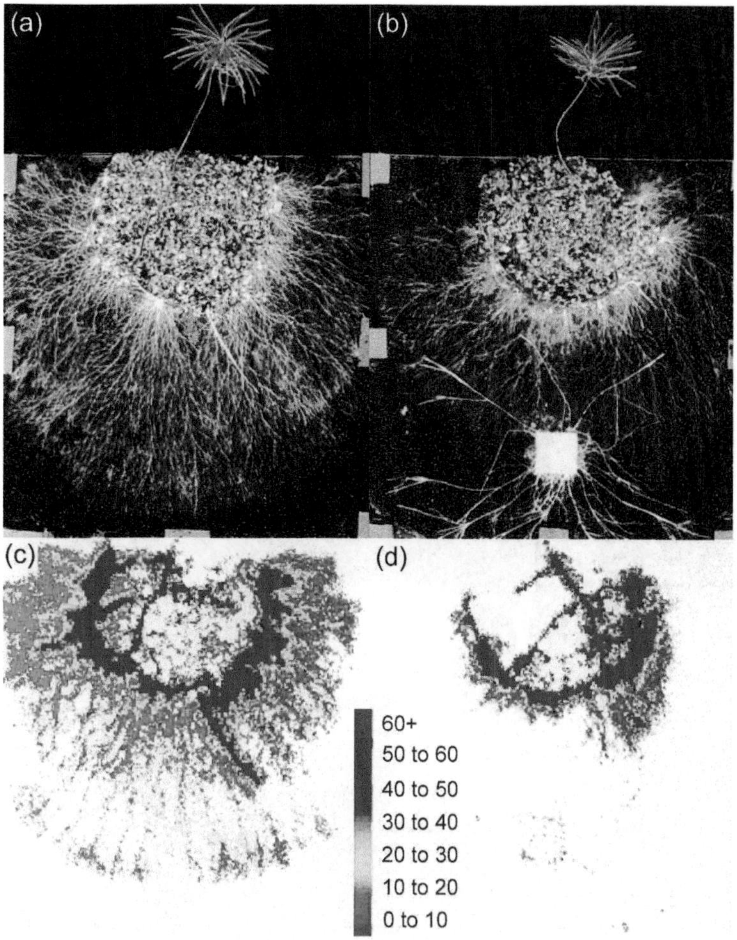

FIGURE 10.8 Saprotrophic basidiomycetes can dramatically affect ectomycorrhizal mycelial spread and allocation of carbon to the extraradical mycelium of the mycorrhizal fungus. Here the wood decay fungus *Phanerochaete velutina*, growing from a piece of wood, is interacting with mycelium of *Suillus bovinus* in association with *Pinus sylvestris* in soil microcosms. (a) and (b) show, respectively, ectomycorrhizal mycelial growth in the absence and presence of the saprotroph. Plants were pulse-labelled with ^{14}C, and this was quantified in a 20 × 24 cm below-ground area by digital autoradiography (c, d). The colour scale represents counts mm^{-2} over 45 min. Very little carbon from the host plant is allocated to ectomycorrhizal mycelium in the area of territorial combat, and its growth is inhibited. There was a 60% reduction in ^{14}C allocated to mycelium of *Suillus bovinus* when interacting with *Phanerochaete velutina*, up to 30 h after pulse labelling. Presence of ^{14}C (0.03 %) was detected in *Phanerochaete velutina* after 5 days. *Source: Modified from: Leake, J.R., Donnelly, D.P., Saunders, E.M., Boddy, L., Read, D.J., 2001. Carbon flux to ectomycorrhizal mycelium following* ^{14}C *pulse labelling of Pinus sylvestris L. seedlings: effects of litter patches and interaction with a wood-decomposer fungus. Tree Physiol. 21, 71–82. By permission of Oxford University Press.*

FIGURE 11.5 Rare and endangered species. (a) The nail fungus *Poronia punctata* on the Red List of many European countries © Martyn Ainsworth; (b) The wax cap *Hygrocybe collucera* endangered in New South Wales Australia © Ray and Elma Kearney; (c) *Ramariopsis pulchella* near threatened in the British Red List © Martyn Ainsworth; (d) *Hericium coralloides* © Martyn Ainsworth; (e) *Hericium erinaceus* protected by law in the UK © Martyn Ainsworth; (f) *Hygrocybe lanecovensis* endangered in New South Wales Australia © Ray and Elma Kearney; (g) *Phellodon melaleucus* © Martyn Ainsworth; (h) young *Piptoporus quercinus* © Martyn Ainsworth; (i) *Zeus olympius* (Ascomycota) found only on Bosnian Pine (*Pinus heldreichii*) in northern Greece © Stephanos Diamandis; (j) *Pleurotus nebrodensis* found only in Sicily © David Minter; (k) *Ophiocordyceps sinensis* (Ascomycota) on a caterpillar (arrowed) © Paul Cannon (CABI/RBG Kew); (l) critically endangered *Erioderma pedicellatum* (boreal felt lichen) © Christoph Scheidegger.

FIGURE 12.1 Cultivated mushrooms. (a) Shiitake, *Lentinula edodes*, on logs. (b) White button mushroom, *Agaricus bisporus*, on beds of compost. *Source: panel (a) http://www.sharondalefarm.com/cultivation/ and (b) http://modernfarmer.com/2014/05/welcome-mushroom-country-population-nearly-half-u-s-mushrooms/.*

FIGURE 12.3 Modern winery with high capacity stainless steel fermenters. *Source: Fedor Kondratenko©123RF.com*